Atomically Dispersed Metallic Materials for Electrochemical Energy Technologies

Atomically dispersed metallic materials (ADMMs) are the most advanced materials used in energy conversion and storage devices to improve their performance for portable electronics, electric vehicles, and stationary power stations. *Atomically Dispersed Metallic Materials for Electrochemical Energy Technologies* aims to facilitate research and development of ADMMs for applications in electrochemical energy devices. It provides a comprehensive description of the science and technology of ADMMs, including material selection, synthesis, characterization, and their applications in fuel cells, batteries, supercapacitors, and $H_2O/CO_2/N_2$ electrolysis to encourage progress in commercialization of these clean energy technologies.

- Offers a comprehensive introduction to various types of ADMMs, their fabrication and characterization, and how to improve their performance
- Analyzes, compares, and discusses advances in different ADMMs in the application of electrochemical energy devices, including commercial requirements
- Describes cutting-edge methodologies in composite ADMM design, selection, and fabrication
- Summarizes current achievements, challenges, and future research directions

Written by authors with strong academic and industry expertise, this book will be attractive to researchers and industry professionals working in the fields of materials, chemical, mechanical, and electrical engineering, as well as nanotechnology and clean energy.

Electrochemical Energy Storage and Conversion

Jiujun Zhang

National Research Council Institute for Fuel Cell Innovation, Vancouver, British Columbia, Canada

PUBLISHED TITLES

Electrochemical Polymer Electrolyte Membranes
Jianhua Fang, Jinli Qiao, David P. Wilkinson, and Jiujun Zhang

Electrochemical Energy: Advanced Materials and Technologies
Pei Kang Shen, Chao-Yang Wang, San Ping Jiang, Xueliang Sun, and Jiujun Zhang

High-Temperature Electrochemical Energy Conversion and Storage: Fundamentals and Applications
Yixiang Shi, Ningsheng Cai, Tianyu Cao, and Jiujun Zhang

Redox Flow Batteries: Fundamentals and Applications
Huamin Zhang, Xianfeng Li, and Jiujun Zhang

Carbon Nanomaterials for Electrochemical Energy Technologies: Fundamentals and Applications
Shuhui Sun, Xueliang Sun, Zhongwei Chen, Yuyu Liu, David P. Wilkinson, and Jiujun Zhang

Proton Exchange Membrane Fuel Cells
Zhigang Qi

Hydrothermal Reduction of Carbon Dioxide to Low-Carbon Fuels
Fangming Jin

Lithium-Ion Supercapacitors: Fundamentals and Energy Applications
Lei Zhang, David P. Wilkinson, Zhongwei Chen, and Jiujun Zhang

Solid Oxide Fuel Cells: From Fundamental Principles to Complete Systems
Radenka Maric and Gholamreza Mirshekari

Atomically Dispersed Metallic Materials for Electrochemical Energy Technologies
Wei Yan, Xifei Li, Shuhui Sun, Xueliang Sun, and Jiujun Zhang

For more information about this series, please visit: https://www.routledge.com/Electrochemical-Energy-Storage-and-Conversion/book-series/CRCELEENESTO

Atomically Dispersed Metallic Materials for Electrochemical Energy Technologies

Edited by
Wei Yan, Xifei Li, Shuhui Sun, Xueliang Sun,
and Jiujun Zhang

CRC Press
Taylor & Francis Group
Boca Raton London New York

CRC Press is an imprint of the
Taylor & Francis Group, an **informa** business

First edition published 2023
by CRC Press
6000 Broken Sound Parkway NW, Suite 300, Boca Raton, FL 33487-2742

and by CRC Press
4 Park Square, Milton Park, Abingdon, Oxon, OX14 4RN

CRC Press is an imprint of Taylor & Francis Group, LLC

© 2023 Taylor & Francis Group, LLC

ISBN: 978-0-367-72098-8 (hbk)
ISBN: 978-0-367-72100-8 (pbk)
ISBN: 978-1-003-15343-6 (ebk)

DOI: 10.1201/9781003153436

Typeset in Times
by codeMantra

Series Preface

The goal of the Electrochemical Energy Storage and Conversion book series is to provide comprehensive coverage of the field, with titles focusing on fundamentals, technologies, applications, and the latest developments, including secondary (or rechargeable) batteries, fuel cells, supercapacitors, CO_2 electroreduction to produce low-carbon fuels, electrolysis for hydrogen generation/storage, and photoelectrochemistry for water splitting to produce hydrogen, among others. Each book in this series is self-contained, written by scientists and engineers with strong academic and industrial expertise who are at the top of their fields and on the cutting-edge of technology. With a broad view of various electrochemical energy conversion and storage devices, this unique book series provides essential reads for university students, scientists, and engineers and allows them to easily locate the latest information on electrochemical technology, fundamentals, and applications.

—Jiujun Zhang
National Research Council of Canada
Richmond, British Colombia

Contents

Preface

Electrocatalytic reactions play critical roles in electrochemical energy storage and conversion devices such as fuel cells, metal-air batteries, lithium batteries and $H_2O/CO_2/N_2$ electrolysis. However, due to the sluggish kinetics of the electrocatalysts used in these devices, their performances are not satisfactory and need to be improved. As a new type of emerging electrocatalysts, the atomically dispersed metallic catalysts (ADMCs) have received great attention and are the frontiers in the field of electrocatalytic research. Different from bulk and nano-sized metals, which are composed of tens to billions of atoms, the ADMCs consist of isolated single metal atoms stabilized by the supporting materials, in which all the catalytic metal atoms can be exposed and be involved in catalytic reactions, whereas the bulk and nano-sized metal catalysts can only give the atoms on the surface for catalytic processes. Since ADMCs are able to achieve up to 100% metal atom utilization, they usually exhibit extraordinary catalytic activity as well as exceptional selectivity because of the uniformity of electronic structure and coordination environment of the metal atoms. Besides the superior catalytic performance, the ADMCs can be recognized as the role-model electrocatalysts for a fundamental understanding of the catalytic mechanisms. With the help of advanced characterization techniques and DFT calculations, it is possible to diagnose the geometric and electronic properties by the low coordination environment and metal-support interaction, and the effects of the activation barriers and adsorption energies of reactants/intermediates/products on the catalytic processes, which will certainly benefit the design of electrocatalysts with prominent catalytic activity and stability.

Based on fruitful and latest research achievements of ADMCs, a book focuses on the fundamental and synthetic technological aspects that should be necessary for facilitating the further research and development toward the practical application of ADMCs. In this book, the latest synthesis methods of ADMCs, their in/ex-situ characterization methods and simulation/modeling are highlighted, and the strategies for stabilizing ADMCs are emphasized. This book also reviews the application of ADMCs in various electrochemical devices.

We have a strong desire to have this book being beneficial to the research and development of academic researchers, graduate/undergraduate students, industry professionals, and manufacturers of electrode/electrolyte systems.

We would like to express our sincere thanks for the contributions of all experts involved in developing and preparing this book. We would also like to express our appreciation to CRC Press for giving us this great opportunity in leading this book project, particularly to Dr. Allison Shatkin and Dr. Vernachio Gabrielle for their guidance and support in the book preparation process.

If there are any errors in the book, we would deeply welcome any constructive comments for further improvement.

Wei Yan, Ph.D., Fuzhou, China
Xifei Li, Ph.D., Xi'an, China
Shuhui Sun, Ph.D., Quebec, Canada
Xueliang Sun, Ph.D., Ontario, Canada
Jiujun Zhang, Ph.D., Fuzhou, China
January 20th, 2021

Editors

Dr. Wei Yan is currently a full professor at Fuzhou University, School of Materials Science & Engineering, Institute for New Energy Materials and Engineering. Dr. Yan's research now focuses on the materials and devices for electrochemical energy storage and conversion application, including electrocatalysts, alkali metal batteries, lead-acid batteries and supercapacitors.

Dr. Xifei Li is currently a full professor at Xi'an University of Technology. He was awarded as 2018, 2019, 2020 and 2021 Highly Cited Researchers of Clarivate Analytics. He is an executive editor-in-chief of Electrochemical Energy Reviews, a vice-president of the International Academy of Electrochemical Energy Science, and a Fellow of the Royal Society of Chemistry. Dr. Li's research group is currently working on optimized interfaces of the anodes and the cathodes with various structures for high-performance rechargeable batteries. Dr. Li has authored/co-authored 310 peer-reviewed articles with 16,400 citations with H index of 68.

Dr. Shuhui Sun is a Full Professor at the Institut National de la Recherche Scientifique (INRS), center for Energy, Materials, and Telecommunications (Canada). He is a member of the Royal Society of Canada's College of New Scholars, a member of Global Young Academy, Vice President of the International Academy of Electrochemical Energy Science (IAOEES). He serves as the Executive Editor-in-Chief of *Electrochemical Energy Reviews*. Dr. Sun's research focuses on Nanomaterials for Energy Conversion and Storage applications, including H_2 fuel cells, green hydrogen, lithium batteries, Zn-air batteries, and metal-ion batteries. He has published over 250 peer-reviewed journal articles with citations of over 14,000 times and an H index of 63. He has edited 3 books and 15 book chapters.

Dr. Xueliang Sun is a Canada Research Chair in Development of Nanomaterials for Clean Energy, Fellow of the Royal Society of Canada and Canadian Academy of Engineering and Full Professor at the University of Western Ontario, Canada. Dr. Sun received his Ph.D. in materials chemistry in 1999 from the University of Manchester, UK, which he followed up by working as a postdoctoral fellow at the University of British Columbia, Canada, and as a Research Associate at L'Institut National de la Recherche Scientifque (INRS), Canada. His current research interests are focused on advanced materials for electrochemical energy storage and conversion, including electrocatalysis in fuel cells and electrodes in lithium-ion batteries and metal-air batteries.

Dr. Jiujun Zhang is a Professor in the School of Materials Science and Engineering/Institute for New Energy Materials and Engineering at FuzhouUniversity, a former Principal Research Officer (Emeritus) at the National Research Council of Canada (NRC). Dr Zhang received his B.S. and M.Sc. in Electrochemistry from Peking University in 1982 and 1985, respectively, and his PhD in Electrochemistry from Wuhan University in 1988. Dr Zhang serves as Editor-in-Chief of Electrochemical Energy Reviews (Springer Nature) and Associate Editor of Green Energy & Environment (KeAi), and Editorial Board Member for several international journals as well as Editor for the book series of Electrochemical Energy Storage and Conversion (CRC). Dr Zhang's expertise areas are electrochemistry, electrocatalysis, fuel cells, batteries, supercapacitors, water/CO_2 electrolysis.

Contributors

Xin Bo
State Key Laboratory of Catalysis
Collaborative Innovation Center of
    Chemistry for Energy Materials
    (iChEM)
Dalian Institute of Chemical Physics
    (DICP)
Chinese Academy of Sciences (CAS)
Dalian, P.R. China

Shuqi Deng
Institute for Sustainable Energy
College of Sciences
Shanghai University
Shanghai, P.R. China

Li Dong
Institute for Sustainable Energy
College of Sciences
Shanghai University
Shanghai, P.R. China
and
Zhaoqing Leoch Battery Technology
    Co. Ltd
Zhaoqing City, Guangdong Province,
    P.R. China

Kieran Doyle-Davis
Department of Mechanical and
    Materials Engineering
The University of Western Ontario
London, Canada

Ejikeme Raphael Ezeigwe
Institute for Sustainable Energy
College of Sciences
Shanghai University
Shanghai, P.R. China

Yi Guan
Department of Mechanical and
    Materials Engineering
University of Western Ontario
London, Canada

Xiao Han
Center of Advanced Nanocatalysis
    (CAN)
Department of Applied Chemistry
University of Science and Technology
    of China
Hefei, Anhui, China

Leiduan Hao
State Key Laboratory of Organic–
    Inorganic Composites
Beijing University of Chemical
    Technology
Beijing, P.R. China

Xun Hong
Center of Advanced Nanocatalysis
    (CAN)
Department of Applied Chemistry
University of Science and Technology
    of China
Hefei, Anhui, China

Rouna Jia
State Key Laboratory of Catalysis
Collaborative Innovation Center of
    Chemistry for Energy Materials
    (iChEM)
Dalian Institute of Chemical Physics
    (DICP)
Chinese Academy of Sciences (CAS)
Dalian, P.R. China

Junjie Li
Department of Mechanical and
    Materials Engineering
University of Western Ontario
London, Canada

Sean Li
School of Materials Science and
    Engineering
University of New South Wales
Sydney, Australia

Wenxian Li
Institute of Materials
School of Materials Science and
    Engineering
Shanghai University
Shanghai, China
and
Institute for Sustainable Energy
Shanghai University
Shanghai, China
and
School of Materials Science and
    Engineering
University of New South Wales
Sydney, Australia
and
Shanghai Key Laboratory of High
    Temperature Superconductors
Shanghai, China

Li-Min Liu
School of Physics
Beihang University
Beijing, China

Revanasiddappa Manjunatha
Institute for Sustainable Energy
College of Sciences
Shanghai University
Shanghai, P.R. China

Yunteng Qu
State Key Laboratory of Photoelectric
    Technology and Functional Materials
International Collaborative Center on
    Photoelectric Technology and Nano
    Functional Materials
Institute of Photonics and Photon-
    Technology, Northwest University
Xi'an, Shaanxi, China

Hariprasad Ranganathan
Énergie Matériaux Télécommunications
    Research Centre
Institut National de la Recherche
    Scientifique (INRS)
Varennes, Canada

Rongbo Sun
Center of Advanced Nanocatalysis
    (CAN)
Department of Applied Chemistry
University of Science and Technology
    of China
Hefei, Anhui, China

Zhenyu Sun
State Key Laboratory of Organic–
    Inorganic Composites
Beijing University of Chemical
    Technology
Beijing, P.R. China

Gilberto Teobaldi
Scientific Computing Department
STFC UKRI, Rutherford Appleton
    Laboratory
Didcot, United Kingdom
and
Department of Chemistry
Stephenson Institute for Renewable
    Energy
University of Liverpool
Liverpool, United Kingdom
and
School of Chemistry
University of Southampton
Highfield, United Kingdom

Mingjie Wu
Énergie Matériaux Télécommunications
    Research Centre
Institut National de la Recherche
    Scientifique (INRS)
Varennes, Canada

Yuen Wu
Hefei National Laboratory for Physical
    Sciences at the Microscale
School of Chemistry and Materials
    Science
University of Science and Technology
    of China
Hefei, China

Jack Yang
School of Materials Science and
    Engineering
University of New South Wales
Sydney, Australia

Xiaozhang Yao
Department of Mechanical and
    Materials Engineering
University of Western Ontario
London, Canada

Wen-Jin Yin
School of Physics and Electronic
    Science
Hunan University of Science and
    Technology
Xiangtan, China
and
School of Physics
Beihang University
Beijing, China

Gaixia Zhang
Énergie Matériaux Télécommunications
    Research Centre
Institut National de la Recherche
    Scientifique (INRS)
Varennes, Canada

Changtai Zhao
Department of Mechanical and
    Materials Engineering
The University of Western Ontario
London, Canada

1 Fundamentals of Atomically Dispersed Metallic Materials

Rongbo Sun, Xiao Han, and Xun Hong
University of Science and Technology of China

CONTENTS

1.1 INTRODUCTION TO ATOMICALLY DISPERSED METALLIC MATERIALS

Catalytic processes are very relevant to our lives in modern industry. More than 90% of production processes of chemical products need to use catalysts, especially in economic development, environmental protection, petrochemical, and energy utilization [1–7]. Catalysts can provide specific selectivity and good reaction rates by adjusting the reaction path and lowering the activation energy. According to the phase state of the reaction system, catalyst can be divided into three categories: heterogeneous catalyst, homogeneous catalyst, and biological (enzyme) catalyst [8,9]. Heterogeneous catalysts usually exist as finely dispersed metal clusters and particles of high surface

DOI: 10.1201/9781003153436-1

area. They could be used under harsh conditions such as high temperature or high pressure and be easily recycled from reactants and products [10–12]. The capability of heterogeneous catalysts depends on the coordination environment and electronic structure of the metal center sites. To lower the cost and improve the utilization of active metals of heterogeneous catalysts, rational design and development of catalytic materials with excellent activity, selectivity, and durability are the most concerning problems at present.

The continuous development of nanomaterial science has led to a wide variety of nanocatalysts [13–20]. Among them, atomically dispersed metallic materials have become star materials in recent years and have received widespread attention in the field of catalysis. Atomically dispersed metallic materials exhibit special physical and chemical properties, and great performance in many catalytic reactions, benefiting from the high dispersion of metal atoms and unique local structure [21–24]. Under the current background of prominent energy and environmental issues, the design of efficient atomic-level dispersed metal catalysts and the realization of clean energy conversion and storage are important research directions in the field of energy catalysis.

1.1.1 DEFINITION

Nanotechnology is revolutionizing the understanding of catalysts in recent decades, resulting in the new concept of "nanocatalysis". Nanocatalysts exhibit high catalytic activity due to their small particle size, high surface volume ratio, and abundant coordination-unsaturated atoms [25–27]. The physical and chemical properties of catalysts can be up to the size of catalytic metal particles. For example, the specific activity of each and every metal atom increases, accompanied by an enhancement to surface free energy when the size of the metal decreases from bulk to nanoscale, and then to the single-atom level (Figure 1.1) [11,28–36]. Nanoparticles with a high ratio of surface area to volume can be prepared by regulating nucleation and growth processes at the nanoscale. What's more, the edges, corners, and steps that have particular morphology of nanoparticles in catalysts are full of abundant coordination-unsaturated atoms, affecting the adsorption, desorption, and activation processes of small molecule reagents, thereby effectively modulating the catalytic efficiency [20]. To increase the utilization rate of atoms, reduce the cost of catalyst production, and improve the activity of the catalyst, the method of decreasing the size of the supported metal substance in catalyst is used to make it in an atomically dispersed state (single, several to hundreds of atoms).

Efforts have been made to reduce the size and promote the durability of metallic nanoparticles. The reduction of the size of metal nanoparticles to atomically dispersed metallic materials changes their catalytic behavior in the following ways:

1. Surface Effect: the percentage of nanoparticles with unsaturated coordination bond increases [37–40].
2. Quantum Size Effects: in which electron confinement leads to an increase in energy levels and a widening of the Kubo HOMO-LUMO gap [41–47].
3. Metal-carrier interactions: the chemical bond between metal and carrier leads to charge transfer [48–53].

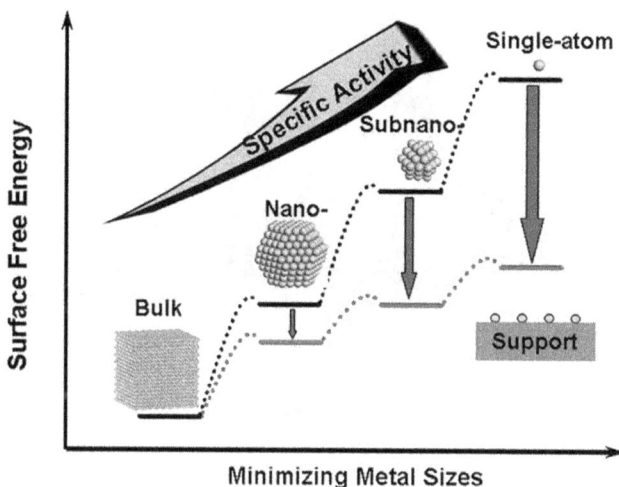

FIGURE 1.1 Schematic illustration of the dramatic increase of surface free energy and specific activity with the decrease of metal sizes. (Reproduced with permission from Ref. [11]. Copyright 2013, American Chemical Society.)

4. Cluster Configuration: specific arrangement of atomic positions and the number of atoms in clusters can greatly change their physical and chemical properties [54–58].

Several definitions of atomically dispersed metallic materials have been proposed in the last decades. In 2013, Jingyue Liu concluded the following terms for four types of atomically dispersed heterogeneous catalysts (Figure 1.2) [59].

1. Single-Atom Catalyst (SAC) [11,25]:
   The catalyst contains only isolated SAs dispersed on the support. No spatial ordering or any other types of appreciable interactions should exist among the isolated single atoms. The active sites generally consist of the individual metal atoms as well as the immediate adjacent atoms of the support surface and/or other functional species. The catalytic properties of each active site can be similar or different, depending on the interaction between the single metal atom and its local surrounding.
2. Atomically Dispersed Metal Catalyst (ADMC) [59–62]:
   The dispersion degree of the supported metal atoms is considered to be nearly 100%. However, the metal form exists as two-dimensional rafts, small clusters, trimers, dimers, monomers, etc. The SAC can be considered as a subset of the ADMC.
3. Single-Site Heterogeneous Catalyst (SSHC) [63]:
   The "single site" may consist of one or more metal atoms. Such single sites are spatially isolated from one another, and no spectroscopic or other cross-link exists between such sites. The interaction energy between each site

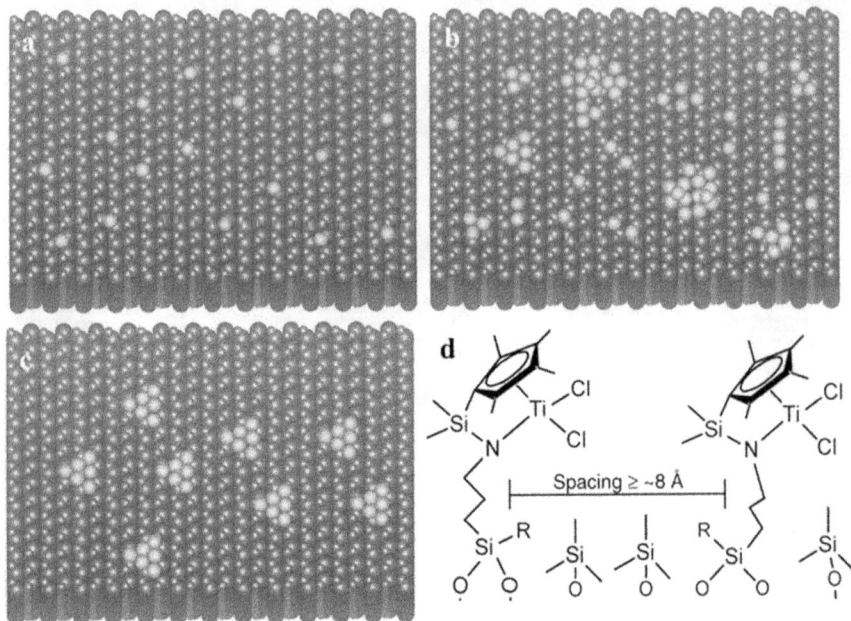

FIGURE 1.2 Schematic illustrations of (a) a supported single metal-atom catalyst (SAC), (b) an atomically dispersed metal catalyst (ADMC), (c) an extreme example of a single-site heterogeneous catalyst (SSHC), and (d) a site-isolated heterogeneous catalyst (SIHC). (Reproduced with permission from Ref. [59]. Copyright 2017, American Chemical Society.)

and a reactant is same, and each such site is structurally well-characterized, which can be treated as a single site in homogeneous molecular catalysts. The SAC can be considered an SSHC when the catalytic behavior of all individual atoms in SAC is the same.

4. Site-Isolated Heterogeneous Catalyst (SIHC) [64–66]:

This term generally refers to a heterogeneous catalyst that contains spatially well-separated organometallic complexes. The organometallic complexes possess well-defined structures and can be anchored to the support surfaces by ligands. A typical example is to graft organometallic complexes on a silica surface. The active sites of SIHCs may contain many metal atoms that are not necessarily atomically dispersed. The unique feature of a SIHC is that the active center is protected by ligands, and all the active centers possess the same catalytic behaviors. When a SIHC has only one type of active center, it can be considered as SSHC. In the reported literature, the terms SSHC and SIHC are frequently used interchangeably.

In general, SSHCs are usually prepared with the existence of ordered microporous/mesoporous supports such as zeolites, whereas functionalization and immobilization of functional groups on the support surfaces are critical to developing SIHCs [63]. SACs generally are not SSHCs because the interactions between isolated metal atoms and the support strongly depend on the surface structures and defects of the supports heterogeneous nature. When single metal atoms are incorporated into well-defined and highly ordered mesoporous structures or when they only interact with structurally well-defined functional groups on the support surfaces, the SAs can be considered as SSHCs [67,68].

The ultimate size for catalytic entities, in principle, is single-atomic. In 2011, the research team led by Tao Zhang achieved the pioneering success of preparing a Pt-based single-atomic site catalyst (denoted as Pt_1/FeO_x) and followed by the explosive development of their concept of "single-atom catalysis" in just a decade [25]. As a new generation of catalytic materials, SACs feature rich superiorities to nanoparticle catalysts, as the active metal species are single-atomically immobilized on the catalyst supports. SACs can readily attain an atom utilization efficiency of 100%, which is of particular advantage for catalysts based on high-cost noble metals [69]. In terms of coordination mode, the metal–metal bonds are absent in SACs, and the unique coordination configurations of single-atomic sites generally lead to exceptionally high catalytic activities in almost all kinds of catalytic reactions such as oxidation reactions [70–74], hydrogenation [75–78], water-gas shift [79,80], photocatalysis [81–84], and electrocatalysis [85–88], among others. In addition, the catalytic sites in SACs are highly uniform and not subject to the influences of crystal facets, vertices, and interfaces (in contrast to nanoparticles), which contribute to achieve higher selectivity. All of these merits qualify SACs as a bridging model between homogeneous and heterogeneous catalysis [11,89].

1.1.2 Development History

In 1925, the concept of active sites in heterogeneous catalysis was introduced by Taylor for the first time (Figure 1.3) [90]. He believes that the steps, cracks, and grooves of monoatomic can catalyze the reaction rather than the entire surface of a solid. Arlman and Cossee reported experimental evidence for the existence of well-defined local active sites in the 1960s [91]. They determined that coordinated unsaturated titanium ions related to chlorine vacancies on α-$TiCl_3$ prisms were used as active sites for Ziegler-Natta polymerization. The research on catalysts with isolated metal atom site can be traced to 1979 when Yates and his colleagues investigated the infrared spectrum of CO chemisorbed on alumina-supported rhodium (Rh) atoms [92]. Based on the C–O stretching frequencies, the adsorbed species can be divided into three types. Species I is formed only with Rh atoms separated from each other, which is assigned as $Rh(CO)_2$. Species II and III are formed on Rh clusters containing two or more Rh atoms, which are assigned as Rh–CO and Rh_2CO, respectively. CO species II and III interact with adjacent CO species, resulting in an increase in wave number as the coverage increases. According to infrared intensity tests for species I, the angle of OC–Rh–CO is ~90°. The single Rh atom is separated from the Al_2O_3 surface through nuclear magnetic resonance (NMR). Besides the metal oxide support, Thomas's group developed titanium single-site nanomaterials by directly grafting metal-organic complexes on the internal wall of ordered mesoporous silica in 1995 [93].

In 1999, Y. Iwasawa et al. synthesized Pt(A)/MgO and $PtMo_6/MgO$ catalysts by calcination of $Pt(C_5H_7O_2)_2$ and $[PtMo_6O_{24}]^{8-}$ and investigated their catalytic activity

	1920s	Taylor's active site definition
	1960s	Cossee's study on active sites
Alumina-supported rhodium atoms	1970s	Atomically adsorbed species distinguished by IR
Pt species at MgO	1990s	Single-atom detection by EELS
Pd$_n$ clusters on MgO	2000s	Single-atom detection by synchrotron XAS
Au and Pt species on CeO$_2$	2003	
Isolated Surface Au^{3+} Ions on ZrO$_2$	2005	
Single-Site Mesoporous Pd/Al$_2$O$_3$	2007	
Single-atom catalyst Pt$_1$/FeO$_x$	2011	The concept of "single-atom catalysis"

The explosive growth in research on atomically dispersed metallic materials

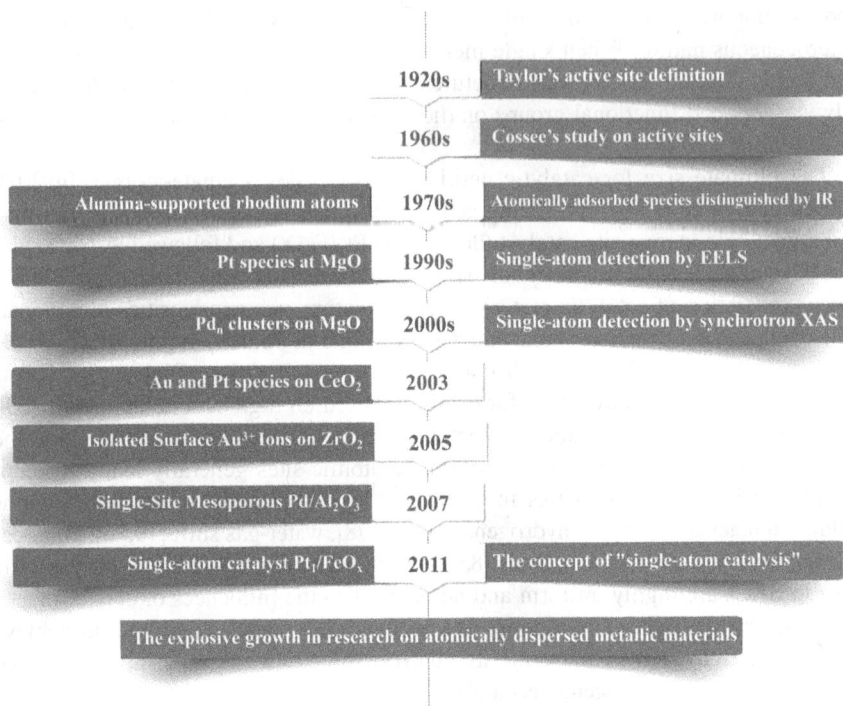

FIGURE 1.3   Chronology of catalysis with isolated atoms, highlighting the development of representative catalysts and their characterization techniques.

for propane $+O_2$ reaction [94]. It was found that the Pt(A)/MgO and PtMo$_6$/MgO catalysts with single Pt sites were as active as MgO-supported Pt nanoparticles for propane combustion. Importantly, by extending the X-ray absorption fine structure, the active Pt species is thought to be a distorted spinel-like structure surrounding Pt ions on the MgO surface. This research report is the first time that the concept of "atomic dispersion" is proposed. Later, Heiz and colleagues used selective soft landing methods to prepare Pd$_{n(1-30)}$ clusters and isolated Pd sites on MgO films, respectively. The atomic-level dispersed materials were characterized through thermal desorption and Fourier transform infrared spectroscopy. In addition, the formation of benzene was discovered by a single-atom Pd at a low temperature (300 K). By using density functional theory calculations to explore the principle, it is proved that free inert palladium atoms are activated by the charge transfer at the defects of the MgO substrate during deposition [95]. Maria Flytzani-Stephanopoulos prepared a nanosized La-doped ceria and deposited Au or Pt on it in 2003 [96]. She reported that for the water-gas shift reaction, nanostructured gold- or platinum-cerium oxide catalysts are used. The catalytic activity was not affected by the removal of metallic Au or Pt particles by cyanide leaching. As a result, Au or Pt exists in an ionic state are expected as the active sites.

In the following years, many scholars reported the successful preparation of atomically dispersed metal materials. For example, in 2005, scientists prepared a small quantity of isolated surface Au^{3+} ions (0.02 Au^{3+}/nm^2) dispersed on a zirconia surface at the atomic level from cheap inorganic raw materials ($HAuCl_4$). This material provides excellent catalytic performance for the selective hydrogenation of 1, 3-butadiene [97]. The extended X-ray absorption fine structure and high-angle annular dark-field scanning transmission electron microscope have developed into the two main and cogent techniques for characterizing SAC. Li et al. produced Pd SAC on ultra-low-mass supported mesoporous Al_2O_3 through the impregnation method, which significantly improved the catalytic efficiency of aerobic selective oxidation of allyl alcohol [98]. These two measurements confirmed the monodisperse structure of Pd on meso-Al_2O_3. In 2009, E. Charles H. Sykes confirmed that the isolated single-atom Pd can promote hydrogen dissociation and spillover onto the specific surface of the metal copper [99].

Tao Zhang and his colleagues reported that single-atom Pt on FeO_x surface without protective organic ligands and linkers are stable and provide excellent catalytic performance for the CO oxidation reaction. They used density functional theory to calculate theoretical evidence that the material produces high catalytic activity. The results show that the partially vacant 5d orbitals of the positively charged high-valent platinum atoms help reduce the CO adsorption energy and the activation barrier for CO oxidation. The curiosity of the laboratory reflects Taylor's early recognition of heterogeneous catalysts with independent active sites.

SACs play an important role in heterogeneous catalysis. As a result of the insufficient resolution of analytical methods, many early studies may have overlooked its importance [100–102]. Although the main focus of atomically dispersed metallic materials was initially on thermal catalysis applications, interest in single-site electrocatalysis has grown promptly and has become an important field in the past few years. This benefited from advances in characterization techniques, including high-resolution electron microscopy, X-ray absorption spectroscopy, and more and more operational studies, supported by thorough theoretical research (Figure 1.4) [103–109]. In 2012, Sykes and colleagues reported that the isolated Pd atoms anchored on Cu surface were constructed to realize the selective hydrogenation of styrene and acetylene, thereby extending from oxides to metals to work as the support of SACs [110]. In 2016, a new strategy for achieving stable Co SAs on nitrogen-doped porous carbon with high metal loading over 4 wt% is reported by Wu and Li's group. The strategy is based on a pyrolysis process of predesigned bimetallic Zn/Co metal-organic frameworks, during which Co can be reduced by carbonization of the organic linker, and Zn is selectively evaporated away at high temperatures above 800°C [111]. This strategy has great development potential in preparing high loading and uniformly coordinated SAC. Another significant study on dual-site SAC was also reported by this team. They constructed the Fe-Co double sites on N-doped carbon support, on which Fe^{3+} species were reduced by carbon and combined with adjacent Co atoms [112]. The construction of dual-site catalysts improved the catalytic performance and further enriched the catalyst library. When exploring other effective single-atom synthesis methods, Wu's group used bulk Cu as the metal source to generate single Cu atoms by means of a gas migration-mediated strategy, and this can meet the needs of mass production [113]. In 2019, reports of SACs used in tumor treatment [114], cancer treatment [115] and enzyme simulation indicated that further attempts to the application of SACs

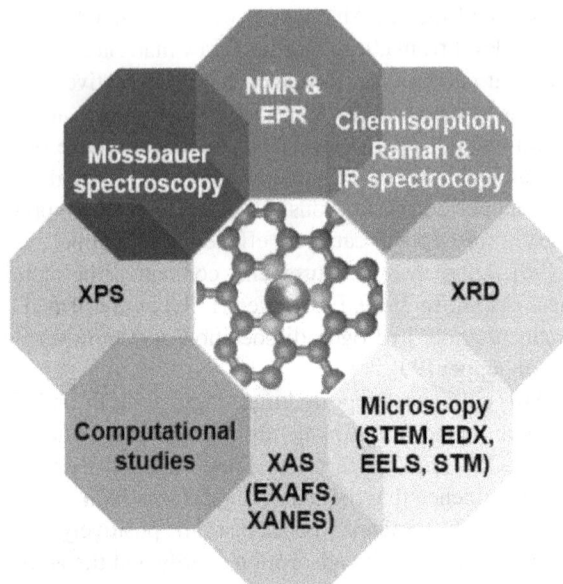

FIGURE 1.4 Overview of complementary techniques to characterize ADMMs. (Reproduced with permission from Ref. [109]. Copyright 2020, American Chemical Society.)

are worthwhile. In 2021, Wang et al. successfully increased the atomic loading of transition metals to as high as 40 wt% or 3.8 wt%, and proposed a general method for synthesizing single-atom catalysts simply. Compared with the reports in the original literature, this method increases the single-atom loading several times [116]. In another important research by Wang and Li et al., the nonmagnetic hexavalent molybdenum (Mo^{6+}) atomically dispersed within oxide lattice and achieve large enhancement in their oxygen reduction reaction activities [117]. In the latest research work, Yao's research team used acidified carbon nanotubes as a carrier and successfully constructed an atomic-level dispersed Pt-modified PtRu alloy catalyst through laser irradiation in liquid (LIL) technology. The development of in-situ synchrotron radiation X-ray absorption spectroscopy combined with theoretical calculations, which revealed the surface structure reorganization of the above-mentioned bimetallic catalyst under electrochemical hydrogen evolution reaction conditions, and clarified the interaction between different kinds of atoms in the reaction state [118]. Sun et al. proposed to combine single-atom catalysis with atomic layer deposition (ALD), using isolated Pt atoms as an aid to prepare Co, Ni, and Fe single-atom catalysts. Through the combination of theoretical calculations and experiments, it is found that on the nitrogen-doped carbon nanosheets (NCNS), Pt single atoms can regulate and promote the dissociation of cobaltocene ($Co(Cp)_2$) during the Co ALD process, so that the dissociative CoCp can be chemically adsorbed to the N_4 site of pyrrole [119] Li's research team reported a novel and efficient strategy to access high-performance nanozymes through direct atomization of platinum nanoparticles (Pt NPs) into single atoms by reversing the thermal sintering process [120].

1.1.3 Unique Features of Atomically Dispersed Metallic Materials

ADMM is composed of isolated single metal atoms or clusters dispersed on the surfaces of the carrier material. By definition, the dispersion of metals needs to be close to 100%. Unlike traditional supported metal nanoparticle catalysts, this catalyst has many unique properties [121–128].

1. The degree of dispersion of metal atoms in ADMMs is higher, and the atom utilization is almost 100%. The higher utilization rate promotes the lower amount of metal used, thus reducing the cost of production and application in the industry.
2. The metal sites in the ADMMs usually have coordination unsaturation. It is conducive to the adsorption of reactant molecules, thereby significantly reducing the adsorption activation energy of the catalytic reaction, so it has a higher intrinsic catalytic activity.
3. The metal sites in the entire catalyst are evenly distributed and the structure is uniform, so it has a high selectivity for the catalytic reaction. The uniform monatomic dispersion and well-defined configuration of ADMMs can also provide a great platform for the optimization of selectivity and activity.
4. The catalytic active centers of ADMMs are usually clear because the coordination structure of their metal centers is usually simple. Therefore, ADMM is a good platform to help understand the structure–reactivity relationships of catalysts at the atomic scale.

ADMMs combine the interfacial dynamics of heterogeneous catalysis with many advantages of homogeneous catalyzes, for example, high selectivity and legible active sites. In recent research directions, the regulation of the stability in nanomaterials has always aroused great interest for researchers, the core of which is the interaction between the supported metal and the carrier. The common sintering mechanisms that affect the stability of ADMMs in synthesis treatment or catalytic reactions are mainly divided into two classifications: Oswald ripening (OR), particle migration and coalescence.

OR process is the main sintering mechanism at present [59]. This process describes that larger nanoparticles will absorb the metal atoms released by smaller nanoparticles. The OR process generally occurs in the catalytic reaction of the supported metal catalyst. As shown in Figure 1.5a, schematic diagram illustrates OR process. Like most supported metal catalysts, Oswald has particle sintering during maturation at high-reaction temperatures. During the sintering process, the smaller metal particles disappear, while the larger metal particles become larger. This process is completed by the diffusion of single metal atoms on the surface of the carrier.

In addition to the above-mentioned OR process, another common agglomeration mechanism is particle migration and coalescence. The high surface energy makes the ADMMs of particularly small size (single atoms or ultrafine nanoparticles) easy to agglomerate into larger particles during the preparation process and long-term catalytic reaction, which affects the catalytic activity and selectivity. Therefore, it is very important to prevent the formation of nanoparticles in the actual experimental

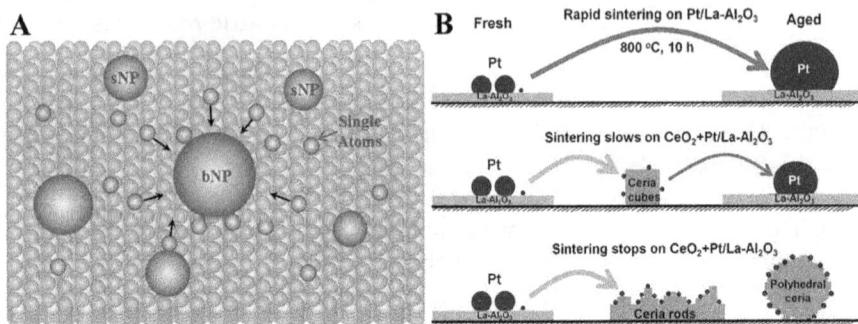

FIGURE 1.5 (a) Schematic diagram of the Oswald ripening process. sNP, small metal nanoparticles; bNP, big metal nanoparticles. (Reproduced with permission from Ref. [59]. Copyright 2017, American Chemical Society.) (b) Illustration of Pt nanoparticle sintering, showing how ceria can trap the mobile Pt to suppress sintering. Cubes appear to be less effective than rods or polyhedral ceria. (Reproduced with permission from Ref. [132]. Copyright 2016, American Association for the Advancement of Science.)

process, but it is a challenge for the current technical requirements to keep a single atom highly dispersed [129]. The key to achieving this goal is to form a strong interaction between the separated metal ions and the catalytic carrier. In this case, optimizing the metal precursor and the catalyst support (especially the anchor location on the support), regulating the interaction between the two, and strictly controlling the preparation process are essential for the stability of the single metal atom in the catalyst [130,131]. Based on the above considerations, the current main preparation methods are to use various methods to utilize and enhance the interaction between the metal atoms and the catalyst support, anchor the highly dispersed metal atoms and clusters on the support, and then select the corresponding preparation method. As shown in Figure 1.5b, Abhaya K. Datye and co-authors chose cerium oxide powders with different exposed surface areas for comparison. When platinum nanoparticles and aluminum oxide catalysts are mixed in the air and aged at 800°C, platinum is transferred to the ceria and is successfully captured. In terms of anchoring platinum, experimental results have shown that polyhedral cerium oxide and nanorods are more effective than cerium oxide cubes [132].

1.2 THE STRUCTURAL SENSITIVITY OF ATOMICALLY DISPERSED METALLIC MATERIALS

In heterogeneous catalysis, the reactions occur on the surface of solid materials. For nanoparticles, the interior atoms do not directly take part in the reaction, and only a small part of surface atoms (commonly featuring unsaturated coordination, such as the atoms at defect sites, edges, and vertices) show high catalytic activities, leading to poor atom utilization efficiencies and, particularly for those noble-metal-based catalysts (e.g., Pt, Pd, Rh, and Ir) [133,134]. Therefore, strategies like downsizing the nanoparticles and elevating surface atom exposure have been adopted to effectively boost atom utilization efficiencies. ADMMs have offered feasible solutions to the common issues associated with traditional heterogeneous catalysts [125]. In essence, the ADMMs can be considered as cluster/SA catalysts with the ultimate small size (i.e., atomic-level size) and thus optimized atom utilization efficiencies, and therefore stand as a perfect model of constructing low-cost catalysts. Besides, every active site in ADMMs is in close contact with the support, maximizing the number of site-support interfaces; therefore, ADMMs display an "interface effect", and the single-atomic sites can be exploited to their full competence by employing the proper support materials [126]. Moreover, the controllable and uniform active sites can endow ADMMs with selectivity comparable with those of homogeneous catalysts. Consequently, ADMMs, integrating the advantages of both homogeneous and heterogeneous catalysts, would definitely play a significant role in the future chemical industry.

In this section, we compare ADMMs with nanoparticles in terms of their catalytic behaviors and systematically interpret the impacts of the structural sensitivity of ADMMs on catalytic performances. To reveal the potential electrocatalytic mechanism and clarify the structure–activity relationship, four engineering effects of size, support, coordination and electronic effects are proposed based on the electrocatalytic applications.

1.2.1 SIZE EFFECT

When the size of nanoparticle is comparable to or even less than the optical wavelength, the de Broglie wavelength, the free path length of electrons, or the coherence wavelength of superconducting state, periodic boundary conditions and the translation symmetry of crystal nanoparticles would be broken down, and atomic density near surface of amorphous nanoparticles decreases under these conditions. Subsequently, the acoustic, magnetic, electronic, thermal, and mechanical characteristics are simultaneously changed so as to cause the appearance of some novel phenomena referred to as small size effect [135]. The most obvious effect of reducing the particle size is the increase of surface and interface areas. The atoms lying at the surfaces or interfaces have different environments in terms of both coordination and chemical properties, since the structure of grain boundaries is significantly different from that of bulk materials [136].

In addition, there are significant differences in the physicochemical properties of the same substance in the macro and micro states, and the main reason for this phenomenon is the change in the size of the substance. When the size of the substance is reduced to a certain value, the electron level near the Fermi level will change from the initial quasi-continuous state to the discrete state, which is known as the quantum size effect. The existence of the quantum size effect leads to many new and original characteristics of ADMMs [137]. For instance, Avelino Corma and co-workers investigated how the geometric and electronic differences between gold clusters and nanoparticles can determine their activity and selectivity. The clusters formed by only small amounts of atoms have discrete molecular-like electronic states, and their chemical reactivity is related to the interaction between the front molecular orbitals of clusters and the molecular orbitals of reactants (Figure 1.6) [138].

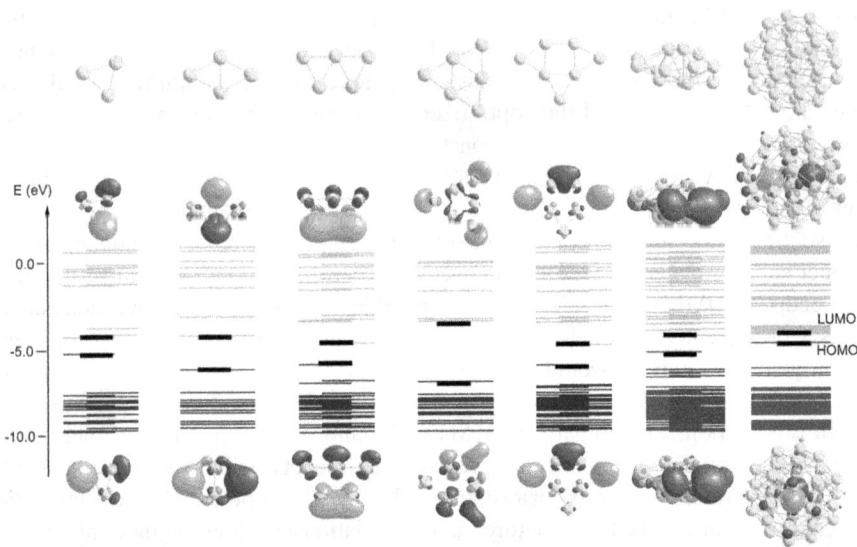

FIGURE 1.6 Optimized structure (top) and calculated isosurfaces of the lowest unoccupied molecular orbital (LUMO, center) and highest occupied molecular orbital (HOMO, bottom) of Au_3, Au_4, Au_5, Au_6, Au_7, Au_{13}, and Au_{38} clusters, together with molecular orbital energy levels. (Reproduced with permission from Ref. [138]. Copyright 2014, American Chemical Society.)

More importantly, the size of catalyst particles is a major influencing factor for catalytic activity. Utilizing the size effect can effectively elevate the atom utilization efficiencies. For instance, Cheng et al. prepared a series of Pt-based catalysts with varying Pt sizes through ALD [139]. By precisely controlling the deposition cycles, the size of Pt species can be tuned from single atoms and subnanometer clusters to nanoparticles. By virtue of the small size and unique electronic structures of single Pt atoms adsorbed on the N-doped graphene support, the single Pt atom catalysts display an exceptionally high hydrogen evolution reaction activity (with mass activity 37-fold higher than that for commercial 20% Pt/C) and high stability in the electrolyte of $0.5 M\ H_2SO_4$.

1.2.2 Support Effect

During the catalysis, the support materials can not only provide a large specific region for anchoring the active sites but also optimize the electronic structure and local geometry of the metal species [140,141]. The synergy between support and metal can further stimulate the reaction intermediates at the interface, thereby bestowing high activity for the catalytic reaction. In addition, the electronic structure of metal atom can be modulated by support materials through charge delocalization, making a slight change in the charge of the metal atom. Selecting a suitable support is a key factor to improve the overall catalytic performance of the supported metal catalysts.

Therefore, it is very important to study the relationship between the metal-support interaction and the corresponding catalytic performance.

Recently, Park et al. investigated the support effect systematically by single Pt atoms on carbon (Pt SA/C) and on tungsten suboxide (Pt SA/m-WO$_{3-x}$), as well as Pt nanoparticles on tungsten suboxide (Pt nanoparticle/m-WO$_{3-x}$) in terms of their hydrogen evolution reaction performances [142]. Scientists further showed that using tungsten suboxide as a support for a single Pt atom maximizes the support effect. The Pt SA/m-WO$_{3-x}$ catalysts showed a much faster hydrogen release at the interface than that of Pt nanoparticle/m-WO$_{3-x}$, leading to a large (10-fold) increase in their hydrogen evolution reaction mass activity. Yang et al. compared the hydrogen evolution reaction performance of single Ru atoms immobilized on different supports, including phosphorus nitride (PN) imide nanotubes, commercial XC-72, and C$_3$N$_4$ [143]. The overpotential of single Ru atoms on PN is 24 mV at 10 mA cm^{-2}, only 14 mV higher than that of commercial Pt/C. By contrast, the ADMMs supported on commercial XC-72 and C$_3$N$_4$ have overpotential values of 191 and 58 mV, respectively. Moreover, the PN-supported catalysts also show a more favorable kinetic with the Tafel slope of 38 mV dec^{-1} (122 mV dec^{-1} for commercial XC-72 and 125 mV dec^{-1} for C$_3$N$_4$). These results elucidate that a pronounced support effect for Ru-based ADMMs to improve their electrochemical hydrogen evolution reaction process. They also demonstrated a more appropriate adsorption-desorption behavior to promote the overall hydrogen evolution reaction performance of Ru SAs@PN by density functional theory calculations.

In addition, Zhang et al. anchored single Pt atoms on the Mo-vacancy defects of Mo$_2$TiC$_2$T$_x$ by the electrochemical exfoliation method during the hydrogen evolution reaction process [144]. The obtained catalyst features isolated Pt ADMMs that occupy the vacancies originally occupied by the Mo atom and coordinate with the adjacent C atoms, forming three Pt–C bonds. The intense interaction between Pt and the support leads to an overpotential as low as 30 mV (at 10 mA cm^{-2} in 0.5 M H$_2$SO$_4$), and 40 times of mass activity more than the commercial 20% Pt/C. Besides, the catalyst also displays outstanding stability, with no obvious activity decline after 10,000 cycles or 100 h of chronoamperometry. Zeng et al. reported a general method for the synthesis of single-atom site catalysts by electrochemical deposition strategy [145]. They synthesized a series of Ir/M (M=MnO$_2$, MoS$_2$, Co$_{0.8}$Fe$_{0.2}$Se$_2$, and N-C) SACs by anodic electrochemical deposition. By fine-tuning the metal type, carrier and electrode, different SACs can be obtained. A-Ir/Co$_{0.8}$Fe$_{0.2}$Se$_2$ exhibited the best oxygen evolution reaction performance.

Chen et al. deployed the site-specific electro-deposition method to fabricate a series of Pt/TMDs (TMDs=MoS$_2$, WS$_2$, MoSe$_2$ and WSe$_2$) to reveal the effect on the electronic metal-support interaction (EMSI) on the hydrogen evolution reaction activity (Figure 1.7). Pt/MoSe$_2$ exhibited the best alkaline hydrogen evolution reaction activity. The authors ascribed this to the EMSI regulation between TMDs supports and Pt single-atom. The electronic structure of single-atom Pt was modulated by 2D TMDs through charge delocalization, enabling the single-atom Pt to take slightly positive charge (Pt$^{\delta+}$), in which the d-band center position of Pt in Pt/WSe$_2$ exhibited a maximum upshift, enabling the stronger Pt–H adsorption energy [146].

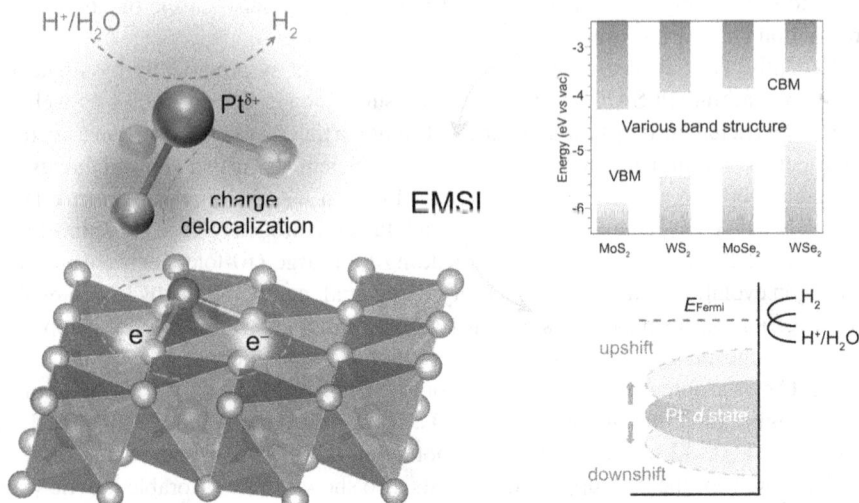

FIGURE 1.7 Electronic metal-support interactions (EMSI) modulation of single-atom Pt for catalyzing hydrogen evolution reaction. Left: schematic structure of single-atom Pt on TMDs material. The structural unit of single-atom Pt was circled by the dashed line and further enlarged above. Top right: schematic diagram of the band edges of TMDs. The schematic band structure showing the electron affinity and ionization potential of various TMDs provide a guideline for rationalizing the EMSI modulation of single-atom Pt. Bottom right: schematic illustrating that the d-state shift of single-atom Pt induced by EMSI regulates the catalytic performance of hydrogen evolution reaction. (Reproduced with permission from Ref. [146], Copyright 2021, Springer Nature.)

1.2.3 COORDINATION EFFECT

Deep understanding of the structure-performance relationship between metal monatomic coordination environment and catalytic performance would be conducive to the development and application of advanced ADMM [147]. The coordination environments of single-site metal atoms can significantly alter the adsorption affinity of the metal centers and, in turn, the catalytic performances. Furthermore, heteroatoms adjacent to metal atom center particularly affect the catalytic properties. In general, coordination engineering can be achieved by precisely adjusting individual metal active centers and coordination atoms. It should be pointed out that precise control of the coordination environment of a single metal active center in the M–N$_x$–C (M = transition metal, N = non-metal) catalyst can effectively improve the catalytic activity and selectivity, but it is extremely challenging [148].

For example, through finely tuning the number of N coordination atoms, efficient electroreduction of CO_2 can be successfully achieved. In 2017, Li et al. designed a ZIF-assisted route to synthesize Ni single atoms anchored nitrogen-doped carbon (Ni/N–C SACs) for CO_2 reduction reaction (Figure 1.8a) [149]. The Fourier transform (FT) k^3-weighted Ni k-edge extended X-ray absorption fine structure spectra for Ni/N–C SACs indicated the active center is NiN$_3$ moiety with Ni atom coordinated

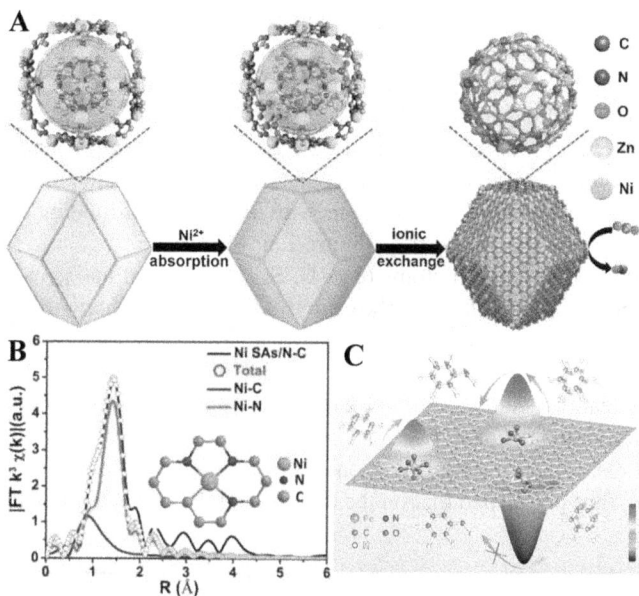

FIGURE 1.8 (a) Schematic diagram of the synthesis process of Ni/N–C SACs. (b) The extended X-ray absorption fine structure fitting curves for Ni/N–C SACs with the proposed Ni–N$_3$ model. (Reproduced with permission from Ref. [149]. Copyright 2017, American Chemical Society.) (c) Coordination effect of different single-atomic-site catalysts in an electrocatalytic benzene oxidation reaction. (Reproduced with permission from Ref. [153], Copyright 2019, Springer Nature.)

by three N atoms (Figure 1.8b). Jiang et al. reported the high-precision construction and manipulation of Fe–N$_4$ active sites by selective destruction of peripheral C–N bonds [150]. Compared with the ideal Fe–N$_4$ model, the Fe–N$_4$ structures located at the edge of the carbon support pores can significantly reduce the reaction free energies of the electron transfer in the first step and the desorption of *OH in the final step during the oxygen reduction reaction process, resulting in the lowest oxygen reduction reaction kinetic barrier. For Fe–N$_4$ located at the edge sites, the E$_{1/2}$ reaches 0.915 V with a merely Fe loading of 0.2 wt%. This work demonstrated Fe–N$_4$ sites located at the pore edges of carbon support possess higher catalytic activities, providing a feasible strategy for rational design and precise regulation of efficient and stable ADMM. Yin et al. synthesized two Pt-based ADMMs with different coordination environments by the coordination effect between the C$_2$–Pt–Cl$_2$ moieties in graphdiyne support and Pt atoms [151]. Compared with the five-coordinated C$_1$–Pt–Cl$_4$ catalyst, the four-coordinated C$_2$–Pt–Cl$_2$ catalyst possesses Pt 5d orbitals with a higher unoccupied state density and exhibits superior performances toward hydrogen evolution reaction. In addition, the mass activity of C$_2$–Pt–Cl$_2$ catalyst was 3.3 times and 26.9 times higher than those of C$_1$–Pt–Cl$_4$ catalyst and commercial Pt/C, respectively. Yao et al. investigated the structure-performance relationships for three Ni-based ADMMs with different coordination configurations (on Di-vacancy, D5775, and perfect hexagons) toward hydrogen evolution reaction and oxygen evolution reaction [152]. They elucidated that the specific coordination configurations containing metal single atoms and different carbon defects lead to different electrochemical activities, which would offer a guidance for the ADMMs configuration-specific construction. Pan et al. prepared a series of Fe-based ADMM with different coordination environments by a polymerization regulated pyrolysis method. For the benzene oxidation reaction, the individual Fe atoms with different coordination environments exhibit different catalytic properties (Figure 1.8c) [153].

Moreover, other non-metal atoms (e.g., N, S, and P) coordinating with the metal centers also show distinct effects on the catalytic activity and selectivity. For example, Zhang et al. fabricated N and S coordinated Cu (S-Cu/SNC) single-atom catalysis via atomic interface engineering for oxygen reduction reaction [154]. X-ray adsorption characterization results unraveled that the moiety configuration in S-Cu/SNC is the unsymmetrically arranged $Cu-S_1N_3$ (Figure 1.9a and b) [155]. By the high-angle annular dark-field scanning transmission electron microscopy and X-ray absorption spectroscopy, they confirmed that on the carbon support with a high S content (HSC,

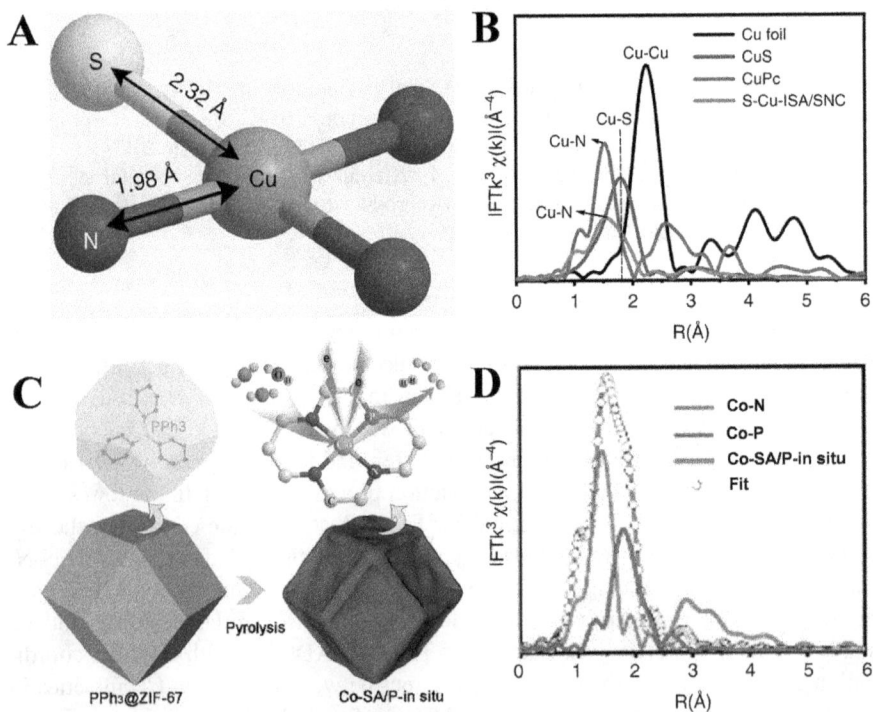

FIGURE 1.9 (a) Schematic model of Cu/S_1N_3C. (b) FT k^3-weighted Cu K-edge extended X-ray absorption fine structure spectra of Cu/S_1N_3C, Cu foil, CuS, and CuPc. (Reproduced with permission from Ref. [154], Copyright 2020, Springer Nature.) (c) Schematic illustration of the synthesis process of Co-SA/P-in situ. (d) k^3-weight extended X-ray absorption fine structure Co K-edge fitting curves of Co-SA/P-in situ. (Reproduced with permission from Ref. [147], Copyright 2020, American Chemical Society.)

17 wt%), the Pt species exists solely in the form of Pt–S_4 sites with 5 wt% Pt loading; on the carbon support with a low S content (LSC; 4 wt%), the Pt species exists in the form of both Pt clusters and Pt single atoms; on the S-free carbon support, Pt exists in the form of nanoparticles (4 nm in size). In the oxygen reduction reaction process, the Pt–S_4/HSC catalyst does not follow the typical 4e⁻ pathway to yield H_2O, while predominantly producing H_2O_2 through a 2e⁻ pathway ($n = 2.1$) without further decomposing. By contrast, the Pt/ZTC and Pt/LSC show mixed 4e⁻ and 2e⁻ pathways ($n = 3.5$ for the former, 2.9 for the latter).

In another case, the phosphorus element was introduced into the carbon matrix to boost the hydrogen reaction kinetics. Specifically, Wang et al. reported a Co-SA/P-in situ single-atom catalysis with Co_1–P_1N_3 interfacial structure through in situ phosphatizing of triphenylphosphine encapsulated within ZIF (Figure 1.9c) [147]. The Co_1–P_1N_3 configuration was verified by the extended X-ray absorption fine structure fitting, evidencing the presence of isolated Co atoms surrounded by three nitrogen atoms and one phosphorus atom (Figure 1.9d). Liu et al. prepared a Co-based single-atom catalysis with Co_1–P_4 sites on a P-doped C_3N_4 support (denoted Co_1–P_4/PCN) by the confinement effect of C_3N_4 [156]. The Co_1–P_4/PCN catalyst has a unique intermediate state in its band structure, which not only drastically enhances the visible light absorption but also effectively suppresses the photogenerated electron-hole pairs recombination (with the lifetime of photocarriers prolonged by 20-fold). As a result, the catalyst achieves a hydrogen production rate of 410.3 mmol $g^{-1}h^{-1}$ under the conditions of simulated solar illumination without using sacrificial agents or noble metals, reached a quantum efficiency of 2.2% at 500 nm. These results further illustrate that modifying the coordination environment of the ADMMs can effectively tune the catalytic selectivity toward different products.

1.2.4 ELECTRONIC EFFECT

On the one hand, the catalytic performance is also affected by ADMMs with different electronic valence states. On the other hand, Benefiting from the synergistic electronic effects between metal and supports, the ADMMs possessed moderate reaction intermediate binding strength and optimized adsorption energy, thus leading to enhanced catalytic activity with excellent durability. The electronic interaction mainly involves the shift of d-band center and the change of charge distribution.

For example, Gu et al. prepared an Fe^{3+}-based ADMM (denoted Fe^{3+}–N–C) capable of converting CO_2 into CO under an overpotential as low as 80 mV [157]. At a high overpotential of 340 mV, the catalyst can deliver a current density of 94 mA cm^2, with an FECO above 90%. By operating XAS, the team determined that Fe^{3+} is the real active site. Fe^{3+} exists in an isolated state coordinated with the pyrrole nitrogen atom in the nitrogen-doped carbon carrier and remains in the valence state during the electrocatalysis process. At the same time, the team also confirmed that the valence state of the metal center is closely related to the catalytic performance and electronic effect. By comparing the Fe^{3+}–N–C catalyst with counterparts with Fe^{2+}–N–C sites (Fe species coordinated with pyridinic N atoms), the team found that the pyrrolic N atoms play a critical role in preserving the Fe^{3+} centers, and the Fe^{3+} centers feature a weaker bonding with CO and faster adsorption to CO_2 with respect to Fe^{2+}, thereby guaranteeing the superior activity and stability. In addition, in the scheme of improving catalytic performance, the doping of heteroatoms (such as sulfur and phosphorus) may also achieve this goal. It is found that the doping of heteroatoms may change the electronic structure of ADMMs [158,159].

Recently, Thomas et al. deployed the transition metal carbides (WC_x) as the substrate to anchor single-atom Fe or Ni, bi-atomic Fe and Ni metal species for oxygen evolution reaction [160]. Combined with a precipitation reaction and pyrolysis process, they obtained a series of atomically dispersed Fe/Ni atoms confined WC_x crystallites supported by the carbon sheet (Figure 1.10a). The calculated overpotentials of all these modulated catalysts are summarized in Figure 1.10b using a three-dimensional volcano-shaped relationship of the free energy difference between OH* and O* (or OOH*) species. In particular, the WC_x-FeNi with C- or W-terminated atoms all provide moderate oxygen-binding strength ($E_{OH*} = 0.94 - 1.13$ eV and $E_{O*} - E_{OH*} = 1.2 - 1.4$ eV) and give near-optimal free energies for each oxygen evolution reaction intermediate, thus resulting in a low theoretical oxygen evolution reaction overpotential (<0.16 V).

1.3   CLASSIFICATION OF ATOMICALLY DISPERSED METALLIC MATERIALS

According to the number of atoms of metal active sites, the atomic-level dispersion catalysts can be divided into SACs, diatomic catalysts and metal nanocluster catalysts.

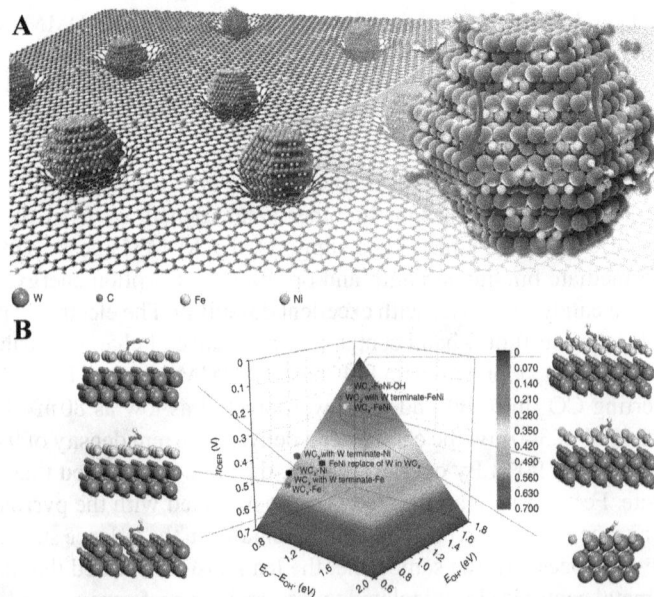

FIGURE 1.10   (a) Schematic illustration of WC_x-FeNi catalyst, consisting of Fe and Ni atoms stabilized on WC_x nanocrystallites (majority component) surrounded by carbon sheets. (b) Volcano plot for all calculated WC_x-based catalysts and corresponding modules. (Reproduced with permission from Ref. [160], Copyright 2021, Springer Nature.)

1.3.1 Single-Atom Materials

A single-atom catalyst is a catalyst formed by dispersing metal atoms in the form of single isolated atoms on the support. There are no metal–metal bonds between atoms of the same metallic element. In 2011, Zhang and his team reported a carbon monoxide oxidation catalyst that uniformly supports monodisperse platinum atoms on iron oxide surfaces. This work was the first to propose the concept of "single-atom catalysis" (SACs) concept [25].

SACs play a huge role as a link of communication between organic and heterogeneous catalysts in the search for advanced, effective and sustainable catalysts. Atomically dispersed supported metal catalysts can provide highly efficient and highly active surface sites for sustainable and benign catalytic transformation. SACs not only maximize the use of active sites, but also improve efficiency, reduce the use of metals, and improve catalytic selectivity in the conversion process [109]. Figure 1.11 summarizes the report on the SACs study, including the fields of catalytic application of SACs by element classification, the type of hosts employed for distinct elemental classes of SACs, and the highest metal loading achieved within all SACs groups.

1. Applications:

   SACs have been widely used in thermalcatalysis during their development, while the application of photocatalysis, especially electrocatalysis, has become prominent only in recent years. A greater exploration of the thermal drive process has been shown in the wider element diversity of the applied SACs (Figure 1.11a). Among all types of catalysis, the most versatile elements are used as SACs, including Pt, Pd, Ir, Ni, and Cu from the late transition metal family, as well as P, S, and Se from nonmetals and B from metalloids. Interestingly, the less-studied SACs of early transition metals, metalloids, and halogens are primarily used as electrocatalysts, while their

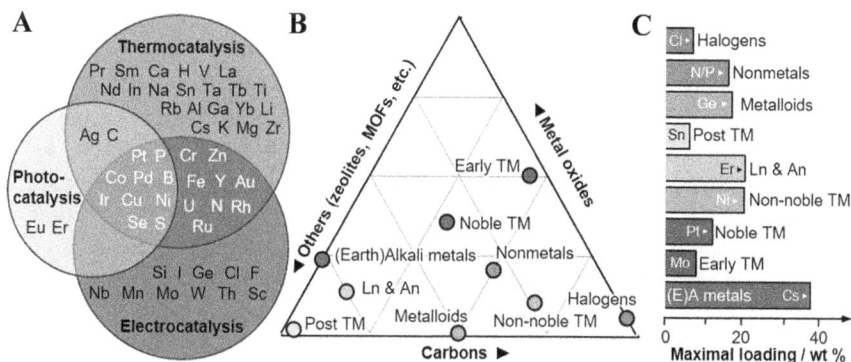

FIGURE 1.11 Comparison of (a) the modes of catalytic application of SACs by element, (b) the type of hosts employed for distinct elemental classes of SACs, and (c) the highest metal loadings achieved within all SACs groups, indicating the corresponding element. (Reproduced with permission from Ref. [109]. Copyright 2020, American Chemical Society.)

potential for thermal and photocatalytic applications remains little explored. Lanthanide elements (such as europium and Erbium) are traditionally used as dopants in photocatalytic materials and are currently mainly used as single-atom photocatalysts, but they have not yet entered other fields. At the same time, the range of elements in photocatalysis for SACs is quite limited and is expected to expand rapidly as interest in the field grows.

2. Hosts:

Among all the categories of elements, the noble transition metals show the largest variety of applied host materials, including reductive and non-reductive metal oxides, carbon-based materials, and many other systems (Figure 1.11b). This diverse metal-host combination is capable of generating a wide range of electron and geometric spectra and opens up a wide range of applications in thermal, electrical, and photocatalysis (e.g. Pt, Pd, Ir). In contrast, most of the other element groups show a strong preference for a particular host system. For example, alkali and alkaline earth metals are almost completely loaded on zeolite and used in thermal catalysis. Similar trends are found in f block elements and metal-like elements, where zeolite and MOF-based support dominate. In contrast, the ADMMs of the known early transition metals were primarily based on metal oxides and carbon. Finally, non-noble transition metals (e.g., Fe, Co, Ni) and halogens (e.g., Cl, F, I) bind primarily to carbon-based bodies, in line with their strong orientation for electrocatalytic applications.

3. Elemental Loadings:

The maximal achieved elemental loadings such as the commonly reported percentage of weight vary widely between SACs classes (Figure 1.11c). In particular, alkali metals (Cs/NaX zeolite), nonmetals (S/C), metal-like metals (Ge/ITQ-22 zeolite), non-noble late transition metals (Ni/NC) and lanthanides (Er/C_3N_4) were stabilized under loadings of ≥ 20 wt%. On the other hand, such high loadings have not been reported for halogens (Cl/C), early transition - (Mo/NC), late transition - (Pt/MOF), and post-transition metals (Sn-β zeolite). It should be noted that there is considerable variation in individual groups. For instance, in noble transition metals, the maximum content is between ~12 wt% (Pt/MOF) and 1 wt% (Au/C or Au/TiO_2). In addition, the hosts and degree of functionalization also have a significant impact on the achievable loading, which has not been studied in detail for most elements. In general, zeolite and doped carbon can accommodate more single atoms than metal oxides. In this regard, with the exploration of zeolite and carbon-based hosts, the possible loading of early transition metal-based SACs is expected to increase.

1.3.1.1 SACs on Oxide Supports

In heterogeneous catalysts, the selection and optimization of support materials play a vital importance in catalytic performance. For one part, the support materials offer a defective coordination environment for metal atoms, to prevent their aggregation effectively and stably. For another part, the electronic structure of as-obtained single metal sites is highly determined by the coordination environment [161,162]. Hence,

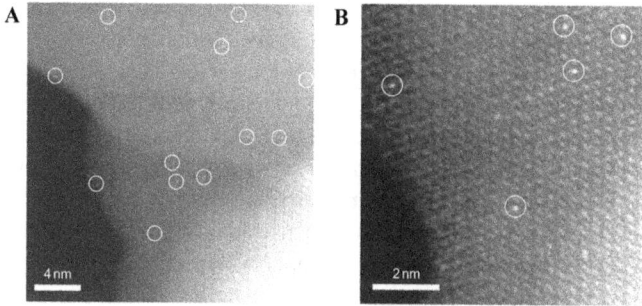

FIGURE 1.12 (a) Pt single atoms (white circles) are seen to be uniformly dispersed on the FeO$_x$ support and occupy exactly the positions of the Fe atoms. (b) Examination of different regions reveals that only Pt single atoms are present in Pt/Fe$_2$O$_3$. (Reproduced with permission from Ref. [25], Copyright 2011, Springer Nature.)

studying the interaction between metal atoms and support materials is essential to improve the catalytic performance of SACs.

In all kinds of substrates, metal oxides are the most common used because of their abundance of metal or oxygen vacancies a method for the synthesis of highly monodisperse Pt/Fe$_2$O$_3$ by co-precipitation synthesis was reported in 2011 by Zhang et al (Figure 1.12) [25]. The extended X-ray absorption fine structure results showed that the coordinated-O with Pt in Pt$_1$/FeO$_x$ is close to 3, which mean every site includes 3 Pt–O bonds. When the Pt atom was in the position of Fe$_2$O$_3$ (001), one of the Fe atoms would be replaced by Pt on the surface of O$_3$ terminal. When the reaction took place in a reducing atmosphere, the FeO$_x$ support would be reduced to Fe$_3$O$_4$, and the coordination number of Pt was reduced to 2. When the Pt atoms on the surface largely aggregated, the Fe$_2$O$_3$ would be transferred to Fe$_3$O$_4$. Since the electrons of Pt were transferred to the surface of FeO$_x$, the Pt atoms became more stable and the positively charged Pt atoms were produced, which contribute to the strong binding energy provided by the vacant d orbital of Pt single atom. In the latest research report, Christophe Copéret and co-authors used the surface organometallic chemistry method to prepare atom-dispersed iridium nanocatalysts on single-crystal MgO(-111) nanosheets [163]. At the metal loading of 1 wt%, the main species is composed of monoatomically dispersed Ir, Ir dimers, and trimers. In the case of low loading, all Ir single atoms have the same coordination structure. These species display unique catalytic properties in the coupling reaction of benzene and ethylene to form styrene, a reactivity that contrasts with conventional homogeneous and heterogeneous iridium catalysts that yield ethylbenzene.

1.3.1.2 SACs on Carbon Supports

Because of the chemical inertness and cheap availability, carbon materials are widely used in numerous catalytic reactions. In particular, most of the commercial catalysts are carbon-supported metal particles, benefiting from their easy synthesis paths, corrosion resistance, and high activities [164,165]. In recent years, carbon supports are becoming the best candidate for the metal SACs substrate because their electronic structures can be easily adjusted by doping N and other heteroatoms, leading to highly stable coordination interactions with metal atoms on the surface [166]. Among all carbon supports, mesoporous polymeric graphitic carbon nitride (mpg-C_3N_4) is well known for its unique tri-s-triazine structure and six-N-N-coordination cavity which is easy to coordinate with metal atoms. The high N content can supply enough coordination sites for anchoring metal atoms. What's more, a two-dimensional structure with van der Waals interaction between layers delivers good stability, high activity, and/or selectivity for photocatalytic and electrocatalytic reactions due to its unique photo-responsive activity and conductive property (Figure 1.13) [167–170].

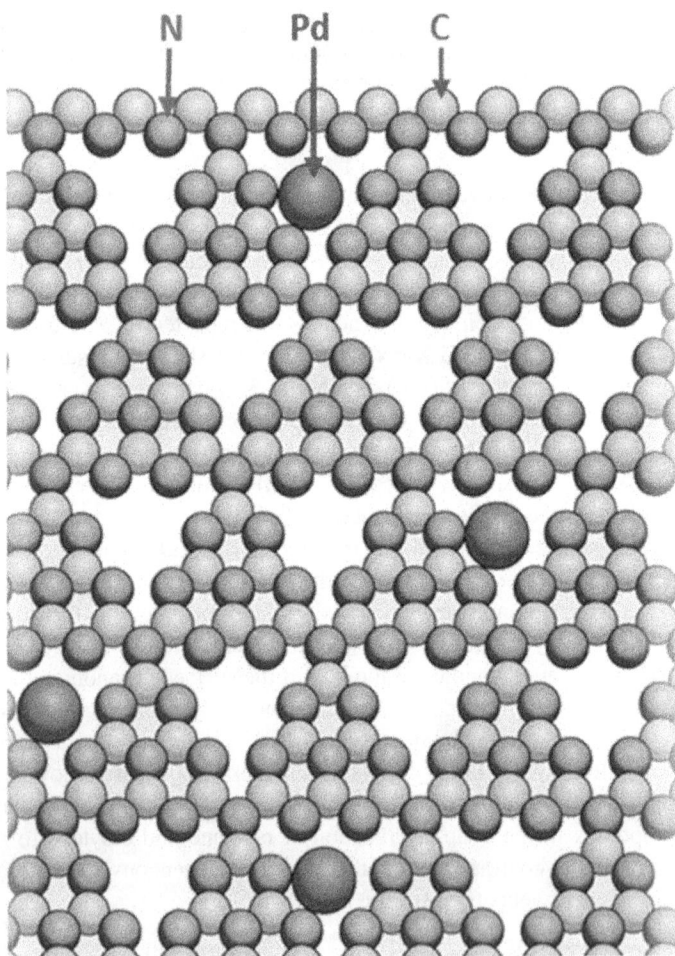

FIGURE 1.13 Isolated Pd atoms anchored onto porous C_3N_4 act as catalysts for hydrogenation reactions. Strong bonds to the nitrogen atoms firmly anchor the Pd atoms inside the roughly triangular pores in the stacked, two-dimensional layers of C_3N_4. Only one layer is depicted, for simplicity. (Reproduced with permission from Ref. [59], Copyright 2017, American Chemical Society.)

In 1964 Jasinski began to study the activity of cobalt phthalocyanine oxygen reduction reaction [171]. Since then, metal-N–C material began to attract attention. By coordinating the monodispersed transition metals with adjacent N atoms, M–N–C catalysts (M = Ni, Fe, Co, Cu, …) exhibit great potential as the alternative catalyst for precious metal-based catalysts in many reactions, especially in electrocatalysis [172–177]. The James M. Tour team synthesized M–N–C using graphene oxide based on graphene's high specific surface area and electronic conductivity. In this strategy, a small amount of cobalt salt is first added to go and then heated in an NH_3 atmosphere to obtain atoms dispersed on n-doped graphene. It was verified that the material showed good catalytic performance and durability for hydrogen evolution reaction in an acidic medium.

1.3.1.3 SACs on Metal Supports

Nørskov and his colleagues discovered that the addition of a small amount of Au on the Ni surface can effectively reduce the activity of the metal, further preventing the sintering of the Ni and improving its steam reforming performance. In another research by the Sykes group, the formation process of single-atom bimetal alloys system (SAAs) was investigated through low-temperature scanning tunneling microscopy and density functional theory calculations. In recent years, a large number of SAAs catalysts have been synthesized, including Pd–Au, Pd–Ag, Pd–Cu, Pt–Cu, Cu–Ag, and so on (Figure 1.14) [178–181]. Hong and Li et al. used a one-pot wet method to disperse ruthenium atoms on the edges or steps of non-atomically flat Pd

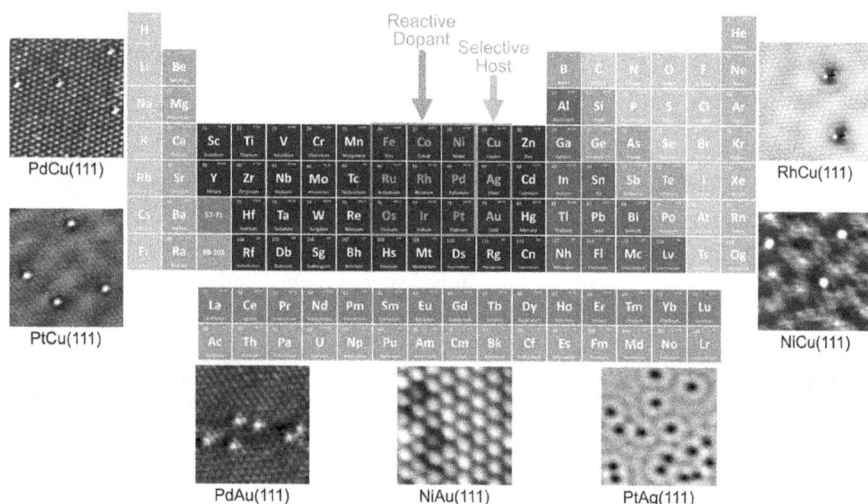

FIGURE 1.14 Periodic table showing the SAA approach of adding small amounts of a reactive dopant into a less reactive, more selective host metal. Atomically resolved scanning tunneling microscopy images of the SAAs pictured are courtesy of Sykes and co-workers. The isolated dopant atom sites appear in scanning tunneling microscopy images as protrusions or depressions depending on the alloy and the imaging conditions. (Reproduced with permission from Ref. [178], Copyright 2020, American Chemical Society.)

nanosheets [182]. This catalyst with a Ru mass loading of 5.9 wt% delivers a good catalytic efficiency in alkene hydrogenation. Through ingenious design and precise control of SAAs, Wu et al. engineered the electronic structure of the atomically dispersed Ru atoms on the bimetallic alloy support through adjusting the compressive strain [183]. Using a sequential procedure including acid etching and electrochemical leaching, they synthesized unique Pt–Cu alloys embedded with Ru atoms and finally realized the PtCu$_x$/Pt skin core-shell structure.

1.3.2 DIATOMIC MATERIALS

Diatomic catalyst is a novel concept in recent years. The metal atoms in the active center of the catalyst appear in pairs. Diatomic catalysts can be divided into homonuclear diatomic catalysts and heteronuclear diatomic catalysts according to the types of the two metal atoms in the active center. Diatomic catalysts maintain the high dispersity of single-atom catalysts and therefore have high atomic utilization and selectivity. Due to the synergy between adjacent active sites, the catalytic activity of diatomic catalysts can be improved to a certain extent, resulting in the effect of "one plus one is greater than two". Especially for some multi-site catalytic reactions, diatomic catalysts have better activity than monatomic catalysts.

Despite of the great success, SACs encountered considerable challenges in some more complicated catalytic reactions involving multiple reactants, intermediates, and products. The limitations generally originate from the simple structure of the active center, which has only one atom. Naturally, scientists came up with the idea of constructing more complex catalytic sites, with multiple metal atoms as active centers (in order to integrate their functions and even take advantage of potential synergies), while being well isolated in SACs (in order to inherit the advantages of high exposure and high atom utilization efficiencies). In recent years, scientists have achieved remarkable success in the design, construction, characterization, and application of metal diatomic site catalysts. While reports on these new systems to date have been sporadic, we believe they are representing an important category of the next generation of advanced catalysts.

Exploring the interactions between adjacent metal atoms in metal diatomic site catalysts will help to further improve catalytic performance and understand catalytic mechanisms at the atomic level. In general, metal diatomic site catalysts have higher metal loadings than SACs because metal atoms are more likely to form diatomic pairs at higher loadings. The metal diatomic site catalysts can be further divided into two sub-classes: (i) Homo-paired metal diatomic site catalysts, with two metal atoms, have the same atomic nucleus; (ii) Different pairs of metal diatomic site catalysts, with different nuclear and chemical environments.

Homo-Paired metal diatomic site catalysts synergistic catalysis by homo-paired metal diatomic site catalysts can be applied to CO_2-related activation reactions. For example, Li et al. reported a diatomic paired Cu_1^0–Cu_1^{x+} catalyst for CO_2 reduction reaction [184]. The paired Cu_1^0–Cu_1^{x+} sites are anchored on one-dimensional $Pd_{10}Te_3$ alloy nanowires (Figure 1.15a). The Cu–Cu bonding is confirmed in the k^3 weighted spectrum of FT; the fitting results on extended X-ray absorption fine structure indicate a Cu–Cu coordination number of 2.1, far lower than that for Cu foil (12),

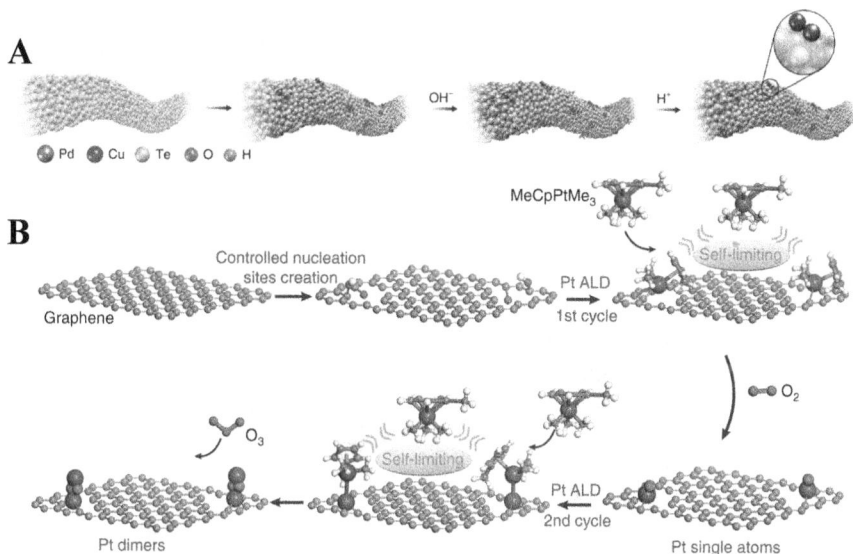

FIGURE 1.15 (a) Synergistic catalysis by homo-paired metal diatomic site catalysts in electrocatalytic CO_2 reduction reaction (the fabrication procedures of atom-pair structured Cu anchored on $Pd_{10}Te_3$ nanowires). (Reproduced with permission from Ref. [184], Copyright 2019, Springer Nature.) (b) Schematic illustration of bottom-up synthesis of dimeric Pt_2/graphene catalysts. Controlled creation of isolated anchor sites on pristine graphene; one cycle of Pt ALD on the anchor sites for Pt single atoms formation by alternately exposing MeCpPtMe$_3$ and molecular O_2 at 250°C; second cycle of Pt ALD on the Pt_1/graphene to selectively deposit the secondary Pt atoms on the preliminary ones for Pt_2 dimers formation at 150°C. (Reproduced with permission from Ref. [185], Copyright 2017, Springer Nature.)

implying that the loaded Cu species exist in the form of well-isolated small clusters. Density functional theory calculations show that the $Cu_1^0-Cu_1^{x+}$ diatomic site has the most favorable kinetics and thermodynamic properties, with Cu_1^0 adsorbing a CO_2 molecule, and the adjacent Cu_1^{x+} adsorbing an H_2O molecule, which is in an ideal configuration to stabilize the chemisorbed CO_2; therefore, the $Cu_1^0-Cu_1^{x+}$ pair works synergistically to facilitate the critical bimolecular step in CO_2 reduction reaction. In brief, this catalytic system follows the "diatomic activating bimolecular" mechanism, so it is a perfect model to demonstrate the unique advantages of metal diatomic site catalysts in complex catalytic reactions.

In addition, Li et al. prepared a Pt/ MoS_2 catalyst with a Pt loading up to 7.5 wt% [185]. Compared with the counterpart with a lower Pt loading (0.2 wt%), the catalyst with high loading gives a turnover frequency 15 times that of the conversion of CO_2 to methanol; in addition, the activation energy of CO_2 hydrogenation is greatly reduced, resulting in a remarkably elevated activity. More interestingly, the synergy within the active sites containing two adjacent Pt atoms not only lowers the kinetic barrier but also alters the reaction pathway. On isolated single Pt atoms, CO_2 is directly converted to CH_3OH without forming HCOOH intermediate; In contrast, CO_2 is converted into HCOOH in the first step on the Pt diatomic sites (Pt–S–Pt), and then further hydrogenated into CH_3OH. Lu group shows that Pt_2 dimers can be fabricated with a bottom-up approach on graphene using ALD, through proper nucleation sites creation, Pt_1 single-atom deposition and attaching a secondary Pt atom selectively on the preliminary one (Figures 1.15b).

1.3.3 NANOCLUSTER MATERIALS

Metal atom clusters are generally composed of a metal core and peripheral ligands. The metal core generally contains several to hundreds of metal atoms, which size between single atoms and nanoparticles, usually less than 3nm. Metal clusters can be categorized into clusters containing multiple metals and single metal clusters containing only one metal pursuant to the metal nuclei classification. Likewise, in supported cluster catalysts exist quantum size effects and metal-interface interactions as well, unlike the discrete energy levels of single-atom catalysts. The electronic structure, physical and chemical properties of atomic cluster catalysts are different from those of macroscopic materials, as well as monatomic and diatomic catalysts, because there are a large number of metal atoms in them, and atomic orbitals still overlap with each other partially. Metal atom cluster curtains are highly susceptible to environmental fluctuations. Its properties will certainly change by altering the number of its atoms, geometric configuration, and coordination environment. Even though the size of metal clusters is between single atoms and nanoparticles, a simple linear combination of the properties of single atoms and metal nanoparticles is not able to obtain their physical and chemical properties. Metal nanoclusters fill the gap between discrete atoms and plasma nanoparticles, providing a unique opportunity to investigate quantum effects and precise structure-property correlations at the atomic level.

In improving the catalytic performance, the synergistic interaction between metal atoms is singular among supported atomic clusters catalysts containing multiple metal atoms. e.g., in many key industrial catalysis processes, especially in multi-substrate catalytic reactions, the active site of a single atom cannot simultaneously adsorb and activate multiple reaction substrates, thereby reducing the catalytic activity. In comparison, clusters containing multiple atoms can provide multiple sites for the adsorption and activation of the reactants to provide a new path for the catalytic reaction and reduce the reaction barrier. In recent years, supported catalysts based on metal clusters have attracted people's attention and been extensively used in many catalytic reactions, which benefit from the development of characterization techniques, particularly in-situ characterization of dynamic catalyst process and the theoretical calculation. These are based on the characteristics mentioned above.

As a special kind of nanoparticle, single crystals can be grown from the unique nanoparticles of atomically precise nanoclusters, and their crystal structures can be determined by X-ray crystallography. This provides a great opportunity for researchers to link their properties to their clearly defined structures.

1. In terms of structure, the fundamental knowledge of the structure-property correlations of nanoparticles is limited by the unknown surface structures and the ambiguity of metal-organic bonding modes. It is an indisputable fact that no two particles are the same. This fact hampers many studies

of their fundamental properties. The precisely characterized structure of ultra-small nanoclusters (typically 1–3 nm in diameter) makes them capable of providing an optimal platform for detailed study of the mechanism of structure-dependent properties at the atomic level.

2. Consider the aspect of size, nanoclusters are a perfect model for establishing a bridge from metal complexes to conventional nanoparticles because nanoclusters with a diameter of 1–3 nm can be placed between metal complexes and plasma metal nanoparticles (Figure 1.16).

3. The strong quantum size effects of nanoclusters are mainly reflected in their chemical and physical characteristics in the property. For instance, discrete energy levels of electrons and multiple absorption bands because of molecular-like electronic transitions. Previous research has established that the change of nanoclusters, even minimal, like adding/subtracting one electron into/from the nanocluster template, will cause non-negligible interference to the electronic structure and geometric structure of nanoclusters, then significantly affect their properties. Nonetheless, in regular nanoparticles, recognizable differences may not result from adding or subtracting several metal atoms.

1.4 FUTURE TECHNICAL CHALLENGES AND PROSPECT

ADMMs are an exploding field of research that is bringing material science into new territory. New catalysts with unique properties and prospects of practical application

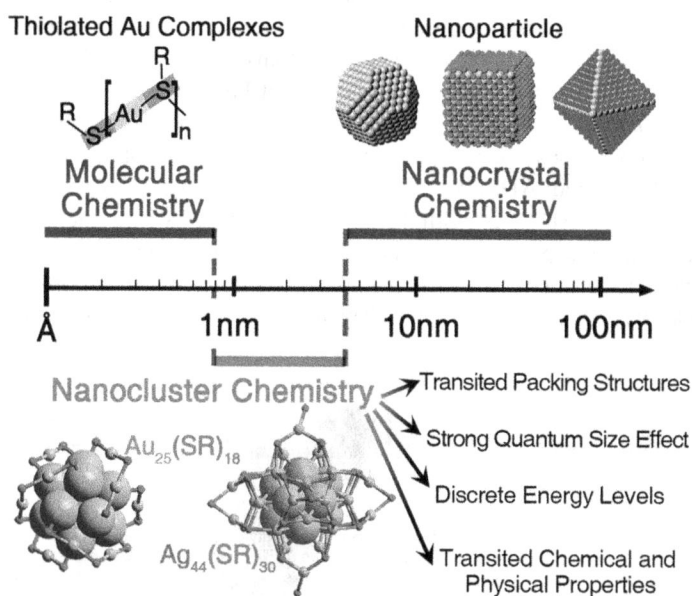

FIGURE 1.16 Illustration of the scope of nanocluster science. Metal nanoclusters bridge metal complexes and plasmonic metal nanoparticles (SR = thiolate). (Reproduced with permission from Ref. [186], Copyright 2020, Royal Society of Chemistry.)

foretell a wide horizon for innovation. Their success can be attributed to a combination of their extensive use in heterogeneous catalysis and their high compatibility for existing characterization methods (e.g., electron microscopy, X-ray absorption spectroscopy, X-ray photoelectron spectroscopy, density functional theory calculation). Advances in the science of ADMMs are stimulating advances in nanomaterials synthesis chemistry, surface single-crystal surface science and characterization techniques. This field is leading material science to increase elegance and rigor, entrenching it more firmly at the center of chemical science. Although some characterization and synthetic challenges still need to be overcome to exploit the potential of ADMMs, it will certainly enrich the field and deepen our understanding of catalytic processes. Furthermore, ADMMs have entered different reaction fields, but there is still a great deal of applications that have not yet been evaluated for their potential. Thus far, still underdeveloped areas encompass tandem/cascade reactions, traditionally homogeneously catalyzed reactions, SAzymes, and energy-related applications. Some timely challenges covering the synthesis, characterization, and application-related challenges are listed in this section (Figure 1.17).

1. Engineering the logical design for large-scale manufacturing of ADMMs is highly desirable for its essential progress in a variety of fields. For ADMMs rational designs, several theoretical and experimental models are provided. Aside from these expected models, comprehending catalytic activity and its relationship to composition is a complicated task. When the catalyst involves a mix of many types of active sites (different active site structures), compositions (support and bimetallic to multimetallic), and size, testing all the necessary catalyst combinations for the required reaction is just intolerable (SAs to a cluster of several single atoms). These variables are highly interdependent, and catalytic activity might be the result of a mixture of all of them. A methodical approach, on the other hand, might lead to the removal of all obstacles and the development of reasonable catalysts.

2. Densely Packed ADMMs: For a variety of industrial applications, the synthesis of ADMMs with high metal loading is extremely difficult. The traditional pyrolysis technique, for example, generally results in the agglomeration of SAs into larger nanoparticles. Several alternative techniques have been described, but they are still confined to metal loadings of

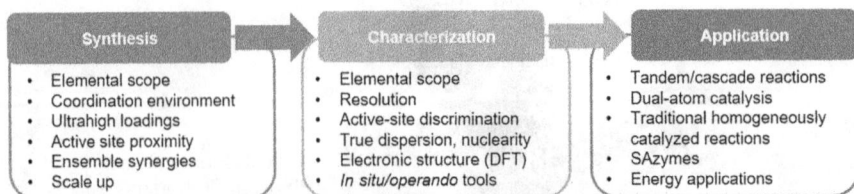

Synthesis	Characterization	Application
• Elemental scope	• Elemental scope	• Tandem/cascade reactions
• Coordination environment	• Resolution	• Dual-atom catalysis
• Ultrahigh loadings	• Active-site discrimination	• Traditional homogeneously
• Active site proximity	• True dispersion, nuclearity	catalyzed reactions
• Ensemble synergies	• Electronic structure (DFT)	• SAzymes
• Scale up	• *In situ/operando* tools	• Energy applications

FIGURE 1.17 Overview of the frontiers in the synthesis, characterization, and application of single-atom catalysts. (Reproduced with permission from Ref. [109] Copyright 2020, American Chemical Society.)

<10% by weight. There are very few reports that have a metal loading more than 10% by weight. Despite significant advances in the synthesis of dense ADMMs, controlling the kind and location of active sites remains difficult.

3. Thermodynamic Stability of ADMMs: Thermodynamic stability of ADMMs is a major concern, and using ADMMs at high temperatures is very desirable. The working temperature is generally kept below 300°C to keep them separate. Few examples of ADMMs being used in industrially important high-temperature (>300°C) gas-phase applications have been recorded. A strong connection through a strong covalent metal-support contact can improve the thermal stability of ADMMs. The SAs was stabilized using a controlled high-temperature shockwave technique, atom trapping, and entropy-driven methods. The creation of thermodynamically advantageous strong metal-defect bonds was enabled by periodic on-off heating for a brief duration (55 ms).

4. Fundamental Knowledge of ADMMs (Atomic-Level Data): ADMMs have been widely used to catalyze a variety of processes, and it now appears that supported metal atoms may successfully catalyze even the most difficult reactions. The majority of the reports to date have focused on synthesis, stability, and increasing metal loading, but it is yet unclear how the coordination environment or structures of active site geometry impact overall performance. Different heteroatom doping and the usage of distinct neighbor metal atoms may be used to modify the coordination environments surrounding the SA, resulting in thousands of different combinations. Homogenization of active sites during catalyst preparation is extremely difficult, and as a result, produced catalysts always include a variety of active sites. Varying molecular sites may have different activity and selectivity for the same processes, which is difficult to predict at the atomic level. Furthermore, support metal contact has a significant influence on catalytic characteristics; in certain cases, support atoms or surface functional groups aid in the stabilization of intermediates, therefore facilitating catalytic cycles. Surface-sensitive methods like X-ray photoelectron spectroscopy may be beneficial, and density functional theory modeling assistance is also crucial.

5. At present so far, a spherical aberration electron microscope is the most prominent characterization tool; nevertheless, it has certain limitations. In the first place, the microscope-based characterization just merely simply furnishes us with visual information about the local structure in the sample. To be more precise, it is almost difficult to simultaneously image the light and heavy elements of a region. If the area of the image acquisition contains two types of atoms with few differences in contrast, data will deviate from the ideal result, reducing the credibility and validity of the data. In addition, the electron microscope can only obtain the projection information of the two-dimensional structure of the sample in a certain direction, resulting in a lack of clear understanding of the three-dimensional structure. Hence, the development of techniques capable of reconstructing the spatial characteristics of ADMMs materials at the atomic level is also of great

significance to spur further research on single-atom catalysts. As another pivotal tool, X-ray absorption fine structure is widely used to characterize the local structure of catalysts. At present, the coordination environment, electronic structure, and valence state of metal centers are mostly concluded by X-ray absorption fine structure. In view of the coordination information as statistical data of all reactive sites in the collected sample, we necessitate a technology that allows us to obtain the actual distribution information of each individual site.

6. We expect to collect in situ and accurate reaction information both in time and space dimensions, including reaction paths, the evolution of catalysts, and reaction kinetics under real conditions to further assist the ADMM design strategy. This is because catalysis is a molecular-level reaction, which usually takes place at the picosecond and nanosecond levels. The research of active centers design strategy is highly dependent on the progress of characterization and instrumentation. Only by clarifying the structural evolution of the active sites of single-atom catalysts can we open the black box and bring about the next explosion in ADMMs research.

7. CO_2 Capture and Conversion at a Hotspot: Continual increases in CO_2 levels in the environment pose a threat to living organisms. Direct air collection and conversion, as well as post-combustion capture and conversion, are in great demand, and millions of dollars have lately been made available to create an efficient and viable CO_2 solution. Even though many companies place a premium on data gathering and storage, this is not a long-term answer. We should trap CO_2 and turn it into value-added goods like gasoline or polymer simultaneously. Many recent publications have highlighted ADMMs for CO_2 conversion, but none of them matched the industrial requirement, which requires further attention.

8. Capturing CO_2 with a Solid Electrochemical System: An electrochemical cell is a solid capture media device with an electrode assembly that catches CO_2 when charged and releases it when discharged. A direct CO_2 collection electrochemical system was recently demonstrated. It was able to collect CO_2 concentrations as low as 0.6% (6,000 ppm). Al Sadat and Archer presented an O_2-assisted Al/CO_2 electrochemical cell as a novel CO_2 collection technique that can also create a significant quantity of electric energy. The cathode is a source of important C_2 species as well as electrical energy. The first cell reduced O_2 to a superoxide intermediate while also reacting with CO_2 to produce aluminum oxalate $Al_2(C_2O_4)_3$, which was the major product. Using ADMMs doped Alcathode, this process may be accelerated.

9. NO_x and SO_x Electrocatalyst: A catalytic converter is a device connected to exhaust that converts hazardous gases including NO_x, CO, and VOC into N_2, O_2, and CO_2. A selective catalytic converter (SCR) is a sophisticated emission control device that injects a liquid-reductant agent into the exhaust stream to lower the quantity of NOx in the exhaust, as described in this series. SCR converts NO_x to N_2 and O_2 by selectively using ammonia. These catalysts are mostly comprised of Rh, Pt, and Pd; however, creating an ADMMs solid electrolyte cell might speed up the catalytic activity even

further. A theoretical model was recently used to examine the direct reduction of NO with CO into N_2 and CO_2 utilizing Ni-SACs.

10. Organics such as chloromethane, methylene chloride, chloroform, and carbon tetrachloride are the most common solvents used by the paint and pharmaceutical industries across the world. Excessive solvent release into the environment not only hurts human health, but also has a significant impact on the ozone layer due to a catalytic chain reaction. For dechlorination of water and air, a heterogeneous catalyst is employed, which may be enhanced further by creating ADMMs. Electrochemical cells can help speed up this process even further.

REFERENCES

1. M. S. Dresselhaus and I. L. Thomas, *Nature*, 2001, 414, 332–337.
2. N. S. Lewis and D. G. Nocera, *Proc. Natl. Acad. Sci. U. S. A.,* 2006, 103, 15729–15735.
3. G. Hutchings, M. Davidson, R. Catlow, C. Hardacre, N. Turner and P. Collier, *Modern Developments in Catalysis*, Vol. 12. World Scientific Publishing Co. Pte Ltd: Singapore, 2017.
4. S. Vajda, M. J. Pellin, J. P. Greeley, C. L. Marshall, L. A. Curtiss, G. A. Ballentine, J. W. Elam, S. Catillon-Mucherie, P. C. Redfern and F. Mehmood. *Nat. Mater.,* 2009, 8, 213–216.
5. J. Lin, B. Qiao, N. Li, L. Li, X. Sun, J. Liu, X. Wang and T. Zhang, *Chem. Commun.,* 2015, 51, 7911–7914.
6. M. Turner, V. B. Golovko, O. P. H. Vaughan, P. Abdulkin, A. Berenguer-Murcia, M. S. Tikhov, B. F. G. Johnson and R. M. Lambert, *Nature*, 2008, 454, 981–983.
7. A. Wang, J. Li and T. Zhang, *Nat. Rev. Chem.,* 2018, 2, 65–81.
8. X. Li, Y. Huang and B. Liu, *Chem*, 2019, 5, 2733.
9. R. Ye, T. J. Hurlburt, K. Sabyrov, S. Alayoglu and G. A. Somorjai, *Proc. Natl. Acad. Sci. U. S. A.,* 2016, 113, 5159–5166.
10. C. M. Wang, Y. D. Wang, J. W. Ge and Z. K. Xie, *Chem*, 2019, 5, 2736–2737.
11. X. Yang, A. Wang, B. Qiao, J. Li, J. Liu and T. Zhang, *Acc. Chem. Res.,* 2013, 46, 1740–1748.
12. B. M. Trost, *Angew. Chem. Int. Ed.,* 1995, 34, 259–281.
13. Y. Luo, Z. Liu, G. Wu, G. Wang, T. Chao, H. Li, J. Liu and X. Hong, *Chin. Chem. Lett.,* 2019, 30, 1093.
14. G. Wu, W. Chen, X. Zheng, D. He, Y. Luo, X. Wang, J. Yang, Y. Wu, W. Yan, Z. Zhuang, X. Hong and Y. Li, *Nano Energy*, 2017, 38, 167–174.
15. X. Wang, Q. Peng and Y. Li, *Acc. Chem. Res.,* 2007, 40, 635–643.
16. Y. Xia, Y. Xiong, B. Lim and S. E. Skrabalak, *Angew. Chem. Int. Ed.,* 2009, 48, 60–103.
17. Y. Wu, D. Wang and Y. Li, *Chem. Soc. Rev.,* 2014, 43, 2112–2124.
18. D. Wang and Y. Li, *Adv. Mater.,* 2011, 23, 1044–1060.
19. Y. Wang, J. He, C. Liu, W. H. Chong and H. Chen, *Angew. Chem. Int. Ed.,* 2015, 54, 2022–2051.
20. Y. Pan, R. Lin, Y. Chen, S. Liu, W. Zhu, X. Cao, W. Chen, K. Wu, W. C. Cheong, Y. Wang, L. Zheng, J. Luo, Y. Lin, Y. Liu, C. Liu, J. Li, Q. Lu, X. Chen, D. Wang, Q. Peng, C. Chen and Y. Li, *J. Am. Chem. Soc.,* 2018, 140, 4218–4221.
21. T. Chao, Y. Hu, X. Hong and Y. Li, *ChemElectroChem*, 2019, 6, 289–303.
22. J. Ge, Z. Li, X. Hong and Y. Li, *Chem. Eur. J.,* 2019, 25, 5113–5127.
23. M. Flytzani-Stephanopoulos, *Chin. J. Catal.,* 2017, 38, 1432–1442.
24. Y. Wang, J. Mao, X. Meng, L. Yu, D. Deng and X. Bao, *Chem. Rev.,* 2019, 119, 1806.

25. B. Qiao, A. Wang, X. Yang, L. F. Allard, Z. Jiang, Y. Cui, J. Liu, J. Li and T. Zhang, *Nat. Chem.*, 2011, 3, 634–641.
26. G. Liu, A. W. Robertson, M. M. Li, W. C. H. Kuo, M. T. Darby, M. H. Muhieddine, Y. C. Lin, K. Suenaga, M. Stamatakis, J. H. Warner and S. C. E. Tsang, *Nat. Chem.*, 2017, 9, 810–816.
27. L. Lin, W. Zhou, R. Gao, S. Yao, X. Zhang, W. Xu, S. Zheng, Z. Jiang, Q. Yu, Y. W. Li, C. Shi, X. D. Wen and D. Ma, *Nature*, 2017, 544, 80–83.
28. J.-H. Zhang, R.-K. Wang, Z. Zhou, L. Zheng and H.-P. Xie, *Rock Soil Mech.*, 2018, 39, 1002–1008.
29. A. Corma, P. Concepcion, M. Boronat, M. J. Sabater, J. Navas, M. J. Yacaman, E. Larios, A. Posadas, M. A. Lopez-Quintela, D. Buceta, E. Mendoza, G. Guilera and A. Mayoral, *Nat. Chem.*, 2013, 5, 775–781.
30. T. He, C. Zhang, L. Zhang and A. Du, *Nano Res.*, 2019, 12, 1817.
31. B. M. Tackett, W. Sheng and J. G. Chen, *Joule*, 2017, 1, S2542435117300090.
32. H. Xu, D. Cheng, D. Cao and X. C. Zeng, *Nat. Catal.*, 2018, 1, 339–348.
33. C. Zhu, S. Fu, Q. Shi, D. Dan and Y. Lin, *Angew. Chem. Int. Ed.*, 2017, 56, 13944.
34. Y. He, S. Hwang, D. A. Cullen, M. A. Uddin, L. Langhorst, B. Li, S. Karakalos, A. J. Kropf, E. C. Wegener and J. Sokolowski, *Energy Environ. Sci.*, 2019, 12, 250.
35. Z. Zhang, J. Sun, W. Feng and L. Dai, *Angew. Chem. Int. Ed.*, 2018, 130, 9038.
36. H. Fei, J. Dong, M. J. Arellano-Jiménez, G. Ye, N. D. Kim, E. L. G. Samuel, Z. Peng, Z. Zhu, F. Qin and J. Bao, *Nat. Commun.*, 2015, 6, 8668.
37. R. Van Hardeveld and F. Hartog *Surf. Sci.*, 1969, 15, 189–230.
38. B. C. Gates, *Chem. Rev.*, 1995, 95, 511–522.
39. N. Lopez, T. V. W. Janssens, B. S. Clausen, Y. Xu, M. Mavrikakis, T. Bligaard and J. K. Nørskov, *J. Catal.*, 2004, 223, 232–235.
40. B. R. Cuenya, *Thin Solid Films*, 2010, 518, 3127–3150.
41. R. J. Kubo, *Phys. Soc. Jpn.*, 1962, 17, 975–986.
42. M. P. Johansson, D. Sundholm and J. Vaara, *Angew. Chem. Int. Ed.*, 2004, 116, 2732–2735.
43. B. von Issendorff and O. Cheshnovsky, *Annu. Rev. Phys. Chem.*, 2005, 56, 549–580.
44. E. Roduner, *Chem. Soc. Rev.*, 2006, 35, 583–592.
45. M. Valden, X. Lai and D. W. Goodman, *Science* 1998, 281, 1647–1650.
46. P. Claus, A. Brückner, C. Mohr, H. J. Hofmeister, *J. Am. Chem. Soc.*, 2000, 122, 11430–11439.
47. J. Li, X. Li, H. J. Zhai and L. S. Wang, *Science*, 2003, 299, 864–867.
48. M. Haruta, *Catal. Today*, 1997, 36, 153–166.
49. B. Yoon, H. Hakkinen, U. Landman, A. S. Worz, J.-M. Antonietti, S. Abbet, K. Judai and U. Heiz, *Science*, 2005, 307, 403–407.
50. S. J. Tauster, S. C. Fung, R. T. K. Baker and J. A. Horsley, *Science*, 1981, 211, 1121–1125.
51. C. T. Campbell, *Surf. Sci. Rep.*, 1997, 27, 1–111.
52. Q. Fu and T. Wagner, *Surf. Sci. Rep.*, 2007, 62, 431–498.
53. J. Liu, *ChemCatChem*, 2011, 3, 934–948.
54. M. Haruta, T. Kobayashi, H. Sano and N. Yamada, *Chem. Lett.*, 1987, 16, 405–408.
55. Z. Xu, F.-S. Xiao, S. K. Purnell, O. Alexeev, S. Kawi, S. E. Deutsch and B. C. Gates, *Nature*, 1994, 372, 346–348.
56. U. Heiz, A. Sanchez, S. Abbet and W.-D. Schneider, *J. Am. Chem. Soc.*, 1999, 121, 3214–3217.
57. G. A. Somorjai and J. Y. Park, *Chem. Soc. Rev.*, 2008, 37, 2155–2162.
58. M. Crespo-Quesada, A. Yarulin, M. Jin, Y. Xia, L. J. Kiwi-Minsker, *Am. Chem. Soc.*, 2011, 133, 12787–12794.
59. J. Liu, *ACS Catal.*, 2017, 7, 34–59.

60. M. Flytzani-Stephanopoulos and B. C. Gates, *Annu. Rev. Chem. Biomol. Eng.* 2012, 3, 545–574.
61. B. C. Gates, M. Flytzani-Stephanopoulos, D. A. Dixon and A. Katz, *Catal. Sci. Technol.*, 2017, 7, 4259–4275.
62. Y. Chen, H. Sun and B. C. Gates, *Small*, 2021, 17, 2004665.
63. J. M. Thomas, R. Raja and D. W. Lewis, *Angew. Chem. Int. Ed.*, 2005, 44, 6456–6482.
64. M. W. McKittrick and C. W. Jones, *J. Am. Chem. Soc.*, 2004, 126, 3052–3053.
65. M. Tada and S. Muratsugu, Site-isolated heterogeneous catalysts. In: *Heterogeneous Catalysts for Clean Technology: Spectroscopy, Design, and Monitoring*, K. Wilson and A. F. Le, Eds, Wiley-VCH Verlag GmbH & Co. KGaA: Weinheim, 2013, pp. 173–191.
66. C. Coperet, A. Comas-Vives, M. P. Conley, D. P. Estes, A. Fedorov, V. Mougel, H. Nagae, F. Nuñez-Zarur and P. A. Zhizhko, *Chem. Rev.*, 2016, 116, 323–421.
67. X.-K. Gu, B. Qiao, C.-Q. Huang, W.-C. Ding, K. Sun, E. Zhan, T. Zhang, J. Liu and W.-X. Li, *ACS Catal.*, 2014, 4, 3886–3890.
68. L. Zhang, A. Wang, W. Wang, Y. Huang, X. Liu, S. Miao, J. Liu and T. Zhang, *ACS Catal.*, 2015, 5, 6563–6572.
69. Y. Y. Jin and P. P. Hao, *Prog. Chem.*, 2015, 27, 1689–1704.
70. B. Long, Y. Tang, and J. Li, *Nano Res.*, 2016, 9, 3868–3880.
71. P. Wu, P. Du, H. Zhang and C. Cai, *Chem. Chem. Phys.*, 2015 17, 1441–1449.
72. B. L. He, J. S. Shen, and Z. X. Tian, *Phys. Chem. Chem. Phys.*, 2016, 18, 24261–24269.
73. J.-X. Liang, X.-F. Yang, A. Wang, T. Zhang and J. Li, *Catal. Sci. Technol.*, 2016, 6, 6886–6892.
74. L. Nie, D. Mei, H. Xiong, B. Peng, Z. Ren, X. I. P. Hernandez, A. DeLaRiva, M. Wang, M. H. Engelhard, L. Kovarik et al., *Science*, 2017, 358, 1419–1423.
75. X. Cao, Y. Ji and Y. Luo, *J. Phys. Chem. C*, 2015, 119, 1016–1023.
76. G. Vile, D. Albani, M. Nachtegaal, Z. Chen, D. Dontsova, M. Antonietti, N. Lopez and J. Perez-Ramırez, *Angew. Chem. Int. Ed.*, 2015, 54, 11265–11269.
77. H. Wei, X. Liu, A. Wang, L. Zhang, B. Qiao, X. Yang, Y. Huang, S. Miao, J. Liu and T. Zhang, *Nat. Commun.*, 2014, 5, 5634.
78. X. Huang, Y. Xia, Y. Cao, X. Zheng, H. Pan, J. Zhu, C. Ma, H. Wang, J. Li, R. You, et al. *Nano Res.*, 2017, 10, 1302–1312.
79. Y. Chen, J. Lin, L. Li, B. Qiao, J. Liu, Y. Su and X. Wang, *ACS Catal.*, 2018, 82, 859–868.
80. X. Sun, J. Lin, Y. Zhou, L. Li, Y. Su, X. Wang and T. Zhang, *AIChE J.*, 2017, 63, 4022–4031.
81. X. Li, W. Bi, L. Zhang, S. Tao, W. Chu, Q. Zhang, Y. Luo, C. Wu and Y. Xie, *Adv. Mater.*, 2016, 28, 2427–2431.
82. G. Gao, Y. Jiao, E. R. Waclawik and A. Du, *J. Am. Chem. Soc.*, 2016, 138, 6292–6297.
83. T. He, C. Zhang, L. Zhang and A. Du, *Nano Res.*, 2019, 12, 1817–1823.
84. X. Fang, Q. Shang, Y. Wang, L. Jiao, T. Yao, Y. Li, Q. Zhang, Y. Luo and H. L. Jiang, *Adv. Mater.*, 2018, 30, 1705112.
85. Z. Zhang, C. Ma, Y. Tu, R. Si, J. Wei, S. Zhang, Z. Wang, J. F. Li, Y. Wang and D. Deng, *Nano Res.*, 2019, 12, 2313–2317.
86. Y. H. He, S. Hwang, D. A. Cullen, M. A. Uddin, L. Langhorst, B. Y. Li, S. Karakalos, A. J. Kropf, E. C. Wegener, J. Sokolowski, et al. *Energy Environ. Sci.*, 2019, 12, 250–260.
87. H. L. Fei, J. C. Dong, M. J. Arellano-Jime ́nez, G. L. Ye, N. D. Kim, E. L. G. Samuel, Z. W. Peng, Z. Zhu, F. Qin, J. M. Bao, et al. *Nat. Commun.*, 2015, 6, 8668.
88. Z. P. Zhang, J. T. Sun, F. Wang and L. M. Dai, *Angew. Chem. Int. Ed.*, 2018, 57, 9038–9043.
89. J. M. Thomas, Z. Saghi and P. L. Gai, *Top. Catal.*, 2011, 54, 588.
90. H. S. Taylor, *Proc. R. Soc. A*, 1925, 108, 105–111.

91. E. J. Arlman and P. Cossee, *J. Catal.*, 1964, 3, 99.
92. J. T. Yates, S. D. Worley, T. M. Duncan and R. W. Vaughan, *J. Chem. Phys.*, 1979, 70, 1225.
93. T. Maschmeyer, F. Rey, G. Sankar and J. M. Thomas, *Nature*, 1995, 378, 159.
94. K. Asakura, H. Nagahiro, N. Ichikuni and Y. Iwasawa, *Appl. Catal. A Gen.*, 1999, 188, 313.
95. S. Abbet, A. Sanchez, U. Heiz, W. D. Schneider and N. Rosch, *J. Am. Chem. Soc.*, 2000, 122, 3453–3457.
96. Q. Fu, H. Saltsburg and M. Flytzani-Stephanopoulos, *Science*, 2003, 301, 935.
97. X. Zhang, H. Shi and B. Q. Xu, *Angew. Chem. Int. Ed.*, 2005, 44, 7132.
98. S. F. Hackett, R. M. Brydson, M. H. Gass, I. Harvey, A. D. Newman, K. Wilson and A. F. Lee, *Angew. Chem. Int. Ed.*, 2007, 46, 8593.
99. H. L. Tierney, A. E. Baber, J. R. Kitchin and E. C. H. Sykes, *Phys. Rev. Lett.*, 2009, 103, 246102.
100. Y. Lei, F. Mehmood, S. Lee, J. Greeley, B. Lee, S. Seifert, R. E. Winans, J. W. Elam, R. J. Meyer, P. C. Redfern, D. Teschner, R. Schlogl, M. J. Pellin, L. A. Curtiss and S. Vajda, *Science*, 2010, 328, 224.
101. J. S. Jirkovsky, I. Panas, E. Ahlberg, M. Halasa, S. Romani and D. J. Schiffrin, *J. Am. Chem. Soc.*, 2011, 133, 19432.
102. Z. Huang, X. Gu, Q. Cao, P. Hu, J. Hao, J. Li and X. Tang, *Angew. Chem. Int. Ed.*, 2012, 51, 4198.
103. X. Guo, G. Fang, G. Li, H. Ma, H. Fan, L. Yu, C. Ma, X. Wu, D. Deng, M. Wei, *Science*, 2014, 344, 616.
104. M. Yang, S. Li, Y. Wang, J. A. Herron, Y. Xu, L. F. Allard, S. Lee, J. Huang, M. Mavrikakis and M. Flytzani-Stephanopoulos, *Science*, 2014, 346, 1498.
105. P. Liu, Y. Zhao, R. Qin, S. Mo, G. Chen, L. Gu, D. M. Chevrier, P. Zhang, Q. Guo and D. Zang, *Science*, 2016, 352, 797.
106. J. Shan, M. Li, L. F. Allard, S. Lee and M. Flytzani-Stephanopoulos, *Nature*, 2017, 551, 605.
107. S. Wei, A. Li, J. C. Liu, Z. Li, W. Chen, Y. Gong, Q. Zhang, W. C. Cheong, Y. Wang, L. Zheng, H. Xiao, C. Chen, D. Wang, Q. Peng, L. Gu, X. Han, J. Li and Y. Li, *Nat. Nanotechnol.*, 2018, 13, 856.
108. L. Liu, M. Lopez-Haro, C. W. Lopes, C. Li, P. Concepcion, L. Simonelli, J. J. Calvino and A. Corma, *Nat. Mater.*, 2019, 18, 866.
109. S. K. Kaiser, Z. Chen, D. F. Akl, S. Mitchell and J Pérez-Ramírez, *Chem. Rev.*, 2020, 120, 11703–11809.
110. G. Kyriakou, M. B. Boucher, A. D. Jewell, E. A. Lewis, T. J. Lawton, A. E. Baber, H. L. Tierney, M. Flytzani-Stephanopoulos and E. C. H. Sykes, *Science*, 2012, 335, 1209.
111. P. Yin, T. Yao, Y. Wu, L. Zheng, Y. Lin, W. Liu, H. Ju, J. Zhu, X. Hong, Z. Deng, G. Zhou, S. Wei and Y. Li, *Angew. Chem. Int. Ed.*, 2016, 55, 10800.
112. J. Wang, Z. Huang, W. Liu, C. Chang, H. Tang, Z. Li, W. Chen, C. Jia, T. Yao, S. Wei, Y. Wu and Y. Li, *J. Am. Chem. Soc.*, 2017, 139, 17281.
113. Y. Qu, Z. Li, W. Chen, Y. Lin, T. Yuan, Z. Yang, C. Zhao, J. Wang, C. Zhao, X. Wang, F. Zhou, Z. Zhuang, Y. Wu and Y. Li, *Nat. Catal.*, 2018, 1, 781.
114. M. Huo, L. Wang, Y. Wang, Y. Chen and J. Shi, *ACS Nano*, 2019, 13, 2643.
115. N. Gong, X. Ma, X. Ye, Q. Zhou, X. Chen, X. Tan, S. Yao, S. Huo, T. Zhang, S. Chen, X. Teng, X. Hu, J. Yu, Y. Gan, H. Jiang, J. Li and X. Lian, *Nat Nanotechnol.*, 2019, 55, 1159.
116. C. Xia, Y. Qiu, Y. Xia, P. Zhu and H. Wang, *Nat. Chem.*, 2021, 13, 887–894.
117. Z. Zhuang, Y. Li, Y. Li, J. Huang, B. Wei, R. Sun, Y. Ren, J. Ding, J. Zhu, Z. Lang, L. V. Moskaleva, C. He, Y. Wang, Z. Wang, D. Wang and Y. Li, *Energy Environ. Sci.*, 2021, 14, 1016–1028.

118. B. Pang, X. Liu, T. Liu, T. Chen, X. Shen, W. Zhang, S. Wang, T. Liu, D. Liu, T. Ding, Z. Liao, Y. Li, C. Liang and T. Yao, *Energy Environ. Sci.,* 2022, 15, 102–108.

119. J. Li, Y. Jiang, Q. Wang, C. Xu, D. Wu, M. N. Banis, K. R. Adair, K. D. Davis, D. M. Meira, Y. Z. Finfrock, W. Li, L. Zhang, T. Sham, R. Li, N. Chen, M. Gu, J. Li and X. Sun, *Nat. Commun.,* 2022, 12, 6806.

120. Y. Chen, P. Wang, H. Hao, J. Hong, H. Li, S. Ji, A. Li, R. Gao, J. Dong, X. Han, M. Liang, D. Wang and Y. Li, *J. Am. Chem. Soc.* 2021, 143, 18643–18651.

121. R. Jin, C. Zeng, M. Zhou and Y. Chen, *Chem. Rev.,* 2016, 116, 10346–10413.

122. S. Liu and Y. Xu, *Sci. Rep.,* 2016, 6, 22742.

123. H. Qian, M. Zhu, Z. Wu and R. Jin, *Acc. Chem. Res.,* 2012, 45, 1470–1479.

124. P. P. Maharjan, Q. Chen, L. Zhang, O. Adebanjo, N. Adhikari, S. Venkatesan, P. Adhikary, B. Vaagensmith and Q. Qiao, *Phys. Chem. Chem. Phys.,* 2013, 15, 6856–6863.

125. X. Cui, L. Wu, R. Pavel, J. Kathrin and B. Matthias, *Nat. Catal.,* 2018, 1, 385–397.

126. X. Qiu, X. Yan, H. Pang, J. Wang, D. Sun, S. Wei, L. Xu and Y. Tang, *Adv. Sci.,* 2019, 6, 1801103.

127. C. Zhu, Q. Shi, B. Z. Xu, S. Fu, G. Wan, C. Yang, S. Yao, J. Song, H. Zhou, D. Du, S. P. Beckman, D. Su and Y. Lin, *Adv. Energy Mater.,* 2018, 8, 1801956.

128. J. Wang, Z. Li, Y. Wu and Y. Li, *Adv. Mater.,* 2018, 30, 1801649.

129. M. B. Gawande, K. Ariga and Y. Yamauchi, *Small,* 2021, 17, 2101584.

130. H. Yang, Z. Chen, S. Kou, G. Lu and D. Chen and Z. Liu, *J. Mater. Chem. A,* 2021, 9, 15919.

131. X. Zheng, P. Li, S. Dou, W. Sun, H. Pan, D. Wang and Y. Li, *Energy Environ. Sci.,* 2021, 14, 2809.

132. J. Jones, H. Xiong, A. T. Delariva, E. J. Peterson and A. K. Datye, *Science,* 2016, 353, 150–154.

133. T. Li, O. Kasian, S. Cherevko, S. Zhang, S. Geiger, C. Scheu, P. Felfer, D. Raabe, B. Gault and K. J. J. Mayrhofer, *Nat. Catal.,* 2018, 1, 300–305.

134. H. Zhang, P. An, W. Zhou, B. Guan, P. Zhang, J. Dong and X. Lou, *Sci. Adv.,* 2018, 4, eaao6657.

135. Y. Pan, C. Zhang, Z. Liu, C. Chen and Y. Li, *Matter,* 2020, 2, 78–110.

136. P. Liu, R. Qin, G. Fu and N. Zheng, *J. Am. Chem. Soc.,* 2017, 139, 2122–2131.

137. C. Zhu, S. Fu, Q. Shi, D. Du and Y. Lin, *Angew. Chem. Int. Ed.,* 2017, 56, 13944–13960.

138. M. Boronat, A. Leyva-Pérez, A. Corma, *Acc. Chem. Res.,* 2014, 47, 834–844.

139. N. Cheng, S. Stambula, D. Wang, M. N. Banis, J. Liu, A. Riese, B. Xiao, R. Li, T. K. Sham and L. M. Liu, *Nat. Commun.,* 2016, 7, 13638.

140. S.-Y. Huang, P. Ganesan, S. Park and B. N. Popov, *J. Am. Chem. Soc.,* 2009, 131, 13898–13899.

141. Z. Xi, D. P. Erdosy, A. Mendoza-Garcia, P. N. Duchesne, J. Li, M. Muzzio, Q. Li, P. Zhang and S. Sun, *Nano Lett.,* 2017, 17, 2727–2731.

142. J. Park, S. Lee, H. Kim, A. Cho, S. Kim, Y. Ye, H. Lee, J. Jang and J. Lee, *Angew. Chem. Int. Ed.,* 2019, 131, 16184–16188.

143. J. Yang, B. Chen, X. Liu, W. Liu, Z. Li, J. Dong, W. Chen, W. Yan, T. Yao, X. Duan, et al. *Angew. Chem. Int. Ed.,* 2018, 57, 9495–9500.

144. J. Zhang, Y. Zhao, X. Guo, C. Chen, C. Dong, R. Liu, C. Han, Y. Li, Y. Gogotsi and G. Wang, *Nat. Catal.,* 2018, 1, 985–992.

145. Z. Zhang, C. Feng, C. Liu, M. Zuo, L. Qin, X. Yan, Y. Xing, H. Li, R. Si, S. Zhou and J. Zengs, *Nat. Commun.,* 2020, 11, 1215.

146. Y. Shi, Z. R. Ma, Y. Y. Xiao, Y. C. Yin, W. M. Huang, Z. C. Huang, Y. Z. Zheng, F. Y. Mu, R. Huang, G. Y. Shi, Y. Y. Sun, X. H. Xia and W. Chen, *Nat. Commun.,* 2021, 12, 3021.

147. J. Wan, Z. Zhao, H. Shang, B. Peng, W. Chen, J. Pei, L. Zheng, J. Dong, R. Cao, R. Sarangi, Z. Jiang, D. Zhou, Z. Zhuang, J. Zhang, D. Wang and Y. Li, *J. Am. Chem. Soc.,* 2020, 142, 8431–8439.

148. C. X. Zhao, B. Q. Li, J. N. Liu and Q. Zhang, *Angew. Chem. Int. Ed.*, 2020, 60, 4448–4463.
149. C. Zhao, X. Dai, T. Yao, W. Chen, X. Wang, J. Wang, J. Yang, S. Wei, Y. Wu and Y. Li, *J. Am. Chem. Soc.*, 2017, 139, 8078–8081.
150. R. Jiang, L. Li, T. Sheng, G. Hu, Y. Chen and L. Wang, *J. Am. Chem. Soc.*, 2018, 140, 11594–11598.
151. X. Yin, H. Wang, S. Tang, X. Lu, M. Shu, R. Si and T. Lu, *Angew. Chem. Int. Ed.*, 2018, 57, 9382–9386.
152. L. Zhang, Y. Jia, G. Gao, X. Yan, N. Chen, J. Chen, M. Soo, B. Wood, D. Yang, A. Du, et al. *Chem*, 2018, 4, 285–297.
153. Y. Pan, Y. Chen, K. Wu, Z. Chen, S. Liu, X. Cao, W. Cheong, T. Meng, J. Luo, L. Zheng, et al. *Nat. Commun.*, 2019, 10, 4290.
154. H. Shang, X. Zhou, J. Dong, A. Li, X. Zhao, Q. Liu, Y. Lin, J. Pei, Z. Li, Z. Jiang, D. Zhou, L. Zheng, Y. Wang, J. Zhou, Z. Yang, R. Cao, R. Sarangi, T. Sun, X. Yang, X. Zheng, W. Yan, Z. Zhuang, J. Li, W. Chen, D. Wang, J. Zhang and Y. Li, *Nat. Commun.*, 2020, 11, 3049.
155. C. H. Choi, M. Kim, H. C. Kwon, S. J. Cho, S. Yun, H.-T. Kim, K. J. J. Mayrhofer, H. Kim and M. Choi, *Nat. Commun.*, 2016, 7, 10922.
156. W. Liu, L. Cao, W. Cheng, Y. Cao, X. Liu, W. Zhang, X. Mou, L. Jin, X. Zheng, W. Che, et al. *Angew. Chem. Int. Ed.*, 2017, 56, 9312–9317.
157. J. Gu, C. Hsu, L. Bai, H. Chen and X. Hu, *Science*, 2019, 364, 1091–1094.
158. K. Wu, X. Chen, S. Liu, Y. Pan, W.-C. Cheong, W. Zhu, X. Cao, R. Shen, W. Chen, J. Luo, et al. *Nano Res.*, 2018, 11, 6260–6269.
159. Q. Li, W. Chen, H. Xiao, Y. Gong, Z. Li, L. Zheng, X. Zheng, W. Yan, W.-C. Cheong, R. Shen, et al. *Adv. Mater.*, 2018, 30, 1800588.
160. S. Li, B. Chen, Y. Wang, M. Ye, P. A. van Aken, C. Cheng and A. Thomas, *Nat. Mater.*, 2021, 20, 1240–1247.
161. J. Li, W. Wang, W. Chen, Q. Gong, J. Luo, R. Lin, H. Xin, H. Zhang, D. Wang, Q. Peng, W. Zhu, C. Chen and Y. Li, *Nano. Res.*, 2018, 11, 4774.
162. R. Qin, P. Liu, F. Gang and N. Zheng, *Small Methods*, 2018, 2, 1700286.
163. P. Liu, X. Huang, D. Mance, C. Copéret, *Nat. Catal.*, 2021, 4, 968–975.
164. B. Liu, H. Shioyama, T. Akita and Q. Xu, *J. Am. Chem. Soc.*, 2008, 130, 5390.
165. B. Chen, X. Zhong, G. Zhou, N. Zhao and H. Cheng, *Adv. Mater.*, 2021, 34, 2105812.
166. Y. Peng, B. Lu and S. Chen, *Adv. Mater.*, 2018, 30, 1801995.
167. X. Wang, K. Maeda, A. Thomas, K. Takanabe, G. Xin, J. M. Carlsson, K. Domen, and M. Antonietti, *Nat. Mater.*, 2009, 8, 76.
168. F. Goettmann, A. Fischer, M. Antonietti and A. Thomas, *Chem. Commun.*, 2006, (43), 4530–4532. DOI: 10.1039/B608532F.
169. E. Kroke, M. Schwarz, E. Horath-Bordon, P. Kroll, B. Noll and A. D. Norman, *New J. Chem.*, 2002, 26, 508.
170. Y.-S. Jun, W. H. Hong, M. Antonietti and A. Thomas, *Adv. Mater.*, 2009, 21, 4270.
171. R. Jasinski, *Nature*, 1964, 201, 1212.
172. W. Liu, L. Zhang, W. Yan, X. Liu, X. Yang, S. Miao, W. Wang, A. Wang, T. Zhang, *Chem. Sci.*, 2016, 7, 5758.
173. H. Wu, H. Li, X. Zhao, Q. Liu, J. Wang, J. Xiao, S. Xie, R. Si, F. Yang, S. Miao, X. Guo, G. Wang, X. Bao, *Energy Environ. Sci.*, 2016, 9, 3736.
174. Y. Chen, S. Ji, Y. Wang, J. Dong, W. Chen, Z. Li, R. Shen, L. Zheng, Z. Zhuang, D. Wang, Y. Li, *Angew. Chem. Int. Ed.*, 2017, 56, 6937.
175. L. Zhang, A. Wang, W. Wang, Y. Huang, X. Liu, S. Miao, J. Liu, T. Zhang, *ACS Catal.*, 2015, 5, 6563.
176. W. Liu, L. Zhang, X. Liu, X. Liu, X. Yang, S. Miao, W. Wang, A. Wang, T. Zhang, *J. Am. Chem. Soc.*, 2017, 139, 10790.

177. P. Chen, T. Zhou, L. Xing, K. Xu, Y. Tong, H. Xie, L. Zhang, W. Yan, W. Chu, C. Wu and Y. Xie, *Angew. Chem. Int. Ed.*, 2017, 56, 610.

178. R. T. Hannagan, G. Giannakakis, M. Flytzani-Stephanopoulos and E. Sykes, *Chem. Rev.*, 2020, 120, 12044.

179. L. Zhang, A. Wang, J. T. Miller, X. Liu, X. Yang, W. Wang, L. Li, Y. Huang, C.-Y. Mou, T. Zhang, *ACS Catal.*, 2014, 4, 1546.

180. F. R. Lucci, M. T. Darby, M. F. Mattera, C. J. Ivimey, A. J. Therrien, A. Michaelides, M. Stamatakis, E. C. Sykes, *J. Phys. Chem. Lett.*, 2016, 7, 480.

181. G. X. Pei, X. Y. Liu, A. Wang, A. F. Lee, M. A. Isaacs, L. Li, X. Pan, X. Yang, X. Wang, Z. Tai, K. Wilson and T. Zhang, *ACS Catal.*, 2015, 5, 3717.

182. J. Ge, D. He, W. Chen, H. Ju, H. Zhang, T. Chao, X. Wang, R. You, Y. Lin, Y. Wang, J. Zhu, H. Li, B. Xiao, W. Huang, Y. Wu, X. Hong and Y. Li, *J. Am. Chem. Soc.*, 2016, 138, 13850.

183. Y. Yao, S. Hu, W. Chen, Z.-Q. Huang, W. Wei, T. Yao, R. Liu, K. Zang, X. Wang, G. Wu, W. Yuan, T. Yuan, B. Zhu, W. Liu, Z. Li, D. He, Z. Xue, Y. Wang, X. Zheng, J. Dong, C.-R. Chang, Y. Chen, X. Hong, J. Luo, S. Wei, W.-X. Li, P. Strasser, Y. Wu and Y. Li, *Nat. Catal.*, 2019, 2, 304.

184. J. Jiao, R. Lin, S. Liu, W. Cheong, C. Zhang, Z. Chen, Y. Pan, J. Tang, K. Wu, S. Hung, et al. *Nat. Chem.*, 2019, 11, 222–228.

185. H. Yan, Y. Lin, H. Wu, W. Zhang, Z. Sun, H. Cheng, W. Liu, C. Wang, J. Li, X. Huang, T. Yao, J. Yang, S. Wei and J. Lu, *Nat. Commun.*, 2017, 8, 1070.

186. X. Kang, Y. Li, M. Zhu and R. Jin, *Chem. Soc. Rev.*, 2020, 49, 6443.

2 Synthesis and Characterization of Atomically Dispersed Metallic Materials

Xin Bo, Rouna Jia, and Dehui Deng
Chinese Academy of Sciences (CAS)

CONTENTS

DOI: 10.1201/9781003153436-2

2.1 INTRODUCTION

Dispersion of the single-atomic-metallic materials (SAMs) onto the various supports is one of the most efficient pathways to realize the maximum utilization of the elemental functions in particular application scenarios. The development of material engineering from bulky to microscale, then from nano-size to quantum dot, offers the basic guidance for the rational design of the next-generation SAMs, which is also the prerequisite for scientific research and practical applications.[1] Tremendous methodologies have been worked out to achieve the well-defined SAMs. As is well-known, the surface free energy of metal increases with the decrease of the metal size, which contributes to the aggregation of the isolated target atoms during both synthesis and application processes.[1] At the same time, a strong interaction (or bonding) between the single-atom site and the sitting substrate is always formed to guard against undesired migration. The principle for SAM fabrication is the competition of the atomically dispersed metal species bonding with surrounded atoms (coordinative bond) against the metal-metal affinity.[1] Therefore, how to artificially build and utilize this coordination with higher anti-aggregation ability is critical for SAMs fabrication.

Generally, the synthesis of SAMs can be categorized into "bottom-up" and "top-down" pathways. In particular for the bottom-up routine, the target metal precursors are dispersed into the media to maintain enough distance and contact with the anchoring sites on the substrate. With the removal of the necessary ligands of the precursors, SAMs can be easily obtained. This routine usually proceeds in liquid media, which gives a diluted target metal source to ensure the well-dispersed single-atom sites, thus leading to the low SAMs loading as a consequence. Therefore, the selection and pretreatment of the proper substrates for efficient SAMs capturing is equally important, such as the porous carbon-based materials and defective oxides with tremendous traps, respectively. Alternatively, the "top-down" pathway mainly conducts that the single-metal sources come from the mother materials via metal bond splitting so that this routine is usually accompanied with a high-energy induction process. The excited single atoms can migrate across both gas-solid and solid-solid interfaces. The continuous energetic input on the mother metal materials can also present successive isolated-atom sources so as to obtain a higher SAM loading. Additionally, the coordinative environment on the support can also be tuned with the carrier atmosphere such as using ammonia to replace the reductive hydrogen. Apart from the "bottom-up" and "top-down" routines, a spatial organic framework assistance access is also pointed out. The well-defined structure through organic ligand combinations generates massive suitable pores, metal-philic sites, coordinative traps, etc., efficiently encapsulating the nomadic single metals, which makes organic frameworks as perfect species for loading the SAMs. Another advantage of organic framework assistance is the in-situ pyrolysis conversion into carbon-based support under annealing treatment. On the one hand, the spatial framework promotes the SAM dispersion with higher content, while carbon support transition further stabilizes the single-site coordination. However, the organic bricks for the framework are expensive, which limits the massive production via this approach. In this section, the synthetic strategies for SAMs on various substrates have been systematically summarized with concrete examples and the comparison for the presentative synthetic approaches for SAMs is also listed in Table 2.1.

TABLE 2.1

Pros and Cons of the Typical Fabrications for SAMs

Methods		Pros	Cons
Liquid	Impregnation	Universal approach, well-dispersed, without specific apparatus	Low mass loading, imprecisely control, pretreatment for supports
	Coprecipitation	Facile approach, well-dispersed, higher mass loading	Low mass loading
	Galvanic Reduction	Metal support available;	Aggregation, imprecisely control, low mass loading
	Etching	Stable, higher mass loading	Strong acid wash
	Photo-induction	Well-dispersed	Low mass loading, imprecisely control
	Electrodeposition	Facile approach, controllable, well-dispersed	Low mass loading;
Gas	ALD; CVD; Sputtering	Well-dispersed, stable, controllable, massive production	Complex manipulation, device
Solid	Ball milling; Bulky transition	Massive production, higher mass loading, stable	High energy, uneven dispersion
Spatial	MOF	Large porous structure, well-dispersed, higher mass loading	Brick materials expensive
	COF	Large porous structure, well-dispersed, higher mass loading	Structure unclear, brick materials expensive

Accompanied by the synthesis of the SAMs, the characterizations of the single-atom dispersion and the relevant coordinative structure are equally vital to both the methodologies and the practical applications. The advanced analytic measurements provide direct evidence to identify whether the successful fabrications of SAMs structure so as to feedback the information for tuning the synthetic parameters. One of the most basic aims is to "see" this SAMs structure as concrete evidence. Fortunately, the development of aberration-correction transition electron microscope (AC-TEM) allows for observative imaging at the atomic-level resolution as well as the coordinative environment. Furthermore, the structural analysis under a statistic account to reflect the spatial dispersion of a single atom is more convincible to get rid of the particular shooting area from an electron microscope. The sensitive and identical X-ray adsorption spectroscope (XAS), providing both electronic information as well as fine structural coordination, can cross-confirm the SAMs existence with statistical results.

At the same time, precise studies on the structure–application relationship are equally important. As in the single-atom system, even with the same central metal atoms, these single-atomic site catalysts usually exhibit different catalytic performances. That means minor changes in coordination atoms can alter the electronic structure of the central atoms, which substantially influences the intrinsic properties. Additionally, the static structural analysis is also insufficient since the intermediate under the reaction-excited state may be derived from the initial materials. Thanks to the development of the spectroscopes under ambient pressure and atmosphere such as XAS, Fourier transform infrared (FTIR), scanning electrochemical microscopy (SECM), ambient-pressure X-ray photoelectron spectroscopy (AP-XPS), etc., the in-situ/operando spectra make it possible to collectively fine-tune structure information on the intermediate under the real-time reaction. Last but not the least, precise structural information, especially from the intermediate, is also helpful for computational modeling construction. Fundamentally, the design of the advanced SAMs can be predictable without relying on massive and unnecessary trial-and-error experiments, thanks to the simulation to abound this database.

2.2 SYNTHETIC METHODS FOR SAMs

2.2.1 WET CHEMISTRY STRATEGY

The fabrication of SAMs is still a great challenge as there is difficulty on the single-atom dispersion from the aggregated particle by the strong surface energy. For example, synthesis proceeding in wet-chemistry media has been regarded as one of the most practical protocols to achieve SAM structures due to ease to handle and feasibility of large-scale manufacturing.[2] The significance of the wet-chemistry method is aimed to disperse and isolate the target metal species/precursors/complexes to the maximum extent and to obtain the single-atomic site steadily implanted into the substrate.[2] A classical wet chemistry technique implies: (i) introduction of the metal precursors onto the support by different methods such as impregnation/ion-exchange, coprecipitation, or deposition-precipitation, etc.; (ii) a drying step and (iii) a calcination and/or reduction step. It is important to stress that every step should be finely controlled to obtain reproducible procedures, and to avoid the by-productions of aggregated particles/clusters.[3] Specific synthetic methods via wet-chemistry routine are introduced as follows.

2.2.1.1 Impregnation Method

Impregnation method has been regarded as universal access to SAMs, in which the metal precursors are anchored onto the immersed support materials suspended in liquid media via physical adsorption, encapsulation or chemical coordination.[4] The critical step is to regulate the interaction between the target metallic ions and the substrate. By selecting the species of the metallic precursors and the coupled substrate, the target ions could be atomically dispersed and isolated under the anchoring effect. Then the further treatment such as calcination/reduction would enhance the isolation and stabilize the SAM structure. However, the mass loading via this method is relatively low since the surface free energy increases the tension, leading to a heavy trend to integrate a cluster structure.[5]

2.2.1.1.1 Carbon-Based Supports

Since the carbon-based supports inherently possess outstanding electrical conductivity, thermal stability, excellent anti-corrosion property, enlarged specific area with exposed pores and diversified structures (*e.g.* 2D graphene and C_3N_4, 1D CNTs), the single-atom confinement is more practical.[1,6,7] Thus, single-atomic metals, such as Co, Cu, Zn, Pd, Ni, Pt, etc., on carbon substrate can be efficiently obtained by impregnation method in aqueous solution through various interactions, *e.g.* covalent bonds, π–π stacking, electrostatic attractions or chelating reactions.[6]

For example, graphene, as a typical 2D carbon-based material with a honeycomb-shaped structure, is constructed with SP^2 hybrid carbon. The symmetry and uniform distribution of charge in the graphene plane present a smaller migration barrier, always resulting in the undesired migration of the conveyed atoms to aggregations.[1] Therefore, the principle to anchor the single atom onto graphene substrate is to reasonably predesign the potential trapping sites and change the uniform charge distribution.[1,8] For example, graphene can be pretreated at a high-temperature range of 700°C~1,050°C mixed with melamine to obtain a defective structure with dense zig-zag carbons and/or arm-chair carbons. Then the defective graphene was mixed with Ni nitrate solution, and the turbid liquid was stirred continuously until the liquid was completely volatilized. After annealing at 750°C, Ni clusters were finally obtained on graphene and the extra Ni clusters were removed with acid wash to achieve the residual single atom Ni (Figure 2.1a).[1,8] The density functional theory (DFT) calculation also predicts the possibility of dopant metals, such as Cr, Mn, Fe, Co, and Cu.[9,10]

Carbon nanotubes (CNTs) are also favorable for being applied as the supports for single metallic atoms with the proper surface functionalization for impregnation. As the scheme shown in Figure 2.1b, CNTs were treated with a consequent acid wash of hydrochloric and nitrate/sulfuric acid. After evaporation the residual $SOCl_2$, thiolated-CNTs were obtained by reaction with $NH_2(CH_2)_2SH$ in dehydrated toluene at 70°C. Then these decorated CNTs were dispersed and mixed with $H_2(PtCl_6)$ in the suspension. After being reduced by $NaBH_4$, the loaded materials were annealed

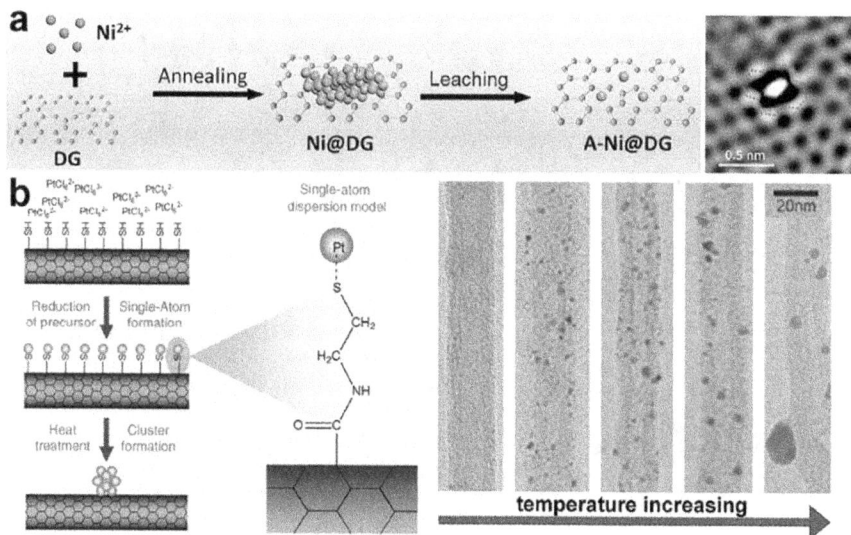

FIGURE 2.1 (a) Scheme of single-atom Ni on defective graphene and the relevant STM.[1] (Copyright ACS.) (b) Scheme of decorated CNTs for single-atom Pt confinement.[11] (Copyright RSC.)

at different temperature ranging from 523 to 873 K in a reductive atmosphere.[12] The SAM Ni_1/CNT was obtained at a relatively low temperature during a short annealing time, unless a cluster structure would be obtained under the thermal creeping.

Thus, it can be implied that the direct capturing of single-atom metals on carbon lattice is relatively unfavorable so that the coordinative atoms such as N, S, O and P have to be introduced to activate the trapping ability. The well-defined carbon-coordination environments have also been displayed in Figure 2.2, showing the possibilities of geometric and electronic features with, thus, different behaviors in catalytic processes.[6] Furthermore, the fabrication of these coordinative single atoms on carbon substrates is limited not only by the impregnation pathway but also by a broader range of achievable routines such as in solid state, in atmosphere, or via spatial organic framework assistance methods, which will be introduced in subsequent sections.[6]

For example, as a result of the strong interactions between metal-nitrogen-carbon (denoted as M–N–C), single atom Co_1–N–C can typically be obtained by impregnation, as shown in Figure 2.3a.[13] Layered C_3N_4 as the substrate was mixed with Co ion source

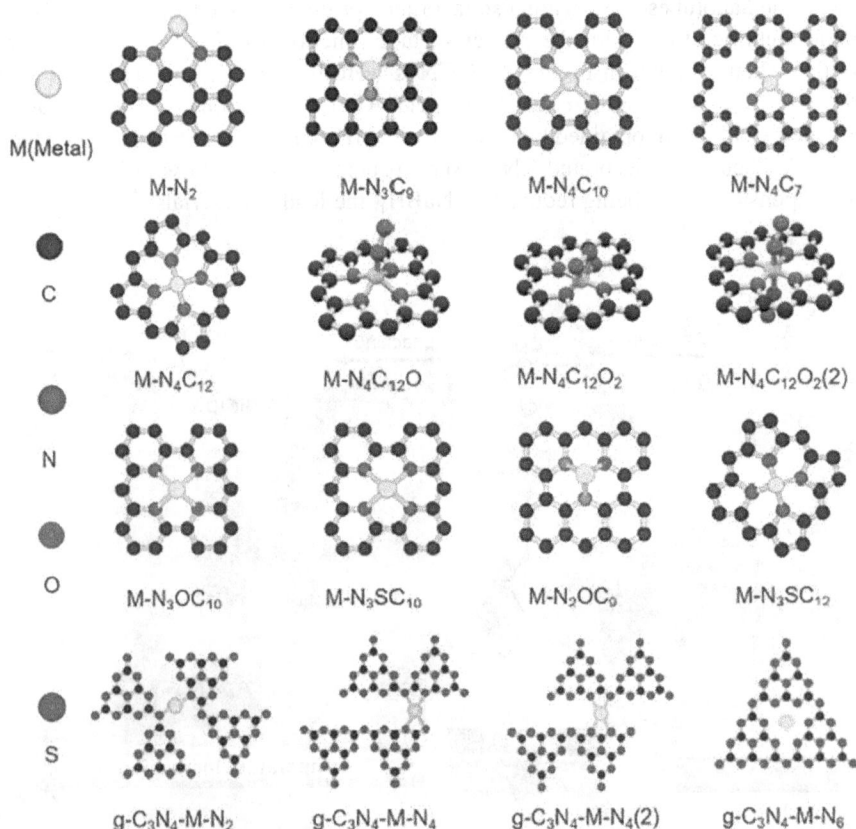

FIGURE 2.2 A variety of carbon-based SAMs with possible coordination environments.[6] (Copyright RSC.)

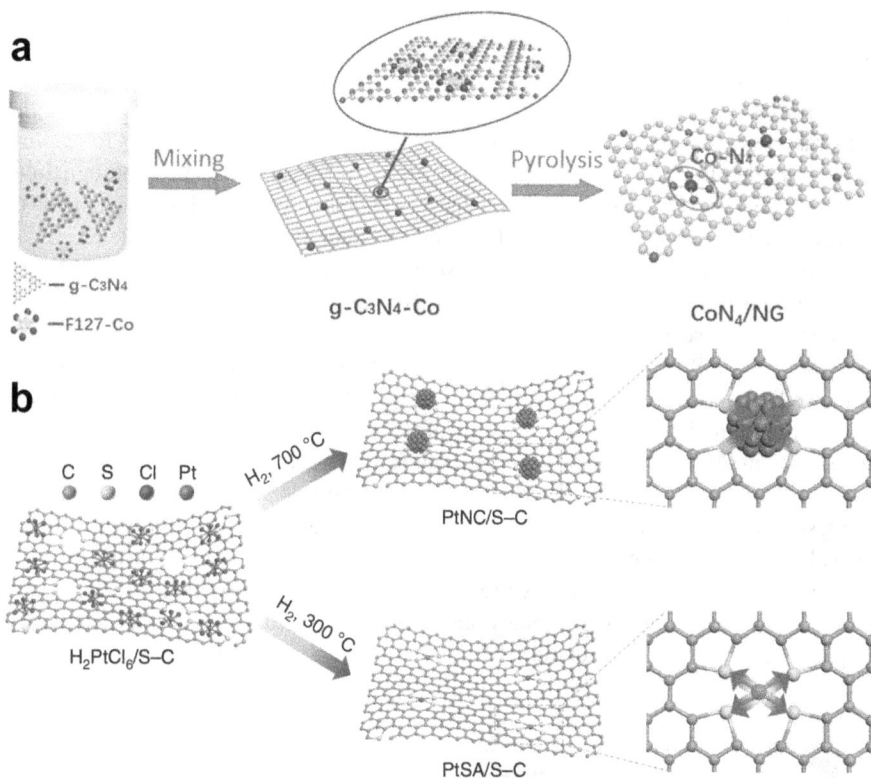

FIGURE 2.3 Schematic illustration of the synthesis of single-atom CoN₄/NG cata-lyst.[15] (Copyright RSC.) (b) Schematic model of the catalyst preparation process and metal size-dependent charge transfer. Single-atom Pt/S–C and nanocluster Pt/S–C were prepared by wet-impregnation of chloroplatinic acid on the S–C supports followed by H₂-reduction at 300 and 700°C, respectively.[14] (Copyright Springer Nature.)

with F127, which is a water-soluble surfactant of polyoxythylene-polyoxypropylene-polyoxyethylene (PEO-PPO-PEO), to guarantee the Co complex penetrated into the 2D substrate. Afterward, the materials were calcinated under an N_2 atmosphere and etched with hydrochloric acid for 24h to dispose the by-product impurities. After the pyrolysis under high-temperature annealing, single-atom Co was successfully trapped with nitrogen coordination in a 2D graphene in-plane structure. Another example comes with S decoration as shown in Figure 2.3b, where the Pt precursors were impregnated into the S-decorated graphite and were anchored stably.[14] After annealing at a relatively low temperature of 300°C, the S-confined-Pt single atoms were obtained.

2.2.1.1.2 Non-Carbon-Based Supports

Apart from the carbon-based supports, which are unstable for SAM aggregation under a high-temperature application scenario, impregnation of single atoms can also be proceeded onto the non-carbon-based supports. According to the support types, the specific processes can be generally classified into metals (alloys), metal oxides, metal hydroxides, metal chalcogenides, nitrides, phosphides, etc., and the corresponding representative structures have been displayed in Figure 2.4, where the substrates also present massive defects and/or potential coordinative traps for capturing the single-atom locations.[16] There are two main processes during impregnation: (i) the dispersion of target metallic precursors and absorption with supports and (ii) removal of the unnecessary ligands (anions) and maintaining the single atom.[17–19] Notably, the fabrications for these representative SAMs on non-carbon-based supports are not only limited to the impregnation method but also included the other methods in later sections.

Among these SAMs on various substrates, metal oxides are one of the most widely employed supports for impregnation. For example, Pt_1/Al_2O_3 was successfully achieved via the impregnation method.[20] Al_2O_3 supports were initially dispersed to form a gel state and then a Pt source was added to the system for ion-exchange and absorption. Then the dried composite was annealed in an ambient atmosphere at 450°C. Accordingly, CeO_2 support can also be utilized to confine the single-atom Au even without a further annealing process.[21] Similarly, a defective NiO substrate was initially achieved by a hydrothermal method using a mixture of nickel nitrate and hexamethylenetetramine (HMT). Then immersed NiO strongly adsorbed the Ir precursors and finally obtained single-atom Ir_1/NiO with an extremely high-mass loading of ~18 wt%.[22]

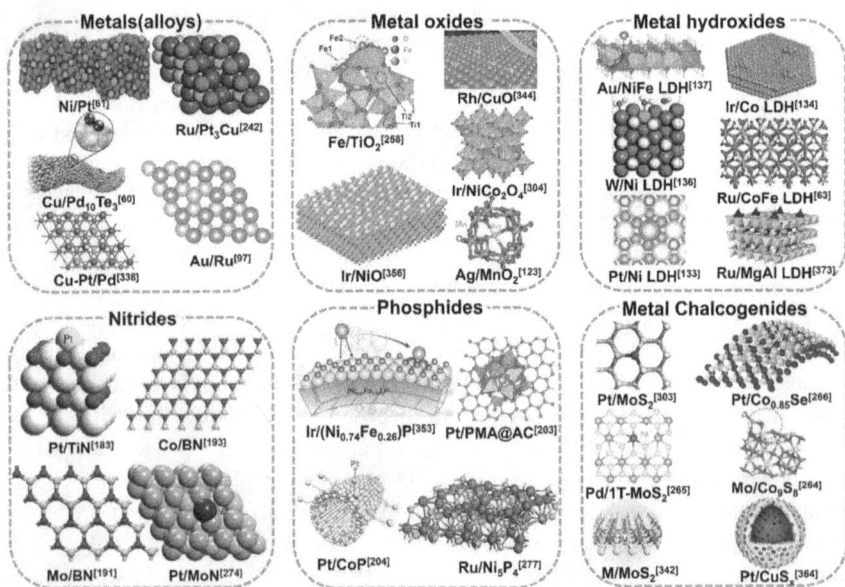

FIGURE 2.4 Representative examples for eight types of non-carbon supported SAMs. The anchoring and coordination environments of the supported isolated transition metal atom are indicated. These examples are mainly derived from applications involving electrochemical reactions.[16] (Copyright RSC.)

Not only for this, the metal oxide substrates with low-dimensional structures such as 2D flakes and 1D rods can be impregnated with SAMs, which is attributed from the enlarged specific area for higher mass loading and catalytic activity. For example, antimony-doped tin oxide (ATO) demonstrates a typical plate structure and can be loaded with $H_2PtCl_6 \cdot 6H_2O$ precursors in suspension. Then the vacuum-dried composite was reduced in H_2/N_2 atmosphere from 100°C to 400°C and the single-atom structure gradually appeared when the temperature arrived at 400°C. It was observed that the Pt single atom was located at the surface of SnSb and SnO_2 while DFT calculation reveals that Pt substitutes steadily for the Sb site position in the SnSb compound.[23] Another instance comes with 1D hollandite-type manganese oxide (HMO) as the substrate. The $Ag(NH_3)_2OH$ complex was mixed with HMO and annealed in a reductive atmosphere. The spatial channel structure of the HMO unit consists of an eight oxygen-atom network, which anchors the Ag for a stable coordination.[24] The ZnO nanowire was also introduced to confine single-atom Ru by the impregnation method.[25]

2.2.1.2 Coprecipitation Method

Alternatively, coprecipitation is another practical method to achieve SAMs. The precursor metallic ions are usually coprecipitated with the introduction of precipitants to induce the formation of single-atom sites onto supports at the same time.[5,26] With the assistance of the aqueous media to control the distance between each metallic-ion species, the target metal precursors can be easily isolated and even be embedded into the bulky body of the support.[1] For example, single-atom Pt on iron oxide was fabricated via a facile coprecipitation method. To be specific, chloroplatinic acid (H_2PtCl_6) and ferric nitrate were used as target metal precursors and substrate precursors, respectively, and dissolved to obtain a homogenous aqueous solution. Then the pH of the mixture was tuned with sodium carbonate solution at 50°C to a value of 8 and the precipitate was thereafter collected. After the calcination of the obtained solid in an ambient or reductive atmosphere, the materials were shown to have single Pt atoms trapped on the vacancies of the FeO_x surface via bonding with O sites, achieving a stable atomic existence.[26] The mass loading of Pt_1 on FeO_x is around 0.17 wt% and a higher content was also achieved around 2.5 wt% with the aggregated cluster structure as the concentrated metal precursors addition. Also, the target metal can be further alternated with Ir following a similar method, whose mass loading of SAM Ir is as low as 0.01 wt%.[27] During this process, H_2IrCl_6 was introduced as a metal precursor and NaOH solution was used as the pH adjustant at 80°C. The SAM Ir_1/FeO_x material was spontaneously achieved without further heat treatment. However, this method cannot precisely control the synthesis process.

2.2.1.3 Galvanic/Successive Reduction Method

Due to its simplicity, galvanic replacement has been used to generate surface alloy coatings without external energy input (e.g. electricity, annealing).[28] This method is aimed at preparing a "mother" nanocluster initially and in-situ reducing a foreign single-atom site on the surface of the cluster via electroless reduction.[29,30] The technique takes advantage of the semi-reduction potential of metal precursors so as to commonly achieve the single-atom alloy (SAA) materials. To evaluate the stability of the SAA formation, many theoretic studies have examined the thermodynamic tendency of the target single-atom dopants to the mother substrate.[28] Stamatakis and co-workers examined the target Pt, Pd, Ni, Rh and Ir atoms in Cu, Ag and Au hosts and disclosed that most of these combinations display a thermodynamic tendency for the dopant atom to be isolated and hence to form a SAA by calculation of the aggregation energy as shown in Figure 2.5a.[28,31–33] To be specific, the combination with positive aggregation energy presents rise to isolated sites in the host substrate while the ones near zero aggregation energy (or lower) tend to appear as aggregated results (just as the insets show the SAA Pt$_1$/Cu vs. Pt$_x$/Au in Figure 2.5a).[28]

For a practical example, Au$_1$/Pd SAA was achieved according to the galvanic principle as shown in Figure 2.5b.[30] The mother Pd nanoparticles was first obtained from PVP-assisted [poly(vinylpyrrolidone)] Pd precursor by an alcohol reduction method. The as-prepared Pd clusters have an average diameter of 1.8 nm and consist

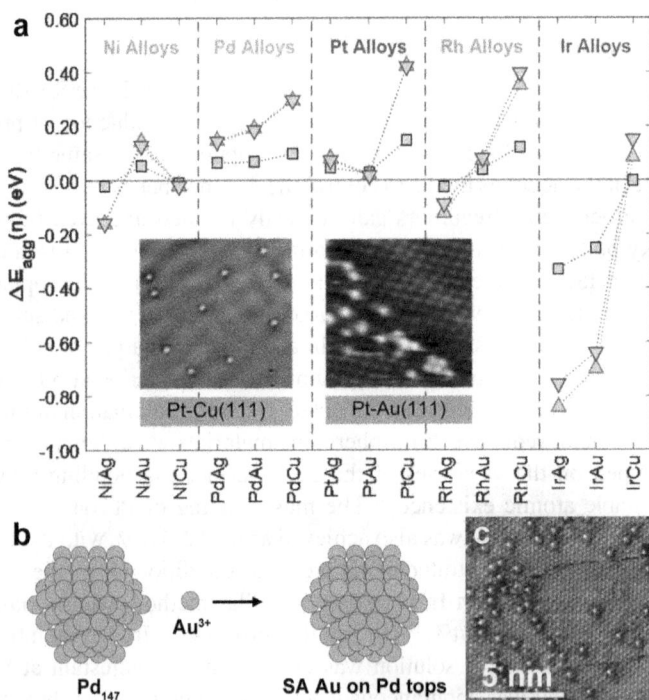

FIGURE 2.5 (a) DFT prediction that metal combinations are thermodynamically disposed to form SAAs. Aggregation energy ΔE_{agg} (n) relative to the SAA phase for the clustering of dopant atoms (Ni, Pd, Pt, Rh, and Ir) in the (111) surface of host metals (Cu, Ag, and Au) into dimers (squares) and trimers where dopants surround fcc (up-triangles) and hcp (down-triangles) sites. Values of ΔE_{agg} (n) that are negative correspond to a preference for dopant clustering, whereas positive values correspond to a preference for dopant atom dispersion to the SAA structure. Insets show examples of when aggregation is not preferred (PtCu) versus when ΔE_{agg}~0 (PtAu) and both single atoms and small clusters of Pt are seen in the surface.[28,33] (Copyright ACS.) (b) Scheme of galvanic reaction to achieve Au$_1$/Pd SAA. (c) STEM of Pd$_1$/Cu SAA. (Copyright RSC.)

of about 147 atoms in a unit. Based on the galvanic replacement principle, $HAuCl_4$ precursors were added to the colloidal dispersion of Pd_{147} clusters, and the obtained Au–Pd colloidal catalysts contain an abundance of top (vertex or corner) Au single atoms.[30] Similarly, single-atomic Au can also be fabricated on a Pd–Ir alloy cluster.[29]

Apart from the precious metallic based mother substrate, in the case that Cu has a relatively lower reduction potential (Cu^{2+}/Cu, $E^0 = 0.340$ V) than that of Pd (Pd^{2+}/Pd, $E^0 = 0.915$ V), it is favorable for Pd to be induced onto the Cu surface.[34] Accordingly, the PVP-Cu^{2+} complex precursors were initially reduced by $NaBH_4$ to achieve the host metal nanoparticles and calcinated in the air and reductive atmosphere to remove the surfactant and surface oxygen, respectively. The galvanic reaction then occurs as a result of the addition of $Pd(NO_3)_2$ source exclusively into Cu substrate suspension, where a simple route for metal exchange occurs on the host metal surface.[34] The typical single-atom dispersed structure on the Pd_1/Cu alloy surface is displayed by STEM in Figure 2.5c. This method can also be applied to single-atom Pt on Cu substrate with the assistance of ultrasonication.[35,36]

These SAA structures are attributed to the strong interbond between the reactive dopants and a less reactive metal host, which leads to thermal stability in terms of keeping the dopant sites isolated.[28] Therefore, galvanic methods can both ensure the successive reduction and load the single atom onto the surface of the host nanoclusters. However, the procedure of this method is relatively complex and hard to control the precise lattice point, which leads to poor durability and a less-industrial prospect.[5] At the same time, although single atom on alloy is a way to tune surface binding energy, this approach is still limited by Bronsted-Evans-Polanyi scaling; thus, there must still be a trade-off between low activation energy and weak binding.[33]

2.2.1.4 Chemical Etching

Fabrication via chemical etching is also a practical approach to achieve a higher mass loading for SAMs. This route mainly contains the chemical etching process from the bulky metal precursor as a soft template to generate a large population of residual single-metal atoms on the unremoved substrate.[1] From another aspect, the hard template could also build the desired spatial framework and then be removed to form the 3D porous channel, which provides abundant coordinative traps for single-atom sitting.

Typically, Ni_1/graphene was achieved via a practical chemical etching method.[37] To be specific, porous nickel was first achieved via the dealloying method from NiMn alloy in $(NH_4)_2SO_4$ solution. Then the porous Ni was annealed in a reductive atmosphere containing benzene and hydrogen and a graphene covered Ni was thereafter obtained. Lastly, the bulky nickel was dissolved in a concentrated HCl solution by controlling the etching time. The mass loading for Ni_1/graphene (Figure 2.6a) reaches around 4~8 at. % and the as-prepared materials are acidic-resistance in proton-rich system for hydrogen evolution reaction (HER) because of the strong binding between Ni and C substrate. For the application of oxygen evolution reaction (OER) in acid media, the Ru_1/Pt_3Cu ternary alloy by acid etching was also developed by a similar etching method.[16,38]

Besides, the favorable coordination between the doped nitrogen and the target single-metal atom (M–N coordination) also permits the higher loading mass for SAMs. Initially, Ni^{2+} as the target metallic precursor was introduced onto the

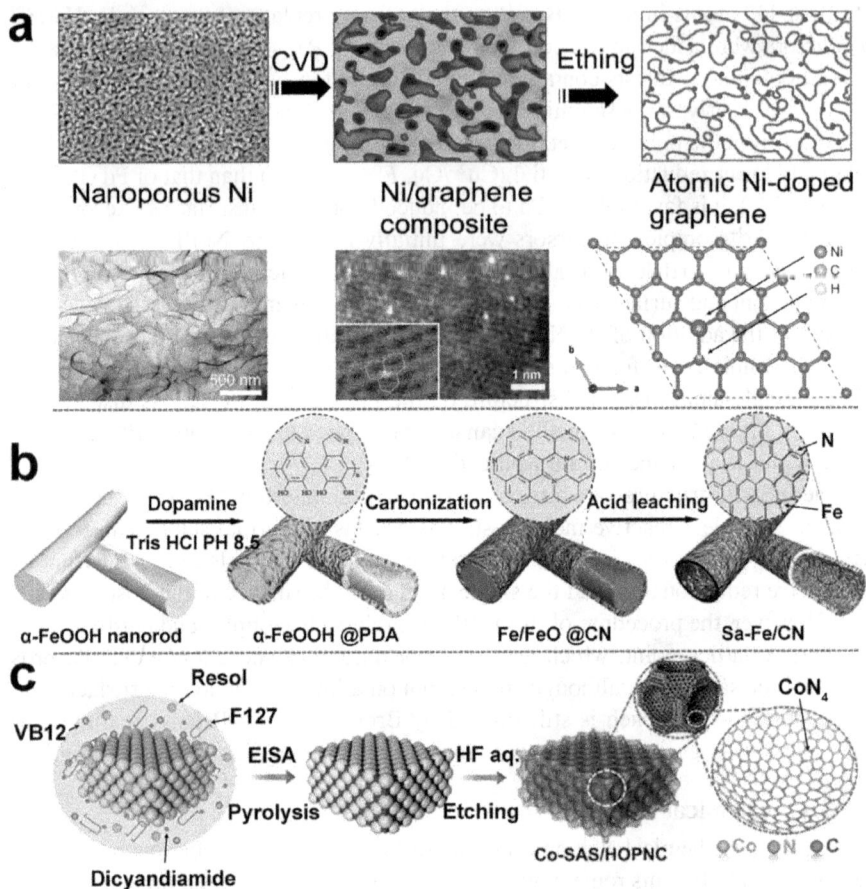

a

CVD → Ething →

Nanoporous Ni   Ni/graphene composite   Atomic Ni-doped graphene

● Ni
● C
○ H

b

Dopamine
Tris HCl PH 8.5 →   Carbonization →   Acid leaching →

N
Fe

α-FeOOH nanorod   α-FeOOH @PDA   Fe/FeO @CN   Sa-Fe/CN

c

Resol
VB12   F127
EISA →   HF aq. →   CoN₄
Pyrolysis   Etching

Dicyandiamide   Co-SAS/HOPNC   ● Co ● N ● C

FIGURE 2.6 (a) Scheme of the Ni₁/graphene by etching bulky nickel from graphene and the relevant SEM/TEM photos.[1] (Copyright ACS.) (b) Scheme of Fe₁/CN using FeOOH nanorod template.[39] (Copyright ACS.) (c) Scheme of porous Co₁/HOPNC using SiO₂ hard template.[40] (Copyright RSC.)

microwave-exfoliated graphene oxide (MEGO) substrate with the addition of urea.[41,42] The dried mixture was then annealed under 800°C in NH_3 atmosphere and also washed with nitric acid to eliminate nickel clusters. The mass loading can reach as high as 6.9 wt%. Not only the N–M bond but the enlarged porous structure stabilizes the single Ni atoms on the edge sites of the 3D porous framework.

To some extent, the etching metal precursor can also be regarded as the template to construct the particular structure with an enlarged surface area and to sufficiently contact with the substrate. Therefore, SAMs with the desired microstructure can also be achieved via etching protocols. For example, FeOOH nanorods were initially obtained by coprecipitation and dispersed into Tris buffer solution to react with dopamine-HC. Consequently, polydopamine-coated FeOOH nanorods were finally annealed in Ar atmosphere by high-temperature pyrolysis to achieve the single-atom

$Fe_1/N–C$ material with a hollow-rod structure as shown in Figure 2.6b.[39] As the single-metal precursors, this route can also be extended into other metal hydroxides or oxides, such as $Co(OH)_2$ and $Ni(OH)_2$ nanoplates and MnO_2 nanorods.[1,39]

From another aspect, the soluble template assistance also provides massive opportunities that permit the metal precursors to make sufficient contact with the substrate by constructing the porous channels. A series of single-atom metal-doped MoS_2 have been fabricated via this template-assistant method.[43–45] To be specific, metal precursors including metallic powders or salt, SiO_2 template and $(NH_4)_6Mo_7O_{24}\cdot4H_2O$ were vigorously stirred and dried to form the templated-complex powder. Then the precursor powder and CS_2 were transferred into an autoclave under a hydrothermal reaction. Afterward, the as-product was finally washed by HF acid to form the porous single-atom doped materials. The target metals include Ni, Co, Fe, Se, Ru, etc. It was further developed into a single-atom Co–N–C structure on hierarchical carbon-based support as the scheme shown in Figure 2.6c.[1,46] Vitamin B12, Pluronic F127 and dicyandiamide were used as the metal, N and C precursors. Then the SiO_2 template was soaked into the precursors solution and the solvent was thereafter removed by evaporation. The pyrolysis took place under a high-temperature treatment in Ar atmosphere. After the strong HF washing, SAM $Co_1–N–C$ with an obviously ordered porous-structure was finally obtained and shows great electrochemical activity for electrochemical hydrogen evolution and oxygen reduction.[46]

2.2.1.5 Electrochemical Deposition

Nanostructured materials from 0D (single atom) to 3D can be efficiently achieved and regulated by electrochemistry method.[47] Most of the transition and precious metal cations in the aqueous solution, driven by the loading electricity, can be reduced into metallic state, which enables electrodeposition as a feasible technology with a real-time response on a large scale.[4,48] However, the momentary procedure also brings severe nucleation and continuous growth, so the craft parameters must be carefully adjusted such as applied potential, current density, electrolyte, concentration, temperature, stirring, cyclic procedure, substrate, additives, etc.[48]

The electrodeposition has been systematically investigated for a series of single-metal moieties on the substrates.[49] The deposited single atoms include Ir, Ru, Rh, Pd, Ag, Pt, Au, Fe, Co, Ni, Zn, V, Cr, Mn and Cu while the substrates such as $Co(OH)_2$ nanosheets, N-doped carbon, MnO_2 nanosheets, MoS_2 nanosheets, and CoFeSe oxide nanosheets are alternated for the practical supports. For instance, in $Ir_1/CoFeSe$ oxide, during the cathodic electrodeposition process, the metallic ions (or positive complexes) migrate onto the substrate and reduced to metallic states (Figure 2.7a). Moreover, the active substrate with high conductivity and specific surface area is often used to induce an underpotential deposition (UPD) process, that is, a metal cation deposits on the active support at a potential more positive than its equilibrium potential (the potential at which it deposits onto itself).[50] These possible metals are Cu, Sn, Pb, Bi, Rh, Pd, Pt, etc.[50] For example, an S and N co-doped graphene oxide substrate enables Cu^{2+} to be reduced into single-atom state with an UPD process.[51] Cu sites can be electrodeposited onto 2D TD_2 (T=Mo, W; D=S, Se, and Sn) materials as an exfoliated support and then the precious metals of Pt, Pd and Rh are galvanically exchanged with Cu sites to achieve the precious metal based SAMs.[50]

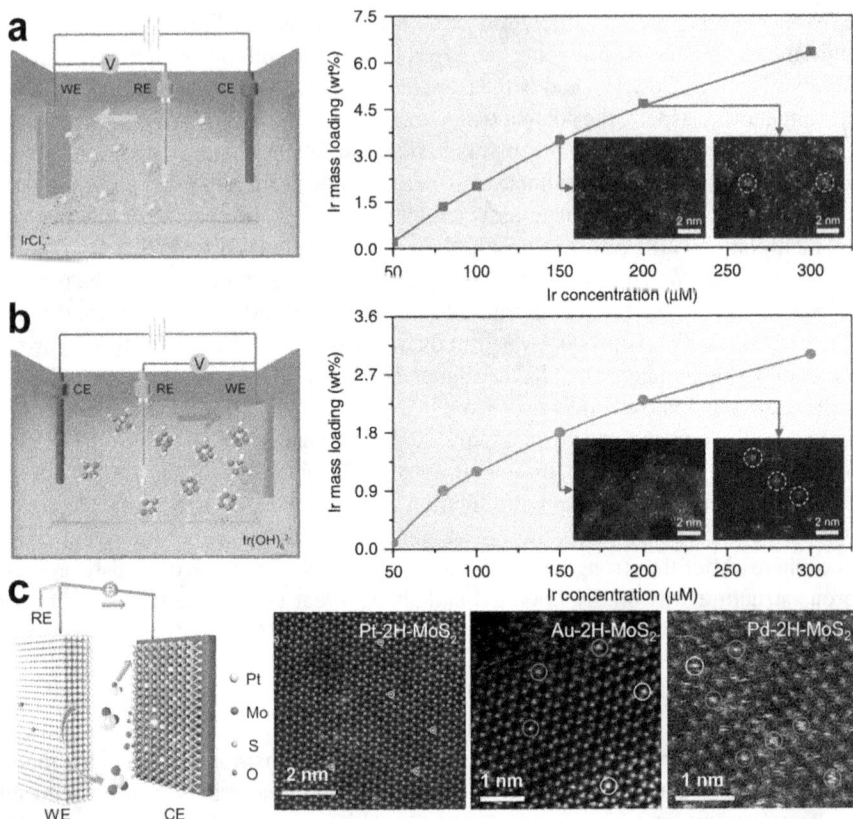

FIGURE 2.7 (a) Scheme of cathodic electrodeposition and the relevant Ir SAMs.[50] (b) Scheme of anodic electrodeposition and the relevant Ir SAMs.[50] (Copyright Springer Nature.) (c) Scheme of electrodeposition with the precursors from anode etching and the relevant Pt, Au and Pd SAMs on MoS$_2$ substrate.[58] (Copyright ACS.)

Notably, the depositing process was not always carried out under a potentiostatic process with a constant applied potential for a period. Since the ordinary coating technique will present a dense film or cluster of islands on the substrate but not the desired single-metal structure. The practical protocol is usually determined by a dynamic process, such as cyclic voltammetric (CV) scan in a redox window. The depositing and peeling-off processes take place consequently on the substrate to prevent the enlargement of the nuclei so as to achieve the single-atom structure.

Atomically dispersed (oxy)hydroxide coatings for OER can also be fabricated via a typical electrodeposition-coprecipitation method. A typical three-electrode system was used as the depositing cell, including a conductive substrate (such as nickel foam, copper foam, glassy carbon, carbon fiber paper, etc.), carbon-based counter and a Ag/AgCl reference. The electrodepositing was applied under −1.0 V *vs.* reference in metal-nitrate aqueous electrolyte (e.g. Ni^{2+}), where the OH$^-$ ions were generated from electro-decompose of H$_2$O and coprecipitated with the local metal cations to form the atomic layer (e.g. NiOOH).[52–57] Through this approach, Fe and Cr atoms were atomically co-dispersed into the oxyhydroxide cage as benchmark OER catalysts in alkaline media.

On the other hand, the metallic cations can also be embraced with the hydroxyl to form the negative-charged complex and transported onto the cathode surface (Figure 2.7b). Therefore, the depositing can, at the same time, proceed on the anodic substrate, where the target single metal in the complex cage will be adsorbed and anchored, such as the octahedral M–OH cage in the hydroxide structure, which presents high activity for OER.[49]

There is also the issue that contamination from anodic etching can also affect the depositing purity. However, this tracing amount of the elemental invasion, from another aspect, is also regarded as a single-atom source. Since the atomic deposition is very sensitive to the concentration of the precursors in the electrolyte, there is a difficulty in controlling the amount of the metal cations addition. As a result, anodic dissolution efficiently provides the request for the ions source for the cathodic electrodeposition with atomic-scaled dispersion. For example, Au, Pt and Pd foils were introduced as the anode and the precursor materials for SAM electrodeposition on MoS$_2$ substrate as shown in Figure 2.7c in 0.5 M H$_2$SO$_4$ electrolyte.[58]

2.2.1.6 Photochemical Induction

The photochemical induction method can also be applied to fabricate SAMs. Since the absorbed energy from light is relatively weak, the subsequent nucleation for single-atom doping is also mild. The central feature of the light-promoted transformations is the involvement of electronically excited states, generated upon absorption of photons, which produce transient reactive intermediates and significantly alter the reactivity of a chemical compound.[1,59,60] Therefore, the selection of the photochemically active substrate that can confine and emit the excited electron is very important. For example, TiO$_2$ has been widely used in the photocatalytic field and can also be applied for a possible support for SAMs via the photochemical induction method.[61] Thus, TiO$_2$ ultrathin sheet was dispersed into H$_2$PdCl$_4$ aqueous solution and ultraviolet (UV) is radiated to age the suspension (Figure 2.8a). The gray Pt$_1$/TiO$_2$ SAMs present a high-mass loading o 1.5 wt%.

Carbon-based support can also be induced for SAMs under UV reduction. To be specific, the nitrogen-doped porous carbon (NPC) was derived from cattle bones under high-temperature annealing of pyrolysis process. Then NPC substrates and H$_2$PtCl$_6$ were mixed and the suspension was dried to form a thin film on the glass plate. Under the UV radiation for 1 h, the adsorbed PtCl$_6^{2-}$ is spontaneously reduced into single-atom state on the substrate as Figure 2.8b shown.[62]

Although the photo-induction presents a mild approach, the separation of the metal precursor is also essential for SAM dispersion.[63,64] Therefore, the precursor solution was initially frozen to maintain the metallic ions distance between each other so that single-atom platinum was efficiently obtained via photochemical reduction under UV light.[63] As Figure 2.8c shows, the H$_2$PtCl$_6$ solution was initially frozen by liquid nitrogen and reacted under UV irradiation for 1 h. After the iced solution was melt at room temperature, the porous carbon was introduced into the mixture and the Pt single atoms were anchored on the substrate. As comparison, the Pt atoms

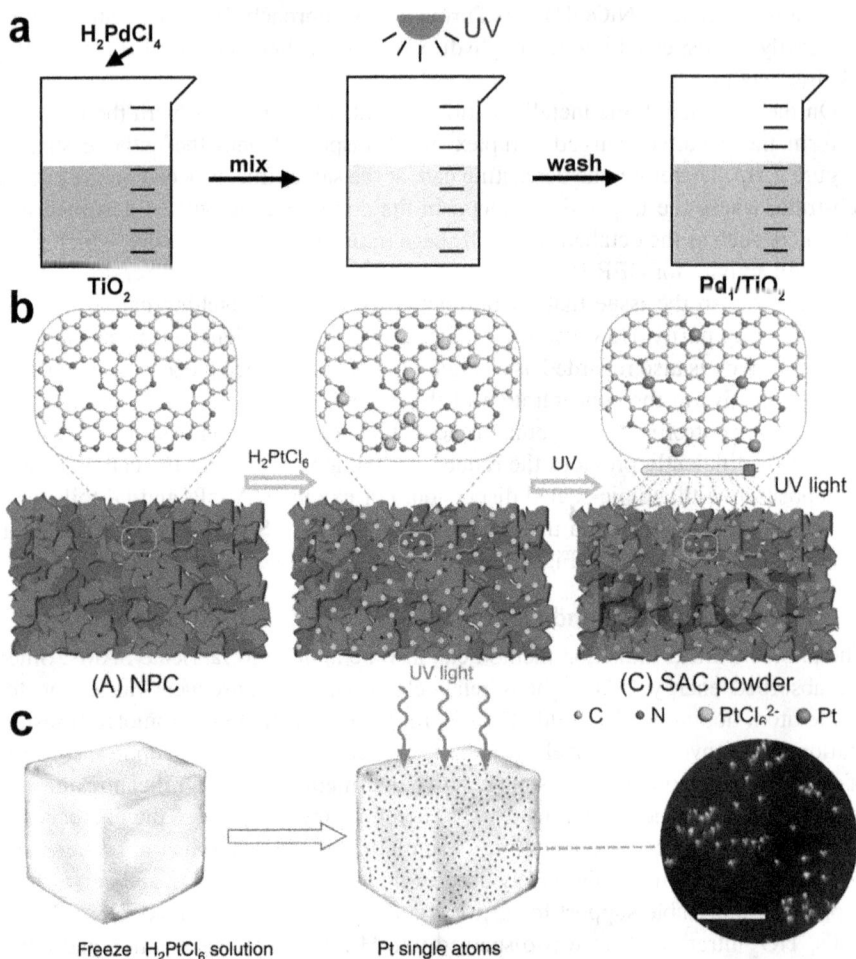

FIGURE 2.8 Scheme for the synthesis of Pd_1/TiO_2. After H_2PdCl_4 was introduced into TiO_2 suspension to allow the adsorption of Pd species thereon. The mixture was then irradiated by Xe lamp with low-density ultraviolet light (UV). After 10 min irradiation, the Pd_1/TiO_2 catalyst was collected and washed thoroughly with water. The resultant product was a light gray dispersion. (b) Schematic illustration of the formation of $Pt_{1/}NPC$ catalyst: (a) resultant NPC substrate, (b) $PtCl_6^{2-}$ ions adsorbed on the NPC, and (c) Pt single atoms anchored on the NPC;[62] (Copyright ACS.) (c) Schematic illustration the iced-photochemical process.[63] (Copyright Springer Nature.)

tended to be nanoparticle structures under the UV-induction approach in liquid state precursor. Technically, this iced-photo-induction method is also working on the above-mentioned TiO_2 substrate.[64] For example, TiO_2 powder and K_2PdCl_4 aqueous solution were mixed and rapidly frozen with the employment of liquid nitrogen. Then Xe light irradiation (300 W) was covered on the iced solution for 5 min till it melted. Finally, the Pd_1/TiO_2 SAMs were achieved.[64]

2.2.1.7 Others Approaches in Liquid Media

There are also some other approaches in the liquid phase to synthesize the SAMs. For example, with the assistance of ionic-liquid (IL), single-atom catalysts employing electrostatic interaction can be stabilized onto the substrate steadily, which was reported as a general stabilization strategy.[65] In particular, hydroxyapatite ($Ca_{10}(OH)_2(PO_4)_6$, HAP) was used as the support and single-atom Pt (or Pd) was introduced by impregnation method. To coat the IL protection, the Pt_1/HAP was mixed with ILs, such as 1-Butyl-3-methylimidazolium bis(trifluoromethylsulfonyl)imide ([Bmim][Tf_2N]), 1-Butyl-3-methylimidazolium trifluoromethanesulfonate ([Bmim][CF_3SO_3]) and 1-Butyl-3-methylimidazolium tetrafluoroborate ([Bmim][BF_4]), and the methanol solvent was finally evaporated. It was found that under the modification of these ILs, single-atom Pt on the support exhibited more stability in isolated state after being applied in propylene hydrogenation reaction (Figure 2.9a). While the SAMs without IL protection tend to aggregate into island-like particles. Besides, the SAMs on CeO_2, TiO_2 and ZrO_2 supports can also be efficiently protected with this strategy.[65] Furthermore, with the IL-induced stabilization of single Ru_1/TiO_2, the leaching of the active Ru species from the support in liquid-phase reactions can also be inhibited.[66]

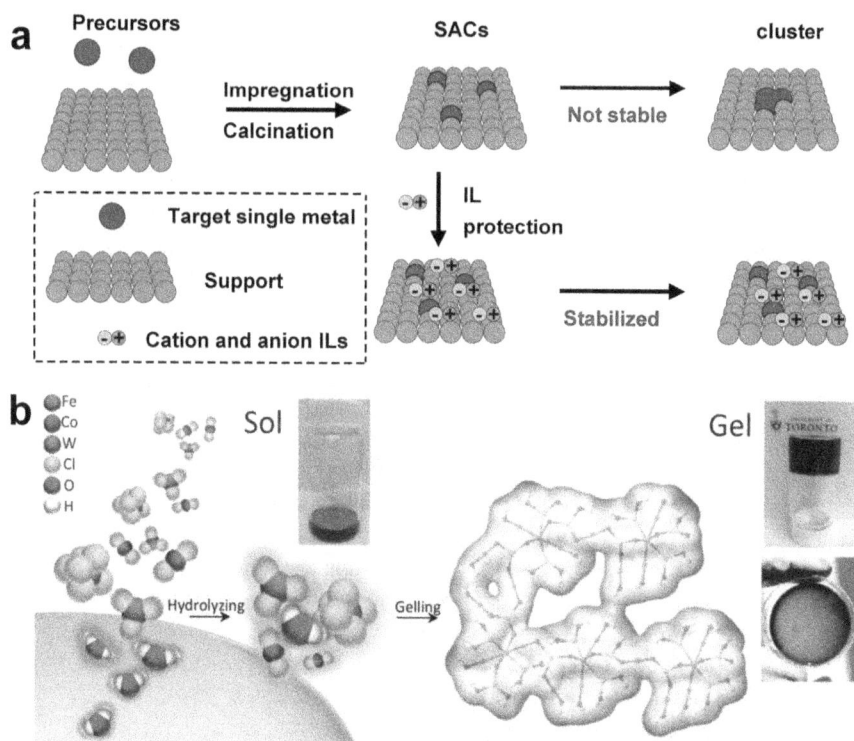

FIGURE 2.9 (a) Schematic illustration of the preparation of SAMs and the stabilization by ILs; (b) schematic illustration of the preparation process for the gelled structure and pictures of the corresponding sol, gel, and gelled film.[54] (Copyright Wiley.)

With a similar principle, the introduction of the complex into the synthesis procedure also plays an important role to ensure the atomic distribution of SAMs. Commonly, Ag cations is more stable in ammonia aqueous as an $[Ag(NH_3)_2]OH$ complex so as to prevent the aggregated nucleation during the SAM fabrication.[24] It was reported a galvanic replacement method with the assistance of adding SR (thiolate) to form the $[Ag_{25}(SR)_{18}]^-$ complex precursor. Then Au can penetrate into the complex center and form the Au_1Ag_{24} SAMs. The use of complex association ensures the stable single-atom state without multiple atoms doping.[67]

Besides, sol-gel approach can also ensure the atomic dispersion for multinary compounds. A series of atomic doping the high-valence elements such as Mo, W were doped into the NiFe and CoFe LDH octahedral cage and performs the benchmark OER activity in alkaline media.[68,69] To be specific, metal chloride precursors such as Ni^{2+}, Co^{2+}, Fe^{3+}, Mo^{4+}, W^{6+} were dissolved in mixture of water and ethanol. After storage in a chilling environment to form the precipitate, the mixture was added with propylene oxide and aged in acetone for 5 days. Then the gel was dried with supercritical CO_2 at room temperature as shown in Figure 2.9b.[54]

2.2.2 ATMOSPHERIC STRATEGY

Proceeding the fabrication of SAMs in liquid phase is, to some extent, hard to control the enough distance and the water molecules may cause hydrated effect, which also influences the atomic dispersion on the substrate. Therefore, using high energy to induce the bulky target metal into the atmosphere, where a longer distance presented between the excited atoms without unnecessary interacted media, thus further improving the dispersion and purities for SAMs. Besides, the pretreatment on the support materials is no longer a necessarily request, as the direct contact between the single-atom metal and the substrate can directly occur and construct a strong covalence bond after the consequent thermal treatment. Last but not the least, gas-phase reaction is also a promising approach that can be potentially promoted in a large-scale production for industrial practice.

2.2.2.1 Chemical Vapor Deposition

Chemical vapor deposition (CVD) is a vapor growth technology that can convert one or more raw materials into volatile substances after heat treatment. The target metal precursors are vaporized and captured by the carrier gas and conveyed to the consequent acceptance as the support, then undergoes a chemical reaction on the surface.[70] Therefore, CVD is aimed to prepare two-dimensional materials, nanoparticles and single-atom catalysts.[70,71] Impressively, the number of layers, lateral size, orientation and defect degree during CVD can be effectively controlled due to the broad tunability of their growth parameters (*e.g.*, precursor, temperature, pressure, flow rate, time, precursor-substrate distance, etc.) and the choice of substrates.[70,71]

To get rid of the tedious fabricating procedures and appropriate interactions between the metal atoms and supports, a simple and practical strategy permits the large-scale synthesis of SAMs via direct atoms emitting from bulk metals, and the subsequent trapping on nitrogen-rich porous carbon with the assistance of ammonia.[72] As shown in Figure 2.10, the copper foam at upstream and carbon-based-substrate powders at

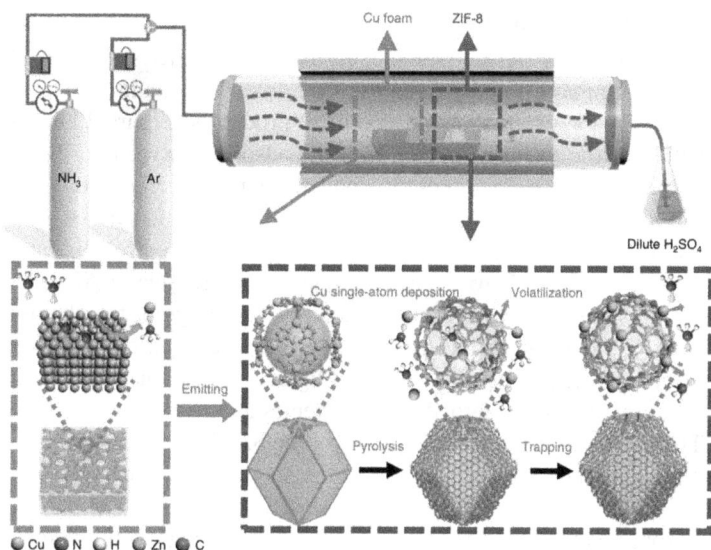

FIGURE 2.10 Schematic of apparatus and reaction mechanism the preparation of Cu_1/-N-C.[72] (Copyright Springer Nature.)

downstream were placed separately in CVD chamber.[72] The system was heated up to 900°C for 1 h in argon atmosphere and then shift to NH_3 atmosphere for another 1 h.[70,72] During this procedure, carbon-based materials underwent a pyrolysis process and included Zn atoms were volatilized so that NPC substrates were obtained.[70,72] With the atmosphere switched into ammonia, ammonia molecules pulled out the surface copper atoms to form the $Cu(NH_3)_x$ species.[70,72] Last the gas-carried Cu complex was trapped by the defect on the carbon substrate to form the Cu_1/N–C SAM final product.[70,72] This method can also be extended into Ni_1/N–C and Co_1/N–C materials by substitution the copper foam with nickel and cobalt foam.[72] Then, this method was updated with the use of ammonia-free atmosphere. CuO powders were introduced as the precursors in the upstream porcelain boat and nitrogen-doped carbon was placed at downstream. The system was heated to 1,000°C for 5 h in N_2 atmosphere.[73] The practical support and target metal can also be extended into carbon nanotubes and MoO_3 and SnO_2, respectively, by this method.[73]

Different from the above-mentioned CVD approach, another route was also introduced to obtain the SAMs in a dual-temperature-zone tube furnace.[74] The 2-methylimidazolate (2-MeIm) was placed in the upstream zone in the CVD furnace and Fe doped ZnO nanosheet was placed in the downstream area. The dual zones were heated to 280°C and 350°C at the same time in argon for 30 min. As the 2-MeIm vaporized, the flow was hit onto Fe-ZnO nanosheet to form the $Fe-Zn(MeIm)_2$ intermediate. With the consequent calcination at 1,000°C for another hour, zinc complex support was transformed into katsenite Fe_1/NC product.[74] This CVD approach, from another aspect, is actually selectively doping the coordinative N into the substrate and thereafter forming the Fe–N–C SAM structure.

2.2.2.2 Atomic Layer Deposition

Atomic layer deposition (ALD) is a special modification derived from CVD with the distinct feature that film growth takes place in a cyclic manner.[78] Generally, the principles for this approach include two critical processes (Figure 2.11a), those are, (i) the reaction of the metal precursor with the adsorbed oxygen on the surface of the substrate and (ii) an oxygen pulse to convert the precursor ligands to M–O species to form a new adsorbed oxygen layer on the M-substrate surface.[1,75] Then this process could be repeated as many times as required to control the quantities of the M–O species on the substrate.[78] To be specific, one recycle of ADL takes place with four main steps: Exposure of the first precursor, purge of the chamber, exposure of the second reactant and a further purge of the reaction chamber. During the cyclic pulse and purge process, the adsorbed ligands are calcinated under particular atmosphere to remove the adsorbed oxygen species and to anchor onto the substrate steadily. Therefore, ALD can precisely control the SAM structure since this method can finely regulate the target population of the single atom or nanocluster onto the substrates.[1,75,76,79]

This approach was initially aimed to produce fine coatings on the electronic component.[1,78,80] In 2013, this strategy was updated (Figure 2.11b) to produce single-atomic or nanomaterials. Trimethyl (methylcyclopentadienyl) platinum (MeCpPtMe$_3$) as the metal source was initially applied onto exfoliated graphene oxide (e–GO) substrate and variated reaction cycles from 50 to 150 to achieve the deposited layers.[76] As the

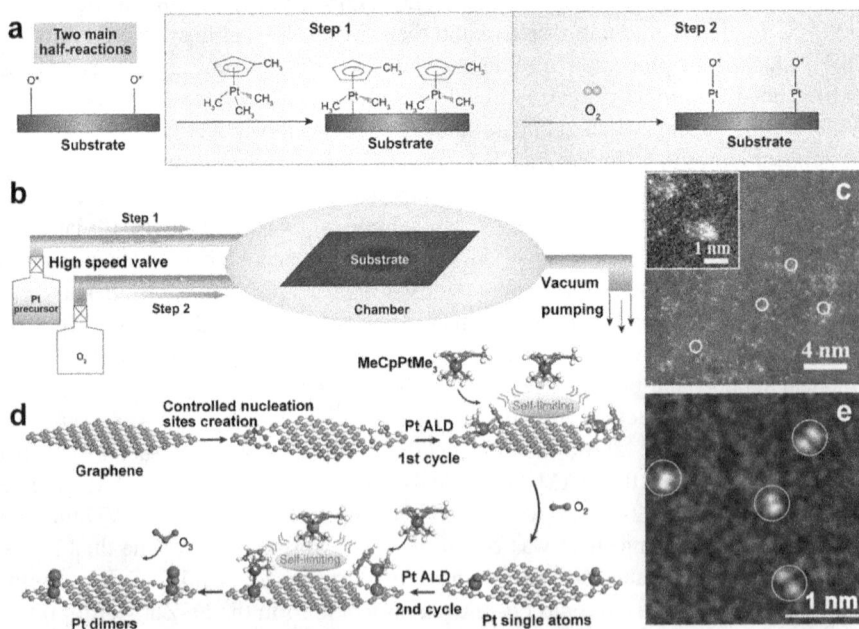

FIGURE 2.11 Schemes of ALD (a) principle and (b) device[75] (Copyright Science Press.) (c) HAADF-STEM images of graphene-supported Pt$_1$-SAMs[76] (Copyright Springer Nature.) (d and e) Scheme and HAADF-STEM of Pt$_2$/GO dimers[77] (Copyright Springer Nature.)

cyclic number increased, Pt–Pt bond intensity gradually enhanced and Pt clusters and Pt nanoparticles appeared gradually. Figure 2.11c shows the Pt single-atomic structure under the cyclic number of 50. Then a nitrogen-doped GO (N–GO) was introduced as an optimal substrate for ALD synthesis of Pt_1/N–GO.[79] Accordingly, the target metal can be replaced with a Pd source by using palladium hexafluoro-acetylacetate (Pd(hfac)$_2$) and formalin as the precursor and complex reagent, respectively, onto acidified exfoliated-GO substrate.[81] The ALD experienced only one cycle in N_2 carrier gas under a mild vapor pressure with a successive time of 120, 120, 60, and 120 s for Pd(hfac)$_2$ exposure, N_2 purge, formalin exposure and N_2 purge, respectively. Furthermore, as the enhanced synergism from multiple-elemental effect, the area-selective ALD method also provides an alternative approach to obtain the heteroatomic structure.[53,54,82] For example, Pt_2 dimers on graphene substrate had been successfully fabricated via a bottom-up ALD approach.[77] Through proper nucleation, Pt_1 single-atom was initially deposited and a secondary Pt site preferentially relocated around Pt_1 site as the scheme and TEM photo shown in Figure 2.11d and e. The typical parameters of ALD synthesis for SAMs are summarized in Table 2.2.[76,77,79,81,83]

However, the mass loading is still very low owing to the relatively diluted source in atmosphere. Further deposition would also lead to the attachment of introduced atom onto the previously existing single-atom sites, which causes the formation of clusters.[1,75] Therefore, alternation of the executive substrates is also essential for the improvement of mass loading and anti-aggregation for SAMs. Not only the pretreated graphene that presents abundant vacancies and defects can be utilized as the anchored convey, the SAMs can also be stabilized onto metal oxide such as CeO_2 to form a steady coordination between each other.[83]

2.2.2.3   Sputtering

Sputtering coating technique is also another efficient protocol to induce a single-atom atmosphere from target metals such as Pt, Au, Ag, etc., which have relatively lower cohesion energy. The procedure is initiated by the impact of energetic particles (~100 eV) and the incident particles cause a multi-atom kinetic collision process, whereby enough energy is transferred to single atoms to overcome the cohesion energy and be emitted from the target.[84] The atmospheric atoms are then captured and coated on the substrate surface.[84] Inert gases such as argon and nitrogen are typically used for the sputtering of metals as an intermediate and the energy of the incident particles can be enhanced by electrical and magnetic fields (Figure 2.12).[84,85] For example, a series of Au decorated carbon and WO_3 catalysts are fabricated by magnetron sputtering method. The Au dispersions are from single-atom (Au_1/C) size to 2.1 nm diameters (Au clusters/WO_3).[86] In a large vacuum chamber, a 5-cm-diameter magnetron sputter source was positioned above stainless steel (SS) cup, tilted 45° from vertical. A second smaller stainless steel cup, filled with support powders, was placed inside the larger one. This rotation of the smaller SS cup causes the support powders to tumble and constantly randomize, exposing a new surface to the sputter source. The Teflon stir bars were used to promote mixing of the powders as they tumbled.[86,87] The instance can also be applied to synthesize the high dense dispersion of single-atom Pt on graphene by plasma sputtering. Particularly, N_2 or Ar gas was introduced into the chamber for 7 min to displace air. The applied voltage was

TABLE 2.2
Parameters of ALD for SAMs Synthesis

SAMs	Precursor	Reactant	Substrate	Pulse 1	Pulse 2	Carrier Gas	Repeat	Temperature
Pt_1/GO	$MeCpPtMe_3$	O_2	Exfoliated-GO	1 s $MeCpPtMe_3$pulse, 65°C, 800 mTorr	5 s O_2pulse, 100°C–150°C	20s N_2pulse	50–150	250°C
Pt_1/N-GO	$MeCpPtMe_3$	O_2	N-GO	1 s $MeCpPtMe_3$pulse, 65°C, 800 mTorr	5 s O_2pulse, 100°C–150°C	20s N_2pulse	50–100	250°C
Pd_1/GO	$Pd(hfac)_2$	HCHO	Exfoliated-GO	120s $Pd(hfac)_2$pulse, 65°C	60s Formalin pulse, 65°C	120s N_2 after each Pulse	1	110°C–150°C
Pt_1/CeO_2	$MeCpPtMe_3$	O_2	CeO_2	50s $MeCpPtMe_3$pulse, 70°C	50s O_2pulse, 100°C–15°C	120s N_2 after each Pulse e	50–100	110°C–150°C
Pt_2/GO	$MeCpPtMe_3$	O_2	N-GO	90s $MeCpPtMe_3$pulse, 65°C	60s O_2pulse, 100°C–150°C	120s N_2 after each Pulse	1–2	110°C

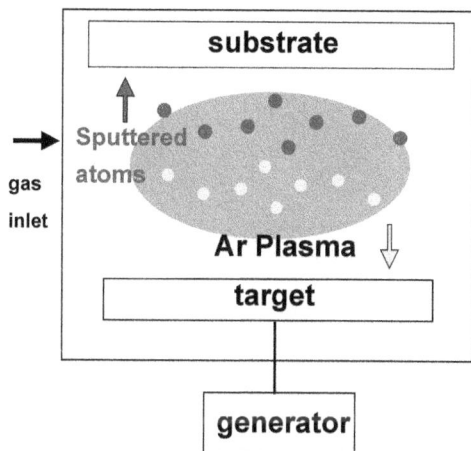

FIGURE 2.12 Overview of plasma sputtering process.

200 V and Pt sputtering was performed in 1 s repetitions at intervals of 10 s. It was finally found the typical single-atom Pt and the integrated clusters are formed as the increase of the depositing time.[85]

2.2.3 SOLID DIFFUSION STRATEGY

With the aim of achieving the desired chemical and physical properties, solid-state diffusion has been widely used to engineer microstructures of functional metals and alloys such as crystal phase, and grain boundary as well as the single-atom composite.[88] For the conventional solid–solid sintering approach, the directional materials and interface flows are affected by differential solid-state diffusion rates of the diffusion couple at high temperature, giving higher migration opportunities of target elements in the solid interfaces. Therefore, the fabrications via solid contact have to be updated for the acceleration of the atom mobility and anti-aggregation under high-temperature driving force. This old method is not generally considered as an effective method of preparing solid materials with sophisticated atomistic structures due to poor control over the drastic reaction at high temperature.

2.2.3.1 Bulky Transition

The mechanism of metal sintering is typically attributed to the migration of atomic metal species between nanoparticles or the coalescence of neighboring metal nanoparticles.[89] Therefore, directly proceeding the solid-state sintering for SAMs seems impossible. The sintering process will significantly reduce the explosion of target metal sites. There are also two solutions, to some extent, to hinder the intergradation phenomenon with the introduction of strong metal-interaction substrates and porous zeolites oxides.[89] Nevertheless, these strategies are still far from satisfactory for atomic isolation.

Recently, an unexpected phenomenon has been found: noble metal nanoparticles (Pd, Pt, and Au) can be transformed into thermally stable single-atom state in a

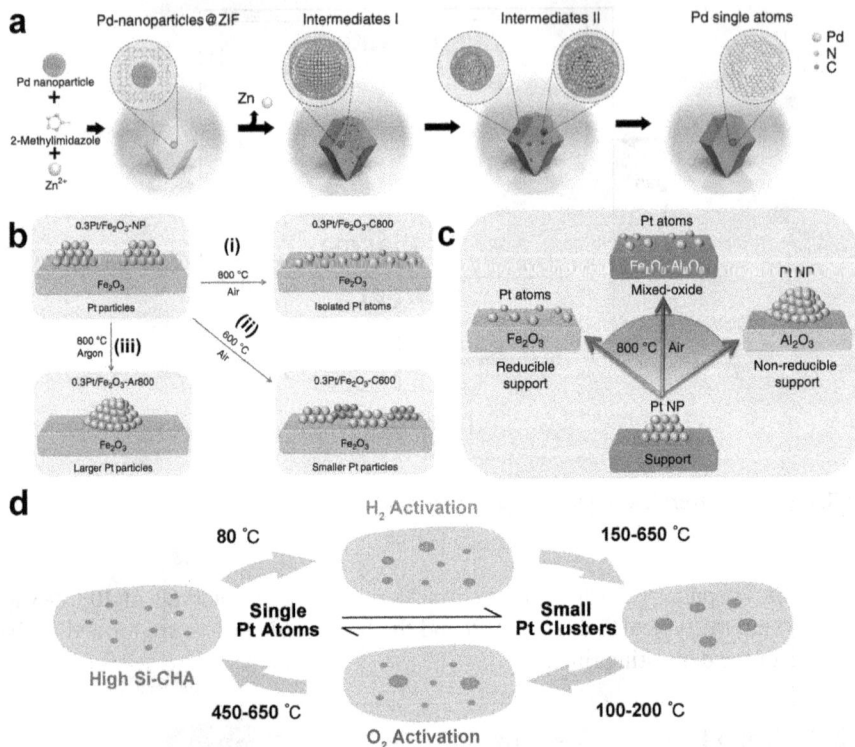

FIGURE 2.13 (a) Scheme of the transformation of Pd nanoparticles to Pd single atoms.[40] (Copyright RSC.) (b) Illustration of thermally induced Pt nanoparticle restructuring. (i, ii) Calcination under oxygen, or under an inert atmosphere (iii), resulting in dispersion as single atoms or particle sintering, respectively (c) Illustration of Pt NP sintering/dispersing on different supports. Metal oxide reducibility dictates the ability of a support to anchor isolated Pt atoms: Fe_2O_3 favors atomically dispersed Pt, whereas Al_2O_3 favors nanoparticle sintering. Doping iron oxide into non-reducible support (Fe_2O_3-Al_2O_3) can adjust noble metal dispersion.[90] (Copyright Springer Nature.) (d) Reversible interconversion of Pt nanoparticles into site-isolated single atoms in reducing and oxidizing atmospheres.[91] (Copyright ACS.)

reversal routine. Different from the in-situ impedance growing up, this bulky transition process, in principle, is very similar to the extraction process. The metallic-philic support grabs the target single atoms and confines them into the lattice trap, thus converting them into single-atom states from bulky mother metal.[89] As shown in Figure 2.13a, Pd nanoparticles were surrounded by ZIF-8 covering and annealed under sintering condition, yielding stable isolated single-atom Pd on the N–C supports.[92] The N defects in the N–C can effectively trap and anchor the mobile Pd atoms emitted from the nanoparticles to form a thermally stable Pd–N_4 structure.[89] This particle-splitting phenomenon via bulky transition was also observed on Pt sites.[90] As shown in Figure 2.13b and c, a robust covalent interaction between metal atoms and reducible supports (Fe_2O_3) enables to anchor and stabilize isolated metal atoms, which is not associated with surface defects of the underlying supports.[92] When the support is replaced with Al_2O_3, the weak interaction between Pt and substrate fails to convert the nanoparticles into a single-atom state (Figure 2.13c). Furthermore, by switching the sintering atmosphere from reductive H_2 to O_2 (or air), the conversion between nanoparticles and single atoms becomes reversible (Figure 2.13d), such as the single-atom Pt, Pd, and dual-atom Pd–Al, Pt–Sn.[91,92]

2.2.3.2 Ball Milling

Another approach, ball milling, via solid state comes without applying high-temperature input. During this process, part of the chemical bonds on the substrate (covalent or coordinative bonds) would be broken, forming the defects sites on the "activated" substrate, which would be employed to capture metal atoms and provide the possibility of preparing single-atomic site catalysts.[1] Also, with the utilization of the ball milling process, the use of dispersion solvent for precursor is no longer needed so that the considerable quantities of liquid wastes can be gotten rid of.[93]

For instance, graphene nanosheet (GN) powder and iron (II) phthalocyanine (FePc) were directly mixed with stainless steel milling balls and proceeded ball milling for 20h.[94] Then the Fe_1N_4/GN structure was successfully obtained and this typical method was further developed by replacing the FePc precursors with other metal-phthalocyanine complexes, such as CoPc, CuPc and NiPc, to form the M_1–N_4 SAM coordinative structure.[95] Besides, the precious SAMs can also fabricated via solid-state ball milling method. For example, cobalt acetylacetonate and 0.2 wt% platinum acetylacetonate were filled into milling chamber and the mixture was ground at the ramp of 400rpm for 60h. Then the obtained powders were annealed in a reductive atmosphere under 200°C to achieve the Pt_1/Co SAA.[96] Also, this series of materials can be extended by replacing the precious metal precursors with palladium acetylacetonate, ruthenium 2,4-pentanedionate, iridium acetylacetonate and rhodium acetylacetonate to fabricate Pd_1/Co, Ru_1/Co, Ir_1/Co and Rh_1/Co, respectively.[96] Apart from the use of metal-complex precursors, potassium tetrachloroplatinate, ruthenium chloride hydrate or rhodium chloride hydrate can also be directly mixed with N-doped carbon (N-C) to achieve the precious metal-based SAMs via ball milling process even without post-annealing treatment.[97]

Most importantly for the massive production of SAMs, ball milling is, to date, one of the most practical approaches, since there are always uneven mass transfer and heat transfer effects that can easily cause uneven high local concentration when the synthesis is scaled up, and thus the agglomeration occurs.[93] It has been reported that the SAMs production by using ball milling for SAMs can reach a massive amount of kilogram-scale, *e.g.* Pd_1/ZnO, Rh_1/ZnO and Pd_1/CuO.[93]

2.2.4 Spatial Framework Construction

Spatial framework assistance to construct SAMs, in this section, is specifically pointed out from the conventional synthetic routes in liquid, solid and gas media. Metal organic frameworks (MOFs) or covalent organic frameworks (COFs), or more widely speaking, coordination polymers, are crystalline compounds consisting of infinite lattices comprised of inorganic secondary building units and organic linkers, connected by coordination bonds of moderate strength.[98,99] Distinct from traditional inorganic materials, organic frameworks can be synthesized from well-defined molecular building blocks and may therefore be understood as molecules arranged in a crystalline lattice, thus presenting several advantages for SAMs fabrications.[98,99] On one hand, the well-defined network structures can be utilized as the SAMs support, with abundant bridgeable sites and isolated by organic linkers. The spatial pores can also capture the nomadic metal precursors for further stabilization. On the other hand, the organic-based composites, including carbon skeleton, metal precursors and low-boiling-point metals, can also participate in the in-situ conversion under pyrolysis into the single metal confined (usually M–N–C) coordinative structure. With the consequent annealing treatment, the stronger bonding experienced after restrict heating results in highly ordered materials with higher stability.[98] Therefore, the typical MOFs and COFs assistance strategies are well classified and introduced in this section.

2.2.4.1   Spatial Metal-Organic Framework

MOFs, as a relatively new type of porous solid materials, have been sufficiently demonstrated to be promising in fabrication SAMs because of their ultrahigh surface area, well-defined structure, accurate designability and flexible tailor-ability.[100,101] The ordered arrangement of the metal nodes and organic ligands construct an ideal framework as single-metal substrates.[102] To be specific, metal ions or metal nodes that are periodically distributed in MOFs can be converted, via high-temperature pyrolysis, into metal (oxide) nanoparticles that are well dispersed in the resultant composites and usually interact with the surrounding S, N, O, P, etc. for a stable location.[100] Therefore, the porous architecture of MOFs favors to spatially separate and encapsulate mononuclear metal precursors via various confinements, which is one of the most efficient methods to achieve atomic separation.[1] The fabricated SAM using MOF-assisted strategy demonstrates excellent catalytic activity, high selectivity and large specific area.[102] In general, there are two schematic approaches to fabricate SAMs via MOF assistance: MOF-stabilized and MOF-derived approaches.[100,101] On one hand, the single atoms directly obtained on MOF substrate without further high-temperature annealing is defined as an MOF-stabilized approach. On the other hand, the MOF-derived pathway is that the metal-MOF precursors experience pyrolysis and/or thermal transformation into a target structure at a high temperature, resulting in the obtained SAMs with higher stability. However, the efficient methods to prevent aggregation in large-scaled produce with higher thermal stability and facile synthesis steps remain a challenge.[102]

2.2.4.1.1   MOF-Stabilized Strategy

For MOF-stabilized strategy, the high-metal loading and thermodynamic stability remain a laborious task.[101] Therefore, tremendous experiments have been carried out, including pore confinement, metal-node bridge and coordination stabilization, to address these issues. The pore confinement strategy mainly consists of the spatial dispersion of the metal precursors in the MOF channels and removal of the ligands of target metal precursors and anchoring the single metal onto the MOF skeleton.[100,101] For example, single-atom Ru can be achieved on ZIF-8 (zeolitic imidazolate frameworks) MOF substrate. $Ru(acac)_3$ precursors were mixed with $Zn(NO_3)_2$ and 2-methylimidazole in methanol (Figure 2.14a, I) under hydrothermal reaction to

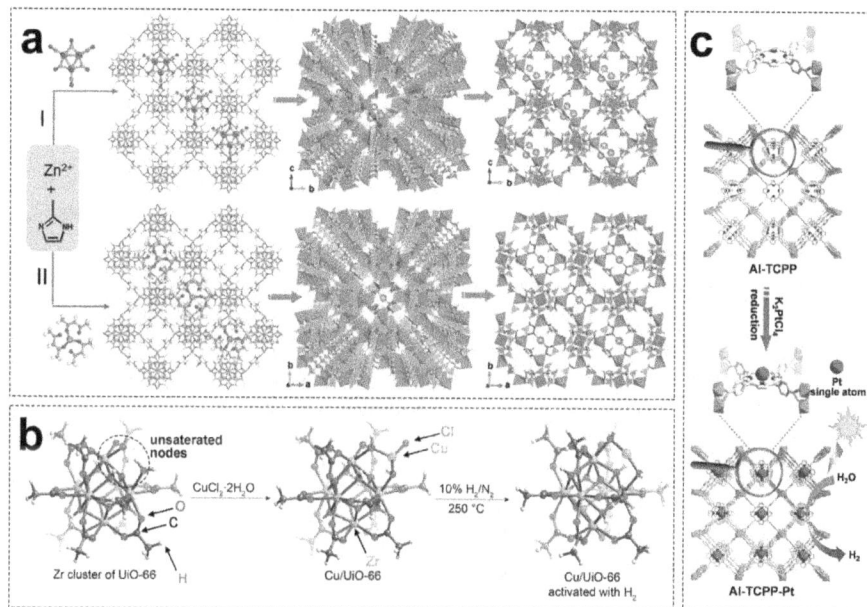

FIGURE 2.14 (a) Preparation process of Ru$_3$@ZIF-8 (I) and Ru$_1$@ZIF-8 (II).[101] (Copyright ACS.) (b) DFT calculated structures of defective UiO-66, Cu/UiO-66 catalyst where Cu is coordinated to a −OH/−OH$_2$ defect site of UiO-66, and Cu/UiO-66 catalyst after activation with H$_2$.[104] (Copyright ACS.) (c) Schematic illustration showing the synthesis of Al-TCPP-Pt for photocatalytic hydrogen production.[101] (Copyright ACS.)

obtain the MOF intermediate and thereafter the intermediate powders were reduced at 400°C to achieve the final Ru$_1$/ZIF-8.[103] As the pore size permission of the MOF substrate, replacement of the Ru(acac)$_3$ with Ru(CO)$_{12}$, a three-atom Ru$_3$/ZIF-8 final product was obtained via the same protocols (Figure 2.14a, II), indicating the significance of metal precursor selection with suitable molecule size based on the cage/pore size of MOF.[101,103]

When the MOF processes abundant coordinatively unsaturated metal sites (metal nodes), which can be readily functionalized with the additional group, this potential adsorption site can capture the single-metal precursors efficiently. Then with a simple reduction, the targeted metal ions, which are bound to the original metal clusters with low coordination numbers, can be readily turned into SAMs.[101] For example, UIO-66 MOF has a missing linker, whose −OH/−OH$_2$ nodes can strongly bond with Cu cations to form a Cu SAMs as shown in Figure 2.14b.[104] Also, single-atom V can be bridged with the active nodes of Hf-MOF-808 and Zr-NU-1000 supports, which present Hf6-type metal nodes and Zr6-type metal nodes, respectively.[105] The V sites were separated and bridged on metal nodes, benefiting from the chelating effect between the vanadium species and metal nodes in the MOF skeletons, the vanadium active sites stay isolated throughout the catalytic oxidation process, achieving long-term catalytic durability.[101] For another instance, TiO$_2$ was further developed into a MOF support with the decoration of $Ti_8(\mu_2\text{-}O)_8(\mu_2\text{-}OH)_4$ node and the resultant Ti-BDC MOF (MIL-125) provides a single-atom capturer for Co site.[106] However, the systematic studies for understanding the detailed fabrication mechanism are still insufficient. This is caused by the fact that some unsaturated sites of metal nodes in MOFs are easily occupied by solvent molecules during the crystallization process of MOFs, which often results in unanticipated and unstable hybrid materials.[101]

Metal-node adsorption is relatively inert to noble metal atoms such as Pt, Pd, Ru, etc., thus hindering the fabrication of noble metal SAMs from metal nodes of MOF.[102] Fortunately, an alternative MOF-assistance approach is developed that the MOFs can be rationally synthesized to provide a wide variety of chelating ligands (*e.g.* $M\text{-}N_4$) to stabilize individual metal atoms.[101] The aggressive coordination sites in the skeleton act as a "claw" to capture and anchor mononuclear noble metal precursors, thus, preventing SAMs from migration and agglomeration by the strong interaction between the metal atoms and the coordination sites.[101] For example, a highly stable Al-based porphyrinnic MOF (Al-TCPP, as shown in Figure 2.14c), in which infinite $Al(OH)O_4$ chains are interconnected by the porphyrin linkers into a 3D microporous framework, was employed for Pt metalation into the porphyrin centers.[107] Following a simple reduction, the Pt_1/Al-TCPP was achieved.[107]

2.2.4.1.2   MOF-Derived Strategy

In addition to the direct immobilization of SAMs via the MOF-stabilized approach, proceeding with the MOF precursors at relatively higher temperature consequently results in pyrolysis reactions, thus capturing the single atoms with stronger coordinative trap. The MOF-derived SAMs can also inherit the high surface area and porosity of MOFs to a large extent, which then contributes to the sufficient exposure of active sites and efficient mass transport.[100] Therefore, the MOF-derived SAMs usually exhibit higher thermal stability than SAMs fabricated under mild conditions and can be applied to reactions under complex scenarios, greatly extending their catalytic application.[108] However, a significant challenge is that the inevitable tendency of the aggregation of metal atoms during the high-temperature pyrolysis must be taken into consideration.[108] Therefore, the predesign of atomically dispersed metal sites that are spatially separated with a certain distance in MOFs, should be systematically optimized for SAMs undergoing pyrolysis.[108] In most cases, abundant N-involved MOFs are constructed to ensure the conversion of N-doped carbon-based materials, which possess particular advantages to anchor and stabilize single-metal atoms by virtue of the metal-N interactions.[108]

Most of MOFs generally have a monolinker node for metal skeleton construction. However, the key design for MOF-derived strategy is the precursor framework that consists (more than) two types of metal nodes-target metal including at least one kind of low-boiling-point metal brick.[101,110] In a subsequent pyrolysis process, the low-boiling-point metals are evaporated, leaving the targeted metal species anchored atomically on ligand-derived N-doped porous carbon matrix by forming strong M–N coordination interactions to prevent the agglomeration of the targeted metal atoms.[101] Therefore, ZIFs are one of the most employed precursors for the fabrication of MOF-derived SAMs since ZIFs perfectly meet the request that contains low-boiling-point metal and target metallic atoms.[111]

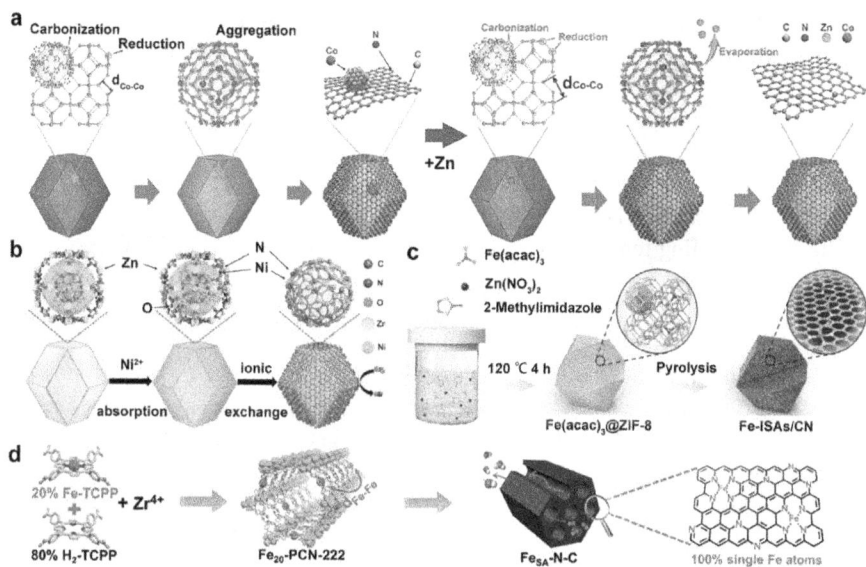

FIGURE 2.15 (a) Fabrication of of Co_1/N-doped carbon via ZIF-67 routine.[40] (Copyright RSC.) (b) Fabrication of of Ni_1/N-doped carbon via ZIF-8 routine.[109] (Copyright ACS.) (c) Fabrication of of Fe_1/N-doped carbon via ZIF-8 routine.[40] (Copyright RSC.) (d) Fabrication of Fe_1–N–C catalyst via PCN-222 MOF routine.[40] (Copyright RSC.)

For example, ZIF-67, which is constructed with metal precursors and 2-methylimidazole (MeIM) in methanol under hydrothermal conditions, was fabricated with both Zn (low-boiling-point metal) and Co (target metal). Then a pyrolysis transition occurred at a high-temperature range of 700°C~1,000°C to obtain Co_1/N-doped-carbon, maintaining the characteristic dodecahedron frame. The control experiment only adding Co metal sites spontaneously formed the aggregated nanoparticles. The absence of low-boiling-point metal Zn confirms the essential distance manipulation for single-metal achievement (Figure 2.15a).[110] Notably, by alternating the annealing temperatures in the argon atmosphere, the saturated Co_1–N_4 coordinations can also be precisely controlled to meet various catalytic functions As the annealing temperature raised from 800°C to 1,000°C, the Co–N coordinative numbers decreased from 4 to 2 as the escape from N elements.[112] A similar dodecahedral MOF frame can also be obtained on ZIF-8 under a relatively mild room temperature by directly using Zn precursor and target Ni precursors as shown in Figure 2.15b.[109] The target metal precursor can also be replaced with $Fe(AC)_3$ (Figure 2.15c) or Fe(PC) to achieve the corresponding single-atomic Fe-confined materials.[113–115] Single-atom Cu can also be installed onto ZIF-8 framework by soaking the Zn-MOF into $CuCl_2$, $CuBr_2$ and $Cu(NO_3)_2$ solutions via metal ionic exchange.[102]

Another porphyrinic MOF (Figure 2.15d), PCN-222, featuring 1D mesochannels, was also chosen as a representative precursor.[116] The assembly of Fe-TCPP into 3D networks of Fe_x-PCN-222 effectively inhibits the molecular stacking, and the mixed-ligand strategy further expands the distance of adjacent Fe-TCPP ligands.[116]

Upon pyrolysis, the optimized Fe-PCN-222 can be converted into single-atom Fe sites implanted in porous N-doped carbon.[116] With the further addition of the SiO_2 template, it is obvious to inhibit the growth of the Fe cluster on the carbon substrate so as to achieve a monodispersive Fe_1–N–C SAMs and higher single-atom Fe loading.[117] Then, with the further development of a bimetallic MgNi-MOF-74 support and polypyrrole (PPy) molecules as the nitrogenous guests filled into the 1D channels of the MOF, single-atom Ni–N_x–C materials ($x=2, 3, 4$) were successfully obtained by tuning the annealing temperature for pyrolysis.[118] Additionally, Fe_1–N–C and Co_1–N–C SAMs can also be achieved via the same protocol.[118]

2.2.4.2   Spatial Covalent-Organic Framework

Similar principle to that of MOF-assistant fabrication for SAMs, COFs also possess a well-defined porous structure to enable the spatial confinement of mononuclear metal precursors.[1] With the strong covalence built between the confined atoms and surrounding networks, COF-precursor units are supposed to be more stable for the maintenance of the atomic dispersion state, since COFs contain more light elements, such as B, C, N and O, that can serve as coordination sites to anchor single-metal atoms against migration and agglomeration.[1,119]

A typical COF-assistance procedure, *e.g.* single-atom Fe_1/COF, is displayed in Figure 2.16a.[120] P-phenylenediamine (Pa) and 1, 3, 5-triformylphloroglucinol (Tp) as building blocks were mechanically ground and mixed in an ambient atmosphere.[120] Then acetic acid as a catalyst and a mixture of mesitylene and dioxane as the solvents were added. Afterward, the obtained material (TpPa) was washed with methanol

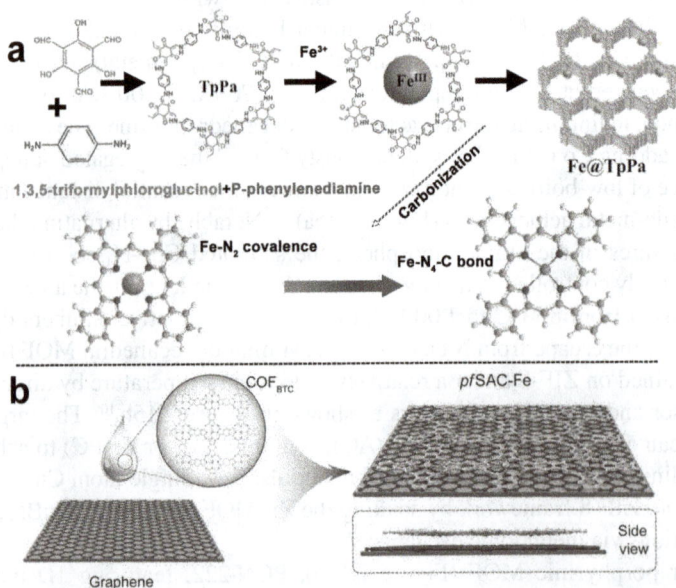

FIGURE 2.16 (a) Illustrative procedure for the fabrication of Fe_1/COF catalysts. (b) Schemes of synthesis route and structure of pyrolysis-free (pf) Fe_1/COF/graphene.[40] (Copyright RSC.)

and CH_2Cl_2 and dried at 180°C to yield a dull red powder (TpPa-COF precursors).[120] $FeCl_3$ and as-prepared TpPa were fully dispersed in methanol and water solution under vigorous magnetic stirring. During this process, Fe^{3+} ions were immobilized within the porous TpPa supports.[120] The resultant co-precipitate was collected and annealed for pyrolysis conversion under a N_2 atmosphere at 700°C.[120] There are typically two $Fe-N_2$ covalent bonds in this COF structure. Then, the pyrolysis transition further improves the Fe–N binding from two coordinations to four. Moreover, this COF pathway also directs a universal transition metal condiments using the same TpPa support, such as Mo, Ni and Zn.[121–123]

However, the precisely controlled and crystallized structure details are still inadequate since the periodic long-term arrangement of COF-assistant SAMs during high-temperature pyrolysis reaction has always been interrupted with the unpredictable structural changes, which is different from MOF structures that achieve various 3D polyhedrons. To solve this problem, pyrolysis-free synthetic approaches toward M–N coordinated SAMs have recently been developed as shown in Figure 2.16b.[124,125] To be specific, a fully closed π-conjugated iron phthalocyanine (FePc)-COF was atomically riveted onto graphene matrix, instead of randomly creating Fe-nitrogen moieties on a carbon matrix (Fe–N–C) through pyrolysis.[125] The resistivity of as-prepared pyrolysis-free Fe_1/COF/graphene was close to that of graphene itself and comparable to that of Fe–N–C on a carbon matrix obtained through pyrolysis.[124,125]

2.3 CHARACTERIZATIONS OF SAMs

For the development of the fabrications on SAMs via various approaches, the advanced characterization techniques to observe and identify the single-atomic sites as a reliable evidence of the correlations between structure and application is equally significant. In particular, it is necessary to analyze the morphology, coordinative environment, structural lattice arrangement, mass loading, etc. Therefore, the analytic techniques with atomic-level resolution such as spherical aberration corrected transmission electron microscope (AC-TEM) provide direct observations on SAM dispersion, morphology, coordination and lattice information. Scanning tunneling microscope (STM) detects the in-situ atomic information on the SAM surface with less damage. In addition, the bulk-sensitive XAS can also disclose the valence state, fine spatial structure and coordination environment of target elements with a statistical belief. Quantification can be precisely provided in conjunction with traditional energy spectroscopic techniques. Apart from those static characterizations of SAMs, *operando* characterizations are also generally introduced to reflect the dynamic interactions during the real-time reactions.

2.3.1 ATOMIC-LEVEL-RESOLUTION ELECTRON MICROSCOPE

With the continuous development of electron microscopic technology, high resolution transmission electron microscope (HR-TEM) has become one of the most widely used characterizations to directly observe the material looking. After all, it has been said by many in the field that "seeing is believing".[28] However, spherical aberration is one of the main reasons that affects the resolution. This shortage cannot be totally eliminated in the TEM system. Referring to the optical lens system that combines

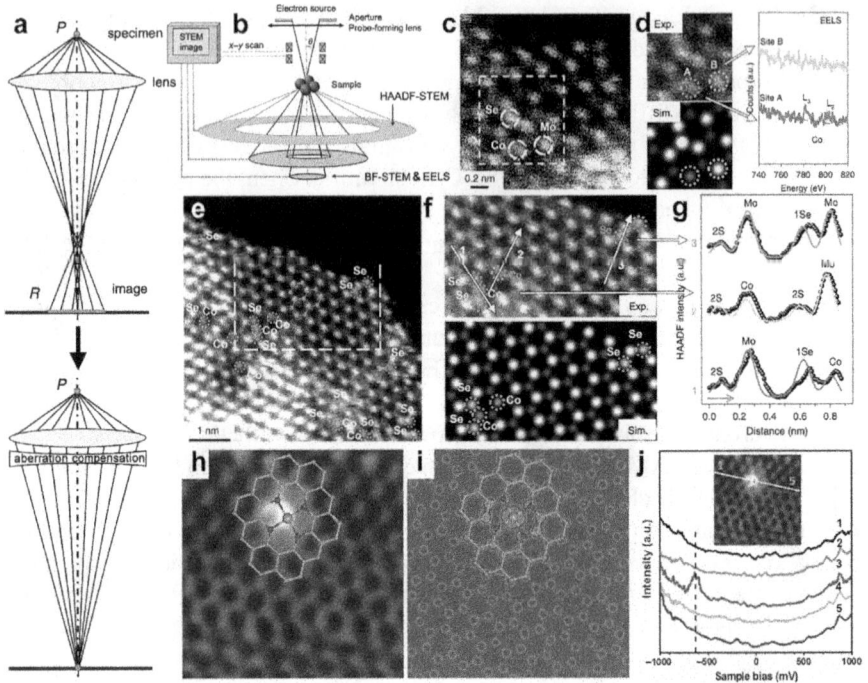

FIGURE 2.17 Schemes of (a) AC-TEM and (b) relevant BF and HAADF model under STEM technique; (c) atomic-resolution HAADF-STEM image of the Co/Se-MoS_2. (d) Enlarged and simulated images of the white dashed line region in (c) and the EELS spectra obtained at the site A and B via line scan. (e) Low pass filtered large area atomic-resolution HAADF-STEM image of the Co/Se-MoS_2. (f) Enlarged and simulated images of the white dashed line in (e). (g) The measured intensity profile along three lines labeled in (f).[43] (Copyright Springer Nature.) (h) LS-STM image of FeN_4 on graphene, measured at a bias of 1.0 V and a current of 0.3 nÅ (2 nm × 2 nm). (i) Simulated STM image for (h). The inserted schematic structures represent the structure of the graphene-embedded FeN_4. The C, N, and Fe atoms in (h) and (i) have been marked, respectively. (j) dI/dV spectra acquired along the white line in the inset image.[94] (Copyright Science.)

convex and concave lenses to reduce the aberrations arising from the uneven convergence ability, AC-TEM was utilized with a spherical aberration corrector device (Figure 2.17a) to act as a concave lens to correct the spherical aberration. The spherical aberration corrector can be installed either at the position of the objective lens or at the position of the condenser lens. There is even an AC-TEM utilized with two correctors. So far, based on the realization of the merits of higher resolution, chemically sensitive and directly interpretable images, the resolution reaches the sub-angstroms scale, which makes it possible to directly observe the distribution of single atoms and atomic structure.

Technically, the conventional TEM mode, where the specimen is illuminated by a near-parallel bundle of electrons and the image is formed by a sequence of lenses equivalent, is seldom used for SAM characterization. Alternatively, scanning transmission electron microscope (STEM) is the most practical mode, where a fine probe is formed by focusing the incident electrons and is then scanned across the specimen (Figure 2.17b).[126] Then the feedback electrons are collected and programmed into image on the screen.[126] Depending on the radial position of the detector, STEM offers two different imaging modes (Figure 2.17b).[126] Bright-field (BF) photo is based on the low-angle scattered electrons, yielding images subject to phase contrast as conventional TEM. On the other hand, the prevailing imaging mode for STEM is high-angle annular dark-field (HAADF), which deals with scattered electrons.[126] The great advantage of HAADF model is that the incoherent intensity distributions are directly related with atomic nature, since the scattered intensity depends on the nuclear charge number Z, allowing different atomic species to be distinguished. The HAADF mode is also referred to as Z-contrast imaging. However, there must be enough of a difference in atomic number for the two metals to be distinguished.[28] Moreover, the low-angle scattered electrons, meanwhile, pass through the annular detector and can be used for spatially resolved electron-energy-loss spectroscopy (EELS) to scan particular electron states localized at or between atoms.

For example, Co and Se were atomically doped into 2D MoS_2 lattice via template-assistance approach to form single-atom confined materials. By removal of the SiO_2 sphere template, the porous framework is constructed.[43] Atomic-resolution HAADF-STEM imaging of a monolayer Co and Se co-doped MoS_2 shows the embedment of Co atoms with a lower contrast at the Mo site and Se atoms being obviously brighter at the S site (Figure 2.17c), as confirmed by simulated HAADF-STEM image in Figure 2.17f. The Co atom is further identified by using atomic-resolution EELS (Figure 2.17d), which shows the feature peaks of L2 and L3 edges of the Co atom at the site A with lower contrast (Figure 2.17f). The distribution of Co and Se atoms being adjacent or separated from each other was observed in the HAADF-STEM image according to the contrast analysis (Figure 2.17e), and was confirmed by the simulated HAADF-STEM images and the line intensity profiles (Figure 2.17f and g).[43]

However, the detective damage to the specific single-atom coordination from the high-energy electron beam should also be taken into consideration because of the unstable state of single-atom confinement. Therefore, observation techniques with less impact in-situ on the target specimen are also necessary. In this regard, STM is an indispensable technique for single-atom observations without heavily damageable impact. To be specific, a small metal tip is brought near enough to the surface that the vacuum tunneling resistance between the surface and the tip is finite and measurable. The tip scans the surface in two dimensions, while its height is adjusted to maintain a constant tunneling resistance.[127,128] The result is essentially a contour map of the surface, thus reflecting the local information for individual atoms in real space.[129] Despite these great advances, the energy resolution remains limited in tunneling-spectroscopy modes by the thermal energy broadening of the electronic tip and sample states.[130] Therefore, proceeding the probe scanning in a low-temperature environment (e.g., liquid He) is a prerequisite.

For example, single-atom FeN_4 confined on graphene was first observed by STM as shown in Figure 2.17h–j. Initially, the sample was dispersed in petroleum ether and further dripped on the surface of HOPG (highly oriented pyrolytic graphite) and then transferred into STM specimen chamber. Before imaging, the sample was degassed at ~450 K to remove surficial impurities and the observation was carried out in liquid He (4 K) at a constant current mode using an electrochemically etched W tip under an ultrahigh vacuum de-pressured below 7.0×10^{-11} mbar.[94,131] In Figure 2.17h, Fe site is a brighter spot, whereas neighboring atoms exhibit a higher apparent height than other carbon atoms in the graphene matrix. STM simulation (Figure 2.17i) displayed the FeN_4 model embedded in the graphene lattice, which is in agreement with the measured STM image (Figure 2.17h). What's more, both the center Fe and coordinative C and N demonstrate brighter signals than carbon atoms located further away. And STM spectroscopic measurements across the FeN_4 center (Figure 2.17j) also show a sharp resonance state at -0.63 eV below the Fermi level, suggesting that the iron center strongly interacts with the graphene lattice and thus introduces a new electronic state near the Fermi level. All these evidence suggest that the FeN_4 center forms stable bonds with neighboring carbon atoms.

2.3.2 X-Ray Absorption Fine Structural Analysis

XAS is another fundamental tool for detecting the structure and electronic state of SAMs because of its high sensitivity and subatomic resolution. XAS is a synchrotron-based characterization technique that requires a tunable, high-intensity X-ray beamline and can be divided into X-ray absorption near-edge spectroscopy (XANES, energy range of 5~150 eV) and extended X-ray absorption fine structure (EXAFS, energy range of 150~2,000 eV) in Figure 2.18a and b.[132–135] As the X-ray energy moves from low to high across the photoemission energy for the atom (usually termed the absorption edge or white line) there is a sharp jump (XANES) in the absorption coefficient, since the core electron in the target atom is excited to an empty state and, as such, X-ray absorption probes the unoccupied part of the electronic structure of the system, and each element has a set of unique absorption edges corresponding to different binding energies of its electrons, giving XAS an "elemental fingerprint" selectivity (Figure 2.18a).[132–134] As the intensity of X-ray scan further increases, the spectrum is followed by a series of oscillations (EXAFS region). This region is the effect of the interference between the fast-moving outgoing and the backscattered photoelectron waves so that analysis this wavelet signals can therefore determine the geometric structure (Figure 2.18b).[133,134,136] The oscillations can be considered as a sum of sine waves, arising from a shell of neighbors of a particular type of atom at a certain radial distance from the target atom, and disclose the structural information on the distance from the target atom to its coordinative neighbors by Fourier transform (FT) of the normalized absorption coefficient, $k \cdot (\chi)k$.[132,134,136] With the introduction of the electronic structure models such as density functional theory to calculate the unoccupied density of states and the interaction during the catalytic process on the active intermediate, XAS fitting and DFT calculation can cross-confirm the accuracy of the built geometric model.[133] Thanks to its high selectivity, sensitivity and precise geometric study for both crystalline and amorphous materials, the applications for XAS cover a wide range, such as structural analysis, electronic state, coordination chemistry and even the biologic field with the development of the soft X-ray source.[134,137] Therefore, the atomic and electronic structure of the SAMs can be precisely characterized by combining the XANES and EXAFS results.

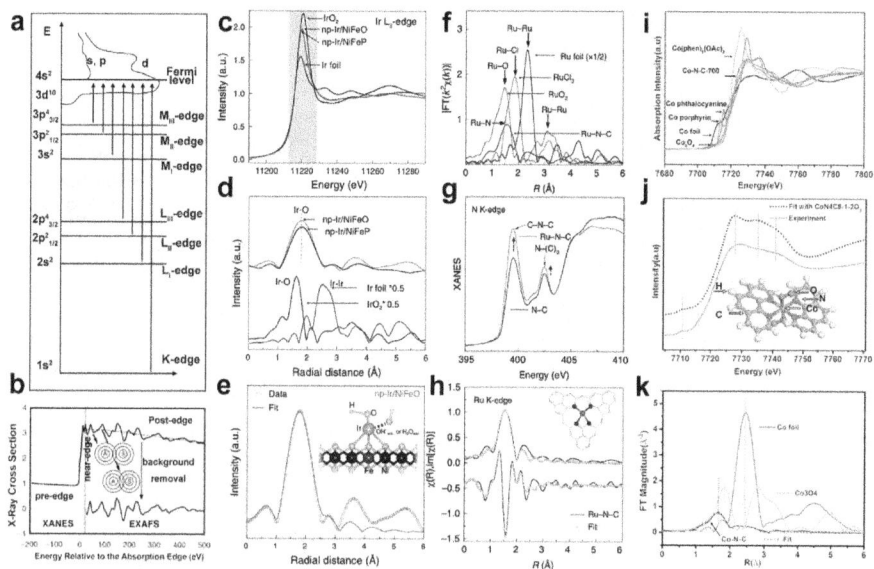

FIGURE 2.18 (a) Principle of X-ray absorption edges; (b) the XANES and EXAFS spectrum of Cu. (c) The XANES at the Ir L₃-edge of Ir foil, IrO₂, Ir₁/NiFeP, and Ir₁/NiFeO. (d) Corresponding to FT-EXAFS spectra from (c). (e) First-shell (Ir-O) fitting of FT-EXAFS spectrum for Ir₁/NiFeO. Inset shows the structure of optimized Ir₁/NiFeO.[138] (Copyright Springer Nature.) (f) The Ru k-edge k²-weighted FT spectra for Ru foil, RuCl₃, RuO₂, and Ru-N-C. (g) N k-edge XANES of pristine N–C and Ru–N–C catalysts. (h) The R-space curve-fitting of ex-situ Ru-N-C. Top and bottom curves are magnitude and imaginary part, respectively. Insert shows the structure of the Ru site in Ru-N-C. The balls in gray, blue, and light green represent C, N, and Ru atoms, respectively.[139] (Copyright Springer Nature.) (i) The normalized XANES spectra at the Co k-edge of Co–N–C structure and different reference samples. (j) Comparison between the k-edge XANES experimental spectrum of Co–N–C (solid line) and the theoretical spectrum (dotted line) based on the inset structure from DFT calculation. The C, H, O, N and Co atoms have been marked, respectively. (k) The k²-weighted FT-EXAFS spectra of the experimental and fitted Co–N–C catalyst as well as the Co foil and Co₃O₄ reference samples.[140] (Copyright RSC.)

For example, single-atom Ir was doped onto nano-porous NiFeP (denoted as np-Ir₁/NiFeP) compound via an electrochemical CV aging approach.[138] The electronic and fine structure of these SAMs was identified with XAS. The valence state of Ir species was measured by Ir L₃-edge XANES. In contrast to most 3D transition metals whose chemical states are related to the shift of the absorption edge position, the oxidation state of some noble metals, such as Ir, Pt and Au, is connected to the white line peak intensity. As shown in Figure 2.18c, the white line intensity of SAM Ir in the XANES spectrum is situated between those of Ir foil and IrO₂, indicating the valence states were between 0 and +4. The FT-EXAFS (Figure 2.18d) of Ir L₃ edge demonstrated that there is only a prominent peak at 1.78 Å corresponding to Ir–O coordination, without Ir–Ir scattering path, revealing the isolated dispersion of Ir atoms. The intensity of Ir–O peak of np-Ir₁/NiFeO was higher than np-Ir₁/NiFeP, indicating that isolated Ir atoms with higher valence and more oxygen ligands became more stable after surface self-reconstruction. Moreover, Ir fine structure was also simulated with the accordant DFT model, where each Ir atom was coordinated by four O atoms from oxygen ligands and two O atoms from absorbed hydroxyl or water group, and the model is perfectly fitting to the FT curves in Figure 2.18e.[138]

For the typical SAM coordinative structure of Metal-N_4, *e.g.* Ru_1–N_4 site anchored on nitrogen-carbon support (Ru–N–C), was investigated by Ru K-edge XAS measurements.[139] The FT-EXAFS was first conducted to distinguish the single-atomic structure of Ru–N–C catalyst with Ru foil and RuO_2 serving as references. As shown in Figure 2.18f, the only dominant peak at 1.5 Å is assigned to the nearest shell coordination of the Ru–N/C bond. In addition, the disappearance of Ru–Ru peaks at ~2.3 Å of Ru foil and 3.1 Å of RuO_2 precludes the existence of Ru-related oxides and clusters. In addition, the significant variations in the peaks' intensity of N K-edge soft XAS indicated the strong interaction between N and Ru atoms (Figure 2.18g), cross-confirming the metal-N coordination. Moreover, the specific coordination distance and numbers of the Ru atoms can be obtained by a least-square curve fitting of the experimental FT-EXAFS data. The best-fitting results for Ru_1–N–C clearly show that the major coordination peak originated from the four Ru–N coordination in the form of Ru_1–N_4 configuration with a mean Ru–N bond length of 2.08 Å (Figure 2.18h).[139]

Not only for the noble metallic single-atom, the single-atomic centers of transition metal-based materials are also widely characterized.[140,141] Co_1–N–C was highly active in converting nitroarenes into aromatic azo compounds. Co K-edge XANES results of Co_1–N–C catalyst is compared with other reference samples (Figure 2.18i) and Co_1–N–C features an absorption edge at 7,720 eV, almost the same as K-edge of Co(II) in its precursor, $Co(phen)_2(OAc)_2$. In addition, the sample lacks the pre-edge peak at the range of 7,714~7,716 eV, suggesting that the Co–N–C catalyst does not form a planar coordination geometry. From the EXAFS results (Figure 2.18j), the coordinative contributions of Co–Co are negligible. As such, various structural models of Co–N–C were built and their corresponding XANES spectra were generated. By comparing the main features of these predicted spectra with the experimental one, the authors determined the most suitable model as shown in Figure 2.18j. The DFT calculated structural data were then used to derive the scattering paths based on EXAFS fitting software in Figure 2.18k. The comparison of total oscillation spectra from calculation and experiment in these spaces further corroborates the proposed structure.[140,141]

2.3.3 OTHER CHARACTERIZATIONS

Apart from the direct observations and coordinative structure of SAMs, the mass loading is also another important index for characterization. However, the low content of single atoms on the support deviates from the real quantifications depending on various analysis resolutions. For example, the relative content can be obtained from energy-dispersive X-ray spectroscopy (EDS or EDX) associated with electron microscope, the illuminated atoms under high-energy electron are ionized to spontaneously generate the holes on the core orbit and emit X-ray as the holes are filled from outer shells. The identified X-ray signals are recorded by detector and simulated for quantification. However, this measured value is not reliable enough since there are many factors influencing the accuracy. The detectable depth varies from different materials so that bulky dispersion of SAMs may be differed. Besides, the emitted X-rays may escape to various directions, thus leading to the collection variables on direction, timing and correction programming.

To obtain a more accurate mass loading content, mass difference method is suggested to use. For carbon-based materials, it is recommended to carry out the TG-DTA (thermogravimetry-differential thermal analysis) measurement to quantify the real single atom contents. By heating to the very amount of SAMs into higher temperature in the air, carbon-based materials are totally burnt and removed. The residual metal content is believed as the noble metal mass. However, this approach is only practical on noble metallic SAMs on carbon-based support. For a universal analysis for mass loading on SAMs, inductively coupled plasma-optical emission spectrometry (ICP-OES) and/or inductively coupled plasma-optical mass spectrometry (ICP-MS) are more accurate. In particular, the SAMs should be dissolved in strong acid such as aqua regia (mixture of hydrochloric acid and nitrate acid) and be diluted to eject into the device chamber. Then component elements are excited by high-energy plasma. On one hand, the excited atoms return to originally low-energy state and emit identical rays (photo signals), which can be collected and simulated to quantify the accounts. On the other hand, the atoms can also be induced into ionic state and measured with mass spectrometry by the charge-to-mass-ratio law.

Based on these advanced analysis devices for single-atom characterizations, the conventional static measurement can be further developed into a dynamic probe, which provides an insight into observing the real-time interactions under particular conditions. For example, XAS under a casual pressure demand gives a great possibility to proceed the measurement in the liquid system. The strong X-ray can pierce the liquid window and feedback the *operando* spectroscopic information. Similar designing principles can also be extended into other spectroscopic measurements such as Raman and Fourier transform infrared (FTIR). However, the relatively lower input energies cannot ensure the permeability. Also, the generated gas phase at the interface also strongly scatters the results. The *operando* spectroscopic measurements can also be designed to apply on the thermal-catalytic process in a customized heated chamber under reactive gas atmosphere. Thus, the side effects from the heat must be taken into consideration and corrected. With the development of the differential pumping pre-lens system, ambient-pressure X-ray photoelectron spectroscopy (AP-XPS) is also possibly applied to *operando* reactions on SAMs.[142,143] This update realizes photoelectron collection under 20 Torr H_2O, allowing the equilibrium vapor pressure of water at room temperature.[142] Combined with a reactive gas supply system and heat-control device, AP-XPS on SAMs can be carried out for thermal reaction under high temperature. Furthermore, this system with heating and ambient pressure can also be installed into an electron microscope system. The so-called environmental electron microscope can observe the dynamic change on the SAMs, thus directly demonstrating the real-time reaction process.

2.4 CONCLUSIONS AND PERSPECTIVE

In conclusion, various synthesis strategies proceeding in liquid, air and solid media as well as by spatial MOF assistance, have been systematically introduced. Notably, the classified synthetic methods, herein, included each other and should not be isolated. The critical step of the rational methods is aiming to stabilize single-metal atoms through coordination, ionic interactions, covalence, etc. with neighboring heteroatoms, thus increasing the metal dispersion and density. Therefore, surface defects (e.g edges, steps, and terraces), lattice vacancy (e.g. cation and anion vacancies), spatial confinement, surfactants-assistance, surface activation/decoration, etc. can be applied to artificially anchor and immobilize the transition metal single atoms on various supports such as porous carbon-based materials, 1D and 2D composites and MOFs.[16] Consequently, the choice of the target metals has been widely broadened, including noble metals, transition metals and even rare earth metals.[4] However, there are still many issues to be further addressed in rational fabrication protocols for SAMs.

First of all (i), even though the reported loading amount can reach as much as 30 wt% for single-atom Fe on carbons, the higher single-metal content for other target elements is still not satisfying enough.[4,144] Second (ii), the uniform dispersion is also difficult to achieve, since the controversial trade-off between high-mass loading and dispersion must be considered seriously. The atomically dispersed atoms in SAMs are also prone to migrate and aggregate in order to reduce their surface energy during the reacting scenario such as in cycling and under high-temperature conditions. Another aspect of this is that it reflects the essential stabilizations for single-atom anchoring. The third challenge (iii) comes to the precisely controllable preparation of SAMs. Since different metal atoms show various complexation properties, the generality of the synthetic method can hardly be expanded to all metal species. The design and fabrication for specific SAMs should depend on the intrinsic properties of target metals, supports and the proposal reactions.[4] The minor changes of coordination atoms can alter the electronic structure of the central atoms, which substantially influences the intrinsic functions.[1] The fourth point (iv) is that, besides the intensive fabrications of single-site SAMs, Dual- and/or Tri- single-atom materials are still in their infancy, which will be utilized with higher active and multifunctionality as the coupling and/or synergistic effects.[4] Last but not least (v), the mature synthetic approach should also meet the industrial request, so it is essential to explore a green, facile strategy to produce SAMs on a large scale. Most of the reported methods are performed at the lab level and the starting materials such as MOF precursors are so expensive that the tedious procedures make the synthetic method neither economical nor environmentally benign.[1,145] In a word, a "perfect" synthetic protocol for SAMs is long-termly desirable, which should be in a facile and universal routine without environmental hazards, and the SAM products should be atomically dispersed at a higher content loading with enough stability.

On the other hand, for the characterizations of SAMs, it is urgent to develop a more efficient atomic-resolution imaging technique to directly "see" the real structures of SAMs such as the first and second coordinative spheres surrounding the confined single-atom.[145,146] The strain tensions caused by the introduction of the foreign

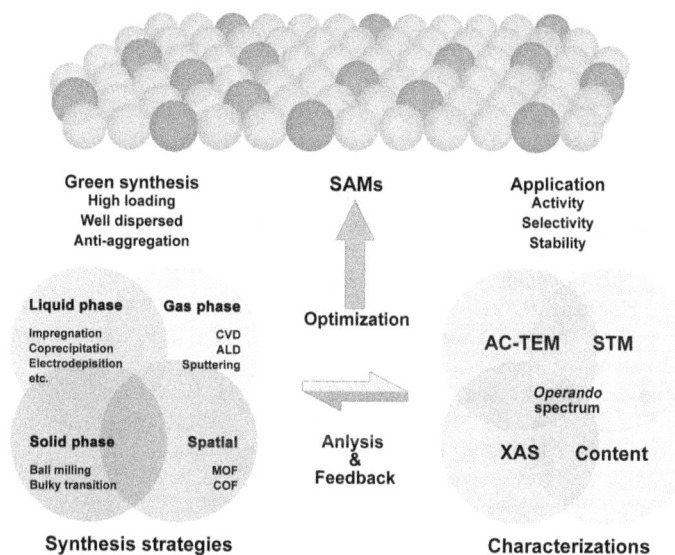

FIGURE 2.19 A summative scheme for SAMs synthesis and characterizations.

single atom and the surrounding coordinative host atoms are still not qualified and quantified. The expansion of the use of in-situ/*operando* techniques such as environmental electron microscope and XAS, and the progress in their applicability to diverse reactions, is a major objective to assess the precise nature of the active sites, resolve potential co-catalytic roles of other elements present and gain further insights into the dynamic nature of SAMs and their behavior under reaction conditions.[145]

Therefore, the understanding of detailed structure and dynamic mechanism can identify the success of the innovative fabrication practice and provide feedback to the synthetic parameters. More importantly, the observation of intermediate during reaction is helpful for building the precise models for computational calculation, which is regarded as a solid reference for the DFT database. This, fundamentally, can predict the advanced SAMs with practical elemental constructions to reduce the low-efficiency trial-and-error experiments. It is believed that both the development of synthetic methods and the characterizations for SAMs are vital for fully elemental utilization and advanced materials design (Figure 2.19).

REFERENCES

1. S. Ji, Y. Chen, X. Wang, Z. Zhang, D. Wang, Y. Li, *Chem. Rev.*, 2020, *120*, 11900–11955.
2. Y. Chen, S. Ji, C. Chen, Q. Peng, D. Wang, Y. Li, *Joule*, 2018, *2*, 1242–1264.
3. C. Rivera-Cárcamo, P. Serp, *ChemCatChem*, 2018, *10*, 5058–5091.
4. J. Xi, H. S. Jung, Y. Xu, F. Xiao, J. W. Bae, S. Wang, *Adv. Funct. Mater.*, 2021, *31*, 2008318.
5. W. Wu, W. Lei, L. Wang, S. Wang, H. Zhang, *Prog. Chem.*, 2020, *32*, 23–32.
6. Y. Shang, X. Xu, B. Gao, S. Wang, X. Duan, *Chem. Soc. Rev.*, 2021, *50*, 5281–5322.
7. M. B. Gawande, P. Fornasiero, R. Zbořil, *ACS Catal.*, 2020, *10*, 2231–2259.

8. L. Zhang, Y. Jia, G. Gao, X. Yan, N. Chen, J. Chen, M. T. Soo, B. Wood, D. Yang, A. Du, X. Yao, *Chem*, 2018, *4*, 285–297.

9. S. Ren, Q. Yu, X. Yu, P. Rong, L. Jiang, J. Jiang, *Sci. China Mater.*, 2020, *63*, 903–920.

10. S. Sahoo, S. L. Suib, S. P. Alpay, *ChemCatChem*, 2018, *10*, 3229–3235.

11. Y. T. Kim, H. Lee, H. J. Kim, T. H. Lim, *Chem. Commun.*, 2010, *46*, 2085–2087.

12. Y. T. Kim, K. Ohshima, K. Higashimine, T. Uruga, M. Takata, H. Suematsu, T. Mitani, *Angew. Chem. Int. Ed.*, 2006, *45*, 407–411.

13. L. Yang, L. Shi, D. Wang, Y. Lv, D. Cao, *Nano Energy*, 2018, *50*, 691–698.

14. Q. Q. Yan, D. X. Wu, S. Q. Chu, Z. Q. Chen, Y. Lin, M. X. Chen, J. Zhang, X. J. Wu, H. W. Liang, *Nat. Commun.*, 2019, *10*, 4977.

15. Z. Liang, H. Zheng, R. Cao, *Sustain. Energy Fuels*, 2020, *4*, 3848–3870.

16. X. Zheng, P. Li, S. Dou, W. Sun, H. Pan, D. Wang, Y. Li, *Energy Environ. Sci.*, 2021, *14*, 2809–2858.

17. Q. Cheng, L. Yang, L. Zou, Z. Zou, C. Chen, Z. Hu, H. Yang, *ACS Catal.*, 2017, *7*, 6864–6871.

18. J. D. Kistler, N. Chotigkrai, P. Xu, B. Enderle, P. Praserthdam, C. Y. Chen, N. D. Browning, B. C. Gates, *Angew. Chem. Int. Ed.*, 2014, *53*, 8904–8907.

19. J. D. Benck, T. R. Hellstern, J. Kibsgaard, P. Chakthranont, T. F. Jaramillo, *ACS Catal.*, 2014, *4*, 3957–3971.

20. M. Moses-DeBusk, M. Yoon, L. F. Allard, D. R. Mullins, Z. Wu, X. Yang, G. Veith, G. M. Stocks, C. K. Narula, *J. Am. Chem. Soc.*, 2013, *135*, 12634–12645.

21. B. Qiao, J. Liu, Y.-G. Wang, Q. Lin, X. Liu, A. Wang, J. Li, T. Zhang, J. Liu, *ACS Catal.*, 2015, *5*, 6249–6254.

22. Q. Wang, X. Huang, Z. L. Zhao, M. Wang, B. Xiang, J. Li, Z. Feng, H. Xu, M. Gu, *J. Am. Chem. Soc.*, 2020, *142*, 7425–7433.

23. J. Kim, C.-W. Roh, S. K. Sahoo, S. Yang, J. Bae, J. W. Han, H. Lee, *Adv. Energy Mater.*, 2018, *8*, 1701476.

24. J. Ding, M. Fan, Q. Zhong, A. G. Russell, *Appl. Catal. B*, 2018, *232*, 348–354.

25. R. Lang, T. Li, D. Matsumura, S. Miao, Y. Ren, Y. T. Cui, Y. Tan, B. Qiao, L. Li, A. Wang, X. Wang, T. Zhang, *Angew. Chem. Int. Ed.*, 2016, *55*, 16054–16058.

26. B. Qiao, A. Wang, X. Yang, L. F. Allard, Z. Jiang, Y. Cui, J. Liu, J. Li, T. Zhang, *Nat. Chem.*, 2011, *3*, 634–641.

27. J. Lin, A. Wang, B. Qiao, X. Liu, X. Yang, X. Wang, J. Liang, J. Li, J. Liu, T. Zhang, *J. Am. Chem. Soc.*, 2013, *135*, 15314–15317.

28. R. T. Hannagan, G. Giannakakis, M. Flytzani-Stephanopoulos, E. C. H. Sykes, *Chem. Rev.*, 2020, *120*, 12044–12088.

29. H. Zhang, L. Lu, K. Kawashima, M. Okumura, M. Haruta, N. Toshima, *Adv. Mater.*, 2015, *27*, 1383–1388.

30. H. Zhang, T. Watanabe, M. Okumura, M. Haruta, N. Toshima, *Nat. Mater.*, 2011, *11*, 49–52.

31. K. G. Papanikolaou, M. T. Darby, M. Stamatakis, *J. Phys. Chem. C*, 2019, *123*, 9128–9138.

32. M. T. Darby, E. C. H. Sykes, A. Michaelides, M. Stamatakis, *Top Catal.*, 2018, *61*, 428–438.

33. M. T. Darby, M. Stamatakis, A. Michaelides, E. C. H. Sykes, *J. Phys. Chem. Lett.*, 2018, *9*, 5636–5646.

34. M. B. Boucher, B. Zugic, G. Cladaras, J. Kammert, M. D. Marcinkowski, T. J. Lawton, E. C. Sykes, M. Flytzani-Stephanopoulos, *Phys. Chem. Chem. Phys.*, 2013, *15*, 12187–12196.

35. M. Mohl, D. Dobo, A. Kukovecz, Z. Konya, K. Kordas, J. Wei, R. Vajtai, P. M. Ajayan, *J. Phys. Chem. C*, 2011, *115*, 9403–9409.

36. Z. Sun, J. Masa, W. Xia, D. König, A. Ludwig, Z.-A. Li, M. Farle, W. Schuhmann, M. Muhler, *ACS Catal.*, 2012, *2*, 1647–1653.

37. H. J. Qiu, Y. Ito, W. Cong, Y. Tan, P. Liu, A. Hirata, T. Fujita, Z. Tang, M. Chen, *Angew. Chem. Int. Ed.*, 2015, *54*, 14031–14035.

38. Y. Yao, S. Hu, W. Chen, Z.-Q. Huang, W. Wei, T. Yao, R. Liu, K. Zang, X. Wang, G. Wu, W. Yuan, T. Yuan, B. Zhu, W. Liu, Z. Li, D. He, Z. Xue, Y. Wang, X. Zheng, J. Dong, C.-R. Chang, Y. Chen, X. Hong, J. Luo, S. Wei, W.-X. Li, P. Strasser, Y. Wu, Y. Li, *Nat. Catal.*, 2019, *2*, 304–313.

39. M. Zhang, Y. G. Wang, W. Chen, J. Dong, L. Zheng, J. Luo, J. Wan, S. Tian, W. C. Cheong, D. Wang, Y. Li, *J. Am. Chem. Soc.*, 2017, *139*, 10976–10979.

40. D. Zhao, Z. Zhuang, X. Cao, C. Zhang, Q. Peng, C. Chen, Y. Li, *Chem. Soc. Rev.*, 2020, *49*, 2215–2264.

41. Y. Zhu, S. Murali, M. D. Stoller, K. J. Ganesh, W. Cai, P. J. Ferreira, A. Pirkle, R. Wallace, K. A. Cychosz, M. Thommes, D. Su, E. A. Stach, R. S. Ruoff, *Sience*, 2011, *332*, 1537–1541.

42. Y. Cheng, S. Zhao, H. Li, S. He, J.-P. Veder, B. Johannessen, J. Xiao, S. Lu, J. Pan, M. F. Chisholm, S.-Z. Yang, C. Liu, J. G. Chen, S. P. Jiang, *Appl. Catal. B*, 2019, *243*, 294–303.

43. Z. Zheng, L. Yu, M. Gao, X. Chen, W. Zhou, C. Ma, L. Wu, J. Zhu, X. Meng, J. Hu, Y. Tu, S. Wu, J. Mao, Z. Tian, D. Deng, *Nat. Commun.*, 2020, *11*, 3315.

44. X. Meng, C. Ma, L. Jiang, R. Si, X. Meng, Y. Tu, L. Yu, X. Bao, D. Deng, *Angew. Chem. Int. Ed.*, 2020, *59*, 10502–10507.

45. X. Meng, L. Yu, C. Ma, B. Nan, R. Si, Y. Tu, J. Deng, D. Deng, X. Bao, *Nano Energy*, 2019, *61*, 611–616.

46. T. Sun, S. Zhao, W. Chen, D. Zhai, J. Dong, Y. Wang, S. Zhang, A. Han, L. Gu, R. Yu, X. Wen, H. Ren, L. Xu, C. Chen, Q. Peng, D. Wang, Y. Li, *Proc. Natl.Acad. Sci.*, 2018, *115*, 12692–12697.

47. L. P. Bicelli, B. Bozzini, C. Mele, L. D'Urzo, *Int. J. Electrochem. Sci.*, 2008, *3*, 356–408.

48. R. Li, Y. Li, P. Yang, D. Wang, H. Xu, B. Wang, F. Meng, J. Zhang, M. An, *J. Energy Chem.*, 2021, *57*, 547–566.

49. Z. Zhang, C. Feng, C. Liu, M. Zuo, L. Qin, X. Yan, Y. Xing, H. Li, R. Si, S. Zhou, J. Zeng, *Nat. Commun.*, 2020, *11*, 1215.

50. Y. Shi, W. M. Huang, J. Li, Y. Zhou, Z. Q. Li, Y. C. Yin, X. H. Xia, *Nat. Commun.*, 2020, *11*, 4558.

51. J. Xu, R. Li, C.-Q. Xu, R. Zeng, Z. Jiang, B. Mei, J. Li, D. Meng, J. Chen, *Appl. Catal. B*, 2021, *289*, 120028.

52. X. Bo, Y. Li, X. Chen, C. Zhao, *Chem. Mater.*, 2020, *32*, 4303–4311.

53. X. Bo, R. K. Hocking, S. Zhou, Y. Li, X. Chen, J. Zhuang, Y. Du, C. Zhao, *Energy Environ. Sci.*, 2020, *13*, 4225–4237.

54. X. Bo, K. Dastafkan, C. Zhao, *ChemPhysChem*, 2019, *20*, 2936–2945.

55. X. Bo, Y. Li, X. Chen, C. Zhao, *J. Power Sources*, 2018, *402*, 381–387.

56. X. Bo, Y. Li, R. K. Hocking, C. Zhao, *ACS Appl. Mater. Interfaces*, 2017, *9*, 41239–41245.

57. X. Lu, C. Zhao, *Nat. Commun.*, 2015, *6*, 6616.

58. N. Xuan, J. Chen, J. Shi, Y. Yue, P. Zhuang, K. Ba, Y. Sun, J. Shen, Y. Liu, B. Ge, Z. Sun, *Chem. Mater.*, 2018, *31*, 429–435.

59. Y. H. Kim, J. S. Heo, T. H. Kim, S. Park, M. H. Yoon, J. Kim, M. S. Oh, G. R. Yi, Y. Y. Noh, S. K. Park, *Nature*, 2012, *489*, 128–132.

60. M. D. Karkas, J. A. Porco, Jr., C. R. Stephenson, *Chem. Rev.*, 2016, *116*, 9683–9747.

61. P. Liu, Y. Zhao, R. Qin, S. Mo, G. Chen, L. Gu, D. M. Chevrier, P. Zhang, Q. Guo, D. Zang, B. Wu, G. Fu, N. Zheng, *Science*, 2016, *352*, 797–801.

62. T. Li, J. Liu, Y. Song, F. Wang, *ACS Catal.*, 2018, *8*, 8450–8458.

63. H. Wei, K. Huang, D. Wang, R. Zhang, B. Ge, J. Ma, B. Wen, S. Zhang, Q. Li, M. Lei, C. Zhang, J. Irawan, L. M. Liu, H. Wu, *Nat. Commun.*, 2017, *8*, 1490.

64. X. Ge, P. Zhou, Q. Zhang, Z. Xia, S. Chen, P. Gao, Z. Zhang, L. Gu, S. Guo, *Angew. Chem. Int. Ed.*, 2020, *59*, 232–236.

65. S. Ding, Y. Guo, M. J. Hülsey, B. Zhang, H. Asakura, L. Liu, Y. Han, M. Gao, J.-Y. Hasegawa, B. Qiao, T. Zhang, N. Yan, *Chem*, 2019, *5*, 3207–3219.
66. S. Ding, M. J. Hülsey, H. An, Q. He, H. Asakura, M. Gao, J.-Y. Hasegawa, T. Tanaka, N. Yan, *CCS Chem.*, 2021, *3*, 1814–1822.
67. M. S. Bootharaju, C. P. Joshi, M. R. Parida, O. F. Mohammed, O. M. Bakr, *Angew. Chem. Int. Ed.*, 2016, *55*, 922–926.
68. B. Zhang, L. Wang, Z. Cao, S. M. Kozlov, F. P. García de Arquer, C. T. Dinh, J. Li, Z. Wang, X. Zheng, L. Zhang, Y. Wen, O. Voznyy, R. Comin, P. De Luna, T. Regier, W. Bi, E. E. Alp, C.-W. Pao, L. Zheng, Y. Hu, Y. Ji, Y. Li, Y. Zhang, L. Cavallo, H. Peng, E. H. Sargent, *Nat. Catal.*, 2020, *3*, 985–992.
69. B. Zhang, X. Zheng, O. Voznyy, R. Comin, M. Bajdich, M. García-Melchor, L. Han, J. Xu, M. Liu, L. Zheng, F. P. G. A. D. Arquer, C. T. Dinh, F. Fan, M. Yuan, E. Yassitepe, N. Chen, T. Regier, P. Liu, Y. Li, P. D. Luna, A. Janmohamed, H. L. Xin, H. Yang, A. Vojvodic, E. H. Sargent, *Science*, 2016, *352*, 333–337.
70. L. Lin, Z. Chen, W. Chen, *Nano Res.*, 2021, *9*, 3412.
71. Q. Wang, Y. Lei, Y. Wang, Y. Liu, C. Song, J. Zeng, Y. Song, X. Duan, D. Wang, Y. Li, *Energy Environ. Sci.*, 2020, *13*, 1593–1616.
72. Y. Qu, Z. Li, W. Chen, Y. Lin, T. Yuan, Z. Yang, C. Zhao, J. Wang, C. Zhao, X. Wang, F. Zhou, Z. Zhuang, Y. Wu, Y. Li, *Nat. Catal.*, 2018, *1*, 781–786.
73. Z. Yang, B. Chen, W. Chen, Y. Qu, F. Zhou, C. Zhao, Q. Xu, Q. Zhang, X. Duan, Y. Wu, *Nat. Commun.*, 2019, *10*, 3734.
74. S. Liu, M. Wang, X. Yang, Q. Shi, Z. Qiao, M. Lucero, Q. Ma, K. L. More, D. A. Cullen, Z. Feng, G. Wu, *Angew. Chem. Int. Ed.*, 2020, *59*, 21698–21705.
75. L. Zhang, M. N. Banis, X. Sun, *Natl. Sci. Rev.*, 2018, *5*, 628–630.
76. S. Sun, G. Zhang, N. Gauquelin, N. Chen, J. Zhou, S. Yang, W. Chen, X. Meng, D. Geng, M. N. Banis, R. Li, S. Ye, S. Knights, G. A. Botton, T.-K. Sham, X. Sun, *Sci. Rep.*, 2013, *3*, 1775.
77. H. Yan, Y. Lin, H. Wu, W. Zhang, Z. Sun, H. Cheng, W. Liu, C. Wang, J. Li, X. Huang, T. Yao, J. Yang, S. Wei, J. Lu, *Nat. Commun.*, 2017, *8*, 1070.
78. M. Leskela, M. Ritala, *Angew. Chem. Int. Ed.*, 2003, *42*, 5548–5554.
79. N. Cheng, S. Stambula, D. Wang, M. N. Banis, J. Liu, A. Riese, B. Xiao, R. Li, T. K. Sham, L. M. Liu, G. A. Botton, X. Sun, *Nat. Commun.*, 2016, *7*, 13638.
80. M. Ritala, K. Kukli, A. Rahtu, P. I. Räisänen, M. Leskelä, T. Sajavaara, J. Keinonen, *Sience*, 2000, *288*, 319–321.
81. H. Yan, H. Cheng, H. Yi, Y. Lin, T. Yao, C. Wang, J. Li, S. Wei, J. Lu, *J. Am. Chem. Soc.*, 2015, *137*, 10484–10487.
82. J. Lu, K. B. Low, Y. Lei, J. A. Libera, A. Nicholls, P. C. Stair, J. W. Elam, *Nat. Commun.*, 2014, *5*, 3264.
83. C. Wang, X.-K. Gu, H. Yan, Y. Lin, J. Li, D. Liu, W.-X. Li, J. Lu, *ACS Catal.*, 2016, *7*, 887–891.
84. D. Hegemann, M. Amberg, A. Ritter, M. Heuberger, *Mater. Tech.*, 2013, *24*, 41–45.
85. K. Yamazaki, Y. Maehara, R. Kitajima, Y. Fukami, K. Gohara, *Appl. Phys. Express*, 2018, *11*, 095101.
86. G. M. Veith, A. R. Lupini, S. J. Pennycook, A. Villa, L. Prati, N. J. Dudney, *Catal. Today*, 2007, *122*, 248–253.
87. G. Veith, A. Lupini, S. Pennycook, G. Ownby, N. Dudney, *J. Catal.*, 2005, *231*, 151–158.
88. C. Zhao, Y. Wang, Z. Li, W. Chen, Q. Xu, D. He, D. Xi, Q. Zhang, T. Yuan, Y. Qu, J. Yang, F. Zhou, Z. Yang, X. Wang, J. Wang, J. Luo, Y. Li, H. Duan, Y. Wu, Y. Li, *Joule*, 2019, *3*, 584–594.
89. S. Wei, A. Li, J. C. Liu, Z. Li, W. Chen, Y. Gong, Q. Zhang, W. C. Cheong, Y. Wang, L. Zheng, H. Xiao, C. Chen, D. Wang, Q. Peng, L. Gu, X. Han, J. Li, Y. Li, *Nat. Nanotechnol.*, 2018, *13*, 856–861.

90. R. Lang, W. Xi, J. C. Liu, Y. T. Cui, T. Li, A. F. Lee, F. Chen, Y. Chen, L. Li, L. Li, J. Lin, S. Miao, X. Liu, A. Q. Wang, X. Wang, J. Luo, B. Qiao, J. Li, T. Zhang, *Nat. Commun.*, 2019, *10*, 234.

91. M. Moliner, J. E. Gabay, C. E. Kliewer, R. T. Carr, J. Guzman, G. L. Casty, P. Serna, A. Corma, *J. Am. Chem. Soc.*, 2016, *138*, 15743–15750.

92. P. Yin, B. You, *Mater. Today Energy*, 2021, *19*, 100586.

93. X. He, Y. Deng, Y. Zhang, Q. He, D. Xiao, M. Peng, Y. Zhao, H. Zhang, R. Luo, T. Gan, H. Ji, D. Ma, *Cell Rep. Phys. Sci.*, 2020, *1*, 100004.

94. D. Deng, X. Chen, L. Yu, X. Wu, Q. Liu, Y. Liu, H. Yang, H. Tian, Y. Hu, P. Du, R. Si, J. Wang, X. Cui, H. Li, J. Xiao, T. Xu, J. Deng, F. Yang, P. N. Duchesne, P. Zhang, J. Zhou, L. Sun, J. Li, X. Pan, X. Bao, *Sci. Adv.*, 2015, *1*, e1500462.

95. X. Cui, J. Xiao, Y. Wu, P. Du, R. Si, H. Yang, H. Tian, J. Li, W. H. Zhang, D. Deng, X. Bao, *Angew. Chem. Int. Ed.*, 2016, *55*, 6708–6712.

96. T. Gan, Y. Liu, Q. He, H. Zhang, X. He, H. Ji, *ACS Sustainable Chem. Eng.*, 2020, *8*, 8692–8699.

97. H. Jin, S. Sultan, M. Ha, J. N. Tiwari, M. G. Kim, K. S. Kim, *Adv. Funct. Mater.*, 2020, *30*, 2000531.

98. S. M. J. Rogge, A. Bavykina, J. Hajek, H. Garcia, A. I. Olivos-Suarez, A. Sepulveda-Escribano, A. Vimont, G. Clet, P. Bazin, F. Kapteijn, M. Daturi, E. V. Ramos-Fernandez, I. X. F. X. Llabres, V. Van Speybroeck, J. Gascon, *Chem. Soc. Rev.*, 2017, *46*, 3134–3184.

99. J. Su, R. Ge, Y. Dong, F. Hao, L. Chen, *J. Mater. Chem. A*, 2018, *6*, 14025–14042.

100. L. Jiao, H.-L. Jiang, *Chem*, 2019, *5*, 786–804.

101. H. Huang, K. Shen, F. Chen, Y. Li, *ACS Catal.*, 2020, *10*, 6579–6586.

102. Z. Song, L. Zhang, K. Doyle-Davis, X. Fu, J. L. Luo, X. Sun, *Adv. Energy Mater.*, 2020, *10*, 2001561.

103. S. Ji, Y. Chen, S. Zhao, W. Chen, L. Shi, Y. Wang, J. Dong, Z. Li, F. Li, C. Chen, Q. Peng, J. Li, D. Wang, Y. Li, *Angew. Chem. Int. Ed.*, 2019, *58*, 4315–4319.

104. A. M. Abdel-Mageed, B. Rungtaweevoranit, M. Parlinska-Wojtan, X. Pei, O. M. Yaghi, R. J. Behm, *J. Am. Chem. Soc.*, 2019, *141*, 5201–5210.

105. K. I. Otake, Y. Cui, C. T. Buru, Z. Li, J. T. Hupp, O. K. Farha, *J. Am. Chem. Soc.*, 2018, *140*, 8652–8656.

106. P. Ji, Y. Song, T. Drake, S. S. Veroneau, Z. Lin, X. Pan, W. Lin, *J. Am. Chem. Soc.*, 2018, *140*, 433–440.

107. X. Fang, Q. Shang, Y. Wang, L. Jiao, T. Yao, Y. Li, Q. Zhang, Y. Luo, H. L. Jiang, *Adv. Mater.*, 2018, *30*, 1705112.

108. Z. Weng, J. Jiang, Y. Wu, Z. Wu, X. Guo, K. L. Materna, W. Liu, V. S. Batista, G. W. Brudvig, H. Wang, *J. Am. Chem. Soc.*, 2016, *138*, 8076–8079.

109. C. Zhao, X. Dai, T. Yao, W. Chen, X. Wang, J. Wang, J. Yang, S. Wei, Y. Wu, Y. Li, *J. Am. Chem. Soc.*, 2017, *139*, 8078–8081.

110. P. Yin, T. Yao, Y. Wu, L. Zheng, Y. Lin, W. Liu, H. Ju, J. Zhu, X. Hong, Z. Deng, G. Zhou, S. Wei, Y. Li, *Angew. Chem. Int. Ed.*, 2016, *2016*, 10800–10805.

111. L. Zou, Y. S. Wei, C. C. Hou, C. Li, Q. Xu, *Small*, 2021, *17*, e2004809.

112. X. Wang, Z. Chen, X. Zhao, T. Yao, W. Chen, R. You, C. Zhao, G. Wu, J. Wang, W. Huang, J. Yang, X. Hong, S. Wei, Y. Wu, Y. Li, *Angew. Chem. Int. Ed.*, 2018, *57*, 1944–1948.

113. Y. Chen, S. Ji, Y. Wang, J. Dong, W. Chen, Z. Li, R. Shen, L. Zheng, Z. Zhuang, D. Wang, Y. Li, *Angew. Chem. Int. Ed.*, 2017, *56*, 6937–6941.

114. A. Han, B. Wang, A. Kumar, Y. Qin, J. Jin, X. Wang, C. Yang, B. Dong, Y. Jia, J. Liu, X. Sun, *Small Methods*, 2019, *3*, 1800471.

115. R. Jiang, L. Li, T. Sheng, G. Hu, Y. Chen, L. Wang, *J. Am. Chem. Soc.*, 2018, *140*, 11594–11598.

116. L. Jiao, G. Wan, R. Zhang, H. Zhou, S. H. Yu, H. L. Jiang, *Angew. Chem. Int. Ed.*, 2018, *57*, 8525–8529.

117. L. Jiao, R. Zhang, G. Wan, W. Yang, X. Wan, H. Zhou, J. Shui, S. H. Yu, H. L. Jiang, *Nat. Commun.*, 2020, *11*, 2831.
118. Y. N. Gong, L. Jiao, Y. Qian, C. Y. Pan, L. Zheng, X. Cai, B. Liu, S. H. Yu, H. L. Jiang, *Angew. Chem. Int. Ed.*, 2020, *59*, 2705–2709.
119. K. Kamiya, *Chem. Sci.*, 2020, *11*, 8339–8349.
120. Y. Yao, H. Yin, M. Gao, Y. Hu, H. Hu, M. Yu, S. Wang, *Chem. Eng. Sci.*, 2019, *209*, 115211.
121. M. Kou, W. Liu, Y. Wang, J. Huang, Y. Chen, Y. Zhou, Y. Chen, M. Ma, K. Lei, H. Xie, P. K. Wong, L. Ye, *Appl. Catal. B*, 2021, *291*, 120146.
122. W. Zhong, R. Sa, L. Li, Y. He, L. Li, J. Bi, Z. Zhuang, Y. Yu, Z. Zou, *J. Am. Chem. Soc.*, 2019, *141*, 7615–7621.
123. Q. Cao, L.-L. Zhang, C. Zhou, J.-H. He, A. Marcomini, J.-M. Lu, *Appl. Catal. B*, 2021, *294*, 120238.
124. L. Peng, L. Shang, T. Zhang, G. I. N. Waterhouse, *Adv. Energy Mater.*, 2020, *10*, 2003018.
125. P. Peng, L. Shi, F. Huo, C. Mi, X. Wu, S. Zhang, Z. Xiang, *Sci. Adv.*, 2019, *5*, eaaw2322.
126. K. W. Urban, *Science*, 2008, *321*, 506–510.
127. J. Tersoff, D. R. Hamann, *Phys. Rev. Let.*, 1983, *50*, 1998–2001.
128. J. Tersoff, D. R. Hamann, *Phys. Rev. B Condens Matter.*, 1985, *31*, 805–813.
129. S. Kano, T. Tada, Y. Majima, *Chem. Soc. Rev.*, 2015, *44*, 970–987.
130. T. S. Seifert, S. Kovarik, C. Nistor, L. Persichetti, S. Stepanow, P. Gambardella, *Phys. Rev. Res.*, 2020, *2*, 013032.
131. J. Yang, D. Wang, Y. Li, *Chemphyschem*, 2020, *21*, 2486–2496.
132. X. Liu, T.-C. Weng, *MRS Bull.*, 2016, *41*, 466–472.
133. F. D. Groot, *Chem. Rev.*, 2001, *101*, 1779–1808.
134. S. L. P. Savin, A. Berko, A. N. Blacklocks, W. Edwards, A. V. Chadwick, *C. R. Chim.*, 2008, *11*, 948–963.
135. J. Yang, W. Li, D. Wang, Y. Li, *Adv. Mater.*, 2020, *32*, e2003300.
136. D. Norman, *J. Phys. C, Solid State Phys.*, 1986, *19*, 3273–3311.
137. Y. Dedkov, C. Karunakaran, C. R. Christensen, C. Gaillard, R. Lahlali, L. M. Blair, V. Perumal, S. S. Miller, A. P. Hitchcock, *PLos One*, 2015, *10*, e0122959.
138. K. Jiang, M. Luo, M. Peng, Y. Yu, Y. R. Lu, T. S. Chan, P. Liu, F. M. F. de Groot, Y. Tan, *Nat. Commun.*, 2020, *11*, 2701.
139. L. Cao, Q. Luo, J. Chen, L. Wang, Y. Lin, H. Wang, X. Liu, X. Shen, W. Zhang, W. Liu, Z. Qi, Z. Jiang, J. Yang, T. Yao, *Nat. Commun.*, 2019, *10*, 4849.
140. W. Liu, L. Zhang, W. Yan, X. Liu, X. Yang, S. Miao, W. Wang, A. Wang, T. Zhang, *Chem. Sci.*, 2016, *7*, 5758–5764.
141. S. Wang, N. Yan, *Synchrotron Radiat. News*, 2020, *33*, 18–26.
142. H. Ali-Löytty, M. W. Louie, M. R. Singh, L. Li, H. G. Sanchez Casalongue, H. Ogasawara, E. J. Crumlin, Z. Liu, A. T. Bell, A. Nilsson, D. Friebel, *J. Phys. Chem. C*, 2016, *120*, 2247–2253.
143. V. M. Varsha, G. Nageswaran, *Front. Chem.*, 2020, *8*, 23.
144. Y. Xiong, W. Sun, P. Xin, W. Chen, X. Zheng, W. Yan, L. Zheng, J. Dong, J. Zhang, D. Wang, Y. Li, *Adv. Mater.*, 2020, *32*, e2000896.
145. S. K. Kaiser, Z. Chen, D. Faust Akl, S. Mitchell, J. Perez-Ramirez, *Chem. Rev.*, 2020, *120*, 11703–11809.
146. Y. Wang, J. Mao, X. Meng, L. Yu, D. Deng, X. Bao, *Chem. Rev.*, 2019, *119*, 1806–1854.

3 Theoretical Modeling and Simulation of Atomically Dispersed Metallic Materials

Wen-Jin Yin
Hunan University of Science and Technology
Beihang University

Gilberto Teobaldi
STFC UKRI, Rutherford Appleton Laboratory
University of Liverpool
University of Southampton

Li-Min Liu
Beihang University

CONTENTS

DOI: 10.1201/9781003153436-3

3.1 INTRODUCTION

The increasing global energy consumption and the fast depletion of fossil fuels demand the development of clean and sustainable energy sources. The abundant usage of fossil fuels increases the emission of CO_2 gas, which has triggered global warming problems and other environmental concerns. To resolve these pressing issues, a variety of methods to clean and sustainable energy sources have been proposed based on thermochemical, biological, electrochemical, and photocatalytic processes.[1–4] Among these processes, water splitting into H_2 and photoreduction of CO_2 are especially promising and appealing strategies. CO_2 molecules can be directly converted into short-chain hydrocarbon renewable fuels, such as CH_4, $HCOOH$, CH_2O, and CH_3OH, which may find use in alleviating the increasingly tense energy crisis.[5] Since pioneering research on the subject,[6] water splitting and CO_2 photoreduction by catalysis have been widely studied. As a result of this work, various kinds of photocatalystes such as TiO_2, CdS, Fe_2O_3, g-C_3N_4, Bi_2WO_6, and Cu_2O[7–9] have been proven to have activity, but low efficiency and poor product selectivity are greatly hindering their further application. Thus, the identification of catalysts with high activity, efficiency, and selectivity for both water splitting and CO_2 photoreduction remains a priority for wider international research in catalysis for sustainable applications.

Zhang et al.[10] successfully synthesized single-atom catalysts (SACs), consisting of only isolated single Pt atoms separated on the surfaces of iron oxide nanocrystallites. Their results showed that these single Pt atoms exhibit excellent stability and high activity for CO oxidation. Further theoretical study demonstrated that the high catalytic activity originates from the partially vacant 5d orbitals of Pt, which is beneficial in reducing both the CO adsorption energy and the activation barriers for CO oxidation. This pioneering research revealed that the high efficiency of the system comes from the dispersed metal atoms, which has attracted extensive attention hereafter.

SACs are defined as isolated metal atoms atomically dispersed on supports without any appreciable interaction between them.[11] Similar to homogeneous catalysts, the single-atom active sites in SACs are atomically distributed on the support, which results in an identical geometric structure for each active center. It has been reported that single-atom modification can greatly change the structural and electronic properties of the catalyst. For example, single transition metal atoms deposited on N-functionalized carbon (TMN_x, TM = transition metal) generally exhibit high catalytic activity.[12] The coordinated TMN_x centers, especially TMN_4 architectures, provide the active sites for catalysis.[13] The N atoms not only act as the anchoring sites to stabilize the single metal atoms but also play an important role in modulating the electronic structures of the active sites.[14] Therefore, by simply altering the electronic structures of the active sites, the electrocatalytic activity can be effectively modified. In particular, doping with hetero-atoms, such as S and P, introduces different coordinated atomistic structures, which can significantly influence the electronic structure of the center single metal atoms, further boosting the electrocatalytic activity.[15] For instance, Li et al.[15] synthesized Fe SACs on N, S-doped carbon support; Qiao et al.[16] investigated the influence of N, C atom on the catalytic activity of Fe_xC/Fe; and Hou et al.[17] studied how well-dispersed molecular $S|NiN_x$ species act as active sites for catalyzing the oxygen evolution reaction. The above studies showed that catalysts with different bindings between the sulfur atoms and the TMN_x centers exhibit superior catalytic activity, which originates from different electronegativity and atomic radius for the S and N atoms.[18]

In addition to peculiar electronic properties, SACs can also have a profound effect on diverse reaction processes and mechanisms. Both experimental and theoretical results demonstrate that the introduction of single metal atoms on an inert substrate can turn it into a highly active and selective catalyst for many reactions. These range from CO oxidation, C–H bond activation, and selective hydrogenation to the N_2 reduction reaction (NRR), the oxygen reduction reaction (ORR), the CO_2 reduction reaction (CRR), the oxygen evolution reaction (OER), and the hydrogen evolution reaction (HER).[19–21] However, the behavior of single metal atoms on different binding sites and reaction processes for distinct supports remains unclear. Therefore, it is essential to design a series of related single metal atoms with tailored center active sites and systematically study the relationship between the orientation of the doping atoms for the SACs and corresponding catalytic activity.

To this end, and to assist further computational and experimental research in this rapidly growing field, we present a brief introduction of recent advances in the investigation and understanding of the structural, electronic, and catalytic properties of SACs from the theoretical side. We first discuss the stability of the single metal atoms on the substrates, including the formation energy and diffusion transfer energy. Then, possible effects on the structural and electronic properties of the single atom are considered. We then present the unique properties of single atoms and their potential for a very diverse range of applications. Finally, we present our conclusions and outlook in terms of unsolved challenges in the field.

3.2 THE CONCEPTION AND MODEL OF SACs

In this section, we mainly address the conception of SACs including their geometric configuration, stability, specific activity, and electronic structure.

3.2.1 CONCEPT AND MODEL

SACs were first proposed by the pioneering work of Zhang et al.[10] SACs are defined as isolated metal atoms atomically dispersed on a substrate without any appreciable interaction between each other, as shown in Figure 3.1.[11,22] Similar to homogeneous catalysts, the single-atom active sites are atomically distributed on the supports, forming the identical active center.[23] In contrast to traditional catalysts, atomically dispersed single metal atoms with the same chemical environment behave as homogeneous catalysts while maintaining the advantages of heterogeneous catalysts. It is well known that homogeneity of single-atom sites offers consistent, excellent selectivity toward a specific product.[24] Therefore, single-atom catalysis has the potential to fill up the gap between homogeneous and heterogeneous catalysts.[25] It is worth mentioning that the active sites in SACs may exist in the immediate neighboring atoms or other functional species apart from the isolated individual metal atoms.[26]

Due to the diversity of substrates, the same single metal atom can form different kinds of SACs.[33] When heteroatomic bonding between the two metals is stronger

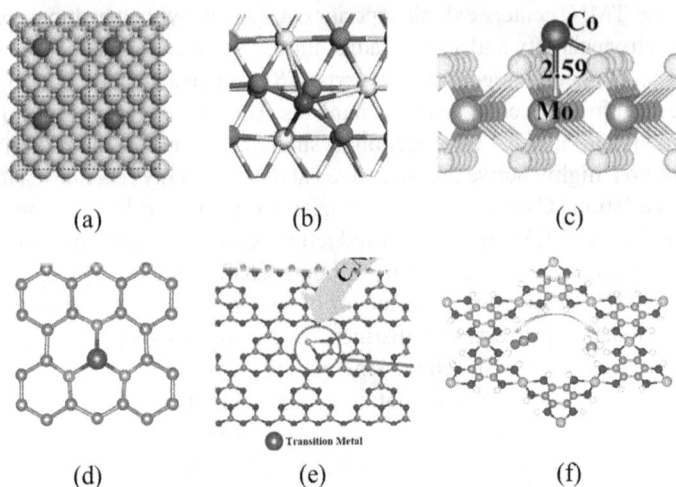

(a)   (b)   (c)

(d)   (e)   (f)

FIGURE 3.1 Typical atomic structures for SACs on the surface of different substrates. (a) Metal Au/Ag,[27] (b) CeO$_2$,[28] (c) single-layer MoS$_2$,[29] (d) defective grapheme,[30] (e) g-C$_3$N$_4$,[31] and (d) MOF surfaces.[32] The single metal atoms are highlighted in different colors relative to the substrate.

than homoatomic bonding, isolated metal atoms can be placed on an array of atoms of a second metal. For example, it has been reported that single-atom alloy catalysts are platinum group metals (PGMs) alloyed with group X and XI metals, yielding the intermetallic Pd-Au, Pd-Ag, Pd-Cu, Pd-Zn, Pd-In, and Pt-Cu as supports.[34,35] Besides metal supports, the single atom can also be dispersed on oxide surfaces (e.g. CeO$_2$), TM dichalcogenides, carbon nitride, and graphene-related materials as shown in Figure 3.1. Furthermore, even on the same support, the single metal atom doping can be extensively varied from noble metals to TM atoms.

3.2.2 Stability

The local geometric structure of the active centers of ideal SACs can be uniform and reproducible, resulting in excellent selectivity compared to pristine support nanoparticles, metal surfaces, and low-dimensional sheets. The unique coordination of single metal atoms with neighboring atoms of support may lead to high activity for specific reactions. It is worth highlighting that all the above advantages are based on the stability of the catalysts. To avoid aggregation under catalytic reaction conditions, single metal atoms should be stabilized by anchoring at specific sites on the supports, including embedding, vacancy, and surface adsorption. When considering complex realistic conditions, the stability of SACs can be altered by the surface condition, reactant species, and finite temperature and pressure. Therefore, it is challenging to design highly stable and reactive SACs.

To know the stability of SACs, both static and dynamic criteria should be considered. The thermodynamic part includes (i) the energetics of supported metal particles, which is based on the Gibbs-Thomson (G-T) relation, considering the adsorbed reactants and (ii) the chemical potential of monomers on supports. The binding energy is a useful factor to estimate the adsorption strength. The binding energy (E_b) between the adsorbed species (A) and supports (B) was calculated according to the following equation:[31]

$$E_b = E_{A/B} - E_A - E_B$$

where E_A and E_B represent the total energy of A and B, and $E_{A/B}$ is the total DFT energy of the system with the adsorbate. According to this definition, a more negative E_b value indicates a stronger adsorption of the adsorbate. On the other hand, the kinetic part includes (i) the diffusion barrier of a single metal atom on perfect surfaces and defects and (ii) the barrier of moving one metal atom from a supported metal nanoparticle to a substrate surface with corresponding sintering rate equations.

In this context, the support effect on the stability of SACs was first discussed in Reference[36]. Ren et al.[31] examined the binding energy of TM (TM = V, Nb, Ta) atoms on a CN sheet to evaluate the overall structural stability. Calculated binding energy of TM/CN ranges from −6.48 eV for V/CN and −7.12 eV for Nb/CN to −7.58 eV for Ta/CN for the single-atom anchoring at the corner vacancy. The larger binding energy demonstrates that the corner vacancy in the CN serves as an ideal anchoring site to stabilize the single metal atoms. On the other hand, surface with different terminations can also affect the relative stability of the adsorbed single metal atoms. It has been reported that the stability sequence of Pt single atoms on different surfaces of CeO_2 follows a $(110) > (100) > (111)$ sequence.[37] Further investigation demonstrated that the high stability of Pt single atoms on the $CeO_2(110)$ surface is attributed to the spontaneous formation of O_2^{2-} species from two surface O atoms that reduce Pt^{IV} to Pt^{II}.

Further work showed that the presence of special defects, such as vacancies or steps, can also improve the stability of SACs. For example, Liu et al. examined Au single atoms on a $CeO_2(111)$ surface, and their calculated stability follows the sequence as cation vacancy > steps > O vacancy > perfect (defect-free) surface.[38] Zhu et al.[39] employed first-principles calculations to investigate how substrate engineering can stabilize SACs by strain tuning the electronic interactions, and they examined two Pd adatoms on a defect-free, single-layer MoS_2 support. They validated that the Pd_2 dimer is prone to dissociate and form highly efficient SACs for CO oxidation due to the enhanced charge transfer and orbital hybridization with the MoS_2 substrate under a suitable tensile strain.

In addition, Fu and Draxl[40] explored the possibility of using hybrid organic-inorganic perovskites as supporting materials for single TM atoms. By means of first-principles calculations, they predicted that single Pt atoms could be incorporated into methylammonium lead iodide surfaces by replacing the methylammonium groups at the outermost layer. The iodide anions at the surface provide potentially uniform anchoring sites for the Pt atoms and donate electrons, generating negatively charged Pt species that allow for preferential O_2 adsorption in the presence of CO.

Due to the unique properties of two-dimensional (2D) systems and their large accessible surface, 2D materials can bring plentiful approaches to enhancing the stability of single metal atoms. For example, Kan et al.[41] studied the stability of Pt atoms on the MXene surface. They calculated the formation energies and diffusion energy barriers

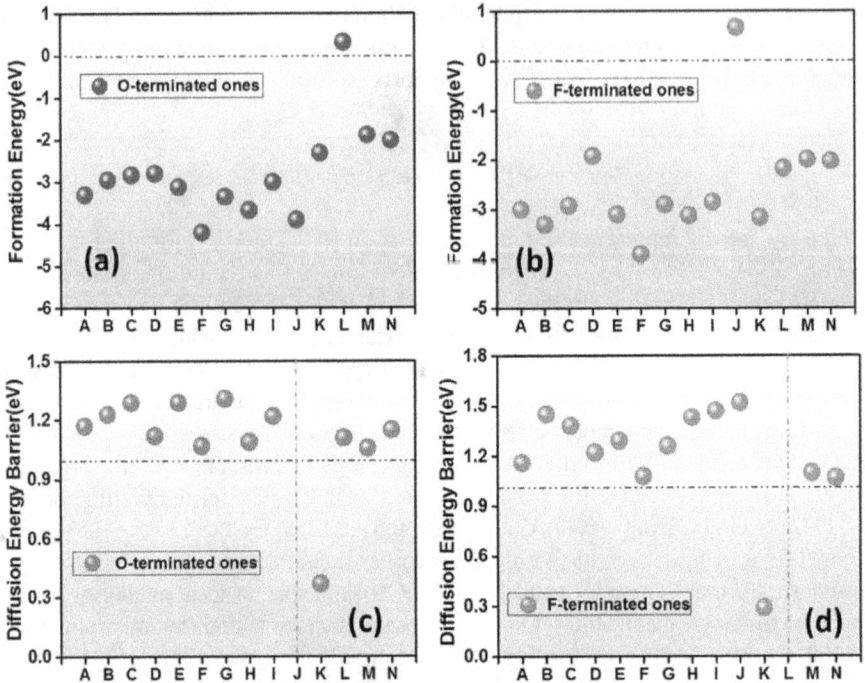

FIGURE 3.2 The binding energy of Pt anchored on O- and F-terminated MXenes (a and b); the diffusion energy barriers of Pt on vacant MXenes (c and d); A-N letters stand for Ti_2CT_2-VT-Pt, V_2CT_2-VT-Pt, Nb_2CT_2-VT-Pt, Mo_2CT_2-VT-Pt, $Ti_3C_2T_2$-VT-Pt, $Zr_3C_2T_2$-VT-Pt, $Ti_3(C,N)_2$-CT_2-VT-Pt, $Ti_3(C,N)_2$-NT_2-VT-Pt, $Nb_4C_3T_2$-VT-Pt, $Ta_4C_3T_2$-VT-Pt, $Ti_4N_3T_2$-VT-Pt, $Cr_2TiC_2F_2$-VF-Pt, $Mo_2TiC_2T_2$-VT-Pt, and $Mo_2Ti_2C_3T_2$-VT-Pt (T=O, F), respectively.[41]

of Pt atoms, as shown in Figure 3.2. The stability of the Pt single atom on the screened structures was studied using the diffusion energy barriers (E_{diff}) of Pt single atoms on the MXene surface. Except for $Ti_4N_3O_2$-Pt and $Ti_4N_3F_2$-Pt, all the other systems exhibited relatively large diffusion barriers ($E_{diff} > 1.0$ eV), suggesting that the Pt single atoms should be stable in such systems. Introducing surface vacancy can further improve the stability of single metal atoms, as suggested by Back et al.[42] in their study of a single TM atoms anchored on defective graphene with single or double vacancies.

In addition to static behavior, the dynamic simulation was also explored to determine the stability of SACs. By means of large-scale *ab initio* molecular dynamic (AIMD) simulations, Liu et al.[43] found that, on reducible oxide-supported Au nanocatalysts, the Au atoms can migrate from the Au nanoparticle to the support to catalyze CO oxidation and reintegrate back to the nanoparticle after completing the reaction. Especially, combining *ab initio* electronic structure and molecular dynamics simulations, as well as a microkinetic simulation, Wang et al.[44] have shown that, on TiO_2-supported Au nanocatalysts, the formation of dynamic SACs is the dominant reaction pathway for CO oxidation under oxidizing conditions and $T < 400$ K. This dynamic formation of single Au atoms under realistic conditions originates from the reducibility of the oxide support, which is strongly coupled with the charge state of

Au. As a result, the single Au atoms achieve an excellent activity due to the dynamic stability. He et al. also found that dynamic single atoms under reaction conditions account for the notorious size effect in Au nanocatalysts.[45]

In all, intrinsic thermodynamic stability and dynamic stability are the key factors in determining the activity of SACs. The stability can be quantitatively evaluated from thermodynamic and dynamic aspects from the theoretical side. The traditional catalytic theory stems mainly from surface science, so the basic theory resorts primarily to band-structure or solid-state theory. However, SACs activity is mainly due to the more local chemical orbitals of single atoms, and thus the local molecular orbitals or the electronic structure of single atoms on the support affects the stability and activity.[46] To achieve the high efficient and stable catalysts, it is vital to explore these intriguing factors on the stability of SACs.

3.2.3 Unique Electronic Property of Single Atom

As shown in Figure 3.3, reducing the catalyst size from metal nanoparticles to SACs causes quantum size effects.[46] Different from the bulk, the SACs exhibit a discrete energy-level distribution and a widening of the Kubo gap (HOMO-LOMO) from valence band maximum to conduction band minimum because of the quantum size effect.[47] The Kubo gap becomes even smaller than those in SCAs and even sub-nanoclusters when the number of metal particles increases to larger than 40 (with a particle size >1 nm).[48] In metal nanoparticles (size larger than 2 nm), a continuous energy level can be formed.[49] As the coordination between isolated atoms and supports affects the unique electronic properties of SACs, metal-support electronic interactions are discussed in the subsequent sections.

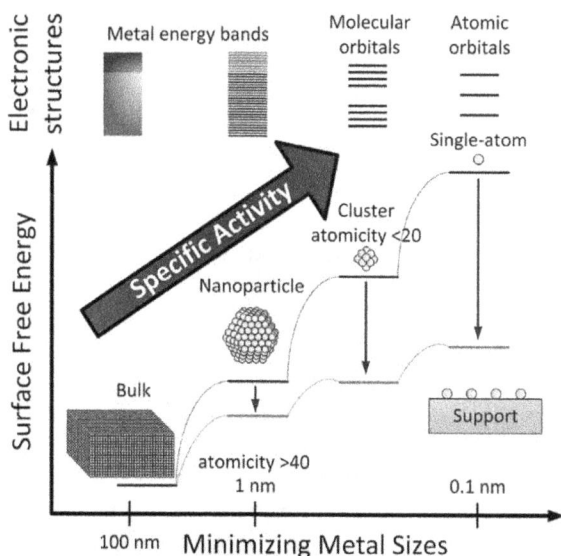

FIGURE 3.3 Evolution of the geometric configuration, electronic structure and surface free energy with the decreasing size of metal nanoparticles.[46]

The catalytic stability, activity, and even selectivity are generally determined by the interaction between the single atoms and their support. Therefore, a suitable support is essential for the application of SACs.[50] Metal single atoms can typically anchor on the support by forming a stable chemical bond. The metal loading density is limited to the anchoring sites of the support.[51] Electron transfer between isolated metal atoms and the support may occur due to their difference in Fermi energy levels. The catalytic performance of metal particles is correlated to the charge transfer between the metal single atoms and the support. It is worth mentioning that the metal-support interaction by the electron-transfer process depends also on the particle size and the ensuing electronic structures. The size effect not only affects the catalyst stability and activity but also the selectivity. The metal atoms may adsorb the reactant strongly enough for subsequent catalysis, resulting in empty nonbonding states. These states are generally vital for the final catalytic activity and selectivity.[46] Thus, the atomic and electronic structures of single atoms on the supports play an important role in the efficiency of SACs.

3.3 THE APPLICATION OF SACs

In this section, a variety of applications of SACs involving photocatalysis and electrocatalysis of different reactions are discussed, including CO_2 reduction, CO oxidation reaction, hydrogenation reaction, and other related reactions. Photocatalytic processes require a semiconductor to absorb a photon with energy higher than the band gap, exciting one electron from the valence into the conduction band, leaving one excited hole in the valence band. These excited electrons and holes then transmit to the surface of the semiconductor encountering adsorbed reactant molecules, thus triggering the reduction and oxidation processes. Different from the photocatalytic case, electrocatalysis implies the presence of an electrochemical reaction on or at the catalyst.[52] This in turn requires that the catalyst can take an electrical current, along with protons/electrons, to achieve the reaction intermediates/products. Good electrocatalysts generally have the following qualities: fast charge transfer capability, high active site density, and high electrical conductivity. SACs have been reported to show excellent performances for several electrochemical reactions, including the ORR, the OER, the HER, the NRR, and the CRR. Table 3.1 summarizes the binding energy and rate-determining steps of some typical electrochemical reactions on different SACs.

3.3.1 CO$_2$ REDUCTION

In CO_2 photocatalytic reduction, CO_2 can be reduced into different kinds of short-chain hydrocarbons, such as CO, CH_3OH, and CH_4, as summarized in Table 3.2. Because of the abundant reaction pathways and effects of the reaction conditions, the selectivity and efficiency of the entire process are quite complex and far from being fully understood. Generally, the activity and product selectivity are mainly influenced by two kinds of factors. One is the thermodynamic kind including photon energy and band edge position, as mentioned above. The other kind includes reaction kinetics factors such as the intensity of the incoming light, surface catalytic active

TABLE 3.1

Effect of Single Metal Atom Anchoring on Different Supports for Diverse Reaction Processes

Single Atom	Supports	Functional	Molecule	E_{bind} (eV)	PDS	E_{bar} (eV) (η(V))	References
Fe	Pt(001)	optB88-vdW	CO	−2.21	CO → *COOH	0.52	Cao et al.[53]
Fe	NG	PBE	O_2		O_2 → H_2O	(0.6)	Yang et al.[54]
Pd	MoS_2	PBE	CO	−1.04	CO → CO_3 LH	0.57	Zhu et al.[39]
V	g-C_3N_4	PBE	N_2	−0.50	Distal *NNH	0.05	Ren et al.[31]
Nb	g-C_3N_4	PBE	N_2	−1.78	Distal *NNH	0.23	
Ta	g-C_3N_4	PBE	N_2	−1.79	Distal *NH_3	0.3	
Ti@N_4	Ru(0001)	RPBE	N_2		*NH_3	0.69	Choi et al.[55]
V@N_4	Ru(0001)	RPBE	N_2		*NNH	0.87	
P	MoS_2	PBE	H_2		*H → *H_2	0.04	Liu et al.[56]
Rn	C_2N	PBE D3	H_2O		OER	(0.37)	Ying et al.[57]
Au	C_2N	PBE D3	O_2		ORR	(0.38)	
Ir	FeO_x	PBE	CO	−1.86	CO → CO_3 ER	1.01	Liang et al.[58]
Pt	$MAPbI_3$	PBE	CO	−2.33	CO → OCOO	0.58	Fu and Draxl[40]
Au	C_3N	PBE	CO		CO → CO_3 LH(ER)	0.32	Fu et al.[59]
						0.47	
Co	N_4 pyridine	DFT+U	H_2O		OER	(0.33)	Hu et al.[30]
Co	N_4 pyridine	DFT+U	O_2		ORR	(0.41)	
Mn	N_2C_2	PBE	H_2O		OER	(0.87)	Shang et al.[60]
RuN_4	Graphene	PBE	H_2O		*H → *H_2HER	(0.2)	Bai et al.[61]
RuN_4OH	Graphene	PBE	O_2		ORR	(0.7)	
RuN_4O	Graphene	PBE	H_2O		OER	(0.6)	
Pt	Ti_2CO_2-V_O	PBE	O_2	−0.65	ORR	0.41	Kan et al.[41]
Rh	C_3N	PBE	O_2		ORR	(0.27)	Zhou et al.[62]

(Continued)

TABLE 3.1 (Continued)

Effect of Single Metal Atom Anchoring on Different Supports for Diverse Reaction Processes

Single Atom	Supports	Functional	Molecule	E_{bind} (eV)	PDS	E_{bar} (eV) (η(V))	References
Pt	FeO$_x$	PBE	CO	−1.27	CO-CO$_2$	0.49	Qiao et al.[10]
Ir@d	TiC	PBE	CO$_2$		*CHOH → *CH	0.09	Back and Jung[63]
Pt@dv	Graphene	PBE	CO$_2$		*CO → *CHO	0.27	Back et al.[42]
Ru@dv	Graphene	PBE	CO$_2$		*HCOOH → *CHO	0.52	
Ti	C$_2$N	PBE	N$_2$	−0.5	NH$_3$	0.88	Qian et al.[54]
Pd	g-C$_3$N$_4$	PBE	CO$_2$		*HCOO → *HCOOH	0.66	Gao et al.[65]
Pt	g-C$_3$N$_4$	PBE	CO$_2$		*HCOOH → *HCO	1.16	
Co-N$_5$	Porous carbon	PBE	CO$_2$		CO$_2$ → CO	0.65	Pan et al.[66]
Cu-N$_4$	Graphene	PBE	CO$_2$		CO$_2$ → *COOH	1.28	Xu et al.[67]
Pt	Ga-CeO$_2$	DFT+U	CO	−2.22	CO → *CO$_3$	0.51	Feng et al.[28]
Fe	Graphdiyne	PBE	O$_2$		*O$_2$ → *OOH	0.21	Gao et al.[68]
Fe	N-graphene	PBE	CO	−0.82	CO → *CO$_3$	0.66	Kropp and Mavrikakis[69]
Rh	V$_N$-g-C$_3$N$_4$	PBE	H$_2$O		OER	(0.32)	Niu et al.[70]
Rh	V$_N$-g-C$_3$N$_4$	PBE	O$_2$		ORR	(0.43)	
Rh	FeO$_x$	PBE	CO	1.82	CO → CO$_3$ LH	0.53	Li et al.[71]

E_{bind}, binding energy of the (reactant) molecule on the catalyst; PDS, potential-determining step; E_{bar}, energy barrier of the rate-determining step. The overpotential, η, for the free energy profiles is shown in brackets.

TABLE 3.2

One, Two, Four, Six, and Eight Electron Reduction Potentials (vs. NHE) of Some Reactions Involved in CO_2 Reduction at pH $= 7$ and Unit Activity[72]

Product	Reaction	E^0_{redox}, V vs. NHE
CO_2^-	$CO_2 + e^- \rightarrow CO_2^-$	-1.90
HCO_2^-	$CO_2 + H^+ + 2e^- \rightarrow HCO_2^-$	-0.49
CO	$CO_2 + 2H^+ + 2e^- \rightarrow CO + H_2O$	-0.53
$HCHO$	$CO_2 + 4H^+ + 4e^- \rightarrow HCHO + H_2O$	-0.48
CH_3OH	$CO_2 + 6H^+ + 6e^- \rightarrow CH_3OH + H_2O$	-0.38
CH_4	$CO_2 + 8H^+ + 8e^- \rightarrow CH_4 + 2H_2O$	-0.24
H_2	$2H^+ + 2e^- \rightarrow H_2$	-0.41
O_2	$H_2O \rightarrow 1/2O_2 + 2H^+ + 2e^-$	0.82

sites, separation of photoinduced electrons and holes, and the adsorption/desorption of intermediates and products.

For instance, single-atom N doping of TiO_2 samples leads to strong activity for CH_4 and CO generation, while g-C_3N_4 and N-TiO_2 composites preferentially result in the formation of CO. Therefore, selectivity depends on the content of g-C_3N_4 in the composites.[73] Cu_2O and Pt co-deposited on TiO_2 can inhibit H_2 production while enhancing the activity and selectivity of CO_2 conversion into CH_4.[74] On the other hand, Gao et al.[65] studied single atoms of Pd and Pt supported on graphitic carbon nitride (g-C_3N_4), i.e., Pd/g-C_3N_4 and Pt/g-C_3N_4, respectively, acting as photocatalysts for CO_2 reduction by density functional theory. They found that the individual metal atoms function as the active sites, while g-C_3N_4 provides the source of hydrogen (H*) from the HER. Further, HCOOH is the preferred product of CO_2 reduction on the Pd/g-C_3N_4 catalyst with a rate-determining barrier of 0.66 eV, while the Pt/g-C_3N_4 catalyst prefers to reduce CO_2 to CH_4 with a rate-determining barrier of 1.16 eV. Therefore, the activity and selectivity of CO_2 photoreduction with H_2O can be greatly affected by single-atom catalysis.

CO_2 reduction is a complex process involving multiple proton/electron transfer and many intermediates, as shown in Table 3.2. To realize the practical application of CO_2 reduction, three key factors need to be considered: the competing HER side reaction, the sluggish kinetics, and the abundant reduction products. As for competing HER side reaction, Back and Jung[63] investigated the catalytic properties of TiC, TiN, and SACs supported on them for CO_2 electrochemical reduction by first-principles calculations. They found that the iridium-doped TiC (Ir@d-TiC) has a low overpotential of -0.09 V, as shown in Figure 3.4. The surface protonation reactions on TiC as a side reaction can be ignored because the overpotential (-0.38 V) is significantly larger than that of the CO_2 electrochemical reduction reaction on SACs (e.g., -0.09 V). On the other hand, Xu et al.[67] revealed that CO_2 reduction is less hindered thermodynamically on Cu-N_4-NG compared to the competing HER due to their limiting potential differences.

To resolve the sluggish kinetics of CO_2 reduction, it is essential to activate the inert CO_2 molecule. To this end, different single metal atoms on various substrates

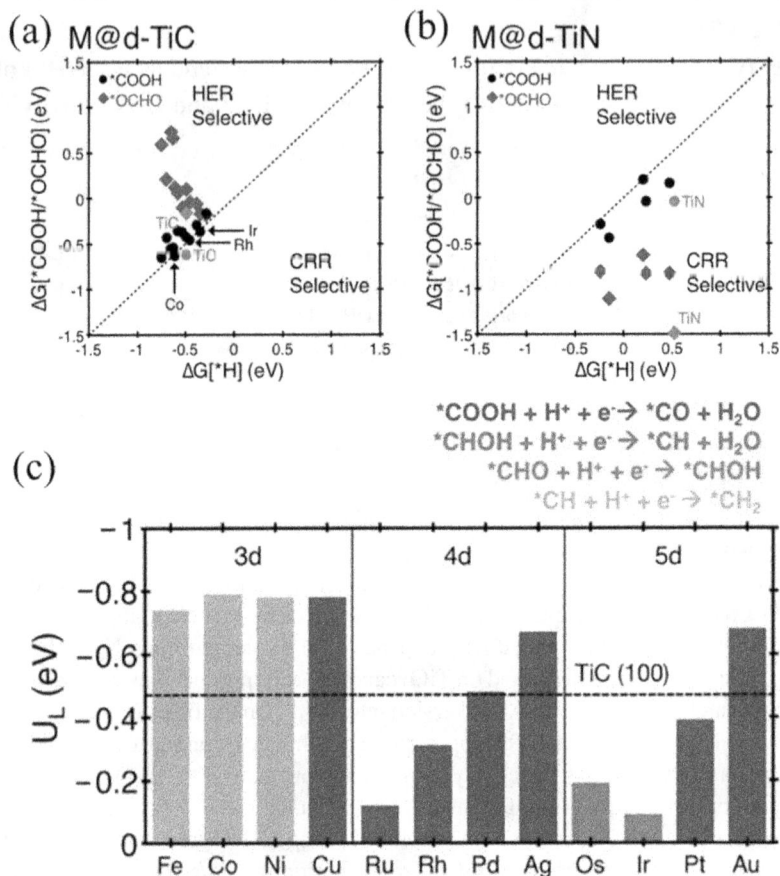

FIGURE 3.4 Gibbs free energy changes (ΔG) of the first protonation step of the CRR and HER on d-TiC (a) and d-TiN (b). Circle and square dots indicate the formation of *COOH and *OCHO, respectively. The reaction on pristine TiC and TiN is marked by the red dots. Catalysts below the dotted line are expected to be CRR-selective. (c) Theoretical limiting potential (U_L) of M@d-TiC(100) for the CRR. The different potential-determining steps (PDS) are marked with different colors. The dotted horizontal line is the U_L of TiC(100) surface, where the PDS is $*CO + H^+ + e^- \rightarrow *CHO$.[63]

have been investigated. Pan et al.[66] developed an N-coordination strategy to design a robust CRR electrocatalyst with atomically dispersed Co–N_5 site anchored on polymer-derived hollow N-doped porous carbon spheres. They found that single-atom Co-N_5 site is the dominating active center simultaneously for CO_2 activation, the rapid formation of key intermediate COOH* as well as the desorption of CO. Furthermore, Lu et al.[29] investigated the CO_2 reduction on the MoS_2 supported single cobalt atom (Co/MoS_2) by first-principles simulation. They found that the preferred CO_2 reduction pathway is the reverse water gas conversion and CO hydrogenation pathway with the rate-limiting step of CO hydrogenation into formyl (HCO). The electronic structures analysis indicates that Co adatom induced gap states play an important role in CO_2 activation and reduction. On the other hand, the single atom can also be anchored in a porous system, which can increase stability and activity. Cui et al.[32] have systemically investigated a family of emerging 2D metal-organic frameworks (MOFs) by DFT calculations. They found that the Mo-based MOF presents the appealing capability of CO_2 activation and reduction under ambient conditions. The value of the energy cost for CO_2 selectively reduced to methane is low, with a value of only 0.42 eV. Moreover, the energy input can be further decreased to 0.27 eV with MoO based on MOF.

For the third aspect, SACs can also play a fundamental role in improving the product selectivity of CO_2 reduction. Cheng et al.[27] used DFT calculations combined with the Poisson–Boltzmann implicit solvation model to study the single-atom alloys for CO_2 reduction, and they found that they are promising electrocatalysts for CO_2 reduction to short-chain hydrocarbons in an aqueous solution. The majority component of the SACs studied is either Au or Ag, in combination with isolated single atoms, M (M = Cu, Ni, Pd, Pt, Co, Rh, and Ir), replacing surface atoms. The results show that the SACs behave as a one-pot tandem catalyst: first Au (or Ag) reduces CO_2 to CO, and the newly formed CO is further reduced to C1 hydrocarbons, such as methane or methanol. The minimum applied voltages to drive the two electrocatalytic systems are −1.01 and −1.12 V RHE for Rh@Au(100) and Rh@Ag(100), respectively. However, CO_2 electrochemical catalysis is limited by scaling relations due to a d-band theory of TMs. Back et al.[42] found that hybridizing the d-orbitals of TM with p-orbitals of main group elements or using naturally hybridized materials such as metal carbides and nitrides is a promising strategy. They studied a single TM atom anchored on defective graphene with single or double vacancies, denoted M@S_v-Gr or M@D_v-Gr, where M = Ag, Au, Co, Cu, Fe, Ir, Ni, Os, Pd, Pt, Rh or Ru, as a CO_2 reduction catalyst. Based on free energy profiles, several promising candidate materials were identified for different products. On the other hand, Xie et al.[75] investigated the CO_2 reduction on Bi-Pd single-atom alloy nanodendrites with Bi atomically dispersed in Pd matrices by DFT. They found that the Faradic efficiencies of CO on the Bi_6Pd_{94}-SAA ND catalyst reach 90.5% and 91.8% in H-type and gas diffusion flow cells with overpotentials of only 290 mV and 200 mV, respectively.

In summary, single metal atoms on diverse supports can effectively regulate the reduction of CO_2 regardless of the specific pathway enabled by the given SAC-support combination. Notably, it turns out that SACs can have a profound effect on the competing HER side reaction, the sluggish kinetics, and the wide product spectrum of the CO_2 reduction, resulting in high activity and product selectivity.

3.3.2 CO OXIDATION

The conversion of CO into nontoxic gases is of great importance in solving the growing environmental and energy problems. The oxidation of CO is often regarded as one of the crucial prototypical reactions, and it has been an object of intense research. Foremost, the reaction mechanisms of CO oxidation should be carefully discussed, since the oxidation reaction involves O_2 in the process. Lu et al.[76] chose the Au-embedded graphene with single vacancy (Au_1/S_v-graphene) as the candidate to illustrate this problem. They found that there are two stages of CO oxidation during the catalytic process, which is

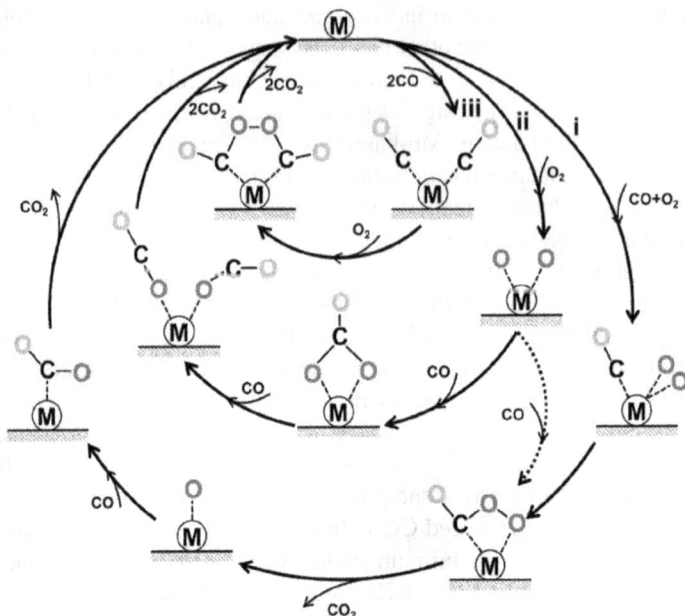

FIGURE 3.5 Proposed reaction mechanisms for CO oxidation on graphene-supported metal single atoms (M1/graphene): (i) L-H+E-R, (ii) E-R+E-R, and (iii) TER.[19]

shown in Figure 3.5. The first CO oxidation stage follows the Langmuir-Hinshelwood (L–H) mechanism with three elementary steps: (a) coadsorption of CO* and O_2*, (b) formation of a peroxo-type OCOO* intermediate with a calculated energy barrier of 0.31 eV, (c) releasing a CO_2 molecule and leaving an adsorbed O* on the Au atom. The second CO oxidation stage follows the Eley-Ridel (E-R) mechanism, where the carbon atom of a CO molecule attacks the left O* directly, leading to the formation of the second CO_2 molecule. Their calculated results indicated that the CO can be oxidized based on two paths, but the activity is still quite low.

Motivated by the successful application of SACs, it is interesting to know the effect of single metal atoms on CO oxidation. To date, several researchers have carried out work aimed at enhancing the activity of CO oxidation reaction from the theory side. Among the single metal atoms, most of the works focused on noble metal anchoring at the metal oxide surface. For instance, Cao et al.[53] indicated that single Pt on $Fe_1(OH)_x$ interfacial sites can readily react with CO and facilitate O_2 activation. Following, Liang et al.[58] give a more detailed work on the catalytic mechanism of CO oxidation on an Ir_1/FeO_x SAC by DFT. They found that the rate-determining step in the catalytic cycle of CO oxidation appears at the formation of the second CO_2 between the adsorbed CO on the surface of Ir_1/FeO_x and the dissociated O atom. Compared with Pt_1/FeO_x catalyst at about 0.52 eV[53], the reaction activation barrier for CO oxidation is higher by 0.62 eV and the adsorption energy for CO molecule is larger by 0.69 eV on Ir_1/FeO_x. To check the effect of other single metal atoms, Li et al.[71] systemically examined M_1/FeO_x (M=Au, Rh, Pd, Co, Cu, Ru and Ti) by means of DFT. They identified five SACs, namely the O-defective Rh_1/FeO_x and Pd_1/FeO_x, Ru_1/FeO_x with or without O_v, and pure Ti_1/FeO_x and Co_1/FeO_x, which exhibit improved overall catalytic performance compared to Pt_1/FeO_x for the CO oxidation via a Langmuir-Hinshelwood (LH) mechanism.

Apart from Fe oxides, the effect of other metal oxides is also examined. Moses-DeBusk et al.[77] reported the CO oxidation activity of monodisperse single Pt atoms supported on a substrate, θ-alumina (Al_2O_3), in the presence of stoichiometric O_2. They found that a single supported Pt atom prefers to bond to O_2 over CO. CO then bonds with the oxygenated Pt atom and forms a carbonate, which dissociates to liberate CO_2, leaving an O atom on Pt. Subsequent reaction with another CO molecule regenerates the SAC. The energetics of the proposed mechanism suggests that the single Pt atoms will get covered with CO_3 unless the temperature is raised to eliminate CO_2. They also found evidence for CO_3 coverage at room temperature. In addition, Long et al.[78] investigated CO oxidation catalyzed by single Au atoms supported on thoria (Au/ThO_2) and doped ThO_2 using DFT with Hubbard-type on-site Coulomb interaction (DFT + U). The calculation results show that the Au-doped $ThO_2(111)$ catalyst exhibits remarkable catalytic activity for CO oxidation via the E-R mechanism in three steps, where the rate-determining step corresponds to the decomposition of the OCOO* intermediate with an energy barrier of 0.58 eV. Moreover, they also revealed a new mechanism of CO oxidation on an Au adatom supported by $ThO_2(111)$, where O_2 is adsorbed only at the Th atom site on the surface, and the gas-phase CO then reacts directly with the activated O_2* to form CO_2 as the rate-limiting step, with a barrier of 0.46 eV.

To understand the role of Pt or Pd on high activity of CO oxidation, Qiao et al.[10] found that the high catalytic activity correlates with the partially vacant 5d orbitals of the positively charged, which will be beneficial to reduce both the CO adsorption energy and the activation barriers for CO oxidation. To describe the electronic property more accurately, Song et al.[79] presented a DFT+U study of CO oxidation for single Pd atoms located on or in the ceria surface as well as a Pd nanorod model on the $CeO_2(110)$ surface. The oxidation of Pd to the 2+ state by ceria weakens the Pd-CO bond for the single Pd models and, in this way, facilitates CO_2 formation. After CO oxidation by O of the ceria surface, Pd relocates to a position below the surface for the Pd-doped model; in this state, CO adsorption is not possible anymore. With Pd on the surface, O_2 will adsorb and dissociate leading to PdO, which can be easily reduced to Pd. The reactivity of the Pd nanorod is low because of the strong bonds of the metallic Pd phase with CO and the O atom derived from O_2 dissociation.

In contrast to metal oxides, other low-dimensional supports, such as two-dimensional graphene, carbon nitride, phosphorene, and other materials, may bring some promising results.[59] For example, Lu et al.[80] investigated the elementary steps of CO oxidation on $Mn-N_4$ porphyrin-like carbon nanotube (MnN_4-CNT). According to the energetic calculations, CO and O_2 prefer to anchor at the MnN_4 site and the adsorption of CO is slightly more favorable than O_2. The three reaction mechanisms of CO oxidation on the MnN_4-CNT are explored, namely, E–R, L–H, and a "new" termolecular E–R TER, respectively. They found that the TER reaction mechanism is the most favorable one and the energy barrier for the rate-limiting step is merely 0.69 eV. Nematollahi and Neyts.[81] investigated the reaction mechanisms of

CO oxidation catalyzed by the Si atom-embedded defective BC_2N nanostructures as well as the analysis of the structural and electronic properties. The results showed that there can be two possible pathways for CO oxidation with O_2 molecule: $O_2 + CO \rightarrow O_2 + CO \rightarrow CO_2 + O$ and $O + CO \rightarrow CO_2$. The first reaction proceeds via the L–H mechanism, while the second reaction goes through the E–R mechanism. On the other hand, Hamid Butt et al.[82] studied Cu-embedded phosphorene as SAC through DFT calculations for CO oxidation. They adopted L–H mechanism, E–R mechanism and TER mechanism for CO oxidation. In the L–H mechanism ($CO + O_2 \rightarrow OOCO \rightarrow CO_2 + O^*$), the activation energy barrier for the formation of peroxy type interme diate OOCO is about 0.16 eV. Furthermore, this intermediate is dissociated into CO_2 and O^* in the rate-limiting step with an energy barrier of 0.06 eV. In the E–R mechanism ($CO + O_2 \rightarrow CO_3 \rightarrow CO_2 + O^*$), a carbonate-like intermediate is formed after the adsorption of the CO molecule over the pre-adsorbed O_2 molecule. The activation energy barrier for the formation of this intermediate is 0.55 eV. This intermediate is dissociated into CO_2 molecule and O^* in the rate-limiting step with an energy barrier of 0.30 eV. In the TER mechanism ($2CO + O_2 \rightarrow OCO\text{-}OCO \rightarrow 2CO_2$), an O_2 molecule is activated over the pre-adsorbed two CO molecules and an intermediate OCO–OCO is formed. The activation energy for this reaction is only 0.02 eV. This intermediate is further dissociated into two CO_2 molecules after the elongation of the bond length of O_2 from 1.41 to 1.57 Å with the rate-limiting energy barrier of 0.08 eV.

To acquire higher stability to keep higher activity, many approaches can be adopted, such as intrinsic defect and external methods. For example, Feng et al.[28] systematically studied CO oxidation reactivity of Pt single atoms supported on $CeO_2(111)$ (Pt/CeO_2) and Ga-doped $CeO_2(111)$ ($Pt/Ga\text{-}CeO_2$). They found that O_v is favored near a surface Ga-doping site ($Pt/Ga\text{-}CeO_2\text{-}O_v$). Significantly, the stability of Pt single atoms anchored on the Ga site was enhanced compared with those on the bare ceria surface. The O_v site plays an important role in activating the O_2 molecule, which then reacts with CO pre-adsorbed on Pt. The calculated energy barrier on $Pt/Ga\text{-}CeO_2\text{-}O_v$ is about 0.43 eV lower than that on the undoped catalyst, suggesting an enhanced reactivity for CO oxidation. Furthermore, Kropp and Mavrikakis[69] studied 14 TMs on pristine and graphene with two vacancies by DFT. They found that for double vacancies, N doping increases the binding strength of harder TMs to the support and reduces their oxygen affinity. Conversely, the oxygen affinity of softer metals increases. Since O_2 binding energies have a great effect on the CO oxidation barrier in a volcano-like trend, doping also affects the activity of the SAC as shown in Figure 3.6. On the other hand, external methods such as stress or strain can also play an important impact. Zhu et al.[39] employed first-principles calculations to investigate Pd on a defect-free MoS_2 monolayer for CO oxidation. Interestingly, their result shows that high CO oxidation activity of SACs can be achieved under a suitable tensile strain, where the tensile strain could greatly enhance charge transfer and orbital hybridization with the MoS_2 substrate. Moreover, low-cost elements, such as Ag, Ni, Cu, and Cr, can also be stabilized into high-performance SACs for CO oxidation with tunable reaction barriers by applying strain.

Generally, the oxidation of CO can produce a variety of products, and SACs can effectively improve the selectivity for the reaction. Han et al.[83] examined two possible reaction pathways related to dimethyl oxalate (DMO) synthesis by employing DFT calculation in combination with microkinetic analysis. Their results showed

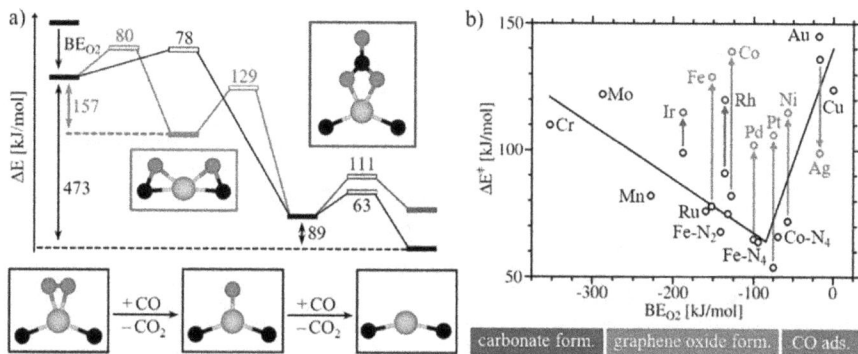

FIGURE 3.6 (a) Energy profile for CO oxidation on Fe/G-V$_2$C starting with pre-adsorbed O$_2$. Minima along the black path are shown below. The gray path refers to O$_2$ dissociation prior to CO oxidation; the path of carbonate formation needs to overcome an energy barrier of 111 KJ/mol. (b) Barriers for the first CO oxidation step (black pathway) are plotted against O$_2$ binding energies for Metal/G-V$_2$C and some metals in N-doped G-V$_2$C. The black lines correspond to a linear fit. If a different reaction is predicted to be rate-limiting, that barrier is shown as well with an arrow connecting both circles.[69]

that COOCH$_3$–COOCH$_3$ coupling pathway is superior to COOCH$_3$–CO on Pd$_4$Cu$_8$/Cu(111) and Pd$_1$–Cu(111). Moreover, the Pd$_1$–Cu(111) surface shows the highest catalytic activity for DMO generation. Additionally, Pd$_1$–Cu(111) surfaces exhibit high DMO selectivity. On the other hand, Aykan Akça et al.[84] studied the mechanism of CO oxidation over iridium (Ir) embedded on both single vacancy graphene and di-vacancy graphene with the aid of DFT. The calculated adsorption energy values of CO and O$_2$ molecules indicate that both molecules can be molecularly adsorbed on the defective graphene. They suggested that the reaction mechanism of CO + O$_2$ → OOCO → CO$_2$ + O* prefers to L-H mechanism with the activation energy of about 0.31 eV. Moreover, the results have confirmed that defective graphene has high catalytic activity and selectivity toward CO oxidation.

To consider the thermodynamic effect on oxidation, Liu et al.[43] have constructed a general thermodynamic model of chemical potentials and applied ab initio electronic structure and molecular dynamics simulations, as well as kinetic Monte Carlo analysis, to probe the dynamical, reactive, and kinetic aspects of metal SAC on oxide support. They examined Au single atoms supported on ceria as a typical example to guide the rational design of highly stable and reactive SACs. Their results showed that the Au single atoms at step sites of ceria support are rather stable, even at temperatures as high as 700 K, and exhibit around 10 orders of magnitude more reactivity for CO oxidation than the terrace sites.

3.3.3 N$_2$ REDUCTION

NH$_3$ as one of the most important chemicals is extensively used in various fields. The traditional synthesis of NH$_3$ is through the well-established Haber-Bosch

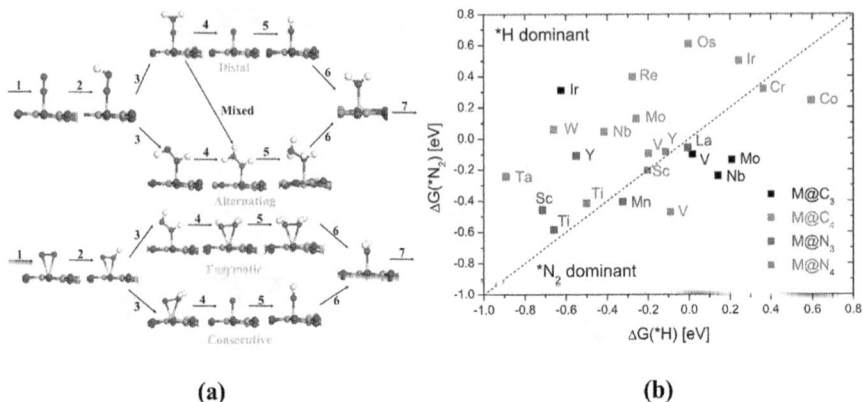

(a)   (b)

FIGURE 3.7 (a) Schematic illustration of distal, alternating, enzymatic, and consecutive mechanisms for NRR. (b) Calculated $\Delta G(*H)$ and $\Delta G(*N_2)$ on SACs that satisfy ΔG (PDS) ≤ 1.0 eV. Dashed line indicates $\Delta G(*H) = \Delta G(*N_2)$. SACs in the $\Delta G(*H) > \Delta G(*N_2)$ region under the dashed line show N_2 adsorption selective.[55]

process, which requires high temperatures, high pressures, and high H_2 consumption. Therefore, the synthesis of NH_3 under mild reaction condition is highly desirable. To this end, electroreduction of N_2 is one of the most promising approaches.[85] The proposed mechanisms for electrochemical NRR include the dissociative pathway and the associative pathways as shown in Figure 3.7a.[55] The whole process involves the transfer of six protons and six electrons. In the dissociative pathway, the first elementary step is the breaking of the N–N triple bond in N_2. In contrast, in the associative pathways, the initial elementary step is the hydrogenation of N_2. The associative pathways can be further divided into the distal, alternating, and enzymatic pathways according to the different N_2 adsorption modes and hydrogenation sequences.

With the aim to design a highly efficient catalyst for ammonia synthesis, numerous research studies have focused on SCAs, especially the single metal on carbon-related materials. For example, Ren et al.[31] explored systematically the potential for N_2 electroreduction of SACs covering V, Nb, and Ta TM centers supported by graphene and g-C_3N_4 substrates. The single Nb-atom embedded on g-C_3N_4 nanosheet possesses outstanding NRR catalytic activity and exhibits better performance than graphene with a small maximum ΔG value (0.05 eV). The single Nb-atom on g-C_3N_4 with more negative valence provides structural advantages for hosting empty d-orbitals for strong N_2 and N_2H adsorption as well as more single d-electrons to further promote back donation to activate the triple bond.

In addition, Qian et al.[64] investigated d-block TM-anchored C_2N single-layer catalyst by DFT. Both single TM-anchored SAC and double TM-anchored double atom catalyst (DAC) exhibit good thermodynamic stability in atomically dispersed catalyst. In the case of SACs, IVB metals (Ti, Zr, Hf) exhibit the highest reactivity and lowest overpotential. While in the case of DACs, the Cr-Cr system leads to the NH_3 formation, but the V-V system leads to the N_2H_4 formation. The SACs show much lower overpotential and stronger activation of N_2 molecule than the DACs due to the different activation mechanisms: traditional σ-donation/π-back donation N_2

activation mechanism is found in SACs, while a new π-donation/π-back donation N_2 activation mechanism is found in the DACs. Besides the prevalent catalysts for natural and artificial N_2 fixation focusing on TM atoms, other metal atoms such as VIIIB atoms are also explored. Liu et al.[86] investigated several TM atoms embedded on boron sheets as N_2 fixation electrocatalysts. Their results revealed that single ruthenium (Ru) atom doped boron sheets exhibit outstanding catalytic activity for ammonia synthesis at ambient conditions, through the distal pathway with small activation barrier of 0.42 eV.

Introducing defects in the substrate can also help to enhance product selectivity. Choi et al.[55] studied electrochemical NRR to ammonia on SACs on defective graphene derivatives by DFT. As shown in Figure 3.7b, they found significantly improved NRR selectivity on SACs compared to that on the existing bulk metal surface due to the strong suppression of the HER on SACs induced by ensemble effects. In addition, several SACs, including Ti@N_4 (0.69 eV) and V@N_4 (0.87 eV), are shown to exhibit lower free energy for the NRR than on the Ru(0001) stepped surface (0.98 eV) due to a strong back-bonding between the hybridized d-orbital metal atom in SAC and the π^* orbital in *N_2.

With the aim to design a highly efficient catalyst for ammonia synthesis, Li et al. investigated the activation behavior of N_2 on Fe_1N_3/S_v-graphene and Fe_1N_4/D_v-graphene by first-principles calculations. They found that Fe_1N_3/SV-graphene is much more efficient for the activation of the inert N-N triple bond of N_2 than Fe_1N_4/DV-graphene due to the higher spin polarization of the Fe_1N_3 center.[87] Among the three reaction pathways, the alternating pathway shows the lowest energy barrier. The key point here is the d-orbital splitting of the central Fe atom as shown in Figure 3.8. It shows that the oxidation state of central Fe atom is Fe(+II) with a d^6 electronic configuration when embedded in N_3-graphene and N_4-graphene. As in Fe_1N_3, the six electrons of Fe d-orbitals occupy five α orbitals and one β orbital, leaving four unpaired electrons. As in Fe_1N_4, the energy level of $d_{x^2-y^2}$ is about 3 eV higher than the other d-orbitals, and thus the electronic configuration is with four α and two β electrons, corresponding to the lower spin of 2 μB. By the systematic investigation on the NRR catalytic performances of SA Fe_1 embedded in the single vacancy S_v-graphene, N_x-S_v-graphene ($x=1$–3), and P_3-S_v-graphene, Guo et al. uncovered a linear scaling relationship between the NRR activity and the magnetic moment of Fe.[88] The NRR activity is promoted by the increased magnetic moment of Fe, which can strengthen the adsorption of N atom and promote the charge transfer between the N_2 molecule and the catalyst surface.

3.3.4 WATER SPLITTING

Water splitting to H_2 product is a promising way to produce clean and renewable energy. The overall reaction consists of two half-reactions: the OER and the HER. It was found that the overall reaction rate of water splitting can be optimized by modulating the coordination environment of metal single atoms. Zhou et al.[89] suggested that a lower-coordinated metal single-atom center exhibits higher HER activity, whereas a higher-coordinated metal SA center is more favorable for the OER. In addition, the N content of the defective graphene support (coordinated with the central metal SA) also affects the HER and OER activity. For early TMs, the fully N-substituted sites

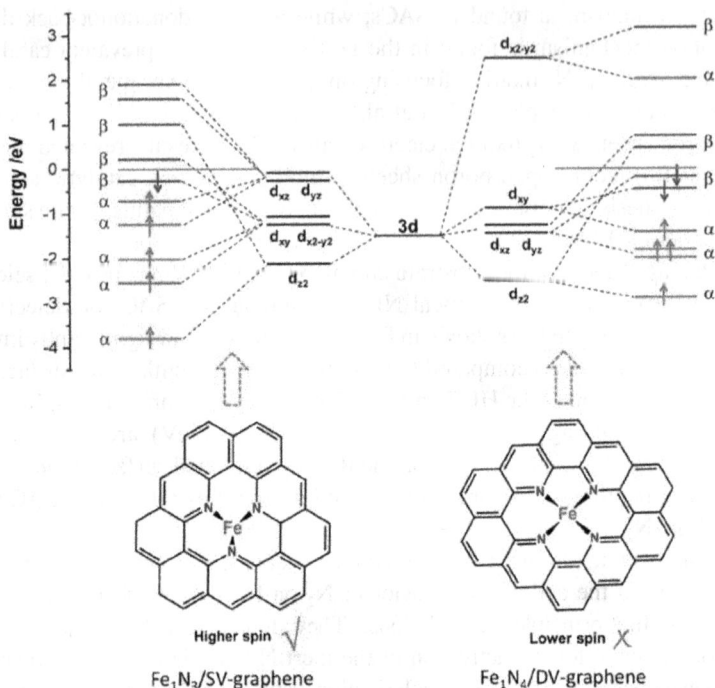

FIGURE 3.8 The structure of Fe_1N_3/S_v-graphene and Fe_1N_4/D_v-graphene, together with the splitting d-orbitals of single Fe atom.[87]

exhibit higher activity; while for later TMs, the partially N-substituted sites are more active.

3.3.4.1 Hydrogen Evolution Reaction

HER can proceed in both photocatalytic and electrocatalytic reaction conditions. It has been reported that single-atom catalysis can greatly impact both these conditions. As for the photocatalytic HER, this reaction suffers from deficient solar light efficiency, high cost of noble metal co-catalysts, and low responses to visible and infrared light. To resolve this problem, Li et al.[90] introduced single-atom Ag into g-C_3N_4 as a low-cost and stable catalyst with higher activities than Ag nanoparticle decorated g-C_3N_4. They found excellent activity due to favorable Gibbs free energy of the adsorbed H atom (ΔG_H^*) and robust structure of the N–Ag bonding between the metal and the support.

As for electrochemical water splitting, graphene was always considered to be a good candidate for SACs. Bai et al.[61] reported a high-temperature annealing strategy to fabricate a highly dispersed ruthenium-based catalyst embedded in N-doped graphene. Their catalyst exhibited high trifunctional electrocatalytic activity and good stability in HER. The electrocatalyst exhibits a low overpotential of only 0.04 and 0.09 V at the current density of 10 mA·cm^{-2} for HER in 1.0 M KOH and 0.5 M H_2SO_4,

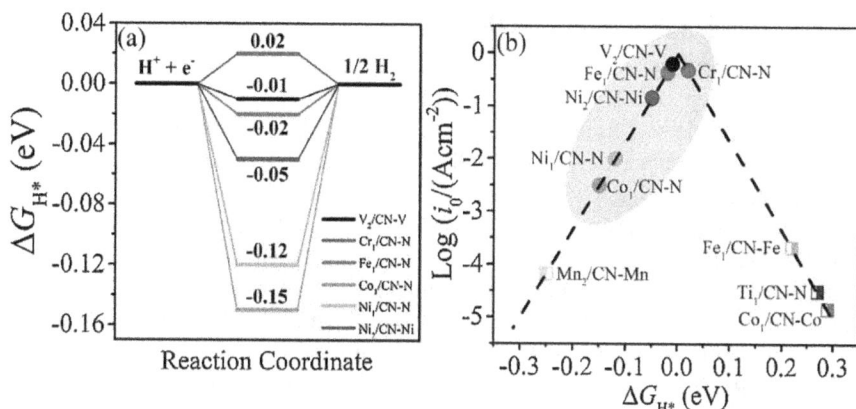

FIGURE 3.9 (a) Gibbs free change for the HER under standard conditions. (b) The HER volcano curve of exchange current (i_0) as a function of the ΔG_{H*} of the H adsorption on TM/g-CN.[91]

respectively. Structural characterizations show that RuN_4C_x is one of the main structures. DFT calculations indicated that the surface states of RuN_4C_x sites evolve in different reaction conditions.

On the other hand, carbon nitride materials are also widely used as substrates. Lv et al.[91] evaluated the catalytic performance of holey g-CN supported SACs (Ti, V, Cr, Mn, Fe, Co and Ni) toward HER, and the corresponding result is shown in Figure 3.9. Co_1/g-CN and Ni_1/g-CN are identified as the most efficient functional SACs for overall water splitting, capable of driving the HER with overpotentials being as low as 0.15 and 0.12 V, outperforming commercial Pt and IrO_2 catalysts. Remarkably, the d-band centers of TM atoms can act as an efficient descriptor for the interaction strength between intermediates and TM/g-CN, which can be tuned to further optimize the catalytic activity. Furthermore, Li et al.[92] investigated the thermodynamic performances during water splitting besides the structural and electronic performances of monolayer $g-C_3N_4$ embedded with single Ni, Pd, Pt, Cu, Ag or Au atom by DFT. Detailed analysis of the electronic redistribution and chemical bond relaxation suggested that the ratio of ionic and covalent bonding parts of chemical bonds could be adjusted by single atoms, and the corresponding free energy and overpotentials of the HER could be changed as well. Interestingly, the overpotential of the HER reduces as the percentage of covalent bonding parts is close to a certain value, which is conducive to improve the HER efficiency.

In addition to carbon material, MoS_2-based TM chalcogenides are also widely explored for the application of electrocatalytic H_2 production. The limited quantity of active sites and poor conductivity are generally believed to hamper the efficiency of H_2 production. Liu et al.[56] demonstrated that P dopants could be new active sites in the basal plane of MoS_2 and help improve the intrinsic electronic conductivity, leading to a significantly improved activity for hydrogen evolution. Furthermore, P-doped MoS_2 nanosheets show enlarged interlayer spacing, facilitating H adsorption and release progress.

3.3.4.2 Oxygen Evolution Reaction

Su et al.[93] reported a strategy to construct the $M_1N_4C_4$ (M = Fe, Co, Ni) structure within the N-doped holey graphene frameworks (NHGFs), and the well-defined $Ni_1N_4C_4$ moiety demonstrates to be a highly active and stable OER catalyst, which presents higher turnover frequency (TOF) values than the reported most active non-precious metal catalysts. Theoretical calculations pointed out that the OER performance of $M_1N_4C_4$ moiety strongly depends on the number of d-electrons of metal atoms. In addition, the OER catalytic behavior of graphene-based SACs is also related to the coordination number, which influences the adsorption strengths of reaction intermediates on the central metal atom.[94] On Ni-NHGF, the metal SA Ni is not the only active site, and the C site plays an important role in the initial reaction steps of OER. The proposed dual-site mechanism extends the understanding of the catalytic mechanisms on graphene-based SACs and provides new ideas for designing catalysts for reactions that require multiple active sites.

To increase the efficiency of the OER reaction, many works have been carried out. Hu et al.[30] used DFT calculations to investigate the OER catalytic activities of 3d single metal atoms coordinated by N atoms on carbon substrates. Among the atom/substrate combinations, the Co atom on the pyridine-N_4 substrate exhibits the lowest theoretical overpotential for the OER reaction. Furthermore, researchers found that the efficiency of the reaction can be greatly modified with a single metal atom and support. Shang et al.[60] rationally designed carbon-based materials named Mn-C_2N_2 to examine its OER catalytic property. By means of operando X-ray absorption fine structure measurements, and they showed that the high-valence Mn^{4+}-N_2C_2 moieties are the catalytic sites for OER, consistent with the related DFT results. The atomic and electronic synergistic effects for the Mn sites and the carbon support are the key for the high catalytic performance. Zhou et al.[89] studied a wide range of TM atoms (Mn, Fe, Co, Ni, Cu, Ru, Rh, Pd, Ir and Pt) embedded into the double carbon vacancy of C_3N monolayers, and the corresponding OER overpotentials for SACs were calculated by DFT. The best catalyst for the OER was Rh on the support with a small overpotential of 0.35 V, followed by Co (0.43 V).

Bai et al.[61] reported a high-temperature annealing strategy to fabricate a highly dispersed ruthenium-based catalyst embedded in N-doped graphene. The catalyst exhibits high-functional electrocatalytic activity and good stability in OER. The electrocatalyst exhibits a low overpotential of 0.372 V for OER in 1.0 M KOH, and their DFT calculations indicated that the surface states of RuN_4C_x sites evolve in different reaction conditions. Kan et al.[41] prepared a series of recombinant SACs by recombining Pt single atoms on 26 representative MXenes by first-principles calculations. They found that F-terminated ones could exhibit high efficiency for the OER.

As for the SAC-support, several works are focused on carbon materials. Li et al.[92] investigated the thermodynamic performance during water splitting besides the structural and electronic performances of monolayer g-C_3N_4 embedded with single Ni, Pd, Pt, Cu, Ag or Au atom. They found that the overpotential of the OER has a correlation with the covalent bonding. Lv et al.[91] found that Co_1/g-CN and Ni_1/g-CN are efficient SACs for overall water splitting with overpotentials of 0.61 and 0.40 V, respectively. The further introduction of defect could increase the stability of the single atom, with benefits to the reaction. Niu et al.[70] clarified that the TM atom supported on defective g-C_3N_4 with N vacancy could be rather stable, and some metal atoms could exhibit low overpotentials for the OER based on the volcano plots and contour maps. They suggested that the OER activity originates from the d-band center, and the number of d-orbital electrons multiplied electronegativity.

Bajdich et al.[95] studied OER on β-CoOOH surfaces to determine the activity trends at experimentally relevant electrochemical conditions. The calculated volume Pourbaix diagram shows that β-CoOOH is the active phase where the OER occurs in alkaline media. By comparing the theoretical overpotentials, they thought that the $(10\bar{1}4)$ surface is the most active one with an overpotential of $\eta = 0.48$ V.

3.3.5 Oxygen Reduction Reaction

The ORR is a pivotal factor for the chemical-electrical energy conversion in fuel cells because it determines the reaction kinetics. The electrocatalytic reduction of O_2 in aqueous electrolytes proceeds through one of two generally recognized pathways according to the acid-base properties, as summarized in Table 3.3.[96]

O_2 can be reduced by the four-electron (4e$^-$) pathway directly or by the two-electron (2e$^-$) process to form H_2O_2. The H_2O_2 intermediate may be further reduced to H_2O or OH$^-$ or undergo disproportionation to regenerate O_2. In the view of theoretical calculations, chemisorption of one O_2 molecule on the catalyst is the first necessary step for the ORR. In the case of the 4e- process, there are three possible reaction paths following the O_2 adsorption: (a) O_2 direct dissociation pathway, (b) OOH dissociation pathway, and (c) HOOH dissociation pathway.[97] The HOOH mechanism has been often ignored, despite the fact that it can sometimes provide superior ORR results than the other two mechanisms.[97]

To date, it is still a challenge to develop a highly efficient and stable nonprecious metal electrocatalyst to replace the Pt-based catalysts for the ORR. Gao et al.[98] systematically studied the relation between a series of TM SACs anchored on NC and the catalytic performance of H2O2 synthesis via the ORR. The corresponding results

---

TABLE 3.3

Electrocatalytic Reduction of O_2 in Different Aqueous Solution Including Acidic Media and Alkaline Media

Aqueous Condition	Reaction Process
Acidic media	$O_2 + 4H^+ + 4e^- \rightarrow 2H_2O$
	$O_2 + 2H^+ + 2e^- \rightarrow H_2O_2$
	$H_2O_2 + 2H^+ + 2e^- \rightarrow 2H_2O$
	$H_2O_2 \rightarrow 2H_2O + 1/2O_2$
Alkaline media	$O_2 + 2H_2O + 4e^- \rightarrow 4OH^-$
	$O_2 + H_2O + 2e^- \rightarrow HO_2^- + OH^-$
	$HO_2^- + HO_2^- + 2e^- \rightarrow 3OH^-$
	$HO_2^- \rightarrow OH^- + 1/2O_2^-$

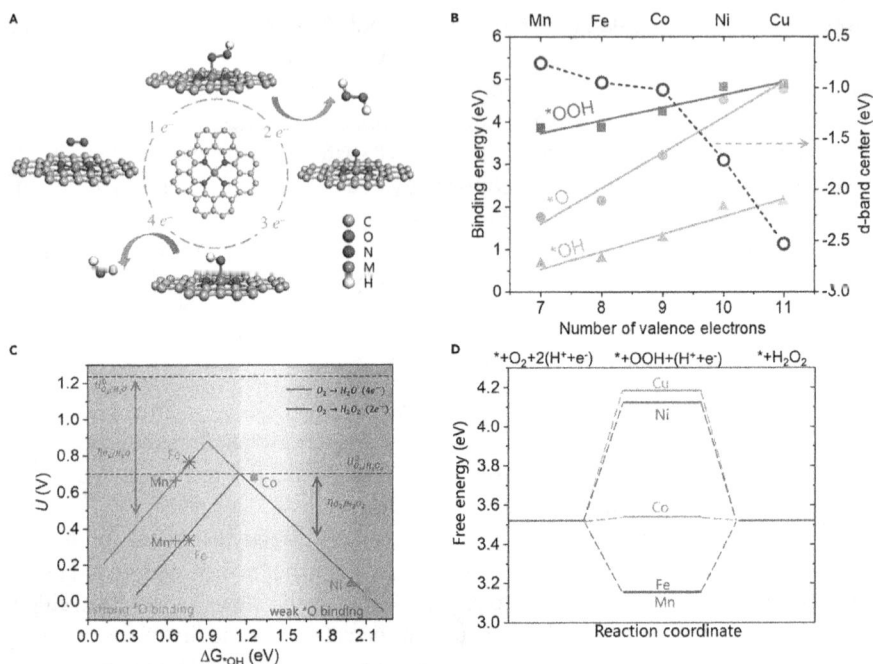

FIGURE 3.10 (a) Schematic of the ORR along the 2e- or 4e- pathway on TM SACs (TM = Mn, Fe, Co, Ni, and Cu) anchored in N-doped graphene. (b) Binding energy of *OOH, *O, and *OH on TM-SAC (TM = Mn, Fe, Co, Ni, and Cu), together with the d-band center (open circle) of Metal atoms in M-SAC (TM = Mn, Fe, Co, Ni, and Cu). (c) Activity-volcano curves of the ORR via the 2e- or 4e- pathway. The limiting potential is plotted as a function of ΔG_{*OH}. (d) Free energy diagrams of the 2e ORR on the SACs at $U = 0.7$ V versus RHE.[98]

are shown in Figure 3.10. It can be found that the Co SAC with the optimal binding energy of 3.54 eV is positioned nearly at the vertex of the activity-volcano map, suggesting that the Co SAC would be highly active for the 2e- pathway. Yang et al.[54] proposed a surfactant-assisted method to synthesize single-atom iron catalysts. Combining experimental results and DFT calculations, they revealed that the origin of high-ORR activity of SA-Fe/NG is from the Fe-pyrrolic-N species. Furthermore, Gao et al.[68] investigated Fe-graphdiyne catalysts, showing high catalytic activity. DFT predicted that the O_2 prefers to bind with a Fe atom, and the ORR process follows the 4e- pathway. To validate the theoretical predictions, the Fe-graphdiyne catalyst was then synthesized. The following electrochemical measurements indicated that the Fe-graphdiyne catalyst facilitates the 4e- ORR while limiting the 2e- transfer reaction, consistent with their DFT predictions. Except for Fe single atom, Ying et al.[57] reported a group of SACs supported on C_2N monolayer as promising ORR catalysts by theoretical calculations. They showed that Au and Pd@C_2N are superior ORR catalysts with the overpotential of 0.38 and 0.40 V. Hu et al.[30] investigated the ORR catalytic activities of 3d single metal atoms coordinated by N atoms on carbon substrates by DFT. They found that the Co atom on the pyridine-N_4 substrate exhibits the lowest theoretical overpotentials for ORR reactions. Zhang et al.[18] developed a template-assisted method to synthesize a series of single metal atoms anchored on porous N, S-codoped carbon matrix as highly efficient ORR catalysts to investigate the correlation between the structure and their catalytic performance. The structure analysis indicates that an identical synthesis method results in distinguished structural differences between Fe-centered SAC and Co-centered/Ni-centered SACs because of the different trends of each metal ion in forming a complex with the N, S-containing precursor during the initial synthesis process. Shang et al.[18,60] designed a carbon-based $Mn-N_2C_2$, and they showed a half-wave potential of 0.92 V versus reversible hydrogen electrode for ORR.

3.4 CONCLUSION

In summary, SACs have recently emerged as promising materials for catalysis and have accordingly attracted considerable interest from various catalytic reaction fields. Compared to conventional catalysts, atomically dispersed SACs behave as homogeneous catalysts while retaining the advantages of heterogeneous catalysts, which bring about unique properties and opportunities for application to different classes of reactions. In this chapter, we have provided a systematic review of recent research progress in investigations of SACs on different substrates mainly from the theoretical side. We have discussed the stability, electronic structure, and activity of SACS for the considered reactions, along with the related computational methods and theory, focusing on the correlation between the electronic structure of SACs and their catalytic activity. Recent theoretical investigations of SACs suggest clearly that the superior activity of some SACs originates from their peculiar electronic structure. Yet, the general factors affecting SACs activity remain unclear and far from fully understood. Since it remains rather challenging to establish such dominant factors for complex reactions either from the experimental or from the theoretical side alone, better combination and integration of more advanced in-situ characterization and computational approaches are urgently needed in the near future for progress in the field to be accelerated. Additional integration of machine learning and database approaches for research in the field is also expected to sustain improved understanding of and advances toward the practical application and large-scale uptake of SACs for relevant chemical reactions.

ACKNOWLEDGMENTS

This work was supported by the National Natural Science Foundation of China (11974037). L.M.L. and G. T. acknowledge support by the Royal Society Newton Advanced Fellowship scheme (grant No. NAF\R1\180242).

REFERENCES

1. K. Zhang, et al., *Journal of the American Chemical Society,* 2020, **142**, 17499–17507.
2. F. Li, L. Chen, M. Xue, T. Williams, Y. Zhang, D. R. MacFarlane, and J. Zhang, *Nano Energy,* 2017, **31**, 270–277.
3. C. Guo, et al., *Physical Review Letters,* 2020, **124**, 206801.

4. D. Dey and A. S. Botana, *Physical Review Materials*, 2020, **4**, 074002.
5. E. S. Sanz-Perez, C. R. Murdock, S. A. Didas, and C. W. Jones, *Chemical Reviews*, 2016, **116**, 11840–11876.
6. T. Inoue, A. Fujishima, S. Konishi, and K. Honda, *Nature*, 1977, **277**, 637–638.
7. W. J. Ong, L. L. Tan, Y. H. Ng, S. T. Yong, and S. P. Chai, *Chemical Reviews*, 2016, **116**, 7159–7329.
8. K. Wenderich and G. Mul, *Chemical Reviews*, 2016, **116**, 14587–14619.
9. W.-J. Yin, B. Wen, C. Zhou, A. Selloni, and L.-M. Liu, *Surface Science Reports*, 2018, **73**, 58–82.
10. B. Qiao, A. Wang, X. Yang, L. F. Allard, Z. Jiang, Y. Cui, J. Liu, J. Li, and T. Zhang, *Nature Chemistry*, 2011, **3**, 634–641.
11. M. Flytzani-Stephanopoulos and B. C. Gates, *Annual Review of Chemical and Biomolecular Engineering*, 2012, **3**, 545–574.
12. A. Zitolo, et al., *Nature Communications*, 2017, **8**, 957.
13. Q. Liu, X. Liu, L. Zheng, and J. Shui, *Angewandte Chemie International Edition in English*, 2018, **57**, 1204–1208.
14. H. Zhang, P. An, W. Zhou, B. Y. Guan, P. Zhang, J. Dong, and X. W. Lou, *Science Advances*, 2018, **4**, eaao6657.
15. Q. Li, et al., *Advanced Materials*, 2018, **30**, 1800588.
16. Y. Qiao, et al., *Advanced Materials*, 2018, **30**, 1804504.
17. Y. Hou, et al., *Nature Communications*, 2019, **10**, 1392.
18. J. Zhang, et al., *Journal of the American Chemical Society*, 2019, **141**, 20118–20126.
19. H. Y. Zhuo, X. Zhang, J. X. Liang, Q. Yu, H. Xiao, and J. Li, *Chemical Reviews*, 2020, **120**, 12315–12341.
20. R. Gusmão, M. Veselý, and Z. Sofer, *ACS Catalysis*, 2020, **10**, 9634–9648.
21. X. Wang, Z. Li, Y. Qu, T. Yuan, W. Wang, Y. Wu, and Y. Li, *Chem*, 2019, **5**, 1486–1511.
22. Z. Kou, W. Zang, P. Wang, X. Li, and J. Wang, *Nanoscale Horizons*, 2020, **5**, 757–764.
23. X.-F. Yang, A. Wang, B. Qiao, J. Li, J. Liu, and T. Zhang, *Accounts of Chemical Research*, 2013, **46**, 1740–1748.
24. Y. Zhai, et al., *Science*, 2010, **329**, 1633.
25. S. Sahu and D. P. Goldberg, *Journal of the American Chemical Society*, 2016, **138**, 11410–11428.
26. M. Yang, L. F. Allard, and M. Flytzani-Stephanopoulos, *Journal of the American Chemical Society*, 2013, **135**, 3768–3771.
27. M.-J. Cheng, E. L. Clark, H. H. Pham, A. T. Bell, and M. Head-Gordon, *ACS Catalysis*, 2016, **6**, 7769–7777.
28. Y. Feng, et al., *The Journal of Physical Chemistry C*, 2018, **122**, 22460–22468.
29. Z. Lu, Y. Cheng, S. Li, Z. Yang, and R. Wu, *Applied Surface Science*, 2020, **528**, 147047.
30. M. Hu, S. Li, S. Zheng, X. Liang, J. Zheng, and F. Pan, *The Journal of Physical Chemistry C*, 2020, **124**, 13168–13176.
31. C. Ren, Q. Jiang, W. Lin, Y. Zhang, S. Huang, and K. Ding, *ACS Applied Nano Materials*, 2020, **3**, 5149–5159.
32. Q. Cui, G. Qin, W. Wang, K. R. Geethalakshmi, A. Du, and Q. Sun, *Applied Surface Science*, 2020, **500**, 143993.
33. A. Wang, J. Li, and T. Zhang, *Nature Reviews Chemistry*, 2018, **2**, 65–81.
34. Q. Feng, et al., *Journal of the American Chemical Society*, 2017, **139**, 7294–7301.
35. F. R. Lucci, J. Liu, M. D. Marcinkowski, M. Yang, L. F. Allard, M. Flytzani-Stephanopoulos, and E. C. H. Sykes, *Nature Communications*, 2015, **6**, 8550.
36. Y. Tang, S. Zhao, B. Long, J.-C. Liu, and J. Li, *The Journal of Physical Chemistry C*, 2016, **120**, 17514–17526.
37. Y. Tang, Y.-G. Wang, and J. Li, *The Journal of Physical Chemistry C*, 2017, **121**, 11281–11289.

38. J.-C. Liu, Y.-G. Wang, and J. Li, *Journal of the American Chemical Society*, 2017, **139**, 6190–6199.
39. Y. Zhu, et al., *ACS Applied Materials and Interfaces*, 2019, **11**, 32887–32894.
40. Q. Fu and C. Draxl, *Physical Review Letters*, 2019, **122**, 046101.
41. D. Kan, R. Lian, D. Wang, X. Zhang, J. Xu, X. Gao, Y. Yu, G. Chen, and Y. Wei, *Journal of Materials Chemistry A*, 2020, **8**, 17065–17077.
42. S. Back, J. Lim, N.-Y. Kim, Y.-H. Kim, and Y. Jung, *Chemical Science*, 2017, **8**, 1090–1096.
43. J. C. Liu, Y. G. Wang, and J. Li, *Journal of the American Chemical Society*, 2017, **139**, 6190–6199.
44. Y.-G. Wang, D. C. Cantu, M.-S. Lee, J. Li, V.-A. Glezakou, and R. Rousseau, *Journal of the American Chemical Society*, 2016, **138**, 10467–10476.
45. Y. He, J.-C. Liu, L. Luo, Y.-G. Wang, J. Zhu, Y. Du, J. Li, S. X. Mao, and C. Wang, *Proceedings of the National Academy of Sciences*, 2018, **115**, 7700.
46. J. Fonseca and J. Lu, *ACS Catalysis*, 2021, **11**, 7018–7059.
47. J. Li, X. Li, H.-J. Zhai, and L.-S. Wang, *Science*, 2003, **299**, 864.
48. L. Liu and A. Corma, *Chemical Reviews*, 2018, **118**, 4981–5079.
49. D. Buceta, Y. Piñeiro, C. Vázquez-Vázquez, J. Rivas, and M. A. López-Quintela, *Catalysts*, 2014, **4**, 356–374.
50. P. Hu, et al., *Angewandte Chemie International Edition in English*, 2014, **53**, 3418–3421.
51. N. Shibata, T. Seki, G. Sánchez-Santolino, S. D. Findlay, Y. Kohno, T. Matsumoto, R. Ishikawa, and Y. Ikuhara, *Nature Communications*, 2017, **8**, 15631.
52. N. Ramaswamy and S. Mukerjee, *Advances in Physical Chemistry*, 2012, **2012**, 491604.
53. L. Cao, et al., *Nature*, 2019, **565**, 631–635.
54. L. Yang, D. Cheng, H. Xu, X. Zeng, X. Wan, J. Shui, Z. Xiang, and D. Cao, *Proceedings of the National Academy of Sciences*, 2018, **115**, 6626.
55. C. Choi, S. Back, N.-Y. Kim, J. Lim, Y.-H. Kim, and Y. Jung, *ACS Catalysis*, 2018, **8**, 7517–7525.
56. P. Liu, J. Zhu, J. Zhang, P. Xi, K. Tao, D. Gao, and D. Xue, *ACS Energy Letters*, 2017, **2**, 745–752.
57. Y. Ying, K. Fan, X. Luo, J. Qiao, and H. Huang, *Journal of Materials Chemistry A*, 2021, **9**, 16860–16867.
58. J.-X. Liang, J. Lin, X.-F. Yang, A.-Q. Wang, B.-T. Qiao, J. Liu, T. Zhang, and J. Li, *The Journal of Physical Chemistry C*, 2014, **118**, 21945–21951.
59. Z. Fu, B. Yang, and R. Wu, *Physical Review Letters*, 2020, **125**, 156001.
60. H. Shang, et al., *Nano Letters*, 2020, **20**, 5443–5450.
61. L. Bai, Z. Duan, X. Wen, R. Si, Q. Zhang, and J. Guan, *ACS Catalysis*, 2019, **9**, 9897–9904.
62. Y. Zhou, G. Gao, J. Kang, W. Chu, and L.-W. Wang, *Journal of Materials Chemistry A*, 2019, **7**, 12050–12059.
63. S. Back and Y. Jung, *ACS Energy Letters*, 2017, **2**, 969–975.
64. Y. Qian, Y. Liu, Y. Zhao, X. Zhang, and G. Yu, *EcoMat*, 2020, **2**, 1.
65. G. Gao, Y. Jiao, E. R. Waclawik, and A. Du, *Journal of the American Chemical Society*, 2016, **138**, 6292–6297.
66. Y. Pan, et al., *Journal of the American Chemical Society*, 2018, **140**, 4218–4221.
67. C. Xu, X. Zhi, A. Vasileff, D. Wang, B. Jin, Y. Jiao, Y. Zheng, and S.-Z. Qiao, *Small Structures*, 2020, **2**, 2000058.
68. Y. Gao, Z. Cai, X. Wu, Z. Lv, P. Wu, and C. Cai, *ACS Catalysis*, 2018, **8**, 10364–10374.
69. T. Kropp and M. Mavrikakis, *ACS Catalysis*, 2019, **9**, 6864–6868.
70. H. Niu, X. Wan, X. Wang, C. Shao, J. Robertson, Z. Zhang, and Y. Guo, *ACS Sustainable Chemistry & Engineering*, 2021, **9**, 3590–3599.
71. F. Li, Y. Li, X. C. Zeng, and Z. Chen, *ACS Catalysis*, 2014, **5**, 544–552.

72. V. P. Indrakanti, J. D. Kubicki, and H. H. Schobert, *Energy & Environmental Science*, 2009, **2**, 745–758.
73. S. Zhou, et al., *Applied Catalysis B*, 2014, **20**, 158–159.
74. R. T. Tung, *Applied Physics Reviews*, 2014, **1**, 011304.
75. H. Xie, et al., *Applied Catalysis B: Environmental*, 2021, **289**, 119783.
76. Y.-H. Lu, M. Zhou, C. Zhang, and Y.-P. Feng, *The Journal of Physical Chemistry C*, 2009, **113**, 20156–20160.
77. M. Moses-DeBusk, M. Yoon, L. F. Allard, D. R. Mullins, Z. Wu, X. Yang, G. Veith, G. M. Stocks, and C. K. Narula, *Journal of the American Chemical Society*, 2013, **135**, 12634–12645.
78. B. Long, Y. Tang, and J. Li, *Nano Research*, 2016, **9**, 3868–3880.
79. W. Song, Y. Su, and E. J. M. Hensen, *The Journal of Physical Chemistry C*, 2015, **119**, 27505–27511.
80. Z. Lu, M. Yang, D. Ma, P. Lv, S. Li, and Z. Yang, *Applied Surface Science*, 2017, **426**, 1232–1240.
81. P. Nematollahi and E. C. Neyts, *Applied Surface Science*, 2018, **439**, 934–945.
82. M. H. Butt, S. H. M. Zaidi, Nabeela, A. Khan, K. Ayub, M. Yar, M. A. Hashmi, M. A. Yawer, and M. A. Zia, *Molecular Catalysis*, 2021, **509**, 111630.
83. B. Han, L. Ling, M. Fan, P. Liu, B. Wang, and R. Zhang, *Applied Surface Science*, 2019, **479**, 1057–1067.
84. A. Akça, O. Karaman, C. Karaman, N. Atar, and M. L. Yola, *Surfaces and Interfaces*, 2021, **25**, 101293.
85. H. Fei, J. Dong, D. Chen, T. Hu, X. Duan, I. Shakir, Y. Huang, and X. Duan, *Chemical Society Reviews*, 2019, **48**, 5207–5241.
86. C. Liu, Q. Li, J. Zhang, Y. Jin, D. R. MacFarlane, and C. Sun, *Journal of Materials Chemistry A*, 2019, **7**, 4771–4776.
87. X.-F. Li, et al., *Journal of the American Chemical Society*, 2016, **138**, 8706–8709.
88. X. Guo and S. Huang, *Electrochimica Acta*, 2018, **284**, 392–399.
89. Y. Zhou, G. Gao, Y. Li, W. Chu, and L.-W. Wang, *Physical Chemistry Chemical Physics*, 2019, **21**, 3024–3032.
90. X. Li, et al., *Applied Catalysis B: Environmental*, 2021, **283**, 119660.
91. X. Lv, W. Wei, H. Wang, B. Huang, and Y. Dai, *Applied Catalysis B: Environmental*, 2020, **264**, 118521.
92. H. Li, Y. Wu, L. Li, Y. Gong, L. Niu, X. Liu, T. Wang, C. Sun, and C. Li, *Applied Surface Science*, 2018, **457**, 735–744.
93. J. Su, R. Ge, Y. Dong, F. Hao, and L. Chen, *Journal of Materials Chemistry A*, 2018, **6**, 14025–14042.
94. G. Gao, S. Bottle, and A. Du, *Catalysis Science & Technology*, 2018, **8**, 996–1001.
95. M. Bajdich, M. Garcia-Mota, A. Vojvodic, J. K. Norskov, and A. T. Bell, *Journal of the American Chemical Society*, 2013, **135**, 13521–13530.
96. J. Masa, W. Xia, M. Muhler, and W. Schuhmann, *Angewandte Chemie International Edition*, 2015, **54**, 10102–10120.
97. M. Yan, Z. Dai, S. Chen, L. Dong, X. L. Zhang, Y. Xu, and C. Sun, *The Journal of Physical Chemistry C*, 2020, **124**, 13283–13290.
98. J. Gao, et al., *Chem*, 2020, 6, 658–674.

4 Fuel Cells Application of Atomically Dispersed Metallic Materials

Yuen Wu
University of Science and Technology of China

Yunteng Qu
Institute of Photonics and Photon-Technology, Northwest University

CONTENTS

DOI: 10.1201/9781003153436-4

4.1   INTRODUCTION

Advances in science and technology have been hampered by energy and environmental concerns. The excessive use of fossil fuels, such as coal and petroleum, as energy sources may increase human productivity while destroying the environment. By 2035, the International Energy Agency (IEA) estimates that global energy consumption will rise to 18 billion tons, and CO_2 emissions will reach 43 gigatonnes per year, resulting in an increase in the world's average temperature and ocean acidification.[1] At the 2019 United Nations Climate Summit, 66 nations committed to achieve carbon neutrality by 2050. As a result, finding new sources of clean energy to replace fossil fuels is a top priority. The zero emissions and reasonable production costs of hydrogen derived from renewable energy make it a viable option.[2,3] Proton-exchange membrane fuel cells (PEMFCs) that utilize hydrogen-oxygen or hydrogen-air as a fuel source have competitive benefits over other hydrogen energy sources, such as high efficiency, quick recharging, and an acceptable working temperature range (80°C –120°C). When hydrogen is oxidized on the anode and oxygen is reduced on the cathode in a conventional hydrogen-oxygen fuel-cell system, electric energy is produced with only clean water as a by-product. In contrast, the oxygen reduction reaction (ORR) at the cathode includes a complex multi-electron and multi-proton coupling mechanism, resulting in slow reaction kinetics, such as difficulties of O_2 activation, O–O bond breakage, and oxide elimination. The hostile electrochemical (high potential) and chemical environment at a cathode, on the other hand, will make catalysts less stable in the long run. For this reason, cathodic catalysts need high standards in terms of ORR activity and long-term robustness.[4]

ORR catalysts often use platinum-based nanoparticles supported on carbon substrates because of their high efficiency and long-term stability.[5] However, the scarcity and high cost of Pt severely limit the use of PEMFCs on a wide scale in the future. With a growing demand for PEMFCs, the breakdown of fuel-cell costs is expected to vary in the next few decades (see Figure 4.1). Fuel-cell stacks now cost a lot, but this is expected to reduce significantly as manufacturing rises. A larger manufacturing rate, on the other hand, will just increase the catalyst cost fraction, which is already rather high.[6] Therefore, it has been generally considered a potential method to create dependable non-noble metal catalysts that can completely replace platinum-based catalysts. PEMFC practicality hinges on the development of low-cost, high-reliability ORR catalysts.[7,8] Nonprecious metal–nitrogen–carbon composite (M–N–C), nonprecious metal oxides, chalcogenides, and oxynitride have all been studied as Pt-based catalyst substitutes in fuel cells during the last several decades.

In particular, nonprecious metal–nitrogen–carbon composites (M–N–C) have been identified as the most promising composites owing to their high ORR activity and long-term stability. The first use of M–N–C composite as ORR catalysts can be traced back to the study of Jasinski,[9] in which cobalt phthalocyanine was discovered to display ORR catalytic activity. Nevertheless, these macrocyclic compounds showed substantially lower activity and stability than Pt-based catalysts, particularly in acidic conditions. Then, enormous efforts have been dedicated to creating novel M–N–C ORR catalysts via the pyrolysis of mixture consisting of non-noble metal, carbon and nitrogen-containing precursors. In 2009, the Dodelet group reported

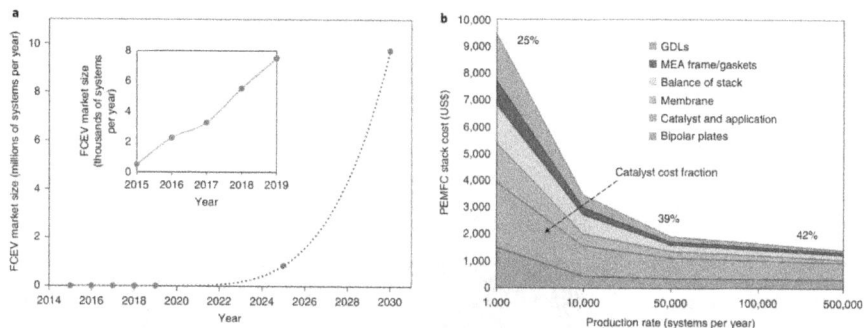

FIGURE 4.1 Evolution of market size for fuel-cell electrical vehicles (FCEVs), and cost breakdown of FCEVs alongside production rate. (a) Predicted market for FCEVs; the market size up to 2019 is shown in the inset. (b) Cost breakdown of fuel-cell stack components as a function of production rate. PEMFC stack components—GDL, MEA frame, balance of stack, membrane, catalyst and application, and bipolar plates are all included. (The scenario data of panel (a) and (b) are Reproduced from the study by Jiantao Fan et al.[6] with permission from Springer Nature.)

iron-nitrogen-carbon catalysts with improved ORR activity.[10] A combination of ferrous acetate, carbon black, and phenanthroline was pyrolyzed in argon and ammonia sequentially after ball-mill. The developed material was used in polymer electrolyte fuel cells as cathode catalysts and was shown to display enhanced intrinsic ORR activity, which is competitive compared with Pt-based catalysts. Subsequently, this group developed another Fe-based catalyst by utilizing ZIF-8 as a microporous host for phenanthroline and ferrous acetate.[11] Compared with the previous one, this catalyst produced an apparent increase in ORR activity in H_2–O_2 fuel cells with a power density of 0.75 W cm^{-2} at 0.6 V (Figure 4.2a), which is near to the goal set by the US DOE (300 A cm^{-3} for non-PGM-based catalysts, Figure 4.2b). However, these as-prepared catalysts failed miserably when used in PEMFCs, with performance dropping by 50% after only 100 h. In 2011, Wu et al. used polyaniline (PANI) as a carbon-nitrogen precursor to create nonprecious metal catalysts containing iron and cobalt contained in multiple C–N shells.[12] Compared to commercial Pt/C, their best catalyst had an increased ORR activity (Figure 4.2c). More significantly, in long-term rotating disk electrode (RDE) and fuel-cell testing, these catalysts proved to be more durable. In addition, the Bao team also noticed a similar kind of issue.[13] Fe nanoparticles were enclosed in pea-pod-like carbon nanotubes (Figure 4.2d), and they discovered that this confinement structure may provide improved endurance during long-term fuel-cell tests. There was an idea put forward that the Fe cluster might become ORR active species by transferring a charge to its neighboring carbon atoms. This would allow ORR to occur without the need for direct contact between O_2 and Fe atoms (Figure 4.2e and f). It is possible that the carbon shells are protecting the iron clusters from acid corrosion, which explains why they are more durable. However, the ORR activity of Fe nanoparticles in encapsulation is still lacking. An H_2–O_2 fuel cell with encapsulated Fe nanoparticles had a voltage of about 0.55 V

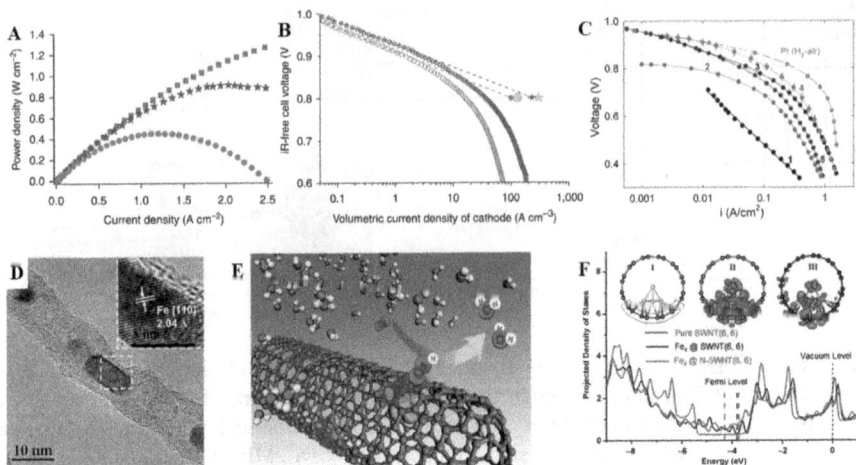

FIGURE 4.2 Investigation of representative M–N–C catalysts. (a) Power density curves of as-prepared catalyst (blue stars), a catalyst reported previously (red circles) and Pt/C reference (green squares). (Reproduced from the study by Eric Proietti et al.[11] with permission from Springer Nature.) (b) Volumetric activity (hollow stars or circles) and extrapolated Tafel plots (dash line) of as-prepared catalyst (blue) and a catalyst reported previously (red). Solid gray circle and star represent US DOE volumetric activity targets for the year 2010 and 2015, respectively. (Reproduced from the study by Eric Proietti et al.[11] with permission from Springer Nature.) (c) H_2–O_2 fuel-cell performance of 1, metal-free PANI-C, 2, PANI-Co-C, 3, PANI-FeCo-C (1), 4, PANI-FeCo-C (2), 5, PANI-Fe-C and Pt-based catalyst (gray dash line). (Reproduced from the study by Gang Wu et al.[12] with permission from the American Association for the Advancement of Science.) (d) TEM images of encapsulated Fe nanoparticles. (Reproduced from the study by Dehui Deng et al.[13] with permission from John Wiley & sons.) (e) Schematic illustration of the ORR at the surface of Fe_4@SWNT. (Reproduced from the study by Dehui Deng et al.[13] with permission from John Wiley & sons.) (f) Projected density of state of the p orbitals of C atoms. The charge transfer in inset II and III is represented by red for increasing and blue for deceasing. (Reproduced from the study by Dehui Deng et al.[13] with permission from John Wiley & sons.)

when fed with current density of 0.10 A cm^{-2}, which is only about 60% of the voltage produced by Pt/C. Despite significant efforts, M–N–C catalysts still fall short of replacing Pt/C due to their low ORR intrinsic activity and stability.

Increasing intrinsic activity and the number of active sites in a catalytic reaction system may significantly increase the reaction rate. Due to limitations in characterization methods, it's difficult to pinpoint exactly where M–N–C catalysts' ORR active sites are. As a result, efforts to improve ORR catalytic activity are hampered. Zhang et al. discovered that the atomically scattered platinum atoms supported on iron oxide (Pt_1/FeO_x) showed excellent CO oxidation performance when using synchrotron radiation and spherical aberration electron microscopy methods.[14] The Pt single atoms served as the real reaction sites. The atomically distributed Fe or Co atoms supported on nitrogen-doped carbon supports developed by Frédéric Jaouen et al. were inspired by this phenomenon and showed similar ORR activity with Pt/C catalysts in RDE experiments when tested.[15] The real ORR active sites of Fe/Co--N–C catalysts are Fe or Co single atoms with Fe/Co–N_4–C_{12} structures, according to X-ray absorption near-edge spectroscopy (XANES) research. Furthermore, these results also shed insight on the synthesis methods for M–N–C single-atom catalysts from the bottom up and highlight the critical function of metal-support interaction in ORR catalysis.

Then, significant efforts were spent to investigating the ORR properties of extremely active single-atom catalysts. Generally, three approaches are used: (i) increasing the intrinsic activity of individual active metal sites in single-atom catalysts; (ii) increasing the loading of metal active sites on supports; and (iii) increasing accessibility by constructing porous or other large structures with a specific surface area. To summarize, in this chapter, we look back at the creation of single-atom catalysts for ORR and show how scientists may improve their catalytic performance by tweaking their component and load parameters as well as electronic structure and support. Significant progress has been made in the design of SACs for ORR as well as their general qualities and unique characteristics. This context also discusses how reasonably designing and constructing MEA to achieve enough excellent performance, such as activity, stability, and durability, eventually propelling SACs assembled PEMFCs into a commercial product. Finally, we look forward to the future of single-atom catalysts and the possibilities they hold.

4.2 ELECTROCATALYTIC ORR ACTIVITY AND PEMFC PERFORMANCE ON SINGLE-ATOM CATALYSTS

Recently, a variety of single-atom catalysts have been developed to effectively catalyze ORR, including non-noble single-atom catalysts (Fe, Co, Ni, Cu, Zn, Mn, Cr, W, Sn, Sb, Mg, etc.) and noble single-atom catalysts (Pt, Ir, Ru, Rh etc.). As practical ORR in PEMFC necessitates the use of single-atom catalysts with high catalytic activity in a given geometric area. This raises three problems. As a result of the high specific surface energy of single-metal atoms, improving metal loading in SACs is one challenge while promoting metal-site accessibility to improve mass transfer in terms of oxygen and water is another. A third challenge is how to improve the intrinsic activity of individual active metal sites. SAC ORR performance may be improved using one of three methods. Creating suitable supports with a bigger particular area can help anchor or expose more single-metal sites, which will in turn supply ORR with more active sites. The second method is to create porous channels with the increased capacity for transporting O_2 and H_2O, which will improve the PEMFC's overall performance. To maximize the inherent catalytic activity, the third strategy is to modify the electronic structure of metal sites by adjusting the coordination environment or doping heteroatoms. In this section, we review the most recent developments in ORR catalysis using SACs and show how scientists are working to solve the problems described above by building rational supports, controlling the regional coordination environment, or doping heteroatoms.

4.2.1 Fe-Based Atom Electrocatalysts for ORR

For Fe–N$_x$/C SACs to function well, the rational design of the synthetic approach is essential for preventing the migration/agglomeration of transition metals while increasing the number of active sites that may be accessed. The adsorption-pyrolysis approach, which includes the adsorption of metal precursors, reduction, and stabilization on defect-rich nitrogen-doped carbon supports, is the most commonly used method for building Fe single-atom catalysts. According to Zelenay et al.,[16] polyaniline, cyanamide, and ferric chloride (III) were pyrolyzed to create an atomically distributed Fe–N–C catalyst, as shown in Figure 4.3a. An electrocatalyst made of single Fe atom produced current densities equal to those of a catalyst containing one active Pt atom per centimeter squared (Figure 4.3b). As exhibited in Figure 4.3c, the Fe SAC displayed superior PEMFC performance, which may be ascribed to the more accessible atomically distributed Fe–N$_4$ active sites as well as a more open framework for enhancing catalytic layer ionomer dispersion. To build Fe-doped zeolitic imidazolate frameworks (ZIF) precursors, Wu and Shao used an ionic exchange technique as a chemical replacement.[17] Fe ions may partly replace Zn ions in this ZIF system and create FeN$_4$ complexes by bonding with imidazolate ligands in 3D frameworks (Figure 4.3d). With a half-wave potential of 0.85 V, the as-prepared Fe catalyst demonstrated good ORR activity in acidic environments (Figure 4.3e), as well as improved stability with a decrease in half-wave potential of just 20 mV after 10,000 cycles of operation (Figure 4.3f). The Fe SAC assembled PEMEC achieved the maximum power density values of 0.94 W cm^{-2} when p$_{O2}$ reached 2.0 bar, which is comparable to commercial Pt/C catalyst. In 2017, Li et al. devised a cage-encapsulated-precursor pyrolysis method to produce an isolated Fe SAC with good ORR performance (Figure 4.3g).[18] Because of its unique pore size and cavity, ZIF-8 was utilized in this study as a molecular cage to isolate and encapsulate the metal precursor Fe(acac)$_3$. Pyrolysis transformed ZIF-8 into nitrogen-doped carbon porous and decreased Fe(acac)$_3$, resulting in isolated Fe sites anchored on N species on the porous carbon. The as-prepared Fe SAC catalyst has a higher Fe loading of 2.16 wt% and is more efficient for ORR in an alkaline medium compared with commercial Pt/C, with a half-wave potential of 0.900 V and a kinetics current density (J_k) of 37.83 mA cm^{-2} at 0.85 V. There is also confirmed that atomically dispersed Fe–N–C electrocatalysts, pyrolyzed by polymer precursor, have superior ORR performance, according to Dai's group.[19] The metallosupramolecular polymer precursor (Fe–TAC) was synthesized by co-ordinating a three-armed catechol (TAC) monomer with either Fe^{3+} or Fe^{2+}, and then pyrolyzing it in an Ar atmosphere. A second annealing was used to produce the final target product (Fe–N/C), after which the superfluous Fe species were removed using an acid etching procedure. Chen's team used the pyrolysis technique to support atomically distributed Fe on N-doped carbon (Fe-NCC), which had good ORR performance in alkaline conditions (Eonset: 0.93 V vs RHE, half-wave potential: 0.82 V vs RHE).[20] When probing the Fe–NCC structure, it's apparent that the elements iron (Fe), nitrogen (N) and carbon (C) were evenly distributed across the whole framework. As an air electrode electrocatalyst, this atomically distributed Fe catalyst provided a high discharge voltage (1.21 V versus RHE) and specific capacity (705 mAh g^{-1} at 5 mA cm^{-2}) due to the porous structure and production of Fe–N$_x$

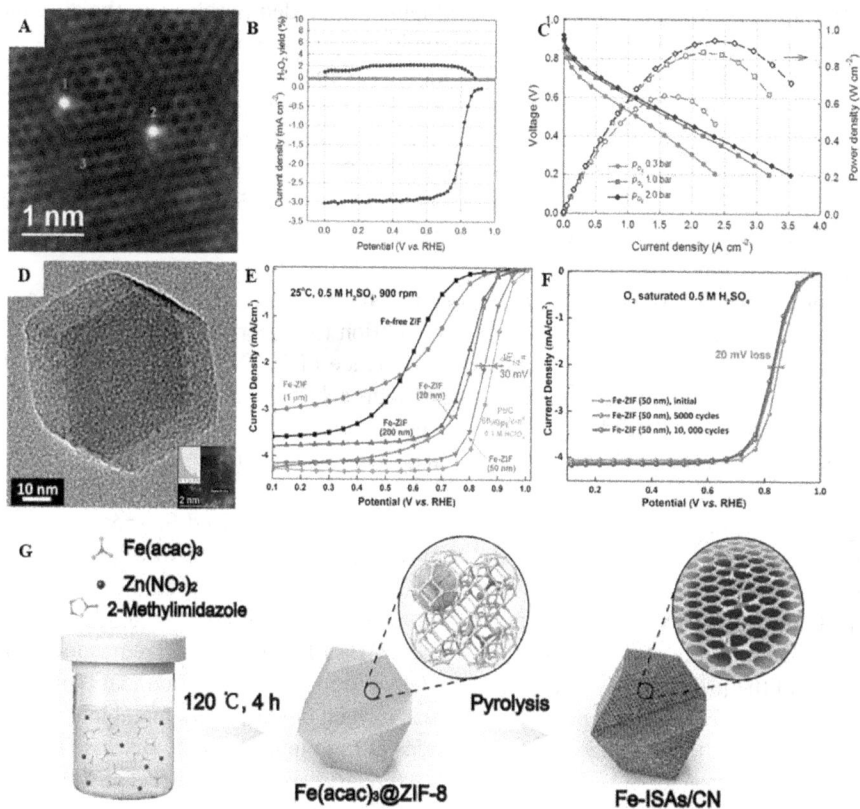

FIGURE 4.3 (a) HAADF-STEM image of individual Fe atoms (labeled 1, 2, and 3) in a few layer graphene sheet. (Reproduced from the study by Hoon T. Chung et al.[16] with permission from the American Association for the Advancement of Science.) (b) ORR performance of Fe–C catalyst. Steady-state RDE polarization plots were obtained by using a 20 mV potential step and 25 s potential hold time at every step. Electrolyte 0.5 M H_2SO_4, temperature 25°C ± 1°C, rotation rate 900 rpm. (Reproduced from the study by Hoon T. Chung et al.[16] with permission from the American Association for the Advancement of Science.) (c) H_2–O_2 fuel-cell polarization plots. Cathode: ~4.0 mg cm^{-2} of (CM + PANI)-Fe-C; O_2 200 mL min^{-1} (40 mL $min^{-1} cm^{-2}$); 100% RH; 0.3, 1.0, and 2.0 bar partial pressures. Anode: 2.0 mg_{Pt} cm^{-2} Pt/C; H_2 200 mL min^{-1}; 100% RH; 1.0 bar partial pressure. Membrane Nafion 211, cell 80°C; 5 cm^2 electrode area. (Reproduced from the study by Hoon T. Chung et al.[16] with permission from the American Association for the Advancement of Science.) (d) Fe-doped ZIF nanocrystal precursors (50 nm). (Reproduced from the study by Gang Wu et al.[17] with permission from American Chemical Society.) (e) H_2SO_4 ORR polarization plots for Fe-ZIF-derived catalysts in 0.5 M H_2SO_4 and Pt/C catalysts (60 $\mu g_{Pt}/cm^2$) in 0.1 M $HClO_4$ at 25°C and 900 rpm. (Reproduced from the study by Gang Wu et al.[17] with permission from American Chemical Society.) (f) Stability AST by cycling the potential (0.6–1.0 V) in O_2 saturated 0.5 M. (Reproduced from the study by Gang Wu et al.[17] with permission from American Chemical Society.) (g) Schematic illustration of the formation of Fe-ISAs/CN. (Reproduced from the study by Yuanjun Chen et al.[18] with permission from John Wiley & sons.)

active species. Xu and coworkers utilized template-guided pyrolysis in the alkaline medium to manufacture single iron atom on N-doped carbon with an atomic Fe–Nx species, which had higher ORR performance than commercial Pt/C catalyst.[21] DFT simulations further showed that the 4e[−] ORR mechanism was prominent in Fe–N–C materials with atomically scattered Fe.[22] Other than cleavage of *OOH to *O on atomic Fe sites, elementary steps in the 4e[−] ORR process were endothermic. The rate-determining step with a free energy change (G) of 0.65 eV was considered the desorption of *OH. This showed the energy barrier for *OH desorption was lowered for ORR on atomic Fe–N$_x$ species.

For Fe–N$_x$/C SACs, optimal support should have a large specific surface area or porous structure for better accessibility in addition to exploring appropriate precursors. This will lead to excellent ORR performance of Fe–N$_x$/C SACs. Figure 4.4a shows how Jiang et al. used a mixed-ligand approach to create a highly effective Fe SA–N–C catalyst in the MOF system with high Fe loading (up to 1.76 wt%), hierarchical pores, and orientated mesochannels, as shown by Jiang et al.[23] The mixed-ligand strategy's large spatial distance from Fe (III) in the MOF framework inhibits Fe aggregation and speeds up the production of atomically scattered Fe species during pyrolysis. In this case, the ideal Fe SA–N–C catalyst had a half-wave potential of 0.776 V RHE in 0.1 M HClO$_4$ and was very long-lasting (4% damping at i-t measurement, and only 6 mV after 5,000 cycles). MOFs with predominant oxygen functional groups and no nitrogen element (e.g. MOF-5, MIL-101 series) have tremendous difficulty in the preparation of SACs when compared to MOFs with abundant nitrogen (e.g. ZIF-8, ZIF-7). This is primarily because oxygen bonding with metal ions would be subjected to the formation of metal oxides at high temperatures. Because iron in the MOF skeleton tends to form less active nanoparticles (Fe oxides) when 2D bimetallic Zn/Fe–MOF formed via coordination of Zn^{2+}, Fe^{2+}, and 1,4,5,8-Naphthalenetetracarboxylic dianhydride($C_{14}H_4O_6$) is directly pyrolyzed, Liao et al. used g-C$_3$N$_4$ as a nitriding (Fe loading is 3.89 wt%).[24] The half-wave potential of Fe/N-PCNs reached up to 0.79 Vs RHE in the 0.1 M HClO$_4$ and followed the 4e[−] ORR process, similar to commercial Pt/C, due to the presence of well-defined porous nanosheets with high surface area and mesoporous features. After a 20,000-s i-t test, Fe/N-PCN ORR performance attenuation is <10% (retention rate of 90.8%), demonstrating outstanding long-term stability.

Additionally, the development of a micro-meso-macroporous structure is critical for enhancing mass transfer and exposing active areas.[25,26] There are two types of pores: micropores (2 nm) that serve as hosts for active sites and contribute to the high specific surface area, but suffer from high mass and charge transfer resistances; and mesopores (>50 nm) that can significantly boost mass and proton transport rates while making quick substance transfer difficult; macropores (>50 nm) that act as reactant and proton buffers and facilitate rapid diffusion, especially long-range diffusion. The micropore-dominated ZIF-8 derived carbon polyhedron is bad for mass transfer, and internal active sites in carbon polyhedra are difficult to make contact with reactants and participate in ORR catalysis, Hu et al. innovated a hierarchical ordered porous carbon catalyst with atomic Fe–N$_4$ sites by hierarchically ordered porosity as a precursor (Figure 4.4b). The FeN$_4$/HOPC-c-1000 has an exciting half-wave potential of 0.80 Vs RHE, approaching commercial Pt/C(0.82 V vs RHE) in 0.5 M H$_2$SO$_4$,

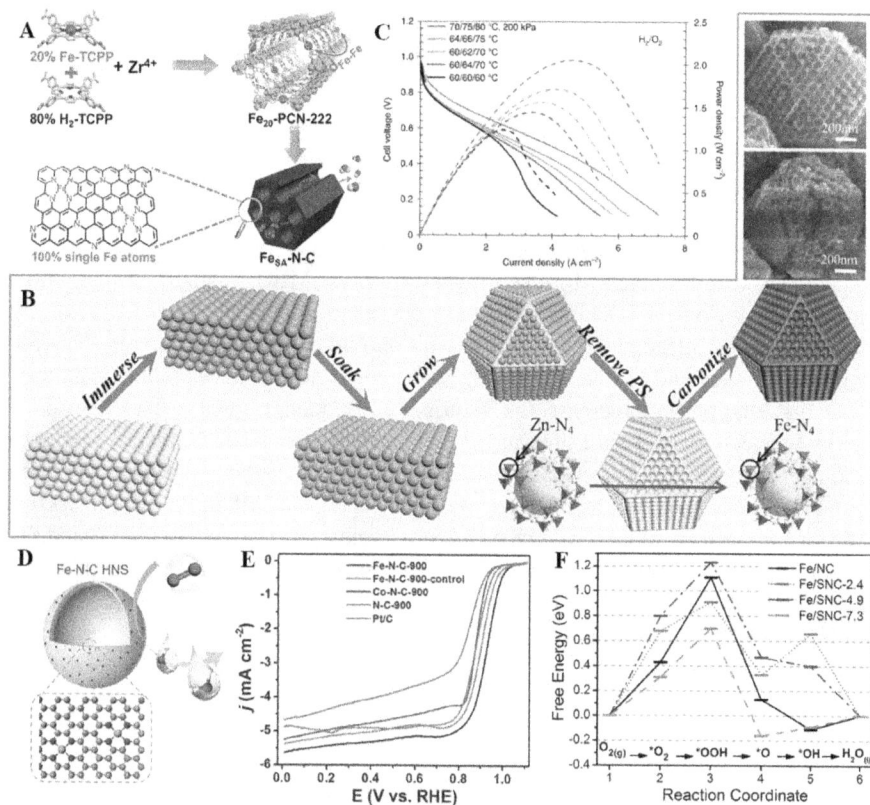

FIGURE 4.4   (a) Illustration of the rational fabrication of single Fe atoms-involved FeSA-N-C catalyst via a mixed-ligand strategy. (Reproduced from the study by Hai-Long Jiang et al.[23] with permission from John Wiley & sons.) (b) Schematic illustration of the synthesis of FeN_4/HOPC-c-1000 and SEM images of OMS-Fe-ZIF-8. (Reproduced from the study by Shuangyin Wang et al.[27] with permission from John Wiley & sons.) (c) Current density–voltage (solid) and current density–power density (dashed) curves for a H_2/O_2 AEMFC with a cathode with 1 mg cm^{-2} Fe–N–C, an anode with 0.6 mg cm^{-2} PtRu and 1 L min^{-1} gas flow. The anode/cathode/cell temperatures and pressurization state are given in the figure. (Reproduced from the study by Horie Adabi et al.[28] with permission from Springer Nature.) (d) Schematic illustration of the synthetic process of Fe–N–C HNS. (Reproduced from the study by Yawen Tang et al.[31] with permission from John Wiley & sons.) (e) LSV curves of Fe(Co)–N–C, N–C, and commercial Pt/C at 10 mV s^{-1} and a rotating speed of 1,600 rpm. (Reproduced from the study by Yuehe Lin et al.[34] with permission from John Wiley & sons.) (f) Free energy diagram for the Fe/NC and Fe/SNC systems during the ORR in acidic conditions at the equilibrium potential of $U_0 = 1.23$ V. The largest change in free energy determines the limiting step and overpotential. (Reproduced from the study by Shaojun Gao et al.[37] with permission from John Wiley & sons.)

which is much higher than FeN_4/C without hierarchically ordered pores due to the increased accessibility of active sites and the quicker mass transfer.[27] Mustain et al. build atomically dispersed Fe-N-C catalysts on carbon-based substrates with larger average pore sizes. Using a pyrolysis reaction including pipemidic acid, metal salt, and a SiO_2 template, carbon-based supports are made by removing the SiO_2 using an alkali solution. Fe–N–C catalysts as produced for anion-exchange membrane fuel cells (AEMFC) were able to achieve extremely high performance, in which the highest peak power density in H_2/O_2 and H_2/air can reach to 2 W cm^{-2} and over 1 W cm^{-2}, respectively (Figure 4.4c). More significantly, the AEMFC was able to maintain a high voltage for over 150 hours without any degradation in performance.[28]

The Sabatier principle may be used to optimize adsorption energy between single-metal sites and O_2/intermediates, which could help to accelerate oxygen molecule dissociation and increase ORR turn-over frequency (TOF). The adsorption energy between single-metal sites and O_2/intermediates may be effectively altered by rationally modifying the electronic structure of Fe SACs, further enhancing ORR performance. Coordination number and heteroatom doping are both important factors that influence the electrical structure of metal single-atom sites, as previously stated.[29,30] Enhancement of the ORR performance was attributed to Fe–N coordination in atomically dispersed Fe–N–C catalysts, and a variety of distinct N-coordination-number $Fe-N_x$ moieties have been observed. $Fe-N_4$ sites in Fe–N–C materials were shown to be the most important active sites for ORR among these materials. For the Fe–N–C HNSs with $Fe-N_4$ active site (Figure 4.4d), Tang and colleagues used the SiO_2-templated technique to manufacture atomically distributed Fe on N-doped carbon nanospheres with excellent ORR performance (onset potential: 1.046 V against RHE, half-wave potential: 0.87 V against RHE) and high durability in alkaline solutions.[31] ZIF-8 and graphitic nitrogen carbon pyrolyzed atomically dispersed Fe catalyst (C-FeHZ8@g-C_3N_4) produced excellent ORR performance.[32] EXAFS spectra showed that the $Fe-N_4$ moiety was present in C-FeHZ8@g-C_3N_4 when the atoms were isolated and distributed on the carbon-based substrate. C-FeHZ8@g-C_3N_4_950 demonstrated excellent ORR in acidic (half-wave potential: 0.78 V vs RHE) and alkaline medium (half-wave potential: 0.845 V vs RHE) as well as strong PEMFC performance with current densities of 400 and 133 mA cm^{-2} at 0.7 and 0.8 V (vs. RHE), respectively. FeN_4/HOPC catalysts were also produced by Wang et al. from pyrolysis of Fe-doped ZIF-8 and showed better ORR performance (half-wave potential: 0.80 V vs. RHE) in PEMFCs, with similar results.[27] Because of the macro-mesoporous structure and $Fe-N_4$ active sites that helped enhance ORR mass transfer, atomic FeN_4-based electrocatalysts had a very high ORR activity. The FeN_2 configuration, in addition to the atomic $Fe-N_4$ site, was shown to be active in ORR. Guo and colleagues synthesized atomically distributed FeN_2 sites on N-doped ordered mesoporous carbon (NOMC) using SBA-15 as a rigid template.[33] Due to its low ORR energy barrier and fast four-electron transport, the FeN_2 moiety was shown to be advantageous in DFT simulations. A Fe–N–C material with better ORR performance (onset potential: 0.99 V versus RHE, half-wave potential: 0.927 V versus RHE) in an alkaline electrolyte was synthesized using the template technique by Lin and coworkers (Figure 4.4e).[34] The FeN_2 active moieties facilitated the release of OH* with a low energy barrier, which resulted in excellent ORR performance. Atomically distributed catalysts with Fe–N$_5$ sites also showed great activity and excellent stability in ORR. Using the pyrolysis technique, Shen and colleagues produced atomically dispersed Fe–N–C materials that showed significant ORR activity with half-wave potentials of 0.908 and 0.795 V (compared to RHE) in 0.1 M KOH and 0.5 M H$_2$SO$_4$, respectively.[35] Highly distributed Fe–N$_5$ sites and high specific surface area of supports may be responsible for the outstanding ORR performance. Fe SAC/N–C was created by Tian and coworkers, and it showed excellent activity (half-wave potential: 0.89 V versus RHE), as well as stability toward ORR in 0.1 M KOH.[36] The Fe–N$_5$ moiety was discovered by XAFS to be a possible active site. DFT studies revealed a low energy barrier for the production of *OOH on Fe–p$_4$N–py (p$_4$N–py: four pyridine nitrogen atoms and one pyridinering), suggesting strong ORR activity on the Fe site coupled with five pyridinic-N atoms.

It has also been demonstrated that heteroatom-doped Fe–N–C may improve the ORR performance of single-atom Fe catalysts. When introducing the C–S–C structure to Fe/SNC catalysts, the introduction of an ORR activity with a full 4e$^-$ ORR process and reduced H$_2$O$_2$ generation in an acidic media were shown by Guo et al.[37] An optimized electronic structure around a Fe center and improved interactions between active sites and oxygen-containing intermediates result in a reduced ORR activation barrier, as shown by the DFT calculation and experimental characterization (Figure 4.4f). Another way to modify the coordination environment and electronic structure of activity sites is to add non-metal elements (such as P, B, F, etc.) into the Fe–N$_x$–C system,[38–40] which successfully increases the ORR activity. It is also extensively used to enhance O$_2$ molecule dissociation via electronic structure optimization by creating defects on the carbon substrate. Using an NH$_4$Cl-assisted approach, Chen et al. created a Fe–N$_x$/GM carbon with numerous FeN$_4$ sites in the edge using a clever design.[41] The formation of pores and nitrogen-doped edges in carbon skeleton is aided by a large amount of released gas (NH$_3$ and HCl) from the decomposition of NH$_4$Cl salt at high temperatures, which increases the number of FeN$_x$ active sites. This has been proven by XAS, STEM, and nitrogen adsorption–desorption, which results in increased ORR activity (half-wave potential = 0.80 Vs RHE, in 0.5 M H$_2$SO$_4$) for e–N$_x$/GM. As a result of the theoretical calculations, the introduction of an adjacent pore defect significantly reduced the adsorption energy of intermediate species (O$_2$* and OOH*), thereby increasing the thermodynamic limiting potential and, as a result, increasing intrinsic activity over the configuration without a pore defect.

Despite the fact that Fe–N–C-based SACs have a high ORR activity and good stability in an acidic medium, certain problems remain.[42–44] Fe–N$_x$ active site undergoes demetalation during ORR, and PEM degradation and carbon oxidation occur as a result of free radical attack from Fenton reactions (particularly Fe with high Fenton reaction activity), both of which result in significant decreases in the active site's stability and activity. M–N–C catalyst applications in MEA of realistic fuel cells would benefit greatly from the development of alternative non-Fe-based atomically dispersed M–N–C catalysts with minimal Fenton reaction activity, promising activity, and excellent acid stability, and recent work in this area is described below.

4.2.2   Co-Based Atom Electrocatalysts for ORR

Co-based SACs with very low Fenton reaction activity have recently been the subject of many research efforts. Co–N$_x$/C ORR performance has been compared to that of Fe SACs in certain investigations.[45] For example, Lou and coworkers used a modular approach to support atomically dispersed Co over a multichannel carbon matrix (Co@MCM), which provided excellent conductivity and a porous structure for atomically distributed Co. (Figure 4.5a).[46] The improved ORR performance was achieved by exposing more Co–N$_4$ active sites and using a carbon support with a high conductivity (Figure 4.5b and c). The pyrolysis technique was used by Shi and colleagues to make an atomically dispersed Co-based catalyst with nonplanar coordination that had good ORR performance owing to the interaction between adsorbates and Co–N$_4$ sites.[47] The ORR mechanism on single-atom Co supported on boron nitride materials (Co/BN) was discovered by Li's team using DFT simulations.[48] In addition, H$_2$/O$_2$ fuel cells have effectively used atomically distributed Co-based catalysts. Co-SACs containing Co-N$_4$ sites, for example, have been used in PEMFCs by Wu and colleagues (Figure 4.5d). It was discovered that by pyrolyzing Co-doped ZIF-8, they could make an atomically dispersed Co catalyst (20Co-NC-1100) with better performance in ORR and PEMFCs (Figure 4.5e and f).[49] Open-circuit voltage was 0.92 V against RHE for this catalyst, with an onset potential of 0.93 V and a maximum power density of 0.87 W cm^{-2}. An F127 layer was used to cover the Co-doped ZIF, and then the carbon shell was created via the carbonization of the F127 layer. This prevents the isolated Co atoms from clumping together, as was previously seen. Template-assistant pyrolysis was used by Li et al. to create a hollow N-doped carbon sphere containing a single scattered Co site (Figure 4.5g and h).[50] The tests and DFT results showed that isolated Co single-atom sites may significantly increase the hydrogenation rate of OH* intermediate. The resulting ISAS-Co/HNCS generated a remarkable ORR activity with a half-wave potential of 0.773 Vs RHE in 0.5 M H$_2$SO$_4$ by integrating highly active Co active sites into the hollow carbon sphere with highly exposed active sites (Figure 4.5i).

Increasing the ORR efficiency by rationally controlling the coordination structure of Co single-atom catalysts is an exciting prospect. The optimum Co$_1$–N$_3$PS active moiety was integrated in a hollow carbon polyhedron (Co$_1$–N$_3$PS/HC) by Chen et al., who developed a MOF-derived Co single-atom catalyst.[51] The as-prepared Co$_1$–N$_3$PS/HC has excellent alkaline ORR activity with a half-wave potential of 0.920 V and an ultralow Tafel slope of 31 mV dec^{-1}, outperforming Pt/C and almost all nonprecious ORR electrocatalysts. The ORR kinetics of Co$_1$–N$_3$PS/HC still outpaces those of Pt/C in acidic conditions. Wei and colleagues demonstrated that Co–N$_2$P$_2$ has a higher affinity for binding O$_2$ ($G = 0.27$ eV) than Co–N$_4$ ($G = 0.13$ eV), which is advantageous for the following processes.[52] With regard to the rate-determining step at $U = 1.23$ V, Co–N$_2$P$_2$ had a lower G than Co–N$_4$ (0.33 eV), indicating that the latter is more active in catalyzing four electrons ORR than the former (Co–N$_4$ has a higher G than Co–N$_2$P$_2$).

4.2.3   Others Non-Noble Metal-Based Single-Atom Catalysts for ORR

Cu-based single-atom catalysts for ORR: atomically distributed Cu catalysts exhibit outstanding ORR performance on Cu-based single-atom platforms.[53–55] Many researchers

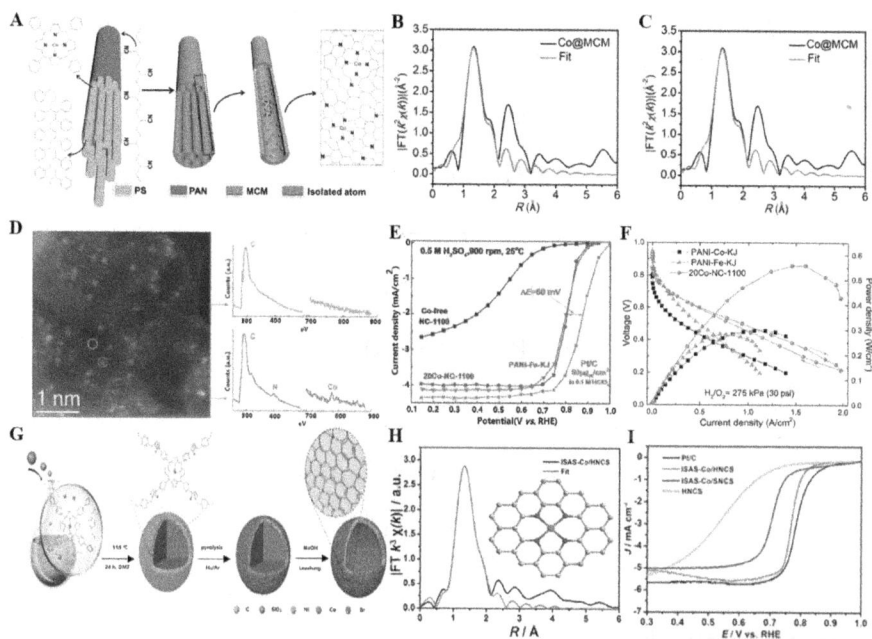

FIGURE 4.5 (a) Synthesis process of Co@MCM. (Reproduced from the study by Huabin Zhang et al.[46] with permission from Royal Society of Chemistry.) (b) Co K-edge EXAFS spectra of Co@PS-PAN (the precursor of Co@MCM), Co@MCM and Co foil. (Reproduced from the study by Huabin Zhang et al.[46] with permission from Royal Society of Chemistry.) (c) The R-space fitting result for Co@MCM. (Reproduced from the study by Huabin Zhang et al.[46] with permission from Royal Society of Chemistry.) (d) Aberration-corrected MAADF-STEM image and The corresponding EELS spectrum. (Reproduced from the study by Gang Wu et al.[49] with permission from John Wiley & sons.) (e) LSV curves of Fe(Co)–N–C, N–C, and commercial Pt/C at 10 mV s⁻¹ and a rotating speed of 1,600 rpm. (Reproduced from the study by Gang Wu et al.[49] with permission from John Wiley & sons.) (f) H_2–O_2 fuel cell polarization plot of samples. (Reproduced from the study by Gang Wu et al.[49] with permission from John Wiley & sons.) (g) Schematic illustration of the synthesis of ISAS-Co/HNCS. (Reproduced from the study by Yunhu Han et al.[50] with permission from American Chemical Society.) (h) Corresponding EXAFS fitting curves of ISAS-Co/HNCS at R-space. Inset: Schematic model of ISAS-Co/HNCS: Co (central atom), N (four adjacent atoms), and C (peripheral atoms). (Reproduced from the study by Yunhu Han et al.[50] with permission from American Chemical Society.) (i) ORR polarization curves S in O_2-saturated 0.5 M H_2SO_4. (Reproduced from the study by Yunhu Han et al.[50] with permission from American Chemical Society.)

are dedicated to using a workable approach to achieve large-scale manufacturing of atomically distributed Cu-based catalysts. For example, the annealing technique was used by Guan and colleagues to produce atomically distributed Cu on N-doped graphene catalyst (Cu@NG).[56] Isolated Cu atoms were found to be distributed across the N-doped graphene substrate in the HAADF-STEM results. The half-wave potential for 0.7% Cu@NG-750 was 0.94 V against RHE. Theoretical modeling by Kulkarni and colleagues predicted ORR activity on a Cu/CTF modified covalent triazine frame-work.[57] The following are two possible ORR processes (associative and dissociative):

Associative mechanism:

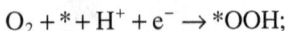

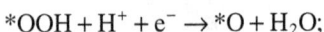

$$O_2 + * + H^+ + e^- \rightarrow *OOH;$$

$$*OOH + H^+ + e^- \rightarrow *O + H_2O;$$

$$*O + H^+ + e^- \rightarrow *OH;$$

$$*OH + H^+ + e^- \rightarrow H_2O$$

Dissociative mechanism:

$$O_2 + * \rightarrow 2 * O;$$

$$*O + H^+ + e^- \rightarrow *OH;$$

$$*OH + H^+ + e^- \rightarrow H_2O$$

There is a strong correlation between ORR activity and the coordination environ-ment and electronic configuration, thus understanding the connection between these two factors is critical for rationally designing atomic Cu-based catalysts. The syn-thesized atomically dispersed $Cu-N_4$ catalyst (Cu–SAs/N–C) showed outstanding ORR performance in 0.1 M KOH (Eonset: 0.99 V versus RHE, half-wave potential: 0.895 V vs. RHE), according to Wu and colleagues' work, in which a gas-migration method to attach single Cu atoms to the support was used.[53] After being pyrolyzed in an Ar environment, ZIF-8 formed Zn defect sites, which could then be used to trap individual Cu atoms. Furthermore, Wang and colleagues synthesized an atomically dispersed Cu-based catalyst (Cu SAC) pyrolyzed from Cu phthalocyanine precursors with onset and half-wave ORR potentials of 0.97 and 0.81 V (vs RHE) in an alka-line solution.[58] Numerous research studies have found a greater ORR activity in the $Cu-N_2$ configuration than in the $Cu-N_4$ configuration. By using the simple pyrolysis technique, the team headed by Bao created an atomically distributed Cu catalyst (Cu–N@C) with an ~8.5 wt% Cu concentration that had better ORR performance.[59] On the carbon-based support, it was discovered that there were isolated Cu atoms (Figure 4.6a and b). This is how $Cu-N_2$ is configured in the Cu–N@C (Figure 4.6c). Based on the results of a theoretical research, the $Cu-N_2$ site has better ORR perfor-mance compared to the $Cu-N_3$ and $Cu-N_4$ ones. Fu and colleagues discovered that a 2D N-doped carbon material with large Cu loadings of 20.9 wt% exposed many Cu active sites.[60] Adopting the pyrolysis of a Cu MOF (Cu(BTC)(H_2O)_3) precursor and followed by an acid-leaching process, the Cu–C–N catalyst was created and showed a half-wave potential of 0.866 V (against the RHE) for ORR in an alkaline media. Because of this, it has been suggested that the coordination number of Cu–N sites is located on the $Cu-N_2$ and $Cu-N_4$ structure (Figure 4.6d–f). DFT computations showed that O_2 and OOH species were more readily adsorbed on the $Cu-N_2$ site

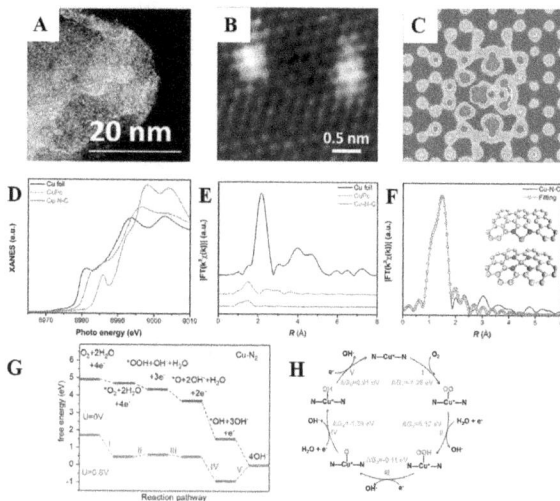

FIGURE 4.6 (a) HAADF-STEM images of Cu-N@C-60 (60 stands for the mass ratio of dicyandiamide over copper phthalocyanine CuPc). (Reproduced from the study by Xinhe Bao et al.[59] with permission from Royal Society of Chemistry.) (b, c) STM and STM simulation images of Cu-N@C-60. (Reproduced from the study by Xinhe Bao et al.[59] with permission from Royal Society of Chemistry.) (d, e) Cu K-edge XANES and FT of the Cu K-edge EXAFS spectrum of Cu–N–C, Cu foil and CuPc. (Reproduced from the study by Feng Li et al.[60] with permission from Royal Society of Chemistry.) (f) The corresponding EXAFS fitting curve of Cu–N–C in r-space. (Reproduced from the study by Feng Li et al.[60] with permission from Royal Society of Chemistry.) (g) Free energy profiles of the ORR steps on Cu–N$_2$ at potential of $U = 0$ and 0.8 V, respectively. (Reproduced from the study by Tao Yao et al.[61] with permission from American Chemical Society.) (h) 4e$^-$ ORR mechanisms on the Cu–N$_2$ site. (Reproduced from the study by Tao Yao et al.[61] with permission from American Chemical Society.)

surface than on the Cu–N$_4$ site. Because of the rapid mass/electron transfer capabilities, the ORR shows better performance. Every basic ORR step has minimal free energy changes due to the Cu–N$_2$ site's strong activity for ORR (Figure 4.6g and h).[61]

Mn-Based Single-Atom Catalysts for ORR: the ORR activity and durability of single-atom Mn supported on graphene matrix were both high, according to the research. For ORR and proton-exchange membrane fuel cells, Wu and coworkers used a two-step synthesis approach to produce an atomically distributed Mn catalyst on graphitic carbon-containing MnN$_4$ active sites that performed better in acidic conditions.[62] An Mn–ZIF precursor was carbonized, and then an acid was added to produce the microporous 20Mn–NC second catalyst. The 20Mn–NC-second catalyst's MnN$_4$ active sites were then multiplied by an additional adsorption step and a heat activation step (Figure 4.7a). There was a significant attenuation in the attenuation of 20Mn–NC-second after 100h at 0.8 V against RHE, suggesting that the ORR activity and stability were strong in the presence of 0.5 M H$_2$SO$_4$ (Figure 4.7b). For a fuel-cell system, the Mn SAC shows a 0.95 V open circuit against RHE and the fuel cell's maximum power density can reach 0.46 W cm^{-2} (Figure 4.7c). There was

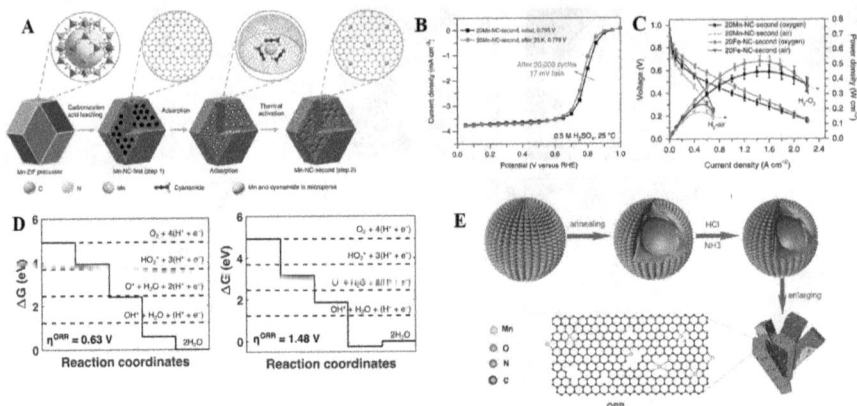

FIGURE 4.7 (a) Schematic of atomically dispersed MnN_4 site catalyst synthesis; (b) steady-state ORR polarization plots before and after potential cycling stability tests (0.6–1.0 V, 30,000 cycles) for the 20Mn-NC-second; (c) Fuel-cell performance of the best-performing 20Mn-NC-second and 20FeNC-second catalysts in both H_2/O_2 and H_2/air conditions. (Reproduced from the study by Jiazhan Li et al.[62] with permission from Springer Nature.) (d) Free energy diagram for ORR on MnN_4–G and MnN_3–G. (Reproduced from the study by Jingqi Guan et al.[63] with permission from Elsevier.) (e) Schematic illustration of the synthetic process for Mn/C–NO. (Reproduced from the study by Qianwang Chen et al.[65] with permission from Wiley.)

also a atomically distributed MnN_4 active site on graphene (Mn@NG) that had 0.95 V as the onset potential and 0.82 V as the half-wave potential against RHE in a solution of 0.1 M KOH.[63] Using DFT calculations to compare the free energy changes of basic steps on MnN_4–G and MnN_3–G, it was shown that ORR was mostly active on MnN_4–G site (Figure 4.7d and e). DFT simulations have recently analyzed the electrochemical ORR performance on Mn–N_x active surfaces.[64] There was evidence to suggest that the nitrogen atoms on Mn–N_4 and Mn–N_3 moieties can attract electrons from adjacent C atoms, thus leading to the better ORR performance. When Mn active centers in graphene frameworks are coordinated with O and N atoms, good ORR performance may be obtained under alkaline conditions, according to Chen and colleagues.[65] Atomically distributed Mn/C–NO electrocatalysts may be made by annealing Mn–BTC, followed by acid etching, and NH_3 activation, as shown in Figure 4.7f. A positive half-wave potential of 0.86 V against RHE was observed for ORR using the Mn/C–NO catalyst in 0.1 M KOH solution.

Sn-Based Single-Atom Catalysts for ORR: Strasser et al. announced the development of an ORR catalyst with high intrinsic activity and TOF based on a p-block Sn single-atom block.[66] The SnN_x sites in the catalyst were distributed on an atomic scale, as shown by HRSTEM-EELS. It was discovered that Sn(iv)N_x is an active site using Mössbauer and XAS, as well as DFT calculations. It was discovered that Sn–N–C catalysts had much better selectivity for four-electron ORR paths than Co–N–C and are similar to Fe–N–C based on the RRDE (rotating-ring-disk electrode) test in 0.1 M $HClO_4$. The results show that the TOF of Sn–N–C is five times more than that of Co–N–C (0.29

e·site^{-1}·s^{-1}), which is comparable to Fe–N–C. Compared with Fe–N–C and Co–N–C, the Sn–N–C had greater current densities at high potential areas ($E \geq 0.7$ V) in the H$_2$/O$_2$ PEMFC test. There is a strong correlation between the density of atomically active sites in SACs and the loading of metals at their respective locations.

Zn-Based Single-Atom Catalysts for ORR: atomically distributed Zn-based catalysts have the potential to replace Pt-based catalysts in the process of converting energy. The Zn–N$_4$ active site was generally accepted as the primary active site, with reports of the Zn–N$_2$ active site also appearing. Single-atom Zn was supported on N-doped carbon (Zn–N–C) with a high Zn loading of 9.33 wt% through a precise control of annealing rate (1°C·min^{-1}). Zn(II) species were highly dispersed over the carbon-based support, and the Zn–N$_4$ configuration was created (Figure 4.8a–d) and found to be crucial in the 4e$^-$ ORR process for single-atom Zn supported on N-doped carbon materials. Zn–N–C catalyst showed high ORR activity and endurance in acidic and alkaline environments (Figure 4.8e–g).[67] The Zn–N$_2$ active site was also suggested for ORR, in addition to the Zn–N$_4$ active site. According to the research

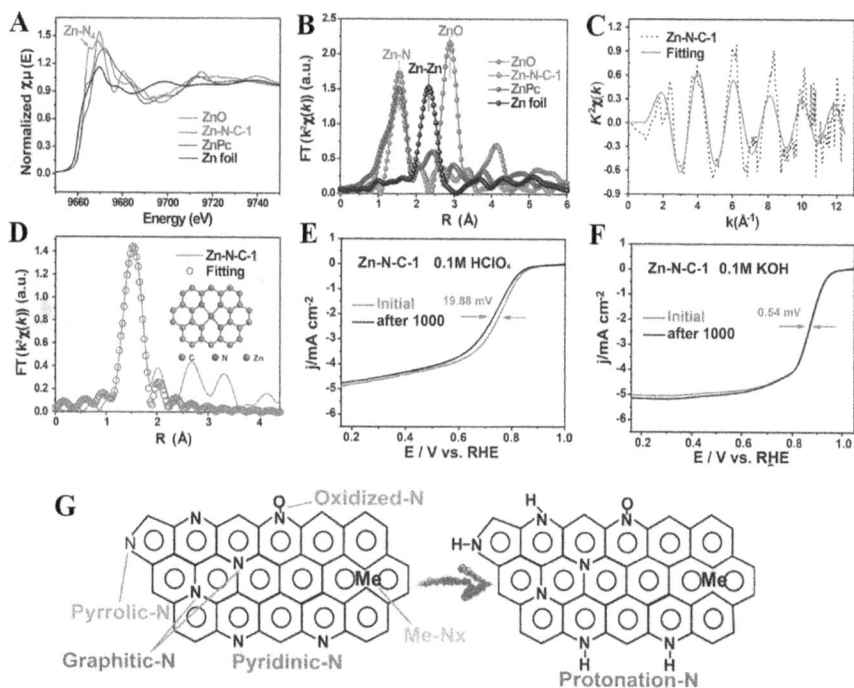

FIGURE 4.8 (a) Zn K-edge XANES spectra of Zn-N-C-1, ZnO, ZnPc, and Zn foil. (b) Fourier transforms (FTs) of the k^2-weighted c(k) functions of the EXAFS spectra for the Zn K-edge. (c, d) Corresponding EXAFS fitting curves in k and R space, respectively; inset shows a model of the Zn environment. (e, f) ORR polarization of the Zn-N-C-1 catalyst before and after 1,000 cycles between 0.6 and 1.1 V versus RHE in 0.1 M HClO$_4$ and 0.1 M KOH. (g) Protonation reaction on Zn-N-C-1 and Fe-N-C-1 in acidic media. (Reproduced from the study by Zidong Wei et al.[67] with permission from Wiley.)

done by Lu and coworkers, the structural features and ORR performance of Zn-N_2 sites were revealed via a combination of DFT calculations and experimental results.[68]

Ce-Based Single-Atom Catalysts for ORR: Wu et al. described metal-organic framework-derived cerium single-atom catalysts. A promising hard-template approach was used to build a hierarchically porous nitrogen-doped carbon support with plenty of Ce sites. To make Ce-doped ZIF-8 precursors, Ce ions and Zn ions were simultaneously injected into a 2-methylimidazole solution together with a SiO_2 hard-template. A hierarchically macro-meso-microporous structure results after the carbonization and leaching processes remove the SiO_2 ball (Figure 4.9a). CeN_4/O_6 may be identified as the main reactive site for Ce SAS/HPNC based on XAFS and

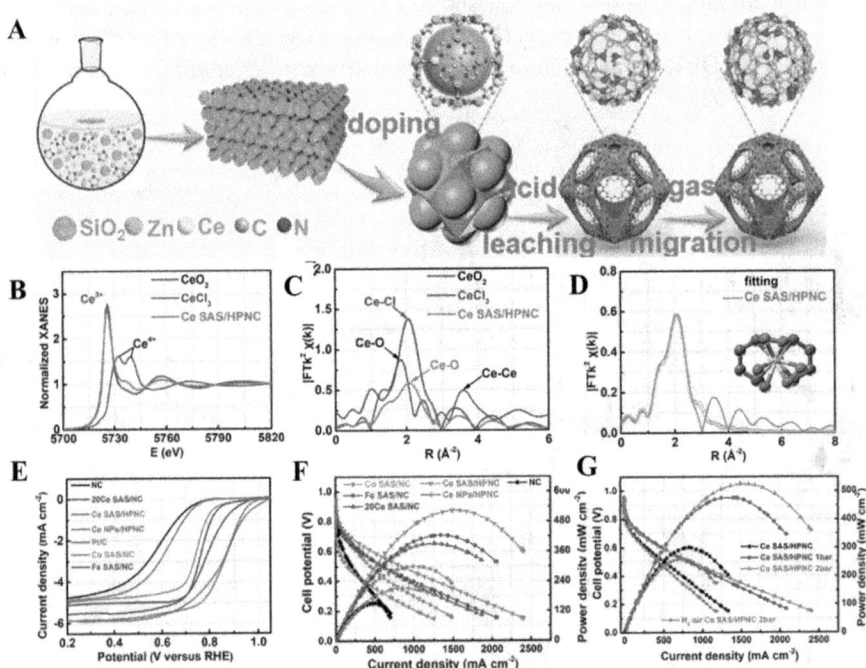

FIGURE 4.9 (a) Fabrication procedure for the Ce sites embedded in a hierarchically macro-meso-microporous N-doped carbon (Ce SAS/HPNC) catalyst. (b) X-ray absorption near-edge structure (XANES) of Ce SAS/HPNC, $CeCl_3$, and CeO_2 at the Ce L_3-edge. (c) Fourier transformed extended X-ray absorption fine structure (FT-EXAFS) spectra of Ce SAS/HPNC, $CeCl_3$, and CeO_2 at the Ce L_3-edge. (d) FT-EXAFS fitting curve of the sample and proposed architectures of Ce−N_4/O_6 (inset). (e) ORR performance of NC, 20Ce SAS/NC, Ce SAS/HPNC, Ce NPS/HPNC, and Pt/C catalysts. Electrolyte 0.1 M $HClO_4$, sweep rate 10 mV s^{-1}, rotating rate 1,600 rpm. (f) O_2 fuel-cell polarization plots of different catalysts. Cathode: 2.0 mg cm^{-2} of catalysts. Anode: 0.15 mgPt cm^{-2} 20% JM Pt/C; 100% RH; membrane Nafion 211; Cell 80°C; back pressure 2 bar; 12.25 cm^2 electrode area. (g) Fuel-cell polarization plots of Ce SAS/HPNC at different conditions. Cathode: 2.0 mg cm^{-2} of Ce SAS/HPNC. Anode: 0.15 mg_{Pt} cm^{-2} JM 20% Pt/C; 100% RH; membrane Nafion 211; Cell 80°C; 12.25 cm^2 electrode area. (Reproduced from the study by Yuen Wu et al.[69] with permission from ACS Publications.)

fitting analysis in R-space, and the ORR four-electron process with high efficiency is initiated (Figure 4.9b–d). The as-prepared Ce single-atom catalyst exhibits outstanding ORR performance in acidic environments, with a half-wave potential of 0.862 V (Figure 4.9e). To obtain the greatest power density in a PEMFC, the components were built at a pressure of 2.0 bar H_2/O_2, achieving a maximum power density of 0.525 W cm^{-2} (Figure 4.9f and g). The hierarchically macro-meso-microporous N-doped carbon support also makes sure that the oxygen gets to the Ce active sites and water stays away from them in a realistic fuel-cell reaction zone.[69]

Sb-Based Single-Atom Catalysts for ORR: Due to the delocalized s/p band, most main-group (s- and p-block) metals are considered catalytically inactive. To efficiently catalyze the oxygen reduction process, we synthesized an antimony single-atom catalyst (Sb SAC) with the SbN_4 structure (ORR). Compared to commercial Pt/C and most transition metal (TM, d-block) based SACs, the newly developed Sb SAC shows better ORR activity and good stability. Theoretical and experimental evidence both point to positively charged SbN_4 single-metal sites with closed d shells as the active catalytic sites. These DOS findings reveal how easily the p orbital of the atomically dispersed Sb cation in Sb SAC interacts with the O_2-p orbital to form hybrid states, which facilitates charge transfer and generates adequate adsorption strength for oxygen intermediates, lowering energy barriers and modulating rate-determining steps. To better understand how p-block Sb metal catalysts are prepared for highly active ORR, this study sheds insight on the atomic level. It also offers useful recommendations for the rational design of additional main-group-metal-based SACs.[70]

Cr-Based Single-Atom Catalysts for ORR: Strong ORR catalysts are urgently needed since most atomically dispersed catalysts have low endurance in ORR owing to the assault of H_2O_2. Stable atomically distributed Cr catalyst with Cr–N_4 active site for ORR was developed by Xing and coworkers (Cr/N/C).[71] The half-wave potential was 0.773 V compared with RHE, and the durability for ORR was exceptional in 0.1 M $HClO_4$. The adsorption of $CrCl_3$ on ZIF-8 produced a Cr-ZIF precursor with microporous structures, as shown in Figure 4.10a, and the Cr/N/C catalyst was made by pyrolyzing the Cr-ZIF precursor and then leaching it in acid. The Cr/N/C had a consistent distribution of C, N, and Cr elements (Figure 4.10b and c), and the EXAFS spectrum and WT plot (Figure 4.10d–g) showed a Cr–N_4 site.

W-Based Single-Atom Catalysts for ORR: ORR has recently been discovered to be catalyzed by single-atom W-based catalysts. Wu and colleagues produced graphitic tungsten carbide with single W atoms on the graphitic layer (WC@C).[72] By structural characterizations and DFT calculations, Zhao and colleagues discovered excellent ORR performance and great stability on W-N_5 sites. In 0.1 M KOH (E_{onset}: 1.01 V versus RHE, half-wave potential: 0.88 V vs RHE) and 0.1 M $HClO_4$ (E_{onset}: 0.87 V vs RHE, half-wave potential: 0.77 V vs RHE), atomically distributed W–N–C material containing W–N_5 sites demonstrated outstanding ORR performance. There was no apparent decrease of ORR activity after 10,000 cycles, suggesting that the W–N–C was stable. This novel and effective W-based atomically dispersed catalyst may pave the way for the development of additional single-atom ORR catalysts based on transition metals in the 4th and 5th dimensions.

Mg-Based Single-Atom Catalysts for ORR: Because of the tight interaction between main-group metal sites and the hydroxyl group intermediate, materials linked to the

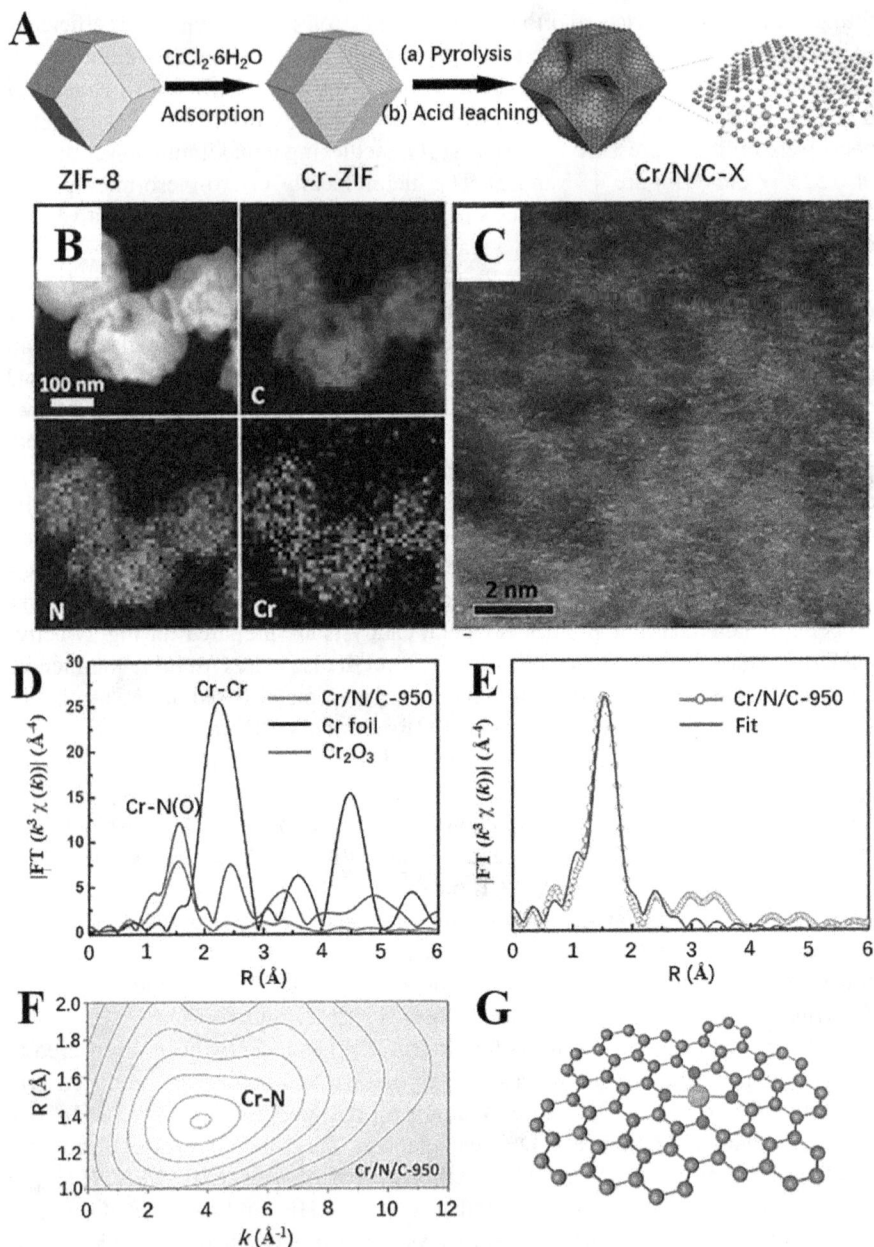

FIGURE 4.10 (a) Schematic depiction of the synthesis of Cr/N/C. (b) STEM, and corresponding elemental mapping images of Cr/N/C. (c) Atomic-resolution AADF-STEM image of Cr/N/C. D) k^3-weighted Cr K-edge Fourier transform EXAFS spectra of Cr/N/C as well as Cr foil and Cr_2O_3 references. (e) Experimental and best-fitting Cr K-edge EXAFS curves of Cr/N/C. (f) WT-EXAFS plots of Cr/N/C ($k = 3$, $s = 1.5$). (g) Schematic model of the Cr-centered configuration in Cr/N/C. Cr, green; N, blue; C, gray. (Reproduced from the study by Prof. Wei Xing et al.[71] with permission from Wiley.)

main-group metals exhibit little catalytic activity for ORR (Figure 4.11a and b). It was shown that atomically distributed Mg single-atom catalysts had better ORR performance, with half-wave potentials as high as 0.91 V in alkaline medium and as low as 0.79 V in acidic media, according to Chen et al (Figure 4.11c–h). When a Mg center is coordinated with two nitrogen atoms in the graphene matrix, DFT calculations show that the higher the p-state location, the weaker the oxygenated species binding strength is at the Mg atom, and the more active an ORR is near the top of the volcano-type activity plots. This is in contrast to other coordination numbers. Based on the results of

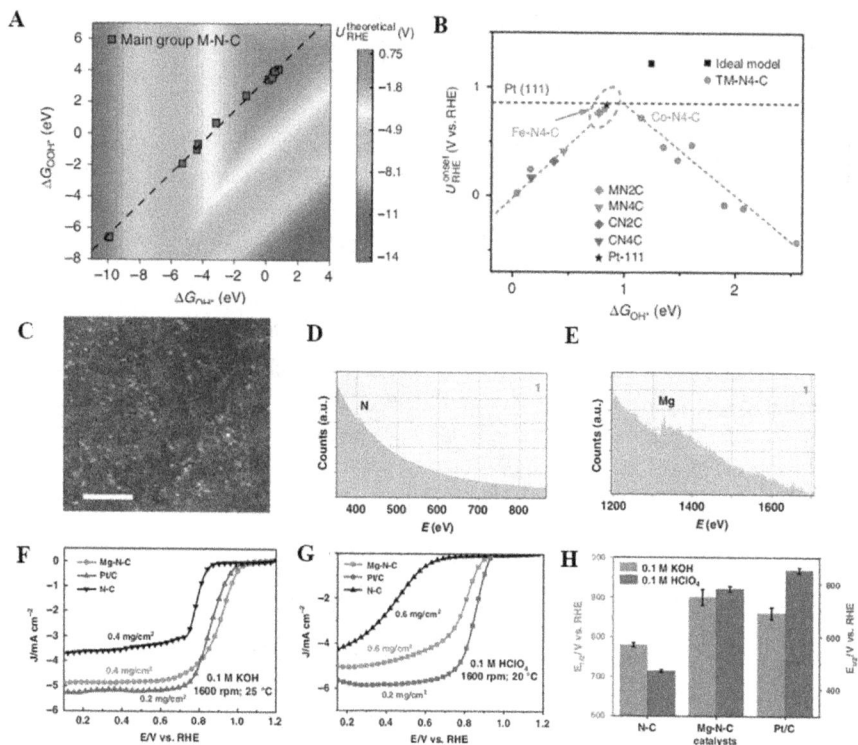

FIGURE 4.11 (a) The two-dimensional volcano map about theoretical onset potential versus adsorption free energies ΔG_{OH}^* and ΔG_{OOH}^* on main group metal cofactors (M–N–C). (b) The zoomed-in view of the onset potential versus ΔG_{OH}^* in the region close to the performance ceiling of models and transition metal cofactors (TM–N$_4$–C). The red dash circles at Mg–N–C represents the active centers. (c) High-angle annular dark-fieldscanning transmission electron microscopy (HAADF-STEM) image of Mg–N–C. Scale bar: 2 nm. The bright dots are Mg atoms, which can be confirmed by electron energy loss spectroscopy (EELS) spectra in (d, e), probe resolution is 0.15 nm. (d, e) The EELS spectrum for atomic site highlighted by red circle in (c). (f) LSV curves after normalization by glass carbon (GC) electrode in O$_2$-saturated 0.1 M aqueous KOH electrolyte solutions at a sweep rate of 5 mV s^{-1}. (g) LSV curves after normalization by GC electrode in O$_2$-saturated 0.1 M aqueous HClO$_4$ electrolyte solution at a sweep rate of 10 mV s^{-1}. (h) $E_{1/2}$ for different catalysts in both alkaline and acidic solutions. (Reproduced from the study by Shuai Liu et al.[73] with permission from Springer Nature.)

this research, it seems that rational materials design may aid in the fabrication of highly active electrocatalytic materials based on metals from the main group, which might provide insight into catalyst research in the future.[73]

4.2.4 NOBLE METAL-BASED SINGLE-ATOM CATALYSTS FOR ORR

ORR on Single-Site Pt-Based Electrocatalysts: Pt catalysts with atomically scattered atoms are currently the most promising ORR choices. Pt was atomistically distributed on S-doped zeolite template carbon, which was synthesized by Choi and colleagues and has a 96% H_2O_2 yield and an onset potential of 0.71 V when compared to RHE when used in the $2e^-$ ORR process.[74] A 5 wt% Pt-loaded catalyst was made using the impregnation technique, in which H_2PtCl_6 was impregnated on a carbon support and then heated under an H_2 environment to soften the catalyst. For ORR, the PtS_4 moiety, which was synthesized by linking the isolated Pt atoms to S sites, was also very active for $2e^-$ ORR process to produce H_2O_2. There is evidence that Pt anchored on TiN nanoparticles (Pt/TiN) may produce significant amounts of H_2O_2 for ORR, and the mass activity of this catalyst can approach 78 A/g_{Pt} at a 50 mV overpotential.[75] After that, they used TiC nanoparticles to attach single Pt atoms, and the resulting elctrocatalytic performance was better to that of the Pt/TiN catalyst in terms of activity and selectivity for H_2O_2 generation.[76] The findings showed the significance of ORR electrocatalysis supports. H_2O_2 generation relied heavily on the coordination settings around the atomic Pt site.

The $4e^-$ ORR performance of single-atom Pt-supported carbon-based materials is better. According to Liu et al., single Pt atoms coordinated with pyridinic-N atoms in atomically distributed Pt supported on N-doped carbon black (Pt_1–N/BP) effectively catalyze ORR.[77] XAFS spectra confirmed the presence of tiny clusters in the Pt_1/BP based on HAADF-STEM images and the distribution of single Pt atoms across the carbon-based support. Half-wave potentials of Pt_1–N/BP were found in $HClO_4$ and KOH to be 0.76 and 0.87 V (vs RHE), respectively, and the material can be built into acidic fuel cells with a maximum power density of 1.02 W cm^{-2} and shows excellent stability. The catalytic performance of a Pt–C_4 catalyst ($Pt_{1.1}$/BP defect) in an acidic ORR and an H_2/O_2 fuel cell was studied by Xu and colleagues.[78] The Pt–C_4 coordination structure in $Pt_{1.1}$/BP defect has been confirmed based on spherical aberration electron microscopy and XAFS analysis (Figure 4.12a–c). On the $Pt_{1.1}$/BP defect in 0.1 M $HClO_4$, excellent ORR activity, good stability, and great methanol tolerance were obtained (Figure 4.12d). The highest power density was 0.52 mW cm^{-2} in fuel cells with the Pt1.1/BP defect (Figure 4.12e). Graphene–Pt (g–Pt, Pt–C_4) active sites formed by DFT calculations showed better electrocatalytic activity (Figure 4.12f).

ORR on Single-Site Ir-Based Electrocatalysts: Researchers led by Chen et al. have developed an Ir–N–C atomically dispersed catalyst (Ir–SAC) for use in ORR and fuel cells that are very efficient.[79] The atomically scattered Ir centers were coordinated with neighboring N atoms, while the isolated Ir atoms were extremely dispersed (Figure 4.12g and h). With an Ir–SAC half-wave potential of 0.864 V against RHE, high mass activity of 12.2 A g^{-1} and a TOF of 24.3 per site and second, the dominating four-electron ORR process was revealed (Figure 4.12i). The Ir–SAC was also shown to have an ORR slope of 41 mV dec^{-1} in 0.1 m $HClO_4$. Due to the appropriate

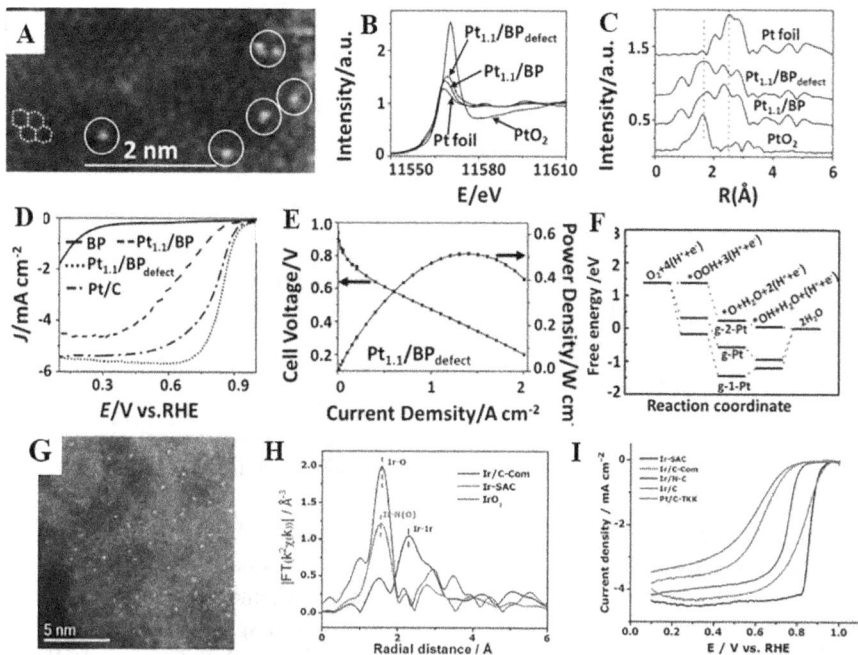

FIGURE 4.12 (a) HAADF-STEM imageto show the distribution of Pt atoms on carbon defects. (b) Pt L_3-edge XANES for Ptfoil, PtO_2, $Pt_{1.1}$/BP and $Pt_{1.1}$/BPdefect. (c) The k^2-weighted R-space FT spectra from EXAFS for Pt foil, PtO_2, $Pt_{1.1}$/BP and $Pt_{1.1}$/BPdefect. (d) RDE polarization curves of BP, Pt1.1/BP and Pt1.1/BPdefect and the commercial Pt/C in O_2-saturated 0.1M $HClO_4$ with scan rate of 5mV s^{-1} and rotation speed of 1,600rpm. (e) Polarization and power density curves of an acidic H_2/O_2 fuel cell with Pt1.1/BPdefect. (f) Free energy diagram for complete O_2 reduction on g-Pt, g-1-Pt, and g-2-Pt substrates in acidic media at 0.83 V, respectively. (Reproduced from the study by Weilin Xu et al.[78] with permission from Wiley.) (g) High-resolution HAADF-STEM image of Ir–SAC. (h) Fourier transforms of k^3-weighted Ir L_3-edge EXAFS data. (i) ORR polarization curves with ascanninprate of 5mV s^{-1} at rotating speed of 900rpm for the synthesized catalysts. (Reproduced from the study by Zhongwei Chen et al.[79] with permission from Wiley.)

adsorption energy of intermediates on the Ir–N_4 site, the better ORR performance of Ir–SAC may be ascribed. This fuel cell has an open-circuit voltage of 0.955 V against RHE and a maximum power density of 932 mW cm^{-2} and may be built into the H_2/O_2 fuel cell.

ORR on Single-Site Ru-Based Electrocatalysts: ORR kinetics may be improved in acidic fuel cells by using single-atom Ru catalysts, which showed great promise. Tour and colleagues used a simple annealing technique to make single-atom Ru supported on N-doped graphene (Ru–N/G).[80] Ru–N/G was very active and long-lasting in acidic medium (Eonset: 0.89 V vs. RHE, half-wave potential: 0.75 V vs. RHE). The superior ORR performance was attributed to Ru–N_4 species with axial oxygen adsorption. DFT studies showed a low energy barrier for ORR at the Ru-oxo-N_4 site, which is likely to cause the rapid 4e⁻ ORR process on Ru–N/G. For this project, the pyrolysis technique was used to produce a Ru–N–C catalyst (Ru–SSC) by Xing's team.[81] Ru–SSC showed an onset potential and half-wave potential (vs RHE) of 0.92 and 0.824 V against RHE, respectively, with an associated Tafel slope of 54.2 mV dec^{-1} in the presence of 0.1 M HClO$_4$. Ru–SSC shows specific activity of 11.95 mA cm^{-2} and mass activity of 4.78 A mg$_{Ru}^{-1}$. DFT simulations revealed that the RuN$_4$OH site was a good candidate for the ORR reaction.

ORR on Single-Site Rh-Based Electrocatalysts: Yang and coworkers investigated the ORR catalytic activity on different single atoms supported on NbC(001) by DFT calculations, and found that atomically dispersed Rh on NbC(001) (Rh/NbC(001)) exhibited high ORR activity.[82] The low workfunction (W–F) and electro-negativity (E–N) of Rh were able to lower the energy barrier of O$_2$ adsorption and activation, thus improving ORR kinetics. In addition, Rh/NbC(001) showed high stability, which was indicated by reasonable agglomeration resistance and oxidation resistance. Interactions between the Rh and O-atom (EM–O) were ideal, which were weaker than those between Rh and substrate (EM–S), implying remarkable stability of Rh/NbC(001).

ORR on Single-Site Pd-Based Electrocatalysts: They created Pd/MnO$_2$-CNT with high mass activity (484 A g^{-1} at 0.9 V compared to RHE) in an alkaline medium for ORR by synthesizing single-atom Pd supported on MnO$_2$ nanowire and carbon nanotube.[83] The isolated Pd atoms were widely distributed throughout the MnO$_2$ nanowires and carbon nanotubes that were used to create them. PdII-O/C and PdII-Mn were found in XAFS spectra of Pd/MnO$_2$-CNT. It was discovered via DFT calculations that the outstanding performance of Pd/MnO$_2$-CNT was owing to the synergistic impact of Pd centers on MnO$_2$ surfaces and to the high conductivity of carbon nanotubes.

4.2.5 DUAL-METAL-SITE ELECTROCATALYSTS

Double transition metal-doped carbon catalysts (M$_1$M$_2$–N–C) with dual-metal sites have been extensively investigated to overcome the catalytic efficiency limitation caused by the linear relationship between reactive intermediates and adsorption energies, resulting in significantly improved catalytic activity and stability.[84] Research on double-atom catalysts is taking off as a result of this breakthrough. It's been shown in the experiments and theoretical calculations that dual transition metal doped double-atom catalysts (heteronuclear DACs) can not only inherit the benefits of SACs but also generate synergistic active sites that can further boost the intrinsic activity and stability of M–N$_x$–C, leading to a superior ORR activity.[85–87]

Using pyrolysis, Li and colleagues created a (Fe, Co)/N–C catalyst with better ORR performance (Eonset: 1.06 V compared to RHE, half-wave potential: 0.863 V compared to RHE) in an acidic medium.[88] A good coordination was found between the metal centers and the N atoms in the support (see Figure 4.13a–f). Highly dense and stable H$_2$/O$_2$ and H$_2$/air fuel cells built with (Fe, Co)/N–C materials were developed (Figure 4.13g and h). The O–O bond was readily broken at the Fe–Co dual sites because to the low dissociation barrier (Figure 4.13i). Furthermore, ORR activity was shown to be greater on the FeCoN$_5$–OH active center than on the FeN$_4$ site by Xing and colleagues.[89] To create FeCoN$_5$–OH active site with a triangular structure and

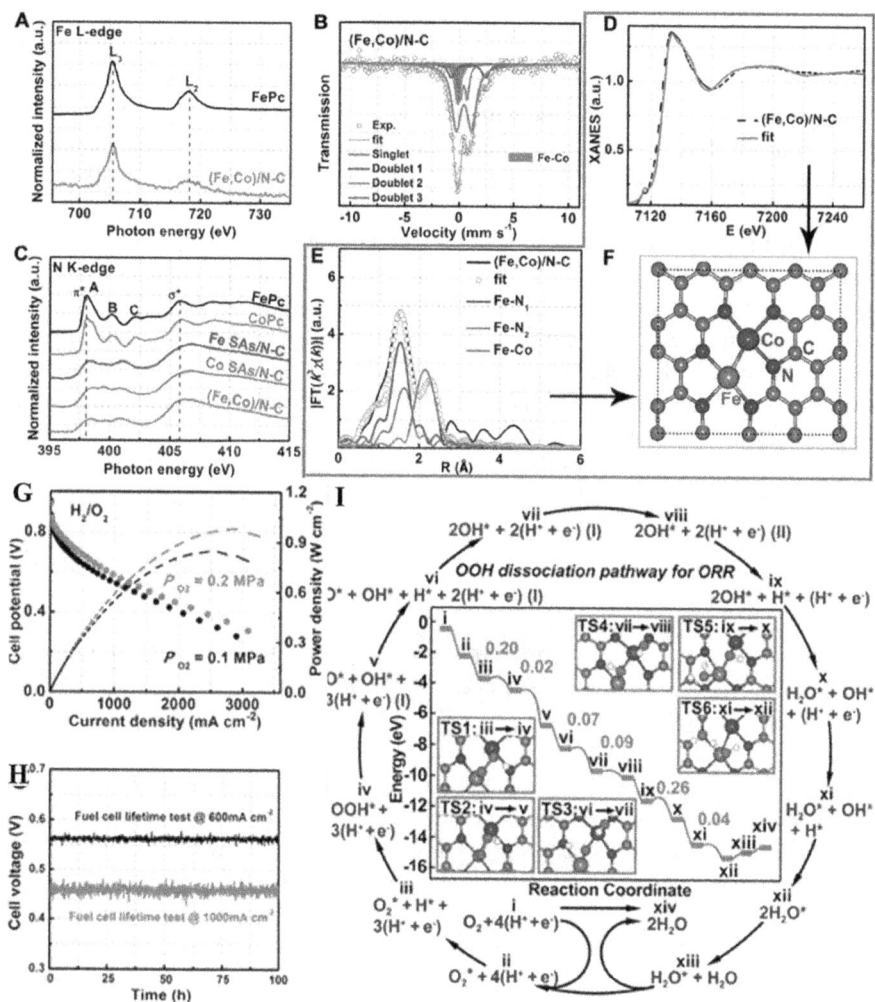

FIGURE 4.13 (a) Fe L-edge XANES spectra of (Fe, Co)/N–C and FePc; (b) experimental 57Fe Mössbauer transmission spectra measured at 298 K for (Fe, Co)/N–C and fittings with spectral components; (c) N K-edge XAS spectra of Fe SAs/N–C, Co SAs/N–C, and (Fe, Co)/N–C; (d) Comparison between K-edge XANES experimental spectrum of (Fe, Co)/N–C (black dashed line) and theoretical spectrum calculated with depicted structure (solid red line); (e) Corresponding Fe K-edge EXAFS fittings of (Fe, Co)/N–C; (f) Proposed architectures of Fe–Co dual sites; (g) H$_2$/O$_2$ fuel-cell polarization plots. Cathode: ≈0.77 mg·cm^{-2} of (Fe, Co)/N–C; O$_2$, 0.1 and 0.2 MPa partial pressures. Anode: 0.1 mg$_{Pt}$·cm^{-2} Pt/C; H$_2$, 0.1 MPa partial pressure. Cell, 353 K; 25 cm^2 electrode area; (h) Stability of (Fe, Co)/N–C in a H$_2$/air fuel cell measured at 600 and 1,000 mA·cm^{-2}; (i) Energies of intermediates and transition states in mechanism of ORR at (Fe, Co)/N–C from DFT. (Reproduced from the study by Jing Wang et al.[88] with permission from American Chemical Society.)

achieve low ORR energy barrier, the OH was attached to the dual-metal $FeCoN_5$ site. A formamide condensation method was utilized by Liu and colleagues to produce the high ORR activity and stability electrocatalyst f-FeCoNC900. FeN_3 and CoN_3 sites synergistically increased the half-wave potential of f-FeCoNC900 to 0.89 and 0.81 V (against RHE) in 0.1 M KOH and 0.1 M $HClO_4$, respectively. When it comes to ORR performance on a hydroxylated NiFe dual-site catalyst, Pei and coworkers utilized DFT simulations to determine that hydroxyl groups may boost ORR activity.[90] One study reported the use of atomically dispersed platinum–copper on an N-doped carbon substrate (A-CoPt NC) with a half-wave potential of 0.96 V against RHE and a massive active concentration (MAC) of 45.47 A mg^{-1} in 0.10 M KOH.[91]

A porous nitrogen-doped carbon ORR catalyst with Fe and Ni dual-metal sites was developed by Zhou et al., and in an acidic media, its active sites and catalytic processes were discovered.[92] The intrinsic activity of the FeNi-N_6 (type I) catalyst was greater than that of the Fe-N_4 (0.11 e s^{-1} $sites^{-1}$) catalyst, with a TOF value of 0.43 e s^{-1} $sites^{-1}$. The ORR activity of the FeNi-N_6 catalyst was found to be superior to that of the Fe-N_4 and Ni-N_4 catalysts in electrochemical tests conducted in 0.1 m $HClO_4$. The FeNi-N_6 catalyst produced much less H_2O_2 (1%–4%) than the Fe-N_4 catalyst (10%–20%). These calculations showed that FeNi-N_6 (type I) is responsible for the catalyst's high activity, and quantitative XANES and DFT calculations confirmed this. It was shown that ORR activity occurs in this order: FeNi-N_6 (type I) > FeNi-N_6 (type II) > Fe-N_4 > Fe_2 – N_6. Fe/Zn DACs (Fe–Zn–SA/NC) were created by Liu et al. as ORR catalysts.[93] Through the pyrolysis of a MOF precursor, Chen and colleagues created a Fe, Mn–N/C catalyst with a uniform distribution of Fe and Mn elements across the porous N-doped carbon substrate.[94] The synergistic impact of the Fe-N_x and Mn-N_x sites on the Fe, Mn–N/C-900 catalyst led to a positive half-wave potential for ORR of 0.904 V against RHE. DFT studies revealed that FeN_x sites were the primary ORR active centers, whereas Mn ions can alter the electronic structure of FeN_x active sites to make ORR kinetics easier to understand.

Other kinds of dual-atom catalysts were investigated as well as various metal atoms utilized in the construction of the catalysts. High-temperature pyrolysis was used by Ye et al. to synthesize Fe_1–N–C, Fe_2–N–C, and Fe_3–N–C (Figure 4.14a–c). Prior to calcining at 800°C, the precursor materials with a predetermined number of Fe atoms were typically encased in a zeolitic imidazolate framework (ZIF-8). Because the Fe–Fe bond at 2.46 did not appear in the FT-EXAFS spectra, this indicates that the isolated Fe atoms are different from those seen in Fe foil and FePc. Fe_2 and Fe_3 clusters, on the other hand, had Fe–Fe bonds, and the strength of these connections rose as the number of Fe atoms in the clusters grew, indicating a stronger tendency in Fe–Fe bonding (Figure 4.14d and e). EXAFS was used to confirm the coordination number of Fe–Fe bonds in Fe_2–N–C and Fe_3–N–C samples were 1.2 and 2.4, respectively (Figure 4.14g and i). For Fe_1–N–C, Fe species primarily existed with Fe-N_4 coordination structure (Figure 4.14h). According to these results, the superoxo-like intermediate species were preferred on Fe_1–N–C, whereas the peroxo-like intermediate were preferred on Fe_2–N–C and Fe_3–N–C.[95] Because of the extended O–O bond, the peroxo-like O_2 had greater adsorption energy and could be activated more readily. Fe_2–N–C had more catalytic sites and better bonding hybridization between the Fe 3d and O 2p orbitals than Fe_3–N–C, according to the study's authors. Graphitization

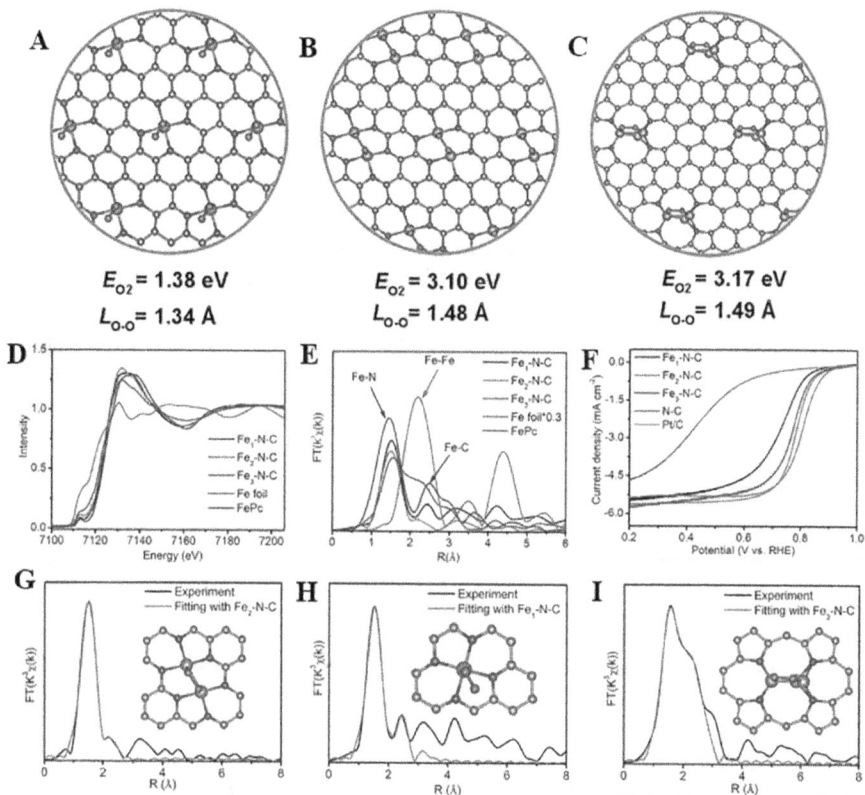

FIGURE 4.14 (a) Superoxo-like adsorption at Fe_1–N–C; (b) Peroxo-like adsorption at Fe_2–N–C; (c) Peroxo-like adsorption at Fe_3–N–C; (d) Normalized Fe K-edge XANES spectra of the Fe_1–N–C, Fe_2–N–C, and Fe_3–N–C in reference to Fe foil and FePc; (e) k^3-Weighted Fourier-transformed Fe K-edge EXAFS spectra. k^3-Weighted Fourier-transform experimental Fe K-edge EXAFS spectrum (black line) and the fitting curve (red line); (f) ORR polarization curves recorded in O_2-saturated 0.5 M H_2SO_4 solution at a sweep rate of 10 mV·s^{-1} and rotation speed of 1,600 rpm; (g) Fe_2–N–C; (h) Fe_1–N–C and; (i) Fe_3–N–C. The insets of (g),(h) and (i) show the optimized structural model of Fe_2–N–C, Fe_1–N–C and Fe_3–N–C (Fe green, N blue, C gray and O red) through theoretical calculations, respectively. (Reproduced from the study by Wei Ye et al.[95] with permission from Elsevier.)

of N-doped carbon was also aided by the presence of Fe_2–N–C, as was the formation of pyridinic-N. Fe_2–N–C was very active in ORR catalysis due to the aforementioned reasons (Figure 4.14f). Fe_2N_6 was also demonstrated to be more active in catalyzing ORR than FeN_4 by Zhang et al.[96]

4.3 INTEGRATING SINGLE-ATOM CATALYSTS INTO CATHODE CATALYST LAYERS

The MEA performance of the H_2–O_2/air PEMFCs using M–N_x/C catalysts is far inferior to that of commercial Pt/C catalysts, despite the fact that M–N_x/C catalysts showed high activity and stability in RDE measurements. The reasons for this phenomenon are as follows: the catalyst loading of M–N$_x$/C catalyst (usually 1–4 mg cm^{-2}) must be much higher than that of commercial Pt/C catalyst (usually 0.2–0.5 mg cm^{-2}) to achieve satisfactory MEA performance, resulting in a thicker catalyst layer (up to 100 μm). In this case, mass transfer resistance increases and water and heat management turns more difficult.[97–99] To achieve competitive ORR performance with commercial Pt/C (generally 100–250 g cm^{-2} for 20 wt% Pt/C),[100–102] most studies often used high loadings of M–N$_x$/C catalyst (400–800 μg cm^{-2}). Fortunately, the disadvantageous mass transfer effect from thick catalyst film is almost eliminated by high-speed rotation in RDE test. In this case, apart from optimization of the M–N$_x$/C catalyst on the RDE level, rational design of the electrode structure based on investigation of structure–activity relationship of MEA is also of great significance for acquiring high-performance MEA, and relevant improvement strategies are as follows.

4.3.1   Optimizing Activity of SACs-Based PEMFCs

4.3.1.1   Optimizing Catalyst Loadings and Ionomer Content

When it comes to CCL catalyst loading and ionomer content, factors such as catalyst activity, density of active sites and surface area (such as tapping density, pore shape and surface area) all have a major effect on MEA performance. After tweaking the CCL to maximize performance, Ye et al. have developed an ultra-high Pmax under H$_2$/air for their NPMC-based MEA.[103] They thoroughly examined the impact of MEA performance on ionomer content and catalyst loadings (Figure 4.15). The MEA with catalyst loadings of 2.5 mg cm^{-2} had the greatest results. In H$_2$–air PEMFCs, lower catalyst loadings result in a higher kinetic overpotential because the number of accessible active sites decreases. Conversely, large catalyst loadings result in unfavorable mass transport at high current densities. It's easier for protons to move through an ionomer with a high concentration than one with a low one, but it increases oxygen diffusion resistance and reduces electrical conductivity. Proton conductivity, electrical conductivity, and O$_2$ transport must all be balanced in the ionomer's composition.

4.3.1.2   Optimizing the Catalyst Slurry Preparation Process

Typically, the catalyst slurry consists of catalyst, ionomer, and solvent. The interfaces between the ionomer and the catalyst particle, the distribution of the ionomer, and the dispersion of the catalyst all have a significant effect on the utilization of active sites and the formation of a miniature three-phase interface, all of which have a significant effect on the performance of CCL (Figure 4.16).[104,105] Consider the following points: (i) a catalyst with a moderate particle size; (ii) a solvent or mixture of solvents with the desired dielectric constant/solubility parameters, which are determined by the physicochemical properties of the catalysts and ionomer; (iii) an effective catalyst ink dispersing method, which requires a powerful dispersion technology (ultrasonication, ball milling, and stirring, etc.) to ensure that the catalyst particles are scattered; (iv) a stable and homogenous catalyst ink, which is influenced by a variety of variables including zeta potential, ionomer, and hydrophilicity/hydrophobicity, pore shape, and particle size.

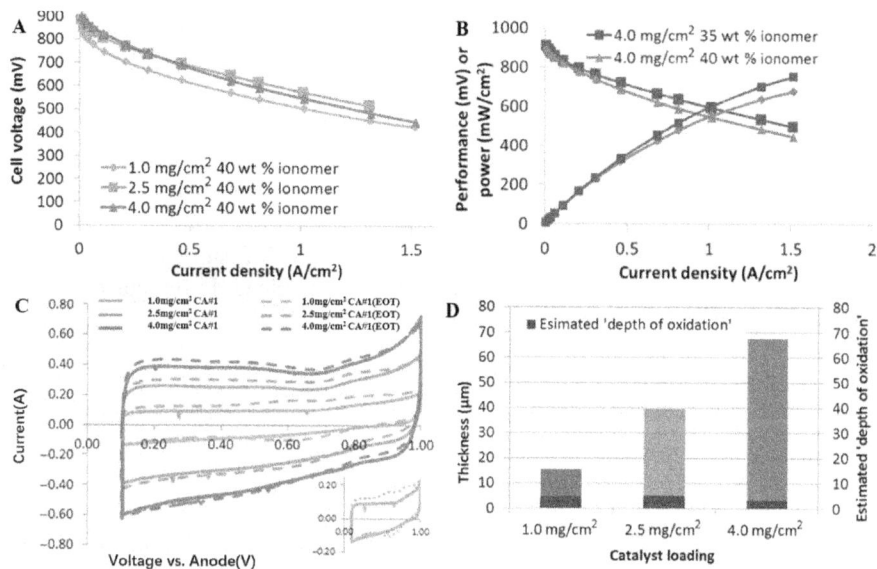

FIGURE 4.15 (a) Polarization curves of MEA (in H_2–O_2 condition) with different catalyst loadings; (b) Polarization and power density curves (in H_2–O_2 condition) for different ionomer content; (c) CVs results before and after stability test; (d) Thickness and estimated depth of oxidation for CCL. (Reproduced from the study by Dustin Banham et al.[103] with permission from Science Advances.)

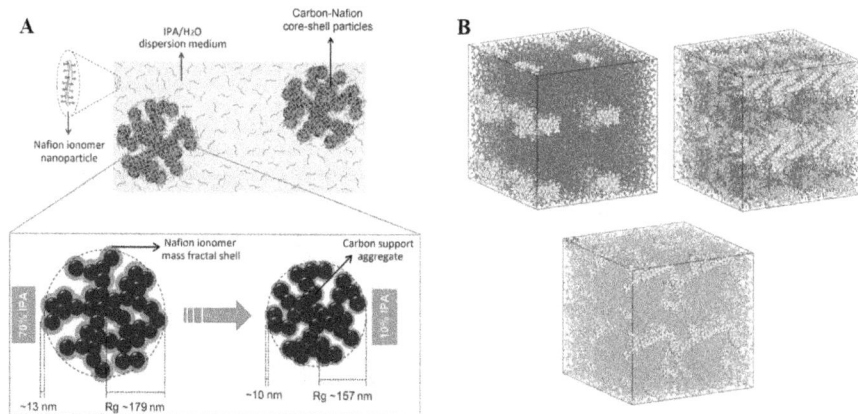

FIGURE 4.16 (a) Illustration for structure of catalyst ink. (Reproduced from the study by Rajkamal Balu et al.[104] with permission from American Chemical Society.) (b) Equilibrium structure of Nafion ionomer (supercells ($2 \times 2 \times 2$)) in three solvents (yellow, cyan, and red represents Nafion ionomer, DPG, water molecules, respectively). (Reproduced from the study by Jihye Lee et al.[105] with permission from Springer Nature.)

4.3.1.3 Cathodic Catalyst Layer Hydrophobicity Improvement

The CCL's MEA performance would plummet if significant amounts of produced water from O_2 reduction couldn't be promptly drained. Hydrophobic materials have recently been shown to enhance CCL water control while also improving MEA performance. The hydrophobicity of CCL was altered by three distinct methods of adding polytetrafluoroethylene (PTFE) into CCL, and the effect of CCL hydrophobicity on MEA performance was examined. MEA with DM-5 percent produced the greatest value (951 mW cm^{-2}) with the optimum quantity of PTFE supplied by the decal technique, and Pmax of PEMFCs was significantly enhanced. Initial current decay was also reduced.[106]

4.3.1.4 Building a 3D Cathode Architecture

A rationally designed electrode with balanced micro-, meso-, and macro-porosity is essential for constructing the triple-phase boundary and speeding up mass and charge transportation.[107,108] Jaouen et al. recently designed a 3D Fe–N–C cathode architecture using an electrospinning technique (E-ZIF-8(Fe)/PAN-Ar) (Figure 4.17).[109] For practical H_2–air PEMFCs, operando X-ray tomography showed that the water-free macroporous gaps may help transport O_2 molecules to active sites and have a favorable proton transport, both of which lead to better MEA efficiency.

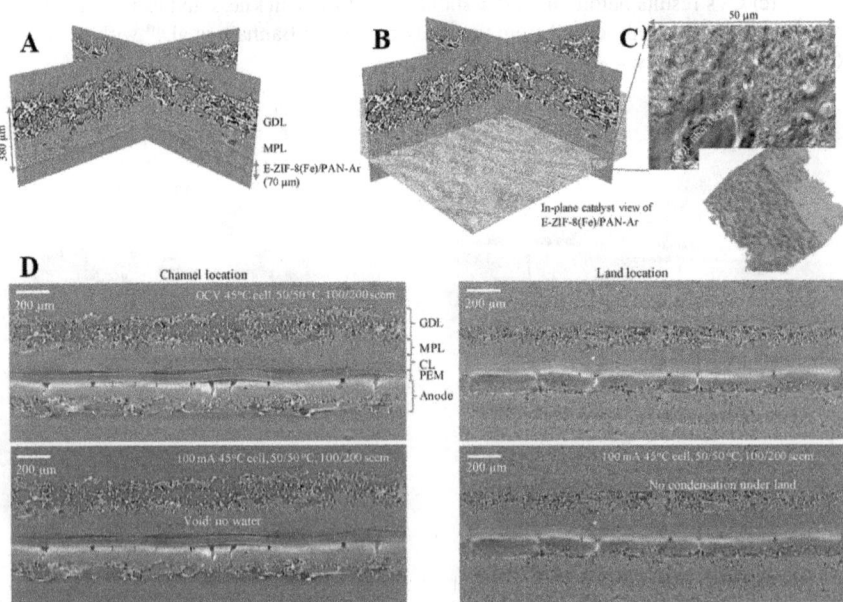

FIGURE 4.17 (a,b) Cross sections obtained by micro-CT, (c) tomograph obtained by nano-CT for E-ZIF-8(Fe)/PAN-Ar electrode analyzed by ex-situ X-ray CT; (d) 2D cross sections from 3D data from operando X-ray CT. (Reproduced from the study by Jingkun Li et al.[109] with permission from American Chemical Society.)

4.3.1.5   Optimizing the GDL's Structure

A well-designed microporous layer (MPL) with a suitable porous structure and wettability plays an essential role in producing a high-performance MEA as part of the GDL's.[110-112] The inclusion of nitrogen-doped CNTs into CCL and GDL has been shown to enhance MEA performance in Pt/C-based PEMFCs by Liao et al.[113] It was shown that adding NCNT to the MPL improved catalyst utilization rates by dispersing the catalyst and forming a profitable porous structure that promotes mass transfer. There have been a few reports to our knowledge on the progress made in optimizing the GDL for M–N$_x$/C-based PEMFCs till now. The optimization of CCL and GDL for M–N$_x$/C-based PEMFCs may be more difficult due to a thick active layer, even if previous research on Pt/C-based PEMFCs may offer some advice. Because of this, M–N$_x$/C-based PEMFCs need immediate research.

4.3.2   Stability and Enhancement Protocols

To date, the practical use of M–N$_x$/C remains a tremendous difficulty due to the low stability of MEA. A number of research studies have indicated that the MEA performance would significantly deteriorate (more than 30%) at high potentials (≥ 0.6 V) during the first few dozen hours of the PEMFCs operating, which is far from the stability objectives established by the U.S. DOE (<10% performance attenuation after 5,000 h).[114] Great efforts have been made to study the major degradation processes that influence M–N$_x$/C during fuel-cell working settings, which has been primarily recognized as: (i) oxidative assault by H$_2$O$_2$ or related free radicals;[115] (ii) demetalation of M–N$_x$ sites;[116-118] (iii) water flooding;[119,120] (iv) protonation/anion binding of active sites;[121,122] and; (v) carbon oxidation.[123,124]

The following sections summarize the methods for increasing stability based on the differentiation of degradation mechanisms.

4.3.2.1   Reducing H$_2$O$_2$ Yield and Combating Radical Attacks

An oxidative assault on the carbon surface and activity sites of the M–N$_x$/C catalyst by H$_2$O$_2$ produced by the unfavorable 2e$^-$ pathway results in a significant decline in intrinsic activity. Additionally, the Fenton reaction between H$_2$O$_2$ and Fe ion may generate ROS that target the PEM and M–N$_x$ sites, decreasing the stability of MEA.[125,126] A new study by Zhang et al. found that the Fenton reaction destroys FeN$_4$ active sites due to significant adsorption of hydroperoxyl (OH), while three hydroxyl free radicals working together causes full destruction of the FeN$_4$ active sites.[127] It has been proposed in many research that improving four-electron selectivity by manipulating the local geometric and electronic structures of M–N$_x$ is a viable strategy for decreasing hydrogen peroxide production. In addition, the introduction of H$_2$O$_2$ scavengers has recently been suggested due to the production of hydrogen peroxide and the inevitable Fenton's reaction encountered by most M–N$_x$/C catalysts. For example, Zou et al. recently revealed that the yield of H$_2$O$_2$ may be effectively decreased when the CeO$_2$ is partnered in the Fe/N/C catalysts, increasing the stability of catalysts and fuel cells.[128] When Pt–Co NPs and Co–N$_4$ sites operate together in synergy, the ORR activity and durability increase, and this is because H_2O_2 is reduced more efficiently on the stretched Pt (111) surface without encountering a thermodynamic barrier.[129]

4.3.2.2 Improving Carbon Materials' Anticorrosion Capability

H_2O_2 or radical chemical assault, as well as thermodynamically advantageous electrochemical corrosion (0.207 Vs standard hydrogen electrode), are the two main origins of carbon corrosion. To accommodate more M–N_x active sites, catalysts are often constructed to have a high specific area with a micropore-dominant porous structure. This, however, results in a low graphitization degree, poor anticorrosion, and unfavorable mass transfer during the catalysis process. As a result, a compromise must be struck between M–N_x/C graphitization and porosity.[130] Kang et al. have developed a stable and effective Fe–N/CNT catalyst that contains numerous atomic Fe–N_x species incorporated in Fe-catalyzed CNT growth.[131] The Fe–N/CNT-2 catalyst showed excellent stability in the cathode of H_2/O_2 PEMFCs, which was attributed to the graphitization degree and anticorrosive capacity improved by the CNTs network. To make M–N_x/C more stable, a simple procedure involves modifying the carbon surface. It was discovered by Zhou et al. that a surface fluorination technique for Fe/N/C decreased attenuation to 15% at 0.6 V after 100 h in H_2–O_2 PEMFCs, considerably better than a non-fluorine Fe/N/C catalyst.[132] A trifluoromethylphenyl (Ar–CF_3) group was covalently grafted onto Fe/N/C, reducing carbon corrosion and promoting mass transfer via increased hydrophobicity.

4.3.2.3 Water Flooding Management

There are two main types of water flooding: (i) micropore flooding caused by carbon oxidation's microporous hydrophily, which results in irreversible performance degradation. In the past, micropore flooding was blamed for a fast decline in early performance; however, Chen et al. recently found that monitoring variations in the double-layer capacitance using cyclic voltammetry shows that micropore flooding is not the main cause of initial instability.[133] Microporous flooding and its impact on instability are still hotly debated topics. (ii) Catalyst layer flooding is caused by water accumulating in the catalyst layer. This kind of flooding may be reversed, and any lost performance can be regained on a regular basis. In addition, research has shown that increasing the hydrophobicity of a catalyst reduces water flooding. For example, Zhou et al. showed that an Ar–CF_3 group decorated Fe/N/C catalyst with hydrophobic properties not only reduced carbon corrosion but also restricted water flooding.[132] When the temperature was raised from 1,050°C to 1,150°C, Dodelet et al. proposed that hydrophobicity of carbon support increased because the amount of heteroatom on carbon surface decreased, which reduced water flooding.[134]

4.3.2.4 Preventing Metal Species from Being Demetalized

In the case of Fe/N/C, Fe demetalation may be ascribed to three different causes. Firstly, in acidic media, iron species (iron carbide, iron clusters, etc.) dissolve in Fe and the enhanced Fenton reaction decomposes the PEM. Demetalation of FeN_x sites occurs during fuel-cell operation, resulting in fast deterioration of performance. This degradation occurs because of the removal of certain unstable species through acid-washing before integrating into MEA.[42,46,55,124,135] Second, when water containing protons and dissolved oxygen passes through a micropore, the thermodynamic stability of $Fe-N_4$-like active sites, as described by Dodelet et al., is disrupted, leading to demetalation of these active sites and rapid degradation of their initial current density. Thirdly, the H_2O_2 assault and/or carbon electro-oxidation may speed up the decomposition of the carbon surface, which in turn destroys FeN_xC_y sites inadvertently. As shown by Choi et al., Fe demetalation from low-activity iron species is not the primary cause of performance deterioration at low potential (0.7 V). But carbon oxidation may damage FeN_xC_y sites indirectly at high potential (>0.9 V).[124] One important factor affecting the stability of SACs is the interaction between single-metal atoms and the anchoring sites on the carbon surface.[46,55] Stability may be improved by creating strong binding $Fe-N_x$ sites on the carbon surface. According to recent research, Shui et al. created a novel kind of $Pt_1@Fe-N-C$ catalyst, which included an active species of $Pt_1-O_2-Fe_1-N_4$. Prior to Fe–N–C-based PEMFCs, the $Pt_1@Fe-N-C$ catalyst shown superior durability in H_2/O_2 PEMFCs.[136] An increase in long-term stability may result from the creation of a stable moiety of Pt_1, O_2, Fe_1, N_4, where the Pt_1-O_2 may stabilize the $Fe-N_4$ moiety in acidic conditions, inhibiting the Fe catalytic Fernton reaction.

4.4 SUMMARY AND OUTLOOK

For ORR and their applications in PEM fuel cells, current work on atomically distributed M–N–C catalysts has been extensively described in this chapter, with a particular focus on MEA performance. To top it all off, the most significant causes of MEA deterioration were uncovered as well as possible solutions to both ORR's activity and MEA's long-term durability. Priorities for the development of highly active and long-lasting atomically distributed M–N–C catalysts include: (i) A well-atomized dispersion increases metal loading. High metal loading in SACs and DACs implies more active sites, which is better for ORR activity. However, owing to metal aggregation at high temperatures, many previously described SACs/DACs have a low loading. As a result, I designing an efficient synthesis technique for SACs/DACs with high metal loading is required; and (ii) increasing intrinsic activity is required. Other heteroatoms (e.g., S, O, P, F, etc.) or dual-metal atoms introduced to the M–N–C catalyst can better regulate the electronic structure of the atomic metal active sites, facilitating the facial cleavage of the O–O bond and improving ORR activity; (iii) increasing active site exposure and mass transfer by constructing a balanced porous structure; (iv) because the local structure of metal active sites changes dynamically in a practical electrochemical environment,[137,138] more efforts should be made to investigate the accurate structure–activity relationship of SACs/DACs using in situ and operando characterization tools, as well as theoretical calculations, in order to provide rational design for high-performance SACs/DACs; (v) the research of DACs is still in its early stages. Controlling the precise amount of metal atoms in homonuclear DACs remains a major problem, and the introduction of homonuclear DACs throughout the synthesis process makes investigating the structure–activity connection in heteronuclear DACs challenging.

As a result, developing improved preparation technologies to regulate the amount and kind of metal dimers is critical. In addition to developing improved ORR catalysts, achieving RDE-level ORR high-performance of SACs/DACs in practical PEMFCs is critical, but it is fraught with difficulties. The combination of catalyst activity and electrode technology is critical for MEA performance. Many published research have attempted to enhance the RDE-level ORR performance of SACs/DACs in recent decades, but less has been spent on investigating MEA, particularly M–N–C generated PEMFCs. As a result, the following suggestions should be considered: (i) developing in-situ characterization techniques for monitoring the water/heat distribution, and surface/interface among Nafion membrane, CCL, and GDL; (ii) establishing in-situ characterization methods for monitoring the water/heat distribution, as well as the surface/interface between the Nafion membrane, the CCL, and the GDL; (iii) To provide a perspective for optimizing ORR catalysts, the structure–activity relationship in MEA should be systematically studied by combining theoretical simulation and advanced characterization; (iv) the stability and durability of SACs/DACs in PEMFCs during practical operation must be further strengthened. It's worth noting that, in addition to the M–N$_x$ sites, pyridine nitrogen is also an active species, but it's susceptible to protonation in an acidic media, resulting in performance deterioration.[139]

Advanced ex-situ and in-situ/operando characterization methods are also highly promoted for distinguishing deterioration processes during fuel-cell operation. (v) most published literature to date have emphasized the highest power density of PEMFCs in the low-voltage range (0.5 V), which is used as an assessment criteria for MEA performance. It is worth mentioning, however, that the actual working voltage of PEMFCs should be kept around 0.6–0.9 V to make water and heat management simpler and to achieve high energy efficiency. To offer an objective and realistic assessment of MEA performance, current densities at the specified voltages (0.6–0.9 V) should also be supplied, in addition to maximum power density. (vi) It is worth mentioning that MEA performance improvement is a massive, systematic engineering project encompassing everything from catalyst design to electrode preparation to effective fuel-cell operation, and it takes a lot of personnel, material resources, and time. Machine learning techniques should be promoted to optimize MEA performance efficiently and correctly.[140] (vii) In future works, more test and technical information, as well as performance assessment criteria, will be made available to better compare findings from various organizations. In summary, single-atom catalysis has opened up a new avenue for developing high-performance M–N–C catalysts that can replace noble metal catalysts, but there is still a long way to go before they can be used in PEMFCs.

REFERENCES

1. S. Chu and A. Majumdar, *Nature,* 2012, 488, 294–303.
2. Z. P. Cano, D. Banham, S. Ye, A. Hintennach, J. Lu, M. Fowler and Z. Chen, *Nature Energy,* 2018, 3, 279–289.
3. M. K. Debe, *Nature,* 2012, 486, 43–51.
4. Y. Nie, L. Li and Z. Wei, *Chemical Society Reviews,* 2015, 44, 2168–2201.
5. M. Shao, Q. Chang, J.-P. Dodelet and R. Chenitz, *Chemical Reviews,* 2016, 116, 3594–3657.
6. J. Fan, M. Chen, Z. Zhao, Z. Zhang, S. Ye, S. Xu, H. Wang and H. Li, *Nature Energy,* 2021, 6, 475–486.

7. B. P. Setzler, Z. Zhuang, J. A. Wittkopf and Y. Yan, *Nature Nanotechnology,* 2016, 11, 1020–1025.
8. W. Yu, M. D. Porosoff and J. G. Chen, *Chemical Reviews,* 2012, 112, 5780–5817.
9. R. Jasinski, *Nature,* 1964, 201, 1212–1213.
10. M. Lefèvre, E. Proietti, F. Jaouen and J.-P. Dodelet, *Science,* 2009, 324, 71–74.
11. E. Proietti, F. Jaouen, M. Lefèvre, N. Larouche, J. Tian, J. Herranz and J.-P. Dodelet, *Nature Communications,* 2011, 2, 1–9.
12. G. Wu, K. L. More, C. M. Johnston and P. Zelenay, *Science,* 2011, 332, 443–447.
13. D. Deng, L. Yu, X. Chen, G. Wang, L. Jin, X. Pan, J. Deng, G. Sun and X. Bao, *Angewandte Chemie,* 2013, 125, 389–393.
14. B. Qiao, A. Wang, X. Yang, L. F. Allard, Z. Jiang, Y. Cui, J. Liu, J. Li and T. Zhang, *Nature Chemistry,* 2011, 3, 634–641.
15. A. Zitolo, V. Goellner, V. Armel, M.-T. Sougrati, T. Mineva, L. Stievano, E. Fonda and F. Jaouen, *Nature Materials,* 2015, 14, 937–942.
16. H. T. Chung, D. A. Cullen, D. Higgins, B. T. Sneed, E. F. Holby, K. L. More and P. Zelenay, *Science,* 2017, 357, 479–484.
17. H. Zhang, S. Hwang, M. Wang, Z. Feng, S. Karakalos, L. Luo, Z. Qiao, X. Xie, C. Wang and D. Su, *Journal of the American Chemical Society,* 2017, 139, 14143–14149.
18. Y. Chen, S. Ji, Y. Wang, J. Dong, W. Chen, Z. Li, R. Shen, L. Zheng, Z. Zhuang and D. Wang, *Angewandte Chemie,* 2017, 129, 7041–7045.
19. T. Wu, Y. Li, Y. Li, J. Hong, J. Wu, J. Mao, Y. Wu, Q. Cai, C. Yuan and L. Dai, *Nanotechnology,* 2019, 30, 305402.
20. N. Jia, Q. Xu, F. Zhao, H.-X. Gao, J. Song, P. Chen, Z. An, X. Chen and Y. Chen, *ACS Applied Energy Materials,* 2018, 1, 4982–4990.
21. Z. K. Yang, C.-Z. Yuan and A.-W. Xu, *ACS Energy Letters,* 2018, 3, 2383–2389.
22. Y. Gao, Z. Cai, X. Wu, Z. Lv, P. Wu and C. Cai, *ACS Catalysis,* 2018, 8, 10364–10374.
23. L. Jiao, G. Wan, R. Zhang, H. Zhou, S. H. Yu and H. L. Jiang, *Angewandte Chemie International Edition,* 2018, 57, 8525–8529.
24. L. Zheng, S. Yu, X. Lu, W. Fan, B. Chi, Y. Ye, X. Shi, J. Zeng, X. Liand, S. Liao, *ACS Applied Materials & Interfaces,* 2020, 12, 13878–13887.
25. C. Tang, H.-F. Wang, J.-Q. Huang, W. Qian, F. Wei, S.-Z. Qiaoand, Q. Zhang, *Electrochemical Energy Reviews,* 2019, 2, 332–371.
26. R. Xing, T. Zhou, Y. Zhou, R. Ma, Q. Liu, J. Luoand, J. Wang, *Nano-Micro Letters,* 2018, 10, 1–14.
27. M. Qiao, Y. Wang, Q. Wang, G. Hu, X. Mamat, S. Zhang and S. Wang, *Angewandte Chemie International Edition,* 2020, 59, 2688–2694.
28. H. Adabi, A. Shakouri, N. Ul Hassan, J. R. Varcoe, B. Zulevi, A. Serov, J. R. Regalbuto and W. E. Mustain, *Nature Energy,* 2021, 6, 834–843.
29. Y. Mun, S. Lee, K. Kim, S. Kim, S. Lee, J. W. Han and J. Lee, *Journal of the American Chemical Society,* 2019, 141, 6254–6262.
30. C. X. Zhao, B. Q. Li, J. N. Liu and Q. Zhang, *Angewandte Chemie International Edition,* 2021, 60, 4448–4463.
31. Y. Chen, Z. Li, Y. Zhu, D. Sun, X. Liu, L. Xu and Y. Tang, *Advanced Materials,* 2019, 31, 1806312.
32. Y. Deng, B. Chi, X. Tian, Z. Cui, E. Liu, Q. Jia, W. Fan, G. Wang, D. Dang and M. Li, *Journal of Materials Chemistry A,* 2019, 7, 5020–5030.
33. H. Shen, E. Gracia-Espino, J. Ma, H. Tang, X. Mamat, T. Wagberg, G. Hu and S. Guo, *Nano Energy,* 2017, 35, 9–16.
34. C. Zhu, Q. Shi, B. Z. Xu, S. Fu, G. Wan, C. Yang, S. Yao, J. Song, H. Zhou and D. Du, *Advanced Energy Materials,* 2018, 8, 1801956.
35. D. Lyu, Y. B. Mollamahale, S. Huang, P. Zhu, X. Zhang, Y. Du, S. Wang, M. Qing, Z. Q. Tian and P. K. Shen, *Journal of Catalysis,* 2018, 368, 279–290.

36. Y. Lin, P. Liu, E. Velasco, G. Yao, Z. Tian, L. Zhang and L. Chen, *Advanced Materials*, 2019, 31, 1808193.
37. H. Shen, E. Gracia-Espino, J. Ma, K. Zang, J. Luo, L. Wang, S. Gao, X. Mamat, G. Hu and T. Wagberg, *Angewandte Chemie*, 2017, 129, 13988–13992.
38. K. Yuan, S. Sfaelou, M. Qiu, D. Lützenkirchen-Hecht, X. Zhuang, Y. Chen, C. Yuan, X. Feng and U. Scherf, *ACS Energy Letters*, 2017, 3, 252–260.
39. M. Karuppannan, J. E. Park, H. E. Bae, Y.-H. Cho and O. J. Kwon, *Nanoscale*, 2020, 12, 2542–2554.
40. J.-C. Li, H. Zhong, M. Xu, T. Li, L. Wang, Q. Shi, S. Feng, Z. Lyu, D. Liu and D. Du, *Science China Materials*, 2020, 63, 965–971.
41. X. Fu, N. Li, B. Ren, G. Jiang, Y. Liu, F. M. Hassan, D. Su, J. Zhu, L. Yang and Z. Bai, *Advanced Energy Materials*, 2019, 9, 1970031.
42. E. Luo, Y. Chu, J. Liu, Z. Shi, S. Zhu, L. Gong, J. Ge, C. H. Choi, C. Liu and W. Xing, *Energy & Environmental Science*, 2021, 14, 2158–2185.
43. W. Wang, Q. Jia, S. Mukerjee and S. Chen, *ACS Catalysis*, 2019, 9, 10126–10141.
44. C. H. Choi, C. Baldizzone, G. Polymeros, E. Pizzutilo, O. Kasian, A. K. Schuppert, N. Ranjbar Sahraie, M.-T. Sougrati, K. J. Mayrhofer and F. Jaouen, *ACS Catalysis*, 2016, 6, 3136–3146.
45. P. Yin, T. Yao, Y. Wu, L. Zheng, Y. Lin, W. Liu, H. Ju, J. Zhu, X. Hong and Z. Deng, *Angewandte Chemie*, 2016, 128, 10958–10963.
46. H. Zhang, W. Zhou, T. Chen, B. Y. Guan, Z. Li and X. W. D. Lou, *Energy & Environmental Science*, 2018, 11, 1980–1984.
47. G. Wan, P. Yu, H. Chen, J. Wen, C. J. Sun, H. Zhou, N. Zhang, Q. Li, W. Zhao, B. Xie, T. Li and J. Shi, *Small*, 2018, 14, 1704319.
48. C. Deng, R. He, W. Shen, M. Li and T. Zhang, *Physical Chemistry Chemical Physics*, 2019, 21, 6900–6907.
49. X. X. Wang, D. A. Cullen, Y. T. Pan, S. Hwang, M. Wang, Z. Feng, J. Wang, M. H. Engelhard, H. Zhang and Y. He, *Advanced Materials*, 2018, 30, 1706758.
50. Y. Han, Y.-G. Wang, W. Chen, R. Xu, L. Zheng, J. Zhang, J. Luo, R.-A. Shen, Y. Zhu, W.-C. Cheong, C. Chen, Q. Peng and Y. Li, *Journal of the American Chemical Society*, 2017, 139, 17269–17272.
51. Y. Chen, R. Gao, S. Ji, H. Li, K. Tang, P. Jiang, H. Hu, Z. Zhang, H. Hao and Q. Qu, *Angewandte Chemie International Edition*, 2021, 60, 3212–3221.
52. X. Wei, D. Zheng, M. Zhao, H. Chen, X. Fan, B. Gao, L. Gu, Y. Guo, J. Qin and J. Wei, *Angewandte Chemie*, 2020, 132, 14747–14754.
53. Y. Qu, Z. Li, W. Chen, Y. Lin, T. Yuan, Z. Yang, C. Zhao, J. Wang, C. Zhao, X. Wang, F. Zhou, Z. Zhuang, Y. Wu and Y. Li, *Nature Catalysis*, 2018, 1, 781–786.
54. X. Ge, A. Sumboja, D. Wuu, T. An, B. Li, F. T. Goh, T. A. Hor, Y. Zong and Z. Liu, *ACS Catalysis*, 2015, 5, 4643–4667.
55. A. Wang, J. Li and T. Zhang, *Nature Reviews Chemistry*, 2018, 2, 65–81.
56. L. Bai, C. Hou, X. Wen and J. Guan, *ACS Applied Energy Materials*, 2019, 2, 4755–4762.
57. A. M. Patel, S. Ringe, S. Siahrostami, M. Bajdich, J. K. Nørskov and A. R. Kulkarni, *The Journal of Physical Chemistry C*, 2018, 122, 29307–29318.
58. L. Cui, L. Cui, Z. Li, J. Zhang, H. Wang, S. Lu and Y. Xiang, *Journal of Materials Chemistry A*, 2019, 7, 16690–16695.
59. H. Wu, H. Li, X. Zhao, Q. Liu, J. Wang, J. Xiao, S. Xie, R. Si, F. Yang, S. Miao, X. Guo, G. Wang and X. Bao, *Energy & Environmental Science*, 2016, 9, 3736–3745.
60. F. Li, G.-F. Han, H.-J. Noh, S.-J. Kim, Y. Lu, H. Y. Jeong, Z. Fu and J.-B. Baek, *Energy & Environmental Science*, 2018, 11, 2263–2269.
61. D. Wang, C. Ao, X. Liu, S. Fang, Y. Lin, W. Liu, W. Zhang, X. Zheng, L. Zhang and T. Yao, *ACS Applied Energy Materials*, 2019, 2, 6497–6504.

62. J. Li, M. Chen, D. A. Cullen, S. Hwang, M. Wang, B. Li, K. Liu, S. Karakalos, M. Lucero and H. Zhang, *Nature Catalysis*, 2018, 1, 935–945.
63. L. Bai, Z. Duan, X. Wen, R. Si and J. Guan, *Applied Catalysis B: Environmental*, 2019, 257, 117930.
64. G. Zhu, F. Liu, Y. Wang, Z. Wei and W. Wang, *Physical Chemistry Chemical Physics*, 2019, 21, 12826–12836.
65. Y. Yang, K. Mao, S. Gao, H. Huang, G. Xia, Z. Lin, P. Jiang, C. Wang, H. Wang and Q. Chen, *Advanced Materials*, 2018, 30, 1801732.
66. F. Luo, A. Roy, L. Silvioli, D. A. Cullen, A. Zitolo, M. T. Sougrati, I. C. Oguz, T. Mineva, D. Teschner, S. Wagner, J. Wen, F. Dionigi, U. I. Kramm, J. Rossmeisl, F. Jaouen and P. Strasser, *Nature Materials*, 2020, 19, 1215–1223.
67. J. Li, S. Chen, N. Yang, M. Deng, S. Ibraheem, J. Deng, J. Li, L. Li and Z. Wei, *Angewandte Chemie International Edition*, 2019, 58, 7035–7039.
68. F. Li, Y. Bu, G.-F. Han, H.-J. Noh, S.-J. Kim, I. Ahmad, Y. Lu, P. Zhang, H. Y. Jeong and Z. Fu, *Nature Communications*, 2019, 10, 1–7.
69. M. Zhu, C. Zhao, X. Liu, X. Wang, F. Zhou, J. Wang, Y. Hu, Y. Zhao, T. Yao, L.-M. Yang and Y. Wu, *ACS Catalysis*, 2021, 11, 3923–3929.
70. T. Wang, X. Cao, H. Qin, L. Shang, S. Zheng, F. Fang and L. Jiao, *Angewandte Chemie*, 2021, 133, 21407–21411.
71. E. Luo, H. Zhang, X. Wang, L. Gao, L. Gong, T. Zhao, Z. Jin, J. Ge, Z. Jiang, C. Liu and W. Xing, *Angewandte Chemie*, 2019, 131, 12599–12605.
72. J. Guo, Z. Mao, X. Yan, R. Su, P. Guan, B. Xu, X. Zhang, G. Qin and S. J. Pennycook, *Nano Energy*, 2016, 28, 261–268.
73. S. Liu, Z. Li, C. Wang, W. Tao, M. Huang, M. Zuo, Y. Yang, K. Yang, L. Zhang and S. Chen, *Nature Communications*, 2020, 11, 1–11.
74. C. H. Choi, M. Kim, H. C. Kwon, S. J. Cho, S. Yun, H.-T. Kim, K. J. Mayrhofer, H. Kim and M. Choi, *Nature Communications*, 2016, 7, 1–9.
75. S. Yang, J. Kim, Y. J. Tak, A. Soon and H. Lee, *Angewandte Chemie International Edition*, 2016, 55, 2058–2062.
76. S. Yang, Y. J. Tak, J. Kim, A. Soon and H. Lee, *ACS Catalysis*, 2017, 7, 1301–1307.
77. J. Liu, M. Jiao, L. Lu, H. M. Barkholtz, Y. Li, Y. Wang, L. Jiang, Z. Wu, D.-J. Liu and L. Zhuang, *Nature Communications*, 2017, 8, 1–10.
78. J. Liu, M. Jiao, B. Mei, Y. Tong, Y. Li, M. Ruan, P. Song, G. Sun, L. Jiang, Y. Wang, Z. Jiang, L. Gu, Z. Zhou and W. Xu, *Angewandte Chemie*, 2019, 131, 1175–1179.
79. M. Xiao, J. Zhu, G. Li, N. Li, S. Li, Z. P. Cano, L. Ma, P. Cui, P. Xu and G. Jiang, *Angewandte Chemie International Edition*, 2019, 58, 9640–9645.
80. C. Zhang, J. Sha, H. Fei, M. Liu, S. Yazdi, J. Zhang, Q. Zhong, X. Zou, N. Zhao and H. Yu, *ACS Nano*, 2017, 11, 6930–6941.
81. M. Xiao, L. Gao, Y. Wang, X. Wang, J. Zhu, Z. Jin, C. Liu, H. Chen, G. Li and J. Ge, *Journal of the American Chemical Society*, 2019, 141, 19800–19806.
82. D. Kan, X. Zhang, Z. Fu, Y. Zhang, Y. Zhao and Z. Yang, *Physical Chemistry Chemical Physics*, 2018, 20, 10302–10310.
83. W. Xiang, Y. Zhao, Z. Jiang, X. Li, H. Zhang, Y. Sun, Z. Ning, F. Du, P. Gao and J. Qian, *Journal of Materials Chemistry A*, 2018, 6, 23366–23377.
84. R. Gao, Y. Yin, F. Niu, A. Wang, S. Li, H. Dong and S. Yang, *ChemElectroChem*, 2019, 6, 1824–1830.
85. D. Zhang, W. Chen, Z. Li, Y. Chen, L. Zheng, Y. Gong, Q. Li, R. Shen, Y. Han and W.-C. Cheong, *Chemical Communications*, 2018, 54, 4274–4277.
86. H. Li, Y. Wen, M. Jiang, Y. Yao, H. Zhou, Z. Huang, J. Li, S. Jiao, Y. Kuang and S. Luo, *Advanced Functional Materials*, 2021, 2011289.
87. X. Han, X. Ling, D. Yu, D. Xie, L. Li, S. Peng, C. Zhong, N. Zhao, Y. Deng and W. Hu, *Advanced Materials*, 2019, 31, 1905622.

88. J. Wang, Z. Huang, W. Liu, C. Chang, H. Tang, Z. Li, W. Chen, C. Jia, T. Yao and S. Wei, *Journal of the American Chemical Society,* 2017, 139, 17281–17284.

89. M. Xiao, Y. Chen, J. Zhu, H. Zhang, X. Zhao, L. Gao, X. Wang, J. Zhao, J. Ge and Z. Jiang, *Journal of the American Chemical Society,* 2019, 141, 17763–17770.

90. X. Zhao, X. Liu, B. Huang, P. Wang and Y. Pei, *Journal of Materials Chemistry A,* 2019, 7, 24583–24593.

91. L. Zhang, J. M. T. A. Fischer, Y. Jia, X. Yan, W. Xu, X. Wang, J. Chen, D. Yang, H. Liu and L. Zhuang, *Journal of the American Chemical Society,* 2018, 140, 10757–10763.

92. Y. Zhou, W. Yang, W. Utetiwabo, Y.-M. Lian, X. Yin, L. Zhou, P. Yu, R. Chen and S. Sun, *The Journal of Physical Chemistry Letters,* 2020, 11, 1404–1410.

93. J. Xu, S. Lai, D. Qi, M. Hu, X. Peng, Y. Liu, W. Liu, G. Hu, H. Xu and F. Li, *Nano Research,* 2021, 14, 1374–1381.

94. S. Gong, C. Wang, P. Jiang, L. Hu, H. Lei and Q. Chen, *Journal of Materials Chemistry A,* 2018, 6, 13254–13262.

95. W. Ye, S. Chen, Y. Lin, L. Yang, S. Chen, X. Zheng, Z. Qi, C. Wang, R. Long and M. Chen, *Chem,* 2019, 5, 2865–2878.

96. N. Zhang, T. Zhou, J. Ge, Y. Lin, Z. Du, W. Wang, Q. Jiao, R. Yuan, Y. Tian and W. Chu, *Matter,* 2020, 3, 509–521.

97. X. Wan, X. Liu, Y. Li, R. Yu, L. Zheng, W. Yan, H. Wang, M. Xu and J. Shui, *Nature Catalysis,* 2019, 2, 259–268.

98. F. Jaouen, D. Jones, N. Coutard, V. Artero, P. Strasser and A. Kucernak, *Johnson Matthey Technology Review,* 2018, 62, 231–255.

99. R. Wu, Y. Song, X. Huang, S. Chen, S. Ibraheem, J. Deng, J. Li, X. Qi and Z. Wei, *Journal of Power Sources,* 2018, 401, 287–295.

100. H. Peng, Z. Mo, S. Liao, H. Liang, L. Yang, F. Luo, H. Song, Y. Zhong and B. Zhang, *Scientific Reports,* 2013, 3, 1–7.

101. W.-J. Jiang, L. Gu, L. Li, Y. Zhang, X. Zhang, L.-J. Zhang, J.-Q. Wang, J.-S. Hu, Z. Wei and L.-J. Wan, *Journal of the American Chemical Society,* 2016, 138, 3570–3578.

102. T. Liu, P. Zhao, X. Hua, W. Luo, S. Chen and G. Cheng, *Journal of Materials Chemistry A,* 2016, 4, 11357–11364.

103. D. Banham, T. Kishimoto, Y. Zhou, T. Sato, K. Bai, J.-I. Ozaki, Y. Imashiro and S. Ye, *Science Advances,* 2018, 4, eaar7180.

104. R. Balu, N. R. Choudhury, J. P. Mata, L. de Campo, C. Rehm, A. J. Hill and , N. K. Dutta, *ACS Applied Materials & Interfaces,* 2019, 11, 9934–9946.

105. J. H. Lee, G. Doo, S. H. Kwon, S. Choi, H.-T. Kim and S. G. Lee, *Scientific Reports,* 2018, 8, 1–8.

106. X. Zhang, Q. Liu and J. Shui, *ChemElectroChem,* 2020, 7, 1775–1780.

107. X. Yang, G. Zhang, L. Du, J. Zhang, F.-K. Chiang, Y. Wen, X. Wang, Y. Wu, N. Chen and S. Sun, *ACS Applied Materials & Interfaces,* 2020, 12, 13739–13749.

108. C. Zhang, Y. C. Wang, B. An, R. Huang, C. Wang, Z. Zhou and W. Lin, *Advanced Materials,* 2017, 29, 1604556.

109. J. Li, S. Brüller, D. C. Sabarirajan, N. Ranjbar-Sahraie, M. T. Sougrati, S. Cavaliere, D. Jones, I. V. Zenyuk, A. Zitolo and F. Jaouen, *ACS Applied Energy Materials,* 2019, 2, 7211–7222.

110. T. Kitahara and H. Nakajima, *International Journal of Hydrogen Energy,* 2016, 41, 9547–9555.

111. J. Zhou, S. Shukla, A. Putz and M. Secanell, *Electrochimica Acta,* 2018, 268, 366–382.

112. J. T. Gostick, M. A. Ioannidis, M. W. Fowler and M. D. Pritzker, *Electrochemistry Communications,* 2009, 11, 576–579.

113. S. Hou, B. Chi, G. Liu, J. Ren, H. Song and S. Liao, *Electrochimica Acta,* 2017, 253, 142–150.

114. X. X. Wang, M. T. Swihart and G. Wu, *Nature Catalysis,* 2019, 2, 578–589.

115. C. H. Choi, H.-K. Lim, M. W. Chung, G. Chon, N. R. Sahraie, A. Altin, M.-T. Sougrati, L. Stievano, H. S. Oh and E. S. Park, *Energy & Environmental Science,* 2018, 11, 3176–3182.
116. M. Ferrandon, X. Wang, A. J. Kropf, D. J. Myers, G. Wu, C. M. Johnston and P. Zelenay, *Electrochimica Acta,* 2013, 110, 282–291.
117. R. Chenitz and U. Kramm, *Energy & Environmental Science* 2018, 11, 365–382.
118. V. Goellner, C. Baldizzone, A. Schuppert, M. T. Sougrati, K. Mayrhofer and F. Jaouen, *Physical Chemistry Chemical Physics,* 2014, 16, 18454–18462.
119. H. Li, Y. Tang, Z. Wang, Z. Shi, S. Wu, D. Song, J. Zhang, K. Fatih, J. Zhang and H. Wang, *Journal of Power Sources,* 2008, 178, 103–117.
120. Y. Li, P. Pei, Z. Wu, P. Ren, X. Jia, D. Chen and S. Huang, *Applied Energy,* 2018, 224, 42–51.
121. D. Banham, S. Ye, K. Pei, J.-I. Ozaki, T. Kishimoto and Y. Imashiro, *Journal of Power Sources,* 2015, 285, 334–348.
122. J. Herranz, F. Jaouen, M. Lefèvre, U. I. Kramm, E. Proietti, J.-P. Dodelet, P. Bogdanoff, S. Fiechter, I. Abs-Wurmbach and P. Bertrand, *The Journal of Physical Chemistry C,* 2011, 115, 16087–16097.
123. G. Zhang, R. Chenitz, M. Lefèvre, S. Sun and J.-P. Dodelet, *Nano Energy,* 2016, 29, 111–125.
124. C. H. Choi, C. Baldizzone, J. P. Grote, A. K. Schuppert, F. Jaouen and K. J. Mayrhofer, *Angewandte Chemie International Edition,* 2015, 54, 12753–12757.
125. A. C. Fernandes and E. A. Ticianelli, *Journal of Power Sources,* 2009, 193, 547–554.
126. K. Hongsirikarn, X. Mo, J. G. Goodwin Jr and S. Creager, *Journal of Power Sources,* 2011, 196, 3060–3072.
127. J. Chen, X. Yan, C. Fu, Y. Feng, C. Lin, X. Li, S. Shen, C. Ke and J. Zhang, *ACS Applied Materials & Interfaces,* 2019, 11, 37779–37786.
128. H. Wei, X. Su, J. Liu, J. Tian, Z. Wang, K. Sun, Z. Rui, W. Yang and Z. Zou, *Electrochemistry Communications,* 2018, 88, 19–23.
129. L. Chong, J. Wen, J. Kubal, F. G. Sen, J. Zou, J. Greeley, M. Chan, H. Barkholtz, W. Ding and D.-J. Liu, *Science,* 2018, 362, 1276–1281.
130. Z. Qiao, S. Hwang, X. Li, C. Wang, W. Samarakoon, S. Karakalos, D. Li, M. Chen, Y. He and M. Wang, *Energy & Environmental Science,* 2019, 12, 2830–2841.
131. D. Xia, X. Yang, L. Xie, Y. Wei, W. Jiang, M. Dou, X. Li, J. Li, L. Gan and F. Kang, *Advanced Functional Materials,* 2019, 29, 1906174.
132. Y.-C. Wang, P.-F. Zhu, H. Yang, L. Huang, Q.-H. Wu, M. Rauf, J.-Y. Zhang, J. Dong, K. Wang and Z.-Y. Zhou, *ChemElectroChem,* 2018, 5, 1914–1921.
133. J.-Y. Choi, L. Yang, T. Kishimoto, X. Fu, S. Ye, Z. Chen and D. Banham, *Energy & Environmental Science,* 2017, 10, 296–305.
134. L. Yang, N. Larouche, R. Chenitz, G. Zhang, M. Lefèvre and J.-P. Dodelet, *Electrochimica Acta,* 2015, 159, 184–197.
135. R. Zheng, S. Liao, S. Hou, X. Qiao, G. Wang, L. Liu, T. Shu and L. Du, *Journal of Materials Chemistry A,* 2016, 4, 7859–7868.
136. X. Zeng, J. Shui, X. Liu, Q. Liu, Y. Li, J. Shang, L. Zheng and R. Yu, *Advanced Energy Materials,* 2018, 8, 1701345.
137. Y. Wang, Y.-J. Tang and K. Zhou, *Journal of the American Chemical Society,* 2019, 141, 14115–14119.
138. M. Xiao, J. Zhu, L. Ma, Z. Jin, J. Ge, X. Deng, Y. Hou, Q. He, J. Li and Q. Jia, *ACS Catalysis,* 2018, 8, 2824–2832.
139. M. Rauf, Y.-D. Zhao, Y.-C. Wang, Y.-P. Zheng, C. Chen, X.-D. Yang, Z.-Y. Zhou and S.-G. Sun, *Electrochemistry Communications,* 2016, 73, 71–74.
140. R. Ding, Y. Ding, H. Zhang, R. Wang, Z. Xu, Y. Liu, W. Yin, J. Wang, J. Li and J. Liu, *Journal of Materials Chemistry A,* 2021, 9, 6841–6850.

5 Water Electrolysis Application of Atomically Dispersed Metallic Materials

Wenxian Li
Shanghai University
University of New South Wales
Shanghai Key Laboratory of High
Temperature Superconductors

Jack Yang and Sean Li
University of New South Wales

Wei Yan and Jiujun Zhang
Shanghai University
Fuzhou University

CONTENTS

DOI: 10.1201/9781003153436-5

5.1   INTRODUCTION

Hydrogen contains about three times the energy density (140 MJ·kg^{-1}) of gasoline (~43 MJ·kg^{-1}) and can be stored in large amounts for many applications including chemical industries and automotive fuel supply such as fuel cells. The hydrogen fuel cell is a possible alternative to batteries in electric vehicle power supplies because of its high energy/power densities, low/zero emission, as well as long recharge mileage. Currently, hydrogen can be produced based either on fossil fuels or on renewable energies [1,2]. The proportion of renewable energy-generated hydrogen is <4% of the total hydrogen product worldwide. The hydrogen produced with renewable electricity energy from water through electrolysis is called green hydrogen. Water electrolysis has been demonstrated as a promising technique for generating green hydrogen using renewable electricity generated from wind, solar, hydro, geothermal, etc., as shown in Figure 5.1. It is believed that green hydrogen would play an important role in the future fuel consumption for decarbonization of the hard-to-electrify sectors of the economy including heavy manufacturing, chemical engineering, food processing, navigation, aviation, and long-haul trucking. It can also be used to produce electricity through fuel cells to power household appliances in a zero-emission manner. However, the practical application of hydrogen has been limited by its handling

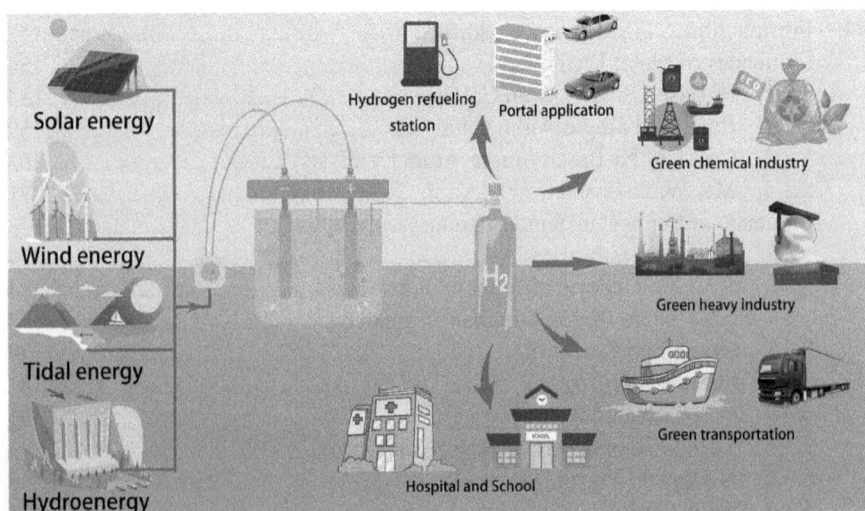

FIGURE 5.1   Production, storage, and application of hydrogen.

processes such as storage, transportation, and suitable utilization. Furthermore, the production cost of green hydrogen is also expensive, which is about €3.50–6 kg^{-1} (US\$ 4.0–7.0 kg^{-1}), much higher than the cost of gray hydrogen, €1.50 kg^{-1} (US\$ 1.75 kg^{-1}) and blue hydrogen, €2–3 kg^{-1} (US\$ 2.3–3.5 kg^{-1}) [3]. Therefore, the innovation of novel hydrogen production will be one of the basic motions to bring down the price of green hydrogen, which depends on electrolysis technology and the mass manufacturing of electrolyzers.

The mechanism of water electrolysis depends on the electrolytes used in electrolyzers. For water electrolysis in acid electrolytes, the cathode reaction is $2H^+ + 2e^- \rightarrow H_2$ (HER) and the anode reaction is $H_2O \rightarrow 2H^+ + 1/2O_2 + 2e^-$ (OER); the corresponding cathode reaction is $2H_2O + 2e^- \rightarrow H_2 + 2OH^-$ (HER) and the anode reaction is $2OH^- \rightarrow H_2O + 1/2O_2 + 2e^-$ (OER) and the overall reaction is $2H_2O \rightarrow 2H_2 + O_2$ in neutral and alkaline solutions. The theoretical cell voltage to drive the water-splitting reaction at standard conditions (25°C, 1.0 atm) is 1.23 V. However, the water-splitting reaction in the electrolyzer needs to conquer the internal resistance of the electrolyte, the circuit, and the interface resistances between the electrodes and the electrolytes, which requires additional energy (i.e. activation energy), leading to higher cell voltage than the theoretical value, resulting in lower energy efficiency of the process. The main extra voltage primarily comes from the cathode and anode overpotentials (expressed by η_c and η_a), which can be partially depressed by the efficient cathode and anode electrocatalysts.

Therefore, electrocatalysts play a critical role in the electrolyzer because they can accelerate the HER and OER processes to increase energy efficiency. For HER, one of the key parameters is hydrogen-adsorption free energy (ΔG_{H*}) defined as $\Delta G_{H*} = \Delta E_{H*} + \Delta E_{ZPE} + T\Delta S_H$ on the active sites of the cathode catalyst surface, in which ΔE_{H*} is the calculated hydrogen adsorption energy, ΔE_{ZPE} is the difference between the adsorption state and the zero-point energy of the gaseous state, and ΔS_H represents the adsorption entropy of $1/2H_2$ [4]. The ΔG_{H*} value is requested to be suitable for both hydrogen adsorption and de-adsorption to fit with the HER process. Three steps are involved in HER in acid electrolyte, i.e., Volmer step ($H^+ + * \rightarrow H^*$), Heyrovsky step ($H^* + H^+ + e^- \rightarrow H_2$), and Tafel step ($H^* + H^* \rightarrow H_2$). In neutral and alkaline electrolytes, the HER process is in different pathways from those in an acidic electrolyte, i.e., Volmer step ($H_2O + * \rightarrow H^* + OH^-$), Heyrovsky step ($H^* + H_2O + e^- \rightarrow H_2 + OH^-$), and Tafel step ($H^* + H^* \rightarrow H_2$). The speeds of hydrogen adsorption and de-adsorption determine the HER mechanisms: Volmer–Heyrovsky or Volmer–Tafel ones.

In general, noble metals, such as Pt, and their oxides are the most active electrocatalysts employed in water electrolysis for high-speed HER and OER. However, some low-loading noble metal electrocatalysts and non-noble metal electrocatalysts are under development to replace noble metal-based electrocatalysts due to their high cost and low nature abundances. In doing so, various microstructures have been researched to expose a high amount of active sites by decreasing the particle size of the catalyst particles. The extreme situation is to achieve atomically dispersed single atomic catalysts, which may decrease the usage amount of noble metals dramatically. For example, Qiao et al. [5] successfully prepared a Pt/FeO$_x$ SAC with dispersed Pt single atoms on FeO$_x$ nanocrystals. The catalyst showed extremely high activity for both the preferential oxidation reaction (PROX) and the CO oxidation reaction, which was 2–3 times high compared with that of the Pt cluster/FeO_x catalyst [6]. The density functional theory simulation results indicated that the Pt atom shares its electrons with FeO_x. The d-orbital of a single Pt atom was more vacant, which could not only lead to the strong bonding and stability of a single Pt atom but also provide a band of positively charged Pt atoms, thus obtaining Pt/FeO_x catalyst with excellent catalytic activity [5,7]. This work led to the research of single atoms in various catalyses, although the Pt single atoms were observed to have significant agglomeration and cluster formation evidenced by the increased loading of Pt atoms from 0.17 to 2.5 wt%.

As achieved, SACs can be synthesized with different types of metals using both noble metals and non-noble metals for catalyzing water electrolysis [8–22]. Some review articles have reported the research progress in the single-atom catalysis of HER and OER [4,8–10,23–30]. In this chapter, we summarize the development of single-atom catalysis based on different active substances including Pt [11–13,31–35], Ru [14,15,36–39], Co [16,17,40–44], Fe [18,19,45,46], Ni [20–22,47–49], Ti [50], W [51,52], and Mo [53]. The catalytic behavior of single atoms depends on the supporting materials greatly because the coordination environment of single atoms comes from the interaction between the single atoms and their supports. In this regard, we also introduce some important supports of the single atoms including graphene [54–60], N-doped graphene [61–65], graphdiyne [66,67], carbon nanoball [68,69], carbon nanotube [70–72], N-doped carbon [48,73–79], metal-organic framework (MOF) derivatives [51,80–84], MXenes [85,86], oxides [87,88], sulfides [89–93], and other composites [11,36,43,94–100]. We will also summarize the current research progress and the basic understanding of the single-atom catalysts (SACs) and their supports with emphasizing their interaction in different coordination environments and their effects on the catalytic performance of water-splitting reactions.

5.2 INFLUENCES OF METAL ELEMENTS

The development of SACs starts from Pt to decrease its loading in catalysts [5]. The atomically dispersed Pt atoms on supports enable the maximum involvement of Pt atoms during the catalytic reactions because all Pt atoms are exposed to the reactants. The other noble metals, Ir and Ru, are also investigated as catalysts for their possible application in water electrolysis [101–104]. The high mass activity enables the noble metals to become competitive when compared with other types of catalysts. However, non-noble metal-based SACs are also developed intensively because of their lower cost [100,103]. The catalytic performance of the non-noble metal-based SACs can be stable even in acidic electrolytes because of the effective metal-support coupling in addition to high activity. The theoretical understanding of the catalytic performance of both noble metals and non-noble metals has advanced their intrinsic catalytic properties and selectivity.

5.2.1 PT-BASED SINGLE-ATOM CATALYSTS

Single Pt atoms immobilized on various supporting materials to form the supported catalyst have received considerable attention for their high utilization of noble platinum and catalytic performance improvement [12,33,105]. Cheng et al. [12]

maximized Pt efficiency by downsizing Pt nanoparticles into single atoms to utilize nearly all Pt atoms. They developed a strategy to produce isolated Pt single atoms supported on nitrogen-doped graphene nanosheets (NGNs) using the atomic layer deposition (ALD) method (Figure 5.2a). It can be seen that three steps are involved in the ALD process as follows: (i) the Pt precursor of MeCpPtMe$_3$ reacts with nitrogen sites in the NGNs; (ii) the Pt precursor under the O$_2$ exposure is oxidized to H$_2$O and CO$_2$, which creates a Pt-containing monolayer; and (iii) Pt surface is covered with a layer of adsorbed oxygen during (ii) step. This layer contains functional groups for the next ALD cycle process that involves both (i) and (ii) steps. The Pt precursor (MeCpPtMe$_3$) can react with NGNs during the ALD growth. The deposited Pt precursor and NGNs support are closely connected by chemical bonds. The HER activities of the commercial Pt/C and ALDPt/NGNs processed with 50 and 100 cycles of deposition were characterized by conducting linear sweep voltammetry (LSV) measurements in N$_2$-saturated 0.5 M H$_2$SO$_4$ solution, as shown in Figure 5.2b. ALDPt/NGNs with 50 and 100 cycles exhibit excellent catalytic activities toward HER, which are superior to both Pt/C and NGNs electrocatalysts. High ALD cycles deteriorate the HER catalytic activity of the ALDPt/NGNs. The mass activity of ALD50Pt/NGNs at 0.05 V is 10.1 A·mg^{-1} (Figure 5.2c). The value is 7.8 times of ALD100Pt/NGNs (2.12 A·mg^{-1}) and 37.4 times of Pt/C catalyst (0.27 A·mg^{-1}). The results indicate that Pt utilization and activity are much higher in single Pt atoms and clusters compared to nanoparticle counterparts. Figure 5.2d exhibits that the N−2p orbitals are mixed with discrete Pt 5d orbitals around the Fermi level. The N atom obtains the electron while the single Pt atoms are positively charged evidenced by the calculated Bader charges, leading to more unoccupied 5d densities of states in single Pt atoms on NGNs. The 5d orbitals of Pt atoms can strongly interact with the 1s orbital of adsorbed H atom, as shown in Figure 5.2e, resulting in the formation of hydride and electron pairing.

Zhang et al. [11] reported a synthesis strategy for SACs, which could be applied for the neutral media HER electrocatalysis. This strategy demonstrated the synthesis of the single Pt atom catalyst supported by CoP nanotube arrays grown on Ni foam (PtSA-NT-NF) through the potential-cycling method in neutral media. As shown in Figure 5.2f, this PtSA-NT-NF has wall thicknesses of about 20 nm and diameters of around 140 nm. The single Pt atoms instead of Pt grains and particles are well dispersed on the CoP-based nanotubes (Figure 5.2g), achieving the maximized Pt efficiency. The as-prepared catalyst supported by a large Ni foam could be used as the binder-free cathodes for HER. The HER activity of a pure NF, an NF with CoP NT arrays (termed NT-NF), a PtSA-NT-NF, and commercial Pt/C were measured. The single Pt atoms could contribute much to the HER performance proved by the 24 mV overpotential at 10 mA·cm^{-2}, which was much more competitive than 337 mV for NF and 46 mV for NT-NF. Notably, the overpotential value of PtSA-NTNF is only 7 mV in neutral media, indicating a high-efficiency HER performance than that of Pt/C. Figure 5.2h shows the Tafel slopes of PtSA-NT-NF (30 mV·dec^{-1}), Pt/C (31 mV·dec^{-1}), NT-NF (38 mV·dec^{-1}), and NF (160 mV·dec^{-1}). The high Tafel slope on NF indicates the rate-limiting step is a slow Heyrovsky reaction while the low Tafel slopes of both Pt/C and PtSA-NT-NF demonstrate that the rate-limiting step is faster.

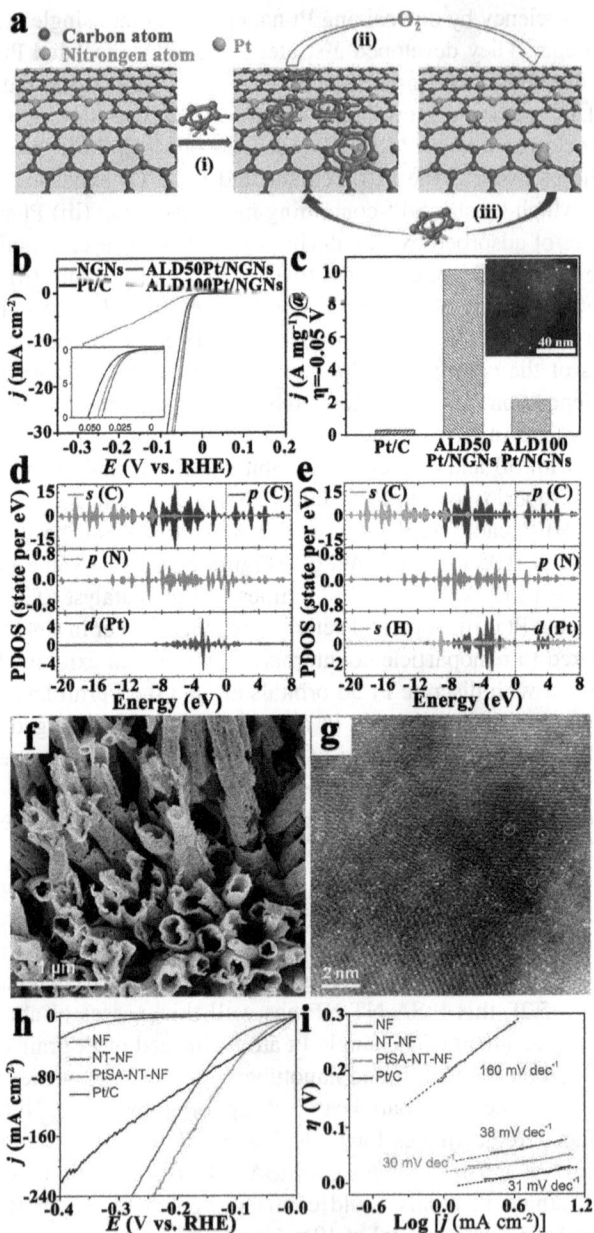

FIGURE 5.2 (a) The Pt ALD mechanism illustration. (b) The HER polarization plots for Pt/C and ALDPt/NGNs catalysts acquired in 0.5 M H_2SO_4 with the enlarged curves at the onset potential region of the HER as inset. (c) Mass activity at 0.05 V (versus RHE) of the Pt/C and the ALDPt/NGNs for HER. Partial density of states (PDOS) before (d) and after (e) hydrogen adsorption for a single Pt atom in ALDPt/NGNs. The Fermi level is shifted to zero. (Ref. [12] Copyright 2016, Springer Nature.) (f) SEM image of PtSA-NT-NF. (g) Atomic-resolution (AR) high-angle annular dark-field (HADDF) image of PtSA-NT-NF. (h) LSV curves and (i) Tafel plots of Pt/C, NF, NT-NF, and PtSA-NT-NF at 5 mV·s^{-1} in 1M PBS at 298 K. (Ref. [11] Copyright 2017 Wiley-VCH Verlag GmbH & Co. KGaA.)

As identified, depending on the different supports, the coordination environment influences the catalytic activity and selectivity of single atoms. Fang et al. [31] reported that a metal-organic framework can confine single Pt atoms to construct fast electron transfer pathways between the MOF photosensitizer and the Pt single-atom acceptor for HER with the aid of visible-light driving. A stable aluminum-based porphyrinic MOF named Al-TCPP was used for Pt atom metalation into the porphyrin centers through a simple reduction. The uniformly dispersed bright spots in Al-TCPP-0.1Pt with Pt loadings of 0.07 wt% represent ultra-small single Pt atoms with sizes between 0.1 and 0.2 nm (Figure 5.3a, inset). Al-TCPP displays a weak activity of $1.5\,\mu mol \cdot g^{-1} \cdot h^{-1}$ owing to the quick electron-hole recombination, as shown in Figure 5.3b. When Pt nanoparticles (NPs) are introduced, the formed Al-TCPP-PtNPs are capable of trapping electrons and acting as proton reduction sites, thus presenting a higher photocatalytic HER rate ($50\,\mu mol \cdot g^{-1} \cdot h^{-1}$). The Al-TCPP-0.1Pt with ultra-small single Pt atoms demonstrates the best photocatalytic efficiency ($129\,\mu mol \cdot g^{-1} \cdot h^{-1}$) on account of the maximized Pt efficiency. The calculated turnover frequency (TOF) of Al-TCPP-0.1Pt in the inset of Figure 5.3b attains $35\,h^{-1}$ which is around 30 times that of Al-TCPP-PtNPs ($1.1\,h^{-1}$). The result explicitly proves that a single-atom Pt cocatalyst can boost photocatalytic electrical activity. Figure 5.3c reveals the Gibbs hydrogen adsorption free energy ΔG_{H*} for photocatalytic HER. The hole-involved Al-TCPP-0.1Pt (Figure 5.3d) exhibits a ΔG_{H*} of 0.05 eV, which is beneficial to both fast hydrogen absorption and desorption. In comparison, the hole-involved Al-TCPP (Figure 5.3e) and ΔG_{H*} of Al-TCPPPtNPs (Figure 5.3f) are −1.09 and −0.37 eV, respectively, suggesting a fast electron–proton acceptance to form H* and slow hydrogen releases. The single Pt atom can optimize H-binding and electronic properties, leading to the superior photocatalytic HER activity of Al-TCPP-0.1Pt when compared with Al-TCPP-PtNPs and Al-TCPP.

Kim et al. [32] designed a platinum single-atom alloy catalyst with high Pt loading and good catalytic durability. Pt single atom supported on antimony-doped SnO_2 (Pt_1/ATO) was fabricated by incipient wetness impregnation. The Pt content can reach 8 wt%. The lattice fringes shown in Figure 5.3g are ascribed to SnO_2 (200) with a lattice space of 0.219 nm. The existence of Pt single atoms is evidenced with magnified high-angle annular dark-field scanning transmission electron microscopy (HAADF-STEM) images, as shown in Figure 5.3h with the dotted box, demonstrating that Pt atoms in white circles are located on the surface of SnO_2. The authors believed that the Pt atoms could substitute the atoms in the SnO_2 because all the Pt atoms are situated on the lattice array. Figure 5.3i illustrates the change in Pt mass activity by performing cyclic voltammetry (CV) after repeated cycles. The Pt mass activities of Pt_1/ATO, Pt NP/ATO, and Pt/C for the direct path of formic acid oxidation reaction (FAOR) after 1,800 cycles are 0.59, 0.35, and 0.003 $A \cdot mg_{Pt}^{-1}$, respectively. The $Pt_1/$ATO remains high activity for FAOR while Pt/C losses most of its electrical activity, which proves that the single atomic nature of Pt in Pt1/ATO could be reserved even after the harsh durability test. The ATO support displays more stable properties at high potentials while the carbon support in Pt/C is severely degraded. This remarkable durability originates from the stable structure of the Pt_1/ATO and the good anti-corrosion property of ATO. The 4 wt% Pt_1/ATO sample was used as an anode to assemble a full cell for direct formic acid fuel cells (DFAFC). Figure 5.3j demonstrates the cell performance based on Pt unit mass, which is 0.04 $mg_{Pt} \cdot cm^{-2}$ in both Pt_1/ATO and Pt/C. The

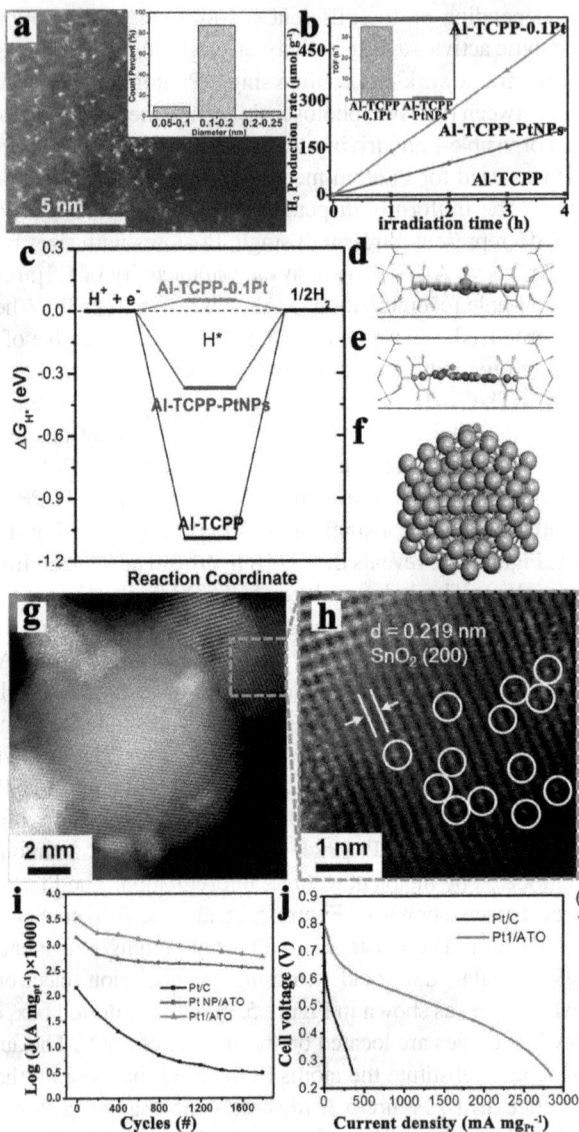

FIGURE 5.3 (a) Aberration-corrected HAADF-STEM images of Al-TCPP-0.1Pt. (b) Photocatalytic H_2 production rates of Al-TCPP-0.1Pt, Al-TCPP-PtNPs, and Al-TCPP, respectively (inset is the calculated TOFs of Al-TCPP-0.1Pt and Al-TCPP-PtNPs). (c) ΔG_{H^*} for photocatalytic H_2 production. The structures of H^* of (d) Al-TCPP-0.1Pt, (e) Al-TCPP, and (f) AlTCPP-PtNPs (Pt147). (Ref. [31] Copyright 2018 WILEY-VCH Verlag GmbH & Co. KGaA, Weinheim. (g) HAADF-STEM images of Pt1/ATO and (h) magnified images of (g). (i) Change in mass activity at 0.6 V during the durability test which was tested by repeating CVs (range from 0.05 to 1.4 V for 1,800 cycles) in 0.1 M $HClO_4$ and 0.5 M formic acid solution with saturated Ar. (j) Cell voltage plots as a function of current density for DFAFC single cells prepared using 4 wt% Pt1/ATO or 20 wt% Pt/C catalysts for the anode. (Ref. [32] Copyright 2017 Wiley-VCH Verlag GmbH & Co. KGaA, Weinheim.)

Pt/C catalyst exhibits an order of magnitude smaller power per Pt unit mass than the Pt_1/ATO after the maximum power density is normalized by the Pt mass, indicating that Pt_1/ATO catalyst has superior catalytic activity.

The coordination environment of Pt single atoms can be tuned by adjusting the bonding elements. For example, Zhou et al. [106] designed the single-atom Pts anchored on NiO/Ni heterostructure (PtSA-NiO/Ni) as an HER catalyst in alkaline electrolyte. Pt single atoms coupled with NiO/Ni heterostructure achieved the tunable binding abilities of hydrogen (H^*) and hydroxyl ions (OH^*), optimized the water dissociation energy and the H^* adsorption for accelerating alkaline HER. The HAADF-STEM image in Figure 5.4a exhibits most of the bright spots located at the interfaces of the NiO/Ni heterostructure. Those bright spots are corresponding to heavy constituent Pt atoms, confirming the immobilization of single Pt atoms atomically dispersed in the NiO/Ni nanosheets. As shown in Figure 5.4b, the PtSA-NiO/Ni illustrates the best HER performance and requires a very low overpotential of 26 mV at 10 mA·cm^{-2} and 85 mV at 100 mA·cm^{-2}, which are significantly superior to Pt/C, NiO/Ni, PtSA-Ni, and PtSANiO catalysts. It is believed that the superior HER performance should originate from the rational design of the interface in PtSA-NiO/Ni, as shown in Figure 5.4c. The NiO/Ni heterostructure-supported single-atom Pt sites can give optimized H-binding ability for the de-adsorption and conversion of H^* to accelerate HER kinetics of PtSA-NiO/Ni.

Zhuang et al. [35] developed a plasma-photochemical method to synthesize atomically coordinated Pt–Co–Se moieties in defect-rich $CoSe_2$ ($CoSe_{2-x}$–Pt). The colored HAADF image of $CoSe_{2-x}$–Pt demonstrates the ordered Pt atom distribution located at Co atop sites on $CoSe_2$ surface, as shown in Figure 5.4d and h. The images prove that single Pt atoms are successfully immobilized at Se vacancies to form atomically coordinated Pt–Co–Se moieties. As shown in Figure 5.4e, $CoSe_{2-x}$–Pt displays a much higher OER activity than $CoSe_{2-x}$, $CoSe_2$-origin-Pt, and commercial Pt/C. This $CoSe_{2-x}$–Pt can optimize the adsorption energy of intermediate. The single Pt atoms and Se vacancies could induce an asymmetrical electron cloud to supply electrons to the intermediates during the OH^* and O^* transformation, and OOH^* desorption process, avoiding excessively strong intermediate binding. As shown in Figure 5.4d, single Pt atoms can theoretically prove to be able to boost the OER activity of $CoSe_2$.

5.2.2 Ru-Based Single-Atom Catalysts

Ru, a transition metal element, has not been widely used in the field of single-atom catalysis until recent years. Researchers have developed various strategies to stabilize Ru single atoms on the support and improve their performance for HER and OER, such as support advantage, coordination advantage, alloy relay, electronic coupling, local electric field, etc. In this section, we introduce the specific applications of these strategies. The choice of support is important in inspiring the catalytic performance of Ru single atoms. Yang et al. [36] anchored Ru metal atoms to HPN (amorphous phosphorus nitride imide nanotubes), in which a strong coordination interaction between the lone pair of electrons of N and the d orbitals of Ru is existed, as shown in Figure 5.5a. Different from traditional C_3N_4 and C supports, PN, composed of the 3D corner-sharing PN_4-tetrahedral units, possesses a twisted spatial structure and polar P–N bonds, so the electron density of PN is extreme inhomogeneity, which will greatly promote the

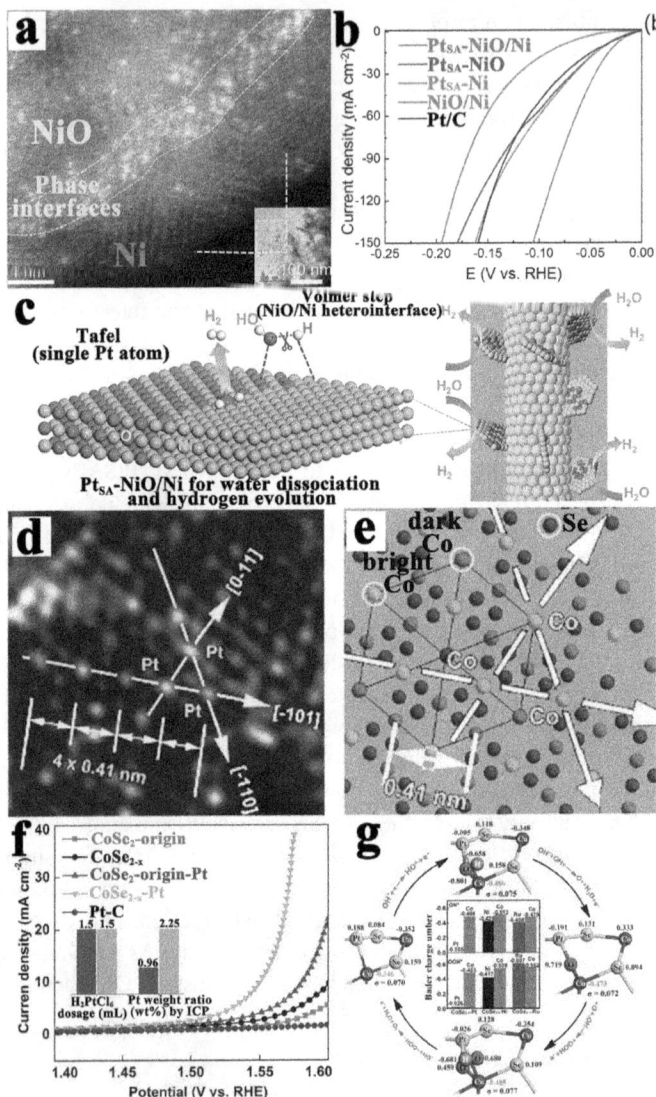

FIGURE 5.4 (a) HAADF-STEM image of PtSA-NiO/Ni. (b) HER plots of Pt/C, PtSA-NiO/Ni, PtSA-NiO, PtSA-Ni, and NiO/Ni in 1-M KOH electrolyte, respectively. (c) The mechanism of PtSA-NiO/Ni as an efficient catalyst toward water dissociation and HER in alkaline media. (Ref. [106] Copyright 2021, Springer Nature.) (d) The processed and colored HAADF-STEM image of CoSe$_{2-x}$-Pt. Bright spots: Pt atoms, dark spots: Co atoms. (e) The 2D atomic arrangement of Co atoms on the (210) plane and Co and Se atoms underneath the (210) plane. (f) OER LSV plots of 20 wt% Pt/C, CoSe$_2$-origin, CoSe$_2$-origin-Pt, CoSe$_{2-x}$, and CoSe$_{2-x}$-Pt, performed in 0.1 m KOH. (g) The Bader charge numbers of atoms of CoSe$_{2-x}$-Pt during the adsorption of OH*, O*, and OOH*, and the desorption of the OOH* process. The inset displays the comparison of Bader charge number of Co atoms and M (Pt, Ni, and Ru in CoSe$_{2-x}$-Pt, CoSe$_{2-x}$-Ni, and CoSe$_{2-x}$-Ru, respectively) after OH* and OOH* adsorption. (Ref. [35] Copyright 2018 Wiley-VCH Verlag GmbH & Co. KGaA, Weinheim.)

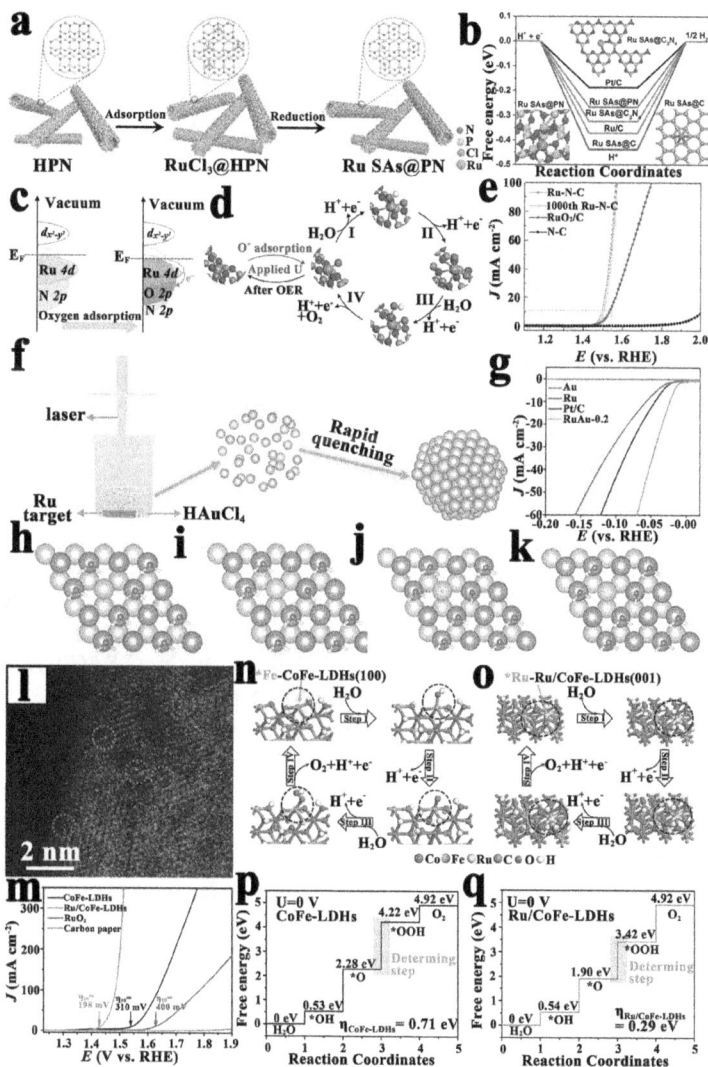

FIGURE 5.5 (a) Synthesis of Ru SAs@PN. (b) The calculated free-energy diagram of HER at the equilibrium potentials for Ru SAs@PN, Ru SAs@C$_3$N$_4$, Ru/C, and Pt/C. (Ref. [36] Copyright 2018, Wiley-VCH Verlag GmbH & Co. KGaA, Weinheim.) (c) Electronic structure variation with the oxygen adsorption for Ru–N–C. (d) OER pathway for Ru–N–C catalyst in the acidic environment. (e) OER performance comparison of the Ru–N–C with that of RuO$_2$/C in 0.5 M H$_2$SO$_4$ solution. (Ref. [14] Copyright 2019, Springer Nature.) (f) Setup of PLAL and SAA nanoparticle synthesis. (g) LSV polarization curves (iR compensated) at 5 mV·s^{-1} scan rate. (h–k) HER pathway on RuAu SAAs. (Ref. [15] Copyright 2019, Wiley-VCH Verlag GmbH & Co. KGaA, Weinheim.) (l) The Cs-corrected STEM image of Ru/CoFe-LDHs nanosheets. (m) Comparison of iR compensated polarization curves of Ru/CoFe-LDHs with CoFe-LDHs, Carbon paper, and the commercial RuO$_2$ catalyst. (n and o) OER pathways on CoFe-LDHs (n) and Ru/CoFe-LDHs (o) for DFT+U simulation. (p and q) ΔG diagrams for OER process on CoFe-LDHs (p) and Ru/CoFe-LDHs (q). (Ref. [37] Copyright 2019, Springer Nature.)

activation of the support. As shown in Figure 5.5b, density functional theory (DFT) shows that compared with Ru SACs@C_3N_4 and Ru SACs@C, the Gibbs free energy of adsorbed H* over the Ru single atoms on PN is much closer to zero, which can greatly improve its HER performance with a low overpotential of 24 mV at 10 mA·cm^{-2} and a Tafel slope of 38 mV·dec^{-1} in 0.5 M H_2SO_4.

To explore the coordination advantage of Ru single atoms, Cao et al. [14] successfully prepared a porphyrin-like Ru_1–N_4 structural configuration by confining Ru single atom within a carbon nitride-derived N–C support as an effective electrocatalyst for OER in acidic electrolyte. The oxygen was pre-adsorbed on Ru_1–N_4 sites forming O–Ru_1–N_4 as revealed by the operando synchrotron radiation X-ray absorption fine structure (XAFS) spectroscopy and Fourier transform infrared (SR-FTIR) spectroscopy under working potential. As shown in Figure 5.5c, the stronger covalent Ru–N/O bond forms due to the downshift of the 4d band of Ru with the adsorption of O. In O–Ru–N–C, the Ru atoms transfer their electrons to the neighboring N and the adsorbed O via orbital hybridization. Therefore, Ru single atom has favorable binding energies with O*, OH*, and OOH* intermediates, which is the reason for its excellent OER performance. Figure 5.5d shows the complete OER reaction mechanism. A single oxygen atom is adsorbed on the Ru site forming O–Ru_1–N_4 coordination. The O–Ru_1–N_4 disassociates water molecules via nucleophilic attack followed by deprotonation to produce OOH*. O_2 is generated with a further proton-coupled electron transfer. The specific Ru_1-N_4 structure makes the Ru–N–C SAC have a highly intrinsic OER activity with a mass activity of 3,571 A g_{metal}^{-1} and a TOF of 3,348 O_2·h^{-1} under an overpotential of 267 mV at a current density of 10 mA·cm^{-2} (Figure 5.5e).

The alloy relay effect of Ru single-alloy-atoms (SAAs) has also been investigated to explore another high-performance source in a water electrolysis application. Chen et al. [15] reported a technique for synthesizing RuAu SAAs by the laser ablation in liquid (LAL) (Figure 5.5f). This process included a strong quenching effect, which could obtain metastable nanostructures with novel properties. As shown in Figure 5.5g, the RuAu SAA has a low overpotential of 24 mV@10 mA·cm^{-2}, and the value is even lower than that of Pt/C (46 mV@10 mA·cm^{-2}) in alkaline media. Its excellent HER performance is mainly derived from the relay catalytic pathway of the host Ru responsible for water dissociation and dopant Au responsible for hydrogen evolution. The immiscibility between Ru and Au forms a unique electronic structure in the RuAu SAAs, that is, electron transfer occurs from Ru to Au, resulting in the decrease in the Ru d-band filling and the increase in the Au d-band filling. In this way, Ru atoms were in positive states and Au atoms were in negative states, which were beneficial to the adsorption of water molecules and hydrogen atoms, respectively, thereby achieving high HER performance under alkaline conditions, as shown in Figure 5.5h–k. The RuAu SAAs show a perfect balance for the adsorption of water molecules and protons.

The electronic coupling between Ru single atoms and supports can be tuned by changing the supporting metals. For example, Li et al. [37] used Ru-O-M (M for Fe or Co) bonds to stabilize Ru single atoms on the surface of cobalt-iron layered double hydroxides (CoFe-LDHs), as shown in Figure 5.5l. The strong electronic coupling between Ru and LDHs achieves a kind of Ru SAC with extremely high electrocatalytic activity and stability and the overpotential of Ru/CoFe-LDHs at the current density of 10 mA·cm^{-2} is 198 mV, and the Tafel slope is greatly reduced to 39 mV·dec^{-1}

in alkaline solutions, as shown in Figure 5.5m. Figure 5.5n–o shows the proposed OER four-electron mechanism and the optimized structures of the intermediates of Ru/CoFe-LDHs and CoFe-LDHs in the free-energy landscape. By comparing the free energy plots in Figure 5.5p–q, it can be found that the Ru atom on the CoFe-LDHs surface shows a lower Gibbs energy of 1.52 eV for the rate-determining step (*O group transforms to *OOH group) than that of the Fe atom on the CoFe-LDHs edge (1.94 eV), which indicates that the Ru/CoFe-LDHs structure has a more favorable OER kinetics and Ru single atoms can serve as highly efficient active sites for catalyzing OER.

The synergistic effect is another factor that improves the catalytic performance of Ru single atoms. Ramalingam et al. [38] stabilized the Ru single atoms on the titanium carbide ($Ti_3C_2T_x$) MXene support through the coordination of nitrogen and sulfur. The obtained Ru SAC exhibited an excellent HER performance with the over-potential as low as 76 mV at the current density of 10 mA·cm^{-2}. Compared to N–S–$Ti_3C_2T_x$, Ru SA–$Ti_3C_2T_x$, and Ru SA–N–$Ti_3C_2T_x$, which had high energy barriers for the formation and desorption of H_2, the Gibbs hydrogen adsorption free energy, ΔG_{H^*}, of Ru SA–N–S–$Ti_3C_2T_x$ achieved an optimal value of 0.08 eV, indicating that Ru SA–N–S–$Ti_3C_2T_x$ had the favorable H adsorption-desorption and subsequent H_2 production characteristics to effectively drive the entire HER process. It was believed that such an excellent HER performance should come from the chemical interaction between the Ru atom and the MXene support. The partial density of states (PDOS) showed that the doping of Ru single-atom could induce the charge transfer between Ru single atom and MXene supports, resulting in non-bonding states near the Fermi level. The decreased intensity of the non-bonding states and difference in density of states indicated that the Ru single atoms could improve the catalytic performance, which was consistent with the changing trend of ΔG_H.

Zhang et al. [39] decorated Ru sites on the edge-rich carbon matrix (ECM) with a precise Ru-N_4 coordination configuration and integrated polydopamine (PDA) particles with Ru species to prepare a highly efficient HER catalyst ECM@Ru. This catalyst showed low overpotentials of 63 mV at a current density of 10 mA·cm^{-2} and 102 mV at 50 mA·cm^{-2}, respectively, which could be derived from the rich edge defects of ECM and the doping effect of Ru atoms. The PODS projected on the C 2p orbital showed that, compared with CM, the band center of ECM was close to the Fermi level, indicating that the ECM had a stronger interaction with adsorbed H*. The intrinsic electronic structure of the doped Ru single atoms had changed, and the decoration of Ru species generate some new hybridized electronic states. This showed that Ru 4d orbitals could hybridize with the coordination non-metal, thereby promoting effective electron transfer and improving conductivity. The charge density distribution of ECM@Ru showed that the defects originating from the carbon matrix edges could induce high reactant concentration and localized electric field around the Ru species, which could enhance the HER kinetics. In addition, the theoretically calculated ΔG_H value of ECM@Ru was the closest to zero.

5.2.3 CO-, FE-, NI-BASED SINGLE-ATOM CATALYSTS

The single-atom catalytic performances of non-noble transition metals have attracted research interest for their application potential. However, these metals suffer from instability in acidic environments or as anodes of water electrolyzers. The coordination environments of single atoms may stabilize the transition metals through the specific coordination environment construction. Fei et al. [16] reported an inexpensive, concise, and scalable method to synthesize a Co SAC based on nitrogen-doped graphene (NG) by heat-treating small amounts of cobalt salts and graphene oxide (GO) under the NH_3 atmosphere. From the peak intensity of the XPS curve, Co species, existing in the form of single atoms, were mainly coordinated with pyridinic N. The N doping of the graphene could provide stable sites for Co incorporation, making it an excellent catalyst for HER in both acidic and alkaline media. In addition, since the precursor suspension of GO with Co addition was very stable and could be formed into a piece of paper, it was used as a free-standing electrode for generating hydrogen. The precursor solution could be coated on conductive supports that were used as a binder-free electrode after post-annealing in NH_3. Therefore, this simple and convenient method increases the versatility of electrode design and construction and makes the catalytic layer easy to integrate with other components in the electrochemical device.

Subsequently, Yin et al. [17] reported a strategy for stabilizing Co single atoms on the N-doped carbon porous framework with metal loading above 4 wt% (Figure 5.6a). During the annealing treatment under an N_2 atmosphere, the organic linker of MOFs could generate N-doped porous carbon through pyrolysis. The metal species could then be reduced by the generated carbon at high temperatures. The pre-existing Zn atoms acted as fences, separating adjacent Co atoms and providing more free N sites to prevent the formation of Co–Co bonds. When the low-boiling Zn atoms were selectively evaporated above 800 °C, the remaining Co atoms were anchored on the N-doped carbon support to form Co SAs/N–C. As shown in Figure 5.6b and c, the extended X-ray absorption fine structure (EXAFS) analysis reveals that the dominant reactive sites in Co SAs/N–C are planar $Co\text{-}N_4$ and $Co\text{-}N_2$ at 800 °C and 900 °C, respectively. The $Co\text{-}N_x$ single sites show excellent oxygen reduction reaction (ORR) property compared with commercial Pt/C and many reported non-precious metal catalysts. The half-wave potential is 0.881 V, which is positive than 0.811 V of the commercial Pt/C (Figure 5.6d). The use of low-boiling metals as fences to avoid the agglomeration of active metal sites is a very promising method for preparing SACs based on MOFs.

Qi et al. [42] used electrochemical CV leaching of Co nanodisk (NDs)-MoS_2 nanosheet hybrids to make Co single-atom array bound covalently on distorted 1T MoS_2 nanosheets via Co–S bonds (SA Co–D 1T MoS_2), as shown in Figure 5.6e. During the hybridization process of Co NDs-MoS_2 nanosheets, both the strain induced by the formation of Co–S covalent bond and lattice mismatch between MoS_2 and Co can cause the phase transformation of MoS_2 from the semiconductive 2H phase to metallic distorted 1T phase (D-1T) (Figure 5.6f). DFT calculation shows that Co atoms tend to bond on 1T MoS_2 than 2H MoS_2. However, the phase transformation of MoS_2 from 2H to 1T is not a thermodynamic spontaneous process, so it needs to be driven by an external force, which precisely comes from the strain induced by the lattice mismatch between the metal Co and the original 2H MoS_2. They also performed EXAFS and X-ray absorption near-edge structure (XANES) spectroscopic analysis to verify the coordination environment of atomically dispersed Co. When the XANES spectrum was simulated by using the SA Co–D 1T

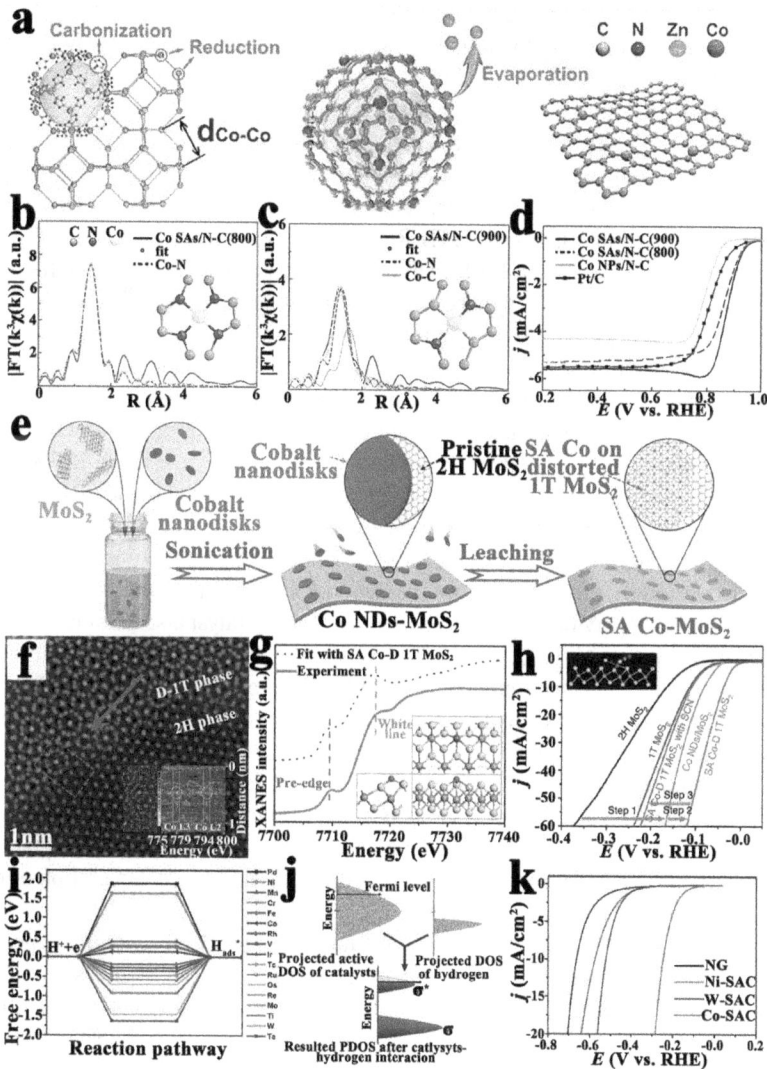

FIGURE 5.6 (a) Fabrication of Co SAs/N–C. The corresponding EXAFS fitting curves for the samples (b) Co SAs/N–C(800) and (c) Co SAs/N–C(900). (d) RDE polarization curves of Co NPs–N/C, Co SAs/N–C, and Pt/C in O_2-saturated 0.1 M KOH with a sweep rate of 10 mV s^{-1} and 1,600 rpm. (Ref. [17] Copyright 2016, Wiley-VCH Verlag GmbH & Co. KGaA, Weinheim.) (e) Synthesis of SA Co-D 1T MoS_2. (f) Aberration-corrected HAADF-STEM image of SA Co-D 1T MoS_2. (g) Co K-edge XANES of SA Co-D 1T MoS_2 and the fitting results. (h) HER polarization plots of MoS_2, 1T MoS_2 prepared by lithiation, Co NDs/MoS_2, and SA Co-D 1T MoS_2 with and without SCN$^-$ ions. (Ref. [42] Copyright 2019, Springer Nature.) (i) ΔG_{H*} diagram for hydrogen adsorption reaction (Volmer reaction). (j) orbital hybridization of catalyst's active sites with hydrogen (σ and σ^* represent for bonding and antibonding state orbitals, respectively). (k) LSV curves for HER alone with the control sample (NG) obtained in 0.5 M H_2SO_4 at a scan rate of 10 mV·s^{-1}. (Ref. [41] Copyright 2019, Wiley-VCH Verlag GmbH & Co. KGaA, Weinheim.)

MoS_2 as a model, two main energy features at 7,710 eV (pre-edge) and 7,718 eV (white line) can fit the experimental results for SA Co–D 1T MoS_2 with a Co concentration of 3.70%, proving that the single Co atom is located at the site directly above the Mo atom and coordinated to three adjacent S atoms (Figure 5.6g). The SA Co–D 1T MoS_2 shows the Pt-like HER catalytic performance in acid environment with 42 mV overpotential, 32 mV·dec^{-1} Tafel slope, and excellent cycling stability at 10 mA·cm^{-2}. The active-site blocking experiment proved that single-atom Co sites give the dominant contribution to Pt-like HER activity (Figure 5.6h) rather than the 2H to D–1T phase transformation of MoS_2. DFT calculations confirmed that the high HER activity was attributed to the synergy effect of Co adatom and S on the MoS_2 (111) support via tuning hydrogen-bonding mode at the interface.

Wu et al. [44] presented a salt-template technique for scalable synthesis of high-density, monodispersed Co single atoms on an N-doped carbon nanosheet (SCoNC), simply by pyrolyzing KCl particles wrapped with a Co-based organic framework (ZIF-67). The site fraction of Co single atoms in the carbon matrix could reach 15.3% and the ECSA was as high as 105.6 m^2·g^{-1}. DFT calculations showed that the Co–N_4 sites were the most possible active sites for ORR and OER bifunctional activity. For the ORR test, the SCoNC catalyst showed a positive half-wave potential (0.91 V vs reversible hydrogen electrode (RHE)) than that of the benchmark Pt/C catalyst (0.88 V vs. RHE) at 1,600 rpm with a 70 mV·dec^{-1} Tafel slope. The SCoNC catalyst also exhibited a current density of 10.0 mA·cm^{-2} ($E_{j=10}$) at 1.54 V versus RHE with a Tafel slope of 74 mV·dec^{-1} for OER measurement. Therefore, the zinc-air batteries based on SCoNC could display good electrochemical performance and long-term stability. In this method, the KCl template played a key role in the formation of SCoNC. The KCl template could strongly support the thin ZIF-67 layer, resulting in little shrinkage during the pyrolysis process, so the Co single atoms could be effectively fixed to prevent aggregation. This simple salt-template method could also be extended to the preparation of other metal SACs on carbon matrix, such as Ni, thereby solving the problem of easy generation of metal particles in the pyrolysis process.

Hossain et al. [41] combined DFT calculations and electrochemical measurements to explore the catalytic performance for many transition metals supported by N-doped graphene as SACs for HER. Figure 5.6i shows that only some of the transition metals (such as Fe, Co, Cr, V, and Rh) as SACs have ΔG_{H*} values between 0.2 and 0.3 eV. Among them, the Gibbs free energy of Co SAC is 0.13 eV, which is the closest to 0. The catalysts' electronic structures are responsible for the activity of SACs for HER, which can be correlated by studying their density of states (DOS) profiles. As shown in Figure 5.6j, two orbital states (blue) will be generated through the interaction of the active valence orbitals (E_p) of catalysts (wooden color) with hydrogen orbital (gray) when H is adsorbed on the surface of catalysts. One is a bonding state (σ) orbital (blue). Another one is a partially filled antibonding state (σ^*) orbital. DOS calculations show that only the valence d_z^2 orbital can participate in the reaction with H actively. The valence d_z^2 energy state and its antibonding orbital are closely related to the descriptor (ΔG_{H*}) of HER performance. The higher energy level of the antibonding state leads to a completely unoccupied orbital, indicating a strong interaction with H, which is not conducive to subsequent H desorption. The location of the Co SAC valence state orbital is near the Fermi level, and its anti-bonding state is only partially filled, which ensures the best hydrogen adsorption strength. In addition, they found that the ΔG_{H*} values of SACs had a linear relationship with their electronic structures (E_p and $E_{\sigma*}$) and the charge transfer along SACs. After that, they synthesized three SACs (Co, Ni, W) to verify the calculation results. EXAFS, XANES, and STEM characterizations show that all the isolated single atoms are uniformly distributed on the surface of NG in the form of $M-N_4C_4$, and the HER activity of Co SAC is significantly higher than that of Ni SAC and W–SAC (Figure 5.6k), which indicates the electrochemical performance of SACs is consistent with the trend of theoretical derivation.

Transition metal single-atom catalysts (SACs) can be constructed by the strong bonding of metal foams (metals = Fe, Co, Ni, and Cu) with suspended bonds supported by GO. As shown in Figure 5.7a, in the case of metallic Fe, charge transfer between Fe^0 and oxygen atoms can generate a large number of isolated $Fe^{\delta+}$ ($0 < \delta < 3$) species by dispersing GO in the foam Fe. Fe–O bonds are formed by coordination between $Fe^{\delta+}$ and oxygen suspension bonds on the GO surface. Finally, the positively charged Fe atoms are pulled out from the surface of the foam iron with the help of ultrasonic treatment to obtain Fe SAs/GO. This provides a generally applicable method for the construction of various transition metal mono atoms. Qu et al. [107] prepared Fe SAs/GO by the above method and subjected them to pyrolysis by poly-dopamine treatment to achieve an isolated Fe–N–C coordination configuration, as shown in Figure 5.7b and c. This was considered the effective active center for catalytic oxygen electroreduction. Meanwhile, the carbon layer produced by the pyrolysis of dopamine could prevent the aggregation of isolated Fe-sites. The obtained SACs showed excellent ORR/OER performance and most importantly this method to create high-density SACs directly from non-precious metal blocks could provide valuable guidance and showed a great potential to meet industrial needs.

In order to synthesize a catalyst that can be multifunctional with single atoms, Chen et al. [108] designed a Janus material (Figure 5.7d), a unique surface that allows two different types of chemical reactions to occur on the same surface. Atomically dispersed Ni and Fe species are located on the inner and outer walls of the hollow graphene (GHSS), respectively. This Janus structure leads to the separation of the ORR and OER active sites; i.e., the $Fe-N_4$ site makes the main contribution to the ORR activity, while the $Ni-N_4$ site is responsible for the OER, which leads to excellent electrocatalytic selectivity and bifunctionality. As shown in Figure 5.7d, the sites of the obtained Ni atoms are on the inner side of the structure and the sites of the Fe atoms are on the outer side of the structure. The $Ni-N_4$ or $Fe-N_4$ planar conformation can be formed by the coordination of Ni or Fe single atoms with four N atoms. The developed $Ni-N_4/GHSS/Fe-N_4$ Janus materials exhibit excellent bifunctional electrocatalytic properties. As shown in Figure 5.7e, the external $Fe-N_4$ can significantly promote high activity in the ORR. As shown in Figure 5.7f, the internal $Ni-N_4$ has an excellent activity for the OER. In addition to this, the versatile catalyst has excellent performance for the Zn-air cell and the specific capacity of

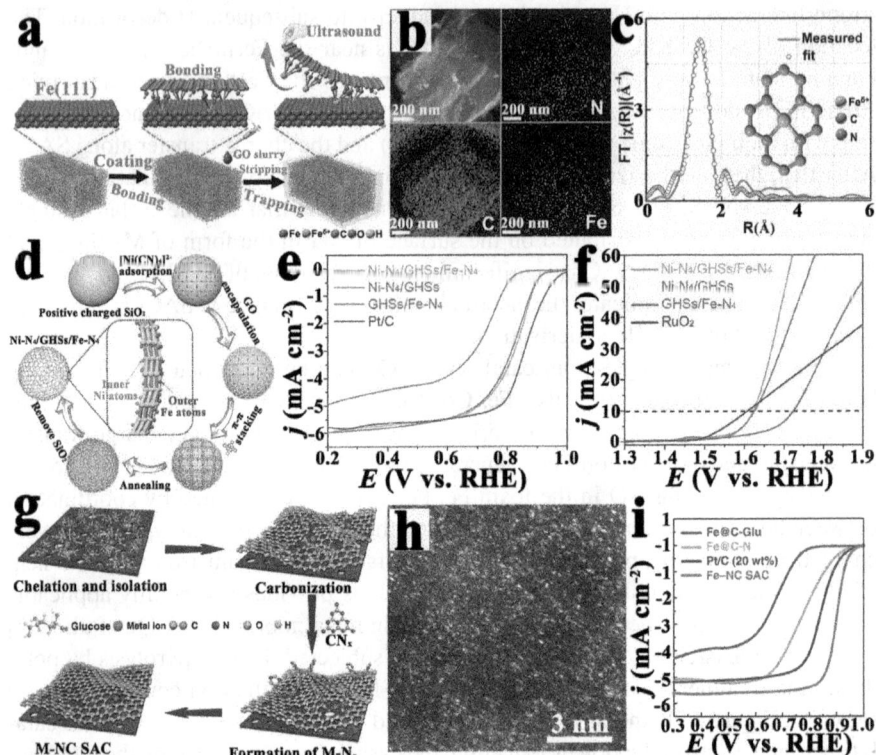

FIGURE 5.7 (a) Schematic illustration for the construction and characterization of Fe SAs/GO. (Ref. [107]. Copyright 2019 Wiley-VCH Verlag GmbH & Co. KGaA, Weinheim.) (b) HAADF-STEM image and corresponding EDS elemental mapping. (Ref. [107]. Copyright 2019 Wiley-VCH Verlag GmbH & Co. KGaA, Weinheim.) (c) k^3-weighted $\chi(k)$ function of the EXAFS spectra and corresponding fitting curve for Fe SAs/N–G. (Ref. [107]. Copyright 2019 Wiley-VCH Verlag GmbH & Co. KGaA, Weinheim.) (d) Synthetic procedure of the Ni–N_4/GHSs/Fe–N_4 catalyst. (Ref. [108]. Copyright 2020 Wiley-VCH Verlag GmbH & Co. KGaA, Weinheim.) (e) ORR polarization curves in O_2-saturated 0.1 m KOH. (Ref. [108]. Copyright 2020 Wiley-VCH Verlag GmbH & Co. KGaA, Weinheim.) (f) OER polarization curves in O_2-saturated 0.1 m KOH. (Ref. [108]. Copyright 2020 Wiley-VCH Verlag GmbH & Co. KGaA, Weinheim.) (g) The cascade anchoring strategy for the synthesis of M–NC SACs. (Ref. [18] Copyright 2019 Springer Nature.) (h) HAADF–STEM images of Fe–NC SAC. (Ref. [18] Copyright 2019 Springer Nature.) (i) Steady-state ORR polarization curves of Fe–NC SAC. (Ref. [18] Copyright 2019 Springer Nature.)

the Ni–N_4/GHSs/Fe–N_4-based Zn-air cell was calculated to be 777.6 mA·h·g_{Zn}^{-1} by normalizing the consumed Zn mass. Zhao et al. [18] also synthesized a SAC loaded on a porous carbon structure, as shown in Figure 5.7g. A generic cascade anchoring technique was proposed for the mass production of a series of M–NC SACs (M = Ni, Co, Mn, Cu, Fe, Mo, Pt) with metal loadings as high as 12.1 wt%. In the case of

Fe, the obtained catalyst HAADF-STEM image is shown in Figure 5.7h. Several single-atom-sized bright spots can be observed, which can be attributed to the Fe atoms in this sample. As shown in Figure 5.7i, Fe–NC SAC exhibits excellent electrocatalytic ORR activity. The onset potential (0.98 V) and half-wave potential (0.90 V) of Fe-NC SAC are significantly positively shifted, 20 and 50 mV respectively higher than the onset potential of Pt/C catalyst.

Ni single atoms can activate catalysis through modifying the electronic structure for HER. For example, during the electrocatalytic water-splitting process, the HER activity of MoS_2 is mainly derived from S atoms at the edge positions, and for MoS_2, the prepared catalysts inevitably contain a large number of non-electrochemical active sites. The introduction of co-catalytic cations has demonstrated its ability to trigger basal inert S atoms. Zhang et al. [20] effectively decorated isolated Ni atoms into the lattice structure of MoS_2, as shown in Figure 5.8a. The Ni single-atom modification produces a large number of activated S atoms at the substrate with an altered electronic structure to facilitate water dissociation and subsequent catalytic processes. Using carbon as a substrate, a thin layer of MoS_2 can be grown on its surface, as shown in Figure 5.8b and c. The introduction of Ni atoms produces shortened Ni-S bonding and distorted conformations on the MoS_2 substrate. However, the decoration of Ni atoms maintains the crystal structure of hexagonal MoS_2. As shown in Figure 5.8d, activation of the S site with decorated Ni atoms results in a large decrease of ΔG_{H*} by 0.15 eV, which is closer to the S atom at the edge site (0.08 eV). In addition, SACs with fine coordination geometry and strong metal-carrier interactions could provide the highest atomic efficiency, and electron coupling between Ni and neighboring N atoms could cause electron redistribution, and lower electron density near the Fermi energy level could improve OER activity [49]. The d-band of the catalyst could strongly interact with the highest occupied molecular orbital of the adsorbate, and the lower electron density near the Fermi energy level would promote the desorption of the adsorbate. The electronic structure of the isolated reaction center controls both the OER activity and the reaction mechanism. As shown in Figure 5.8e, the effective electronic coupling via Ni–N coordination can reduce EF and lower the adsorption energy of intermediates, thus promoting OER kinetics.

The introduction of active single atoms into catalysts is one of the most widely employed methods to tailor the electronic structure of catalysts to improve their intrinsic activity, the density of active sites, electron transport, and stability. For fuel cells and metal-air cells, Han et al. [109] designed synergistic catalysis with highly reactive uniformly isolated monoatomic and bimetallic Co–Ni sites to reduce the energy potential barrier and accelerate the reaction kinetics, which greatly contributed to the electrocatalytic ORR/OER activities. The atomically dispersed binary Co–Ni configuration could be successfully synthesized in CoNi–SAs/NC, which had a unique electronic structure and coordination environment that facilitates the electrocatalytic process. This catalyst provided an excellent bifunctional activity, long-term durability, and reversibility for bifunctional ORR/OER and rechargeable zinc-air batteries. CoNi–SAs/NC provided the same potential gap (ΔE) value of 0.81 V as the state-of-the-art Pt/C–IrO_2 noble metal catalyst, and as shown in Figure 5.8f and g, the

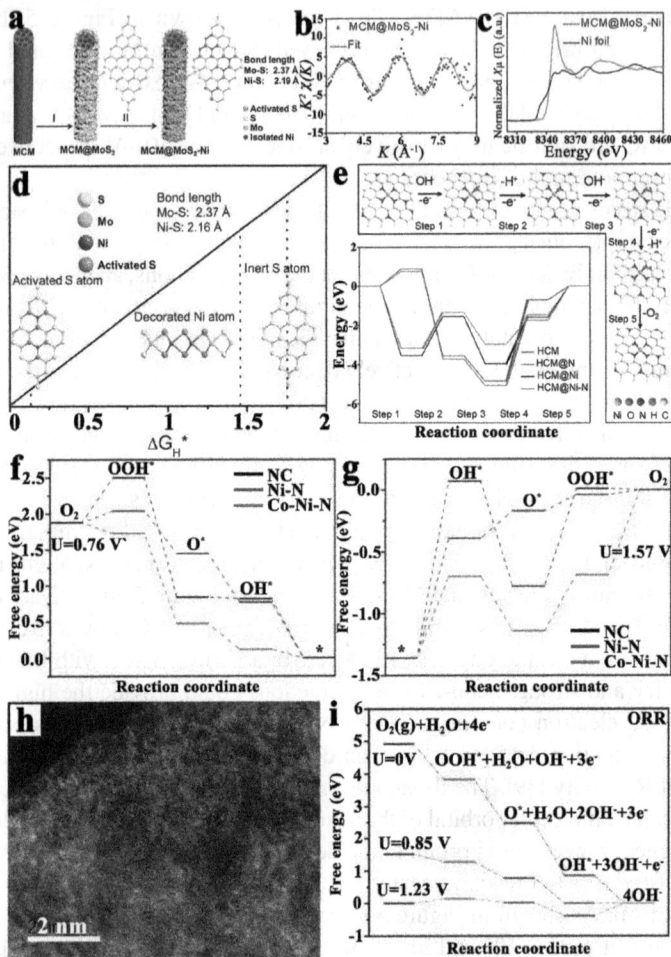

FIGURE 5.8 (a) Schematic illustration of the synthetic process for MCM@MoS$_2$–Ni. (b) Experimental and best-fitted EXAFS spectra in K space for MCM@MoS$_2$–Ni. (c) Experimental XANES spectra for MCM@MoS$_2$–Ni and Ni foil. (d) Theoretical calculations for the effects of Ni decoration on the HER activity of MoS$_2$. (Ref. [20]. Copyright 2018 Wiley-VCH Verlag GmbH & Co. KGaA, Weinheim.) (e) Proposed mechanism of O$_2$ evolution over HCM@-Ni–N.and free energy diagram at 0 V for OER over HCM, HCM@N, HCM@Ni, and HCM@-Ni–N. (Ref. [49]. Copyright 2019 Wiley-VCH Verlag GmbH & Co. KGaA, Weinheim.) Free energy diagram of (f) ORR, and (g) OER processes on NC, Ni–N, and Co–Ni–N. (h) HAADF-STEM of CoNi-SAs/NC. (Ref. [109]. Copyright 2019 Wiley-VCH Verlag GmbH & Co. KGaA, Weinheim.) (i) Free energy diagram of ORR on Co–N$_4$/C. (Ref. [22]. Copyright 2019 Wiley-VCH Verlag GmbH & Co. KGaA, Weinheim.)

favorable adsorption configuration can be optimized by pre-adsorption on the Ni site of CoNi–SAs/NC on the OH* surface. HAADF-STEM of CoNi–SAs/NC is shown in Figure 5.8h, which also demonstrates that the catalyst is in the form of single atoms. Compared to catalysts in the presence of clusters, single-atom coordination sites exhibit a lower catalytic barrier to oxygen catalysis, which greatly facilitates ORR and OER in alkaline media [22]. The layered porous structure of Co SA@NCF/CNF allows high accessibility to single-atom sites, and the binder-free design allows control of the active site and surface morphology of the electrode, as shown in Figure 5.8i, resulting in better electrode performance.

5.2.4 Mo, W, Ti-Based Single-Atom Catalysts

The rapid development of single-atom catalysis covers more elements rather than noble metals (Pt, Ru, Pd et al.) and common transition metallic elements (Fe, Co, Ni et al.). Some metals with high melting points, such as Mo, Ti, and W, have also been explored to extend the understanding of SACs. Titanium (Ti) as one typical transition metal has plenty of significant applications and has long been theoretically predicted to be a good candidate to modify the properties of carbon materials when chemically bonded with the substrate of isolated Ti single atoms. For example, Liang et al. [53] synthesized Ti SACs (Ti_1/rGO) which were used in non-Pt cathode material in dye-sensitized solar cells (DSCs) and it exhibited high electrocatalytic activity toward the tri-iodide reduction reaction (T-IRR), as shown in Figure 5.9b. Ti atoms were anchored by oxygen atoms on the reduced graphene oxide (rGO). Its structure was determined to be five coordinated local structures TiO_5, which could be characterized by HAADF-STEM and X-ray absorption fine structure spectroscopy (XAFS), as shown in Figure 5.9a. Based on this structure, DFT calculation was also employed to reveal possible orbital interaction between the Ti atom and the support atoms (O and C). The calculation results, the total density of states (TDOS), and PDOS are shown in Figure 5.9c. The top and bottom on the left are PDOS images of rGO and Ti_1/rGO, respectively. It can be found that Ti 3d orbits overlap with O 2p and C 2p at three positions, which indicates that there is a strong electron interaction between Ti atoms and support atoms. It can also be found that the electronic structure of Ti_1/rGO is well modified by comparison to rGO in the TDOS curves, as shown in Figure 5.9c. rGO has a small bandgap of around 0.3 eV, while the orbits are more continuous near the Fermi level for Ti_1/rGO. The above results indicate that Ti_1/rGO has a higher conductivity than rGO, which is favorable to charge transfer in T-IRR.

Wang et al. [50] produced the Mo SACs based on surface oxidized Co_9S_8 nanoflakes (Mo–Co_9S_8@C) by a simple solvothermal method. The atomically dispersed Mo sites can be identified via HAADFSTEM, XPS, and synchrotron radiation-based XAFS. Particularly, XPS and synchrotron radiation-based XAFS also support the conclusion that monodispersed Mo atoms species are stabilized by the surrounding O, rather than Mo–S or Mo–C bonds, as shown in Figure 5.9d. Mo–Co_9S_8@C was studied in different electrolytes, illustrating excellent electrocatalytic performance for both HER and OER in a wide pH range. Based on such an excellent performance, it can be concluded that the catalyst has the potential to be used as both cathode and anode for overall water splitting. The Mo–Co Co_9S_8@C couple can drive the overall water splitting with a cell voltage of 1.56 V (1.0 M KOH), and 1.68 V (0.5 M H_2SO_4) at 10.0 mA·cm^{-2} current density (Figure 5.9e). Unexpectedly, in both acidic and alkaline electrolytes, the overall water splitting of Mo–Co9S8@C exceeded those of precious

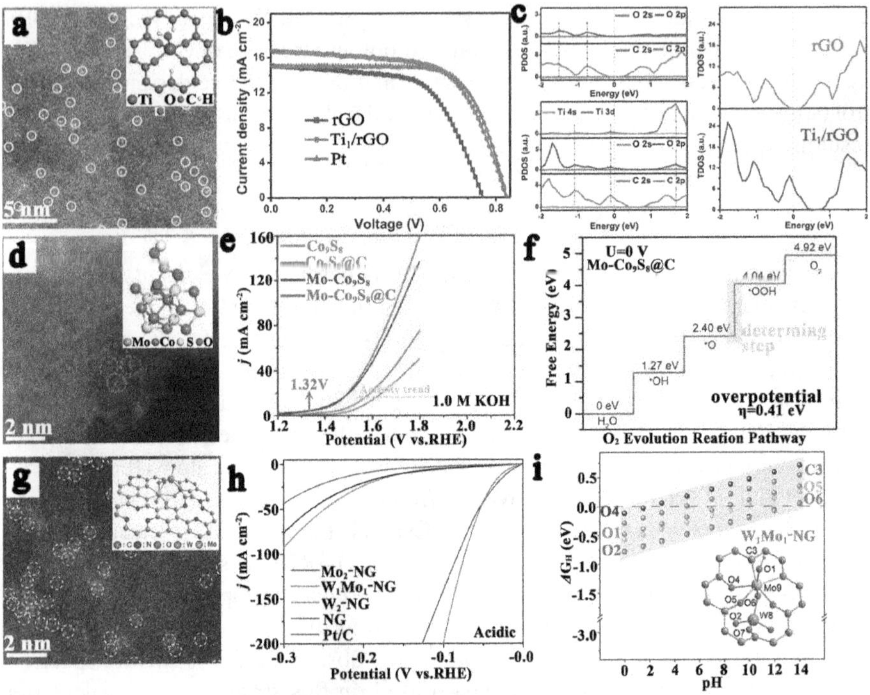

FIGURE 5.9 (a) HAADF-STEM image of Ti_1/rGO (inset: the most possible configuration of Ti_1/rGO). (b) Current-voltage (J–V) curves of DSCs using different cathodes (rGO, Ti_1/rGO, Pt). (c) PDOS of rGO and Ti_1/rGO (on the left) and TDOS of rGO and Ti_1/rGO (on the right), the Fermi energy is set at 0 eV. (Ref. [53] Copyright 2020 Wiley-VCH Verlag GmbH & Co. KGaA, Weinheim.) (d) Atomic-resolution HAADF-STEM image. (e) LSV curves for OER of Mo–Co_9S_8@C and the controlled samples in 1.0 M KOH. (f) Gibbs free-energy diagram for four steps of acidic OER on Co_9S_8 and Mo–Co_9S_8@C. The shaded region is the rate-determining step and η stands for overpotential. (Ref. [50] Copyright 2019 Wiley-VCH Verlag GmbH & Co. KGaA, Weinheim.) (g) AC HAADF-STEM images of W_1Mo_1–NG (inset: the schematic models of W_1Mo_1–NG). (h) The polarization curves and the corresponding Tafel plots of W_1Mo_1–NG and the controlled samples in 0.5 M H_2SO_4. (i) ΔG_H diagrams of W_1Mo_1–NG on different sites (inset: optimized geometries and possible active sites for H adsorption on W_1Mo_1–NG). (Ref. [52] Copyright 2020 AAAS.)

metals, which was the first case that a non-noble catalyst could be better than the Pt/C‖IrO_2 for overall water splitting, especially in an acidic electrolyte. To get further insight into the effects of single Mo atoms on performance, DFT calculations were utilized to understand the underlying mechanism of the OER process. The OER computational models adopt a four-layered (311) Co_9S_8 crystal plane and then a single Mo atom is added on the basal plane of Co_9S_8 via O atom, as shown in Figure 5.9f because $4e^-$ processes greatly restrict the progress of overall water splitting. Compared with pure Co_9S_8, the introduced alien elements could promote the catalyst's activity. The synergistic effect between single-site Mo and Co site endowed Mo-Co_9S_8@C to have optimal intermediates adsorption free energy, which led to a lower theoretical OER overpotential of 0.41 V (0.74 eV in pure Co_9S_8). This means that single-site Mo active species play an important role in the acidic water oxidation reaction.

In general, the SACs can enable the reasonable use of metal resources, maximize the efficiency of metal atoms utilization, and achieve atomic economy [100]. However, the isolated single-metal sites on the surface are likely to increase the surface free energy of the catalyst and cause agglomeration. Compared with the SACs, the electronic environment of the multiatom catalyst can be adjustable, which is very conducive to the intrinsic activity and stability of the catalyst. In the multiatom catalyst, there is a strong interaction between adjacent atoms, which allows the species to remain relatively independent and prevents the occurrence of agglomeration. In this case, by adjusting the ligand atom, coordination number, and structural distortion to adjust the electronic structure, the metal atoms can be dispersed as much as possible on the support, and finally, binuclear sites or extremely small clusters can be formed. The point to be emphasized is that the heteronuclear metal atom catalyst can optimize the activity, stability, and selectivity of the catalyst by adjusting the metal active center, but this adjustable characteristic is rarely seen in a single-metal single-atom system [77]. Thus, the atom utilization and catalytic activity are expected to be improved in the dual-atom catalysts (DACs) with interacted bi-atomic metal cores. For example, Yang et al. [52] proposed a unique DACs structure (W—O—Mo—O—C) with strong interaction between the atoms. The synthesis process mainly included hydrothermal, freeze-drying, and thermal treatment. The SEM images showed abundant wrinkles and ripples, which can be anchoring sites to fix the Mo and W atoms. The bright dots, Mo and W atoms, are obvious in the AC HAADF-STEM image (Figure 5.9g). It demonstrates the existence of a large proportion of isolated heteronuclear W-Mo atoms. Combining with XPS, XANES, and EXAFS, the DAC with W—O—Mo—O—C configuration can be confirmed. In the end, the catalyst was used in HER in 0.5 M H_2SO_4 and 1.0 M KOH solutions. Impressively, as shown in Figure 5.9h, W_1Mo_1–NG produces cathodic geometric current density (j) of 10 mA·cm^{-2} at an overpotential of 24 mV ($\eta_{10} = 24$ mV), which is much lower than earth-abundant HER catalysts in the acid electrolyte, such as CoP/NiCoP nano-tadpoles (125 mV) [110], Ni_2P/Ni@C (149 mV) [111], Fe_3C–Co/NC (298 mV) [112], and even commercial Pt/C catalyst. To understand that the O-coordinated W_1Mo_1–NG system promotes the catalytic reaction over a wide range of pH 0 to pH 14, a theoretical calculation was carried out. Based on EXAFS fitting results, W—O—Mo—O—C configuration was found in the DFT geometry optimization indicating that the O-bridged W–Mo atoms are anchored in NG vacancies through oxygen atoms in a heteronuclear model of W_1Mo_1–NG, as shown in the bottom right corner of Figure 5.9i, including the possible active sites, namely, O_1, O_2, C_3, O_4, O_5, O_6, O_7, W_8, and Mo_9. The calculated ΔG_H on the abovementioned sites for DACs and the results are summarized in Figure 5.9i. O_4, O_5, and O_6 are preferred for proton adsorption, indicating that the W—O—Mo—O—C configuration can redistribute the electronic structure to construct an improved electron environment for HER.

5.3 INFLUENCE OF SUPPORT OF SINGLE-ATOM

In principle, SACs represent the maximum utilization efficiency of the metal atom. However, the high surface energy of single-metal atoms makes them prone to agglomeration during the preparation process. Therefore, using suitable support to anchor single-metal atoms to prevent them from agglomerating into nanoparticles is a great challenge in this field. The strong interaction between the support and single-metal atoms is the core principle for the construction of SACs. Researchers have found that in addition to surface-active atoms as active centers, the interface between active atoms and substrates is also where some catalytic reactions occur. It can be said that the interaction between active atoms and substrates can promote catalytic activity [113]. Therefore, the choice of substrates is particularly important. With respect to this, various types of substrates have been developed based on the understanding of the substrates and metal atoms, such as graphene [54,55,60], nanocages [68,114], nanotubes [70–72], graphdiyne [66,67], MXene [85,86], oxides [87,88], sulfides [91,92], and borides [94,95].

5.3.1 GRAPHENE SUBSTRATES

Graphene, which has a large specific surface area, excellent conductivity, and electrochemical stability, has been considered to be the ideal substrate for single atoms [59,60]. The 2D structure of graphene contains a great number of defects, which can be used as natural sites to trap single atoms and supply stable coordination environments for single atoms. The intrinsic physical and chemical properties of graphene can be modulated by doping different elements in the flexible 2D sheet lattice, which also generates new types of defects. Both the dopants and the defects are possible trapping centers for single atoms [65,88]. However, it is still a challenge to synthesize graphene materials with high single-atom loading. This is mainly due to the poor thermal stability of metal precursors and the high surface energy of metal single atoms, which can cause the aggregation of metal atoms and the growth of metal particles in the traditional pyrolysis process [115].

5.3.1.1 Undoped Graphene Substrates

At present, there are two bonding methods between graphene substrates and single atoms: the embedded type and the supported type [116]. The embedded type is inserting single atoms into substrates with defects, and the supported type is a simple connection between single atoms and atoms on substrates. At present, the more extensive research studies focus on SACs with an embedded type. For example, Yao et al. [57] prepared a single-atom defect graphene-based catalyst (A–Ni@DG) with 1.24 wt% of Ni atoms by a simple acid leaching method, as shown in Figure 5.10a. By comparing the catalytic activity of A–Ni@DG and perfect graphene-supported SAC (A–Ni@D), the interaction between defects and single atoms was further proved. As shown in Figure 5.10b, compared to A–Ni@D, A–Ni@DG has better HER and OER activities. The lower Tafel slope of A–Ni@DG shows a faster electron transfer rate. The higher Ni content indicates that the defects can capture more Ni atoms, thereby forming a unique connection configuration and well-dispersed metal atoms without aggregation. The defects can optimize the electronic structure of Ni atoms and reduce the reaction barriers of HER and OER. As shown in Figure 5.10c, D5775(D5775 represents an ordinary Stone-Wales graphene defect) plays a huge role in HER, while vacancy defects play a huge role in OER. The desorption energy of the intermediate product of A–Ni@DG with D5775 is

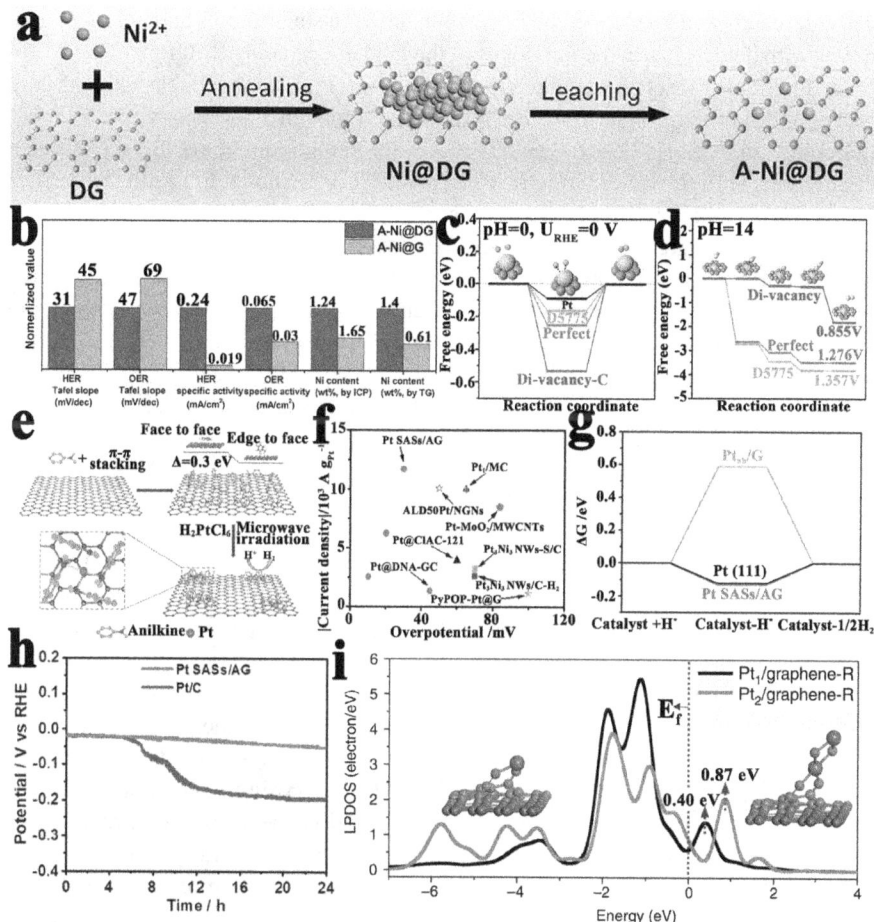

FIGURE 5.10 (a) Schematic diagram of the preparation method of A–Ni@DG. (b) Comparison graph of the Tafel slope and catalytic specific activity of HER and OER of A–Ni@DG and A–Ni@G, and the comparison graph of their Ni content. The changes in the free energy of three special atomic configurations during (c) HER and (d) OER. (Ref. [57] Copyright 2018 Nature Publishing Group.) (e) Schematic diagram of the preparation method of Pt–SASs/AG. (f) The HER catalytic activity of Pt–SASs/AG, Pt/C, and other advanced Pt-based catalysts in 0.5 M H_2SO_4. (g) The HER free energy diagram of Pt(111), Ptab/G, and Pt–SASs/AG were calculated by DFT. (Ref. [117] Copyright 2019 The Royal Society of Chemistry.) (h) Chronopotentiometric curves of Pt SASs/AG and Pt/C in a 0.5 M H_2SO_4 solution at 10 mA cm^{-2}. (i) The local projected density of states (LPDOS) of the electron orbitals on the top Pt atoms of Pt$_1$/graphene-R and Pt$_2$/graphene-R. (Ref. [54] Copyright 2017 Nature Publishing Group.)

close to that of Pt, indicating a better HER catalytic activity. As shown in Figure 5.10d, A–Ni@DG with vacancy defects has the lowest desorption energy and exhibits the lowest potential of 0.855 V during the dissociation process, showing that it has the best OER catalytic activity. The embedded metal atoms form local ligands through ionic and covalent bonds with the support atoms, and the metal-support interaction will also make the catalyst more stable. The interaction between single atoms and graphene substrates with defects has come into focus in the field of designing SACs.

In addition to inserting single atoms into defects, another reliable method is to anchor and stabilize single atoms by introducing some other atoms or molecules. For example, Sun et al. [117] developed a microwave reduction method to anchor single Pt atoms on aniline-stacked graphene (Pt-SASs/AG), as shown in Figure 5.10e. The maximum protonation of aniline and the gentle microwave reduction method together can prevent the aggregation of Pt atoms. As shown in Figure 5.10f, Pt-SASs/AG has better catalytic activity than other advanced Pt catalysts due to the special structure and combined action between aniline and surrounding Pt atoms. As shown in Figure 5.10g, three typical special structures of Pt-based catalysts are compared, they are Pt(111), single Pt atom adsorbed on graphene(Pt_{ab}/G), and Pt SASs/AG. Through the analysis of the free energy changes of the three structures in the HER process, it can be found that the ΔG_{H*} of Pt_{ab}/G reaches 0.587 eV, which indicates that the single-atom catalytic HER activity in Pt_{ab}/G is very low. The ΔG_{H*} of Pt SASs/AG is similar to that of Pt(111), indicating that the coordination of aniline and Pt atoms is the key factor for high catalytic HER activity. As shown in Figure 5.10h, after a long-term catalytic reaction, Pt SASs/AG still shows good stability, further indicating that Pt atoms do not agglomerate. It is believed that the aniline molecule can optimize the electronic structure of the single Pt atom and inhibit the aggregation of single atoms, and the dispersed single atoms increase the utilization rate of the catalytic center. Aniline can also optimize the hydrophilicity of the catalytic surface of graphene with excellent conductivity. In general, the excellent catalytic performance is mainly derived from the common role of single atoms, graphene substrates, and modifying groups. In addition, based on single-metal atoms, single atoms can be modified to form metal dimers to modify the adsorption properties of reaction intermediates. This catalyst has flexible active sites and synergistic effects between atoms, which has the potential of catalyzing multistep reactions and optimizing the catalytic activity and selectivity. Lu et al. [54] deposited secondary Pt atoms on the initial Pt atoms of the graphene substrate by depositing Pt atoms on the phenol-related oxygen using a bottom-up method to form Pt_2 dimer on the graphene support (Pt_2/graphene). Through experimental research, it was found that in the hydrolysis reaction of ammonia borane (AB), Pt_2/graphene had excellent catalytic HER activity ($2,800\,mol_{H_2} \cdot mol_{Pt}^{-1} \cdot min^{-1}$ at room temperature). This activity was 17 times and 45 times those of the graphene-supported initial Pt SAC (Pt_1/graphene) and Pt nanoparticle catalyst, respectively. Pt_2/graphene also had excellent reaction stability under current reaction conditions. This excellent catalytic performance was largely due to the lower adsorption energy of Pt_2/graphene in AB and H_2O. As shown in Figure 5.10i, Pt atoms and O atoms are connected to form a special configuration. From the projected density of states of the 5d orbital of the top Pt atom of Pt_1/graphene and Pt_2/graphene, it can be found that the orbital energy level of the Pt atom at the top of the Pt_2/graphene catalyst is 0.87 eV above the Fermi level, which is higher than the energy position of Pt_1/graphene (0.410 eV). This shows that Pt_2/graphene is uneasy to accept electrons, so the adsorption capacity of the intermediate will not be very strong. This weak adsorption characteristic is an important factor for the excellent catalytic activity of Pt_2/graphene.

5.3.1.2 N-Doped Graphene Supports

Although the current research on single-atom-supported graphene catalysts has been in-depth, these kinds of catalysts still have a great potential for improvement. In addition to introducing defects into the graphene substrate, doping with heteroatoms is also the most effective strategy for optimizing catalytic performance [118]. Nitrogen (N) atoms have higher electronegativity and smaller atomic radius, making N atoms a potential dopant [119]. For example, Lou et al. [41] used DFT to study the relationship between the catalytic activity and electronic structure of a series of metal SACs based on N-doped graphene and explored their HER activities. The surface configuration of SACs synthesized on the N-doped graphene substrate is composed of a single atom and several active nitrogen sites. As shown in Figure 5.11a, the active sites on

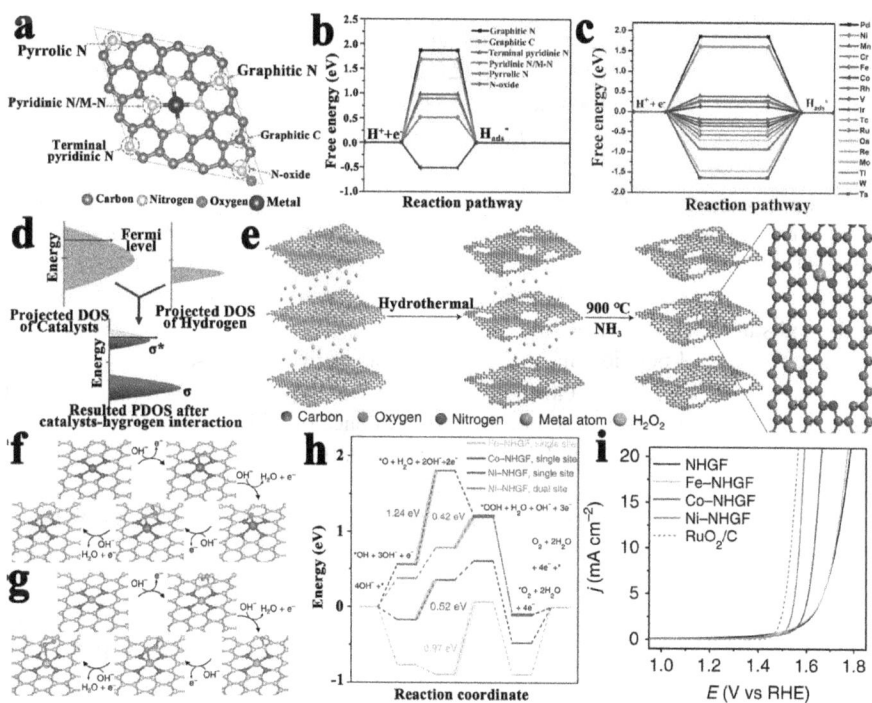

FIGURE 5.11 (a) The molecular model of the non-metallic sites where hydrogen adsorption occurs in the N-doped graphene substrate is calculated by Gibbs free energy. (b) The Gibbs free energy ΔG_{H*} changes during the hydrogen adsorption reaction at each non-metallic site. (c) For the hydrogen adsorption reaction (Volmer reaction), the free energy ΔG_{H*} diagram of different types of metal SACs. (d) Electron orbital hybridization of the adsorbed hydrogen and the active site of the catalyst. (Ref. [41] Copyright 2019 Wiley-VCH.) (e) Schematic diagram of the preparation method of M–NHGFs. (f) The single-site and (g) the dual-site mechanisms of metal sites in the adsorption and reaction process of OER intermediates. (h) Free energy diagrams of the OER process of Ni–NHGF, Fe–NHGF, and Co–NHGF with single-center mechanism and dual-center mechanism. (i) The OER activity of NHGF, Fe–NHGF, Co–NHGF, Ni–NHGF, and RuO$_2$/C. (Ref. [61] Copyright 2018 Nature Publishing Group.)

the N-doped graphene substrate could be graphite-N, graphite-C, pyridine-N, terminal pyridine-N, pyrrole-N, and N-oxide species. These models take into account all possible N atom configurations. As shown in Figure 5.11b, the calculation of the free energy of these active sites can be used to initially determine the source of activity. Through the calculation and comparison of the free energy of these active sites, it can be found that these possible active sites do not have high-efficient electrochemical activity due to their higher free energy. Graphite carbon (1.69 eV) and graphitic nitrogen (1.89 eV) have the highest ΔG_{H*}, showing their weak effect on the hydrogen adsorption step in HER. This also illustrates from the side that the interaction between a single-metal atom and an N atom is the key to a catalyst with high catalytic activity. The transition metal atoms coordinate with the four nitrogen atoms to form a special structure embedded in the graphene structure. The lone pair electrons existing in the metal single-atom part of the empty orbital and nitrogen are combined to affect the catalytic HER activity. As shown in Figure 5.11c, the Gibbs free energy exhibited by the combination of different types of metal single atoms and N-doped graphene can have strict selectivity to HER. Co SAC ($\Delta G_{H*} = 0.13$ eV) shows the best HER activity compared to other kinds of single atoms. The higher ΔG_{H*} values of Pb SAC and Ni SAC indicate the weak adsorption between the catalyst and hydrogen. Re, Mo, Ti, and other SACs have negative free energy, which indicates the strong adsorption between them and hydrogen. The in-depth mechanism of hydrogen adsorption on the SAC is shown in Figure 5.11d. After the interaction between hydrogen and single atoms, some of the valence orbitals hybridized with hydrogen on the metal atoms are converted into new bonded orbitals and anti-bonded orbitals. The occupancy rate of the anti-bond state is closely related to the energy level. The empty anti-bond state has a higher energy level position and will have a stronger interaction with hydrogen. The low-price orbital leads to a weaker interaction, while the high-price orbital ensures a stronger interaction. The position of the active valence orbital close to the Fermi level produces a partially filled anti-bond state, which is the ideal case for HER catalysis ($\Delta G_{H*} \approx 0$).

The special structure formed by the coordination of metal atoms and N atoms can optimize the catalytic HER activity of catalysts [62]. But for OER catalysis, the optimization mechanism of single-atom catalysis is more complicated. Sun et al. [61] embedded a series of different kinds of transition metal atoms in the framework of nitrogen-doped porous graphene (M–NHGFs, M=Fe, Co, or Ni), and explored the correlation between the atomic structure of the catalyst and the electrocatalytic OER activity. As shown in Figure 5.11e, the researchers used the doped N atoms as the binding sites of the introduced M single atoms through two steps of hydrothermal and annealing, so that the M single atoms are embedded in the lattice of N-doped graphene. The N atom acts as an intermediate to connect the C atom and the M atom and does not serve as a binding site for the reaction intermediate. After the doping of N atoms, the local coordination environment of the catalyst has changed. Both the M and C atoms around the N atom have the energy conditions to serve as intermediate reaction sites. As shown in Figure 5.11f, the adsorption and reaction of the reaction intermediates on the M atom site connected with N are simulated by DFT. The entire OER process is the process of continuously adsorbing OH^* and finally dissociating into O_2. As shown in Figure 5.11g, the site of this reaction may not only exist on the

M atom but also exist on the surrounding C atoms. The participation of the C atom depends on the number of d electrons of an M single atom. For Fe and Co, which have a low number of d electrons, the binding strength of the intermediate at the M atom site is higher than that at the C site, so the reaction only proceeds at the metal atom site (a single-site mechanism). On the contrary, for Ni atoms with a higher number of d electrons, O^* and OH^* tend to be adsorbed on C, and OOH^* tends to be adsorbed on M atoms (a two-site mechanism). The free energy of the OER process of different M-NHGFs is shown in Figure 5.11h. Compared with Fe-NHGF and Co-NHGF, the two-site mechanism of Ni-NHGF has the smallest limiting reaction barrier for OOH^* formation of 0.42 eV. The single-site mechanism Ni–NHGF exhibits a larger limiting reaction barrier of 1.24 eV. This illustrates the importance of the C atom as the active site of OER and the great potential of the dual-site mechanism for optimizing the catalytic OER activity. As shown in Figure 5.11i, the catalytic OER activity of M-NHGFs is consistent with the results of the above DFT simulation. The catalytic activity of Ni–NHGFs is higher than that of Co, Fe, and metal-free corresponding catalysts, similar to the precious metal RuO_2/C.

In short, it is particularly important to clarify the active sites in the catalyst and to rationally regulate the local environment and atomic connection configuration of the active sites [65]. To improve the catalytic activity and stability of SACs, three aspects need to be optimized: single atoms, substrates, and the interaction between the single atoms and substrates. For single atoms, the choice of metal atom species, the way of introduction, and the modification with different atoms are the major areas worth optimizing. For substrates, the selection of substrate types, such as graphene with high conductivity and excellent stability, and modification of defects, groups, and doping on the surface of the support, are all conducive to single atoms to anchor on the substrate with good stabilization and high dispersion ratio. The interaction between single atoms and substrates is extremely important and has become the key to the study of SACs, which affects the Fermi energy level of single atoms and the adsorption of intermediates. The design and construction of SACs need to fully consider the synergistic efficiency of single atoms and substrates.

5.3.2 GRAPHDIYNE SUBSTRATES

Graphdiyne (GYD) is a new kind of 2D carbon material with uniform 18 C-hexagonal pores formed by three butadiyne linkages ($-C\equiv C-C\equiv C-$) between the benzene rings. The rich carbon chemical bonds, high π-conjunction, and excellent chemical stability make GYD a great application potential in the field of catalysis as a novel carbon-support material. Yin et al. [66] successfully prepared two kinds of Pt SACS (Pt-GYD$_1$ and Pt-GYD$_2$) on GYD through the reaction of K_2PtCl_4 and GYD (Figure 5.12a). The FT-EXAFS curve results indicate that the Pt atoms are anchored on GDY forming C_1-Pt-Cl$_4$ in Pt-GDY$_1$ and C_2-Pt-Cl$_2$ in Pt-GDY$_2$ (Figure 5.12b). The authors found that the total unoccupied density of states of Pt 5d in Pt-GYD$_2$ is higher than that in Pt-GYD$_1$ with the XANE analysis, indicating that Pt 5d orbital could easier interact with the 1s orbital of H atom to accelerate electrons transfer to H atom with a C_2-Pt-Cl$_2$ coordination environment on GDY in Pt-GYD$_2$ (Figure 5.12c). The DFT calculations show the Gibbs free energy for hydrogen adsorption ΔG_{H^*} is

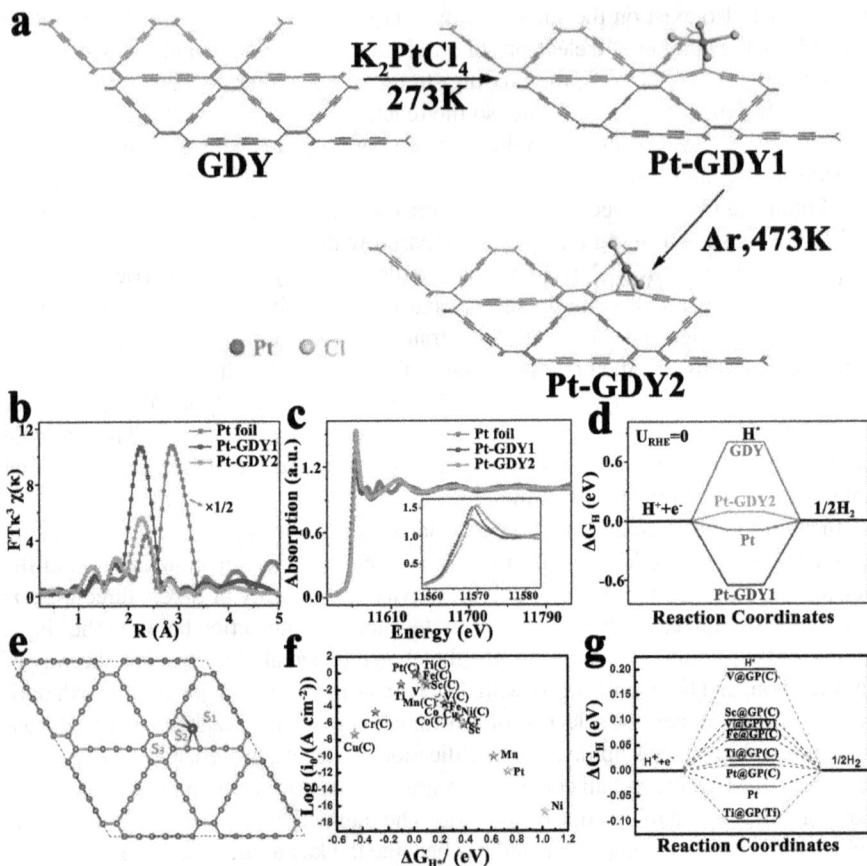

FIGURE 5.12 (a) Synthesis illustration of Pt-GDY1 and Pt-GDY2. (b) Pt L_3-edge k^3-weighted FT-EXAFS spectra of Pt-GDY1, Pt-GDY2, and Pt foil. (c) Pt L_3-edge XANES spectra of Pt-GDY1, Pt-GDY2, and Pt foil. The inset is the enlarged spectra at Pt L_3-edge. (d) The calculated Gibbs free energy diagram for hydrogen evolution on different catalysts. (Ref. [66] Copyright 2018 Wiley-VCH.) (e) Different possible adsorption sites for the single TM atoms are supported on GDY. (f) The volcano curve of exchange current related to the Gibbs free energy for hydrogen binding on different active sites on the GDY monolayer. (g) The corresponding ΔG_{H*} for the TM and C atoms in TM@GDY systems. (Ref. [67] Copyright 2019 Wiley-VCH.)

0.092 eV in the Pt-GYD$_2$, which further confirms that the Pt active site anchored on GDY with a C_2-Pt-Cl$_2$ coordination environment can contribute to the excellent catalytic HER activity in Pt-GYD$_2$ catalysts (Figure 5.12d). He et al. [67] investigated a series of transition metal atoms anchored on a GYD monolayer (TM@GYD, where transition metal (TM) represents a transition metal from Sc to Zn and Pt) using first-principle calculations. They found that the single-metal atom was most stable at the corner of the acetylenic ring (S$_2$) compared with those at the center of the holes (S$_1$) or the center of the hexatomic ring (S$_3$) (Figure 5.12e). The ΔG_{H*} on the TM and the C atoms of TM@GYD were further calculated to compare the HER performance (Figure 5.12f and g). The C atoms that bond to TM in Sc@GYD, Ti@GYD, V@GYD, Fe@GYD, and Pt@GYD exhibit the close values of ΔG_{H*}(C) to 0 eV, which are much smaller than that in pristine GYD of 0.73 eV, indicating that C atoms could be highly active sites for HER in Sc@GYD, Ti@GYD, V@GYD, Fe@GYD, and Pt@GYD. Especially, the ΔG_{H*}(Ti) and ΔG_{H*}(V) are also very close to zero in Ti@GYD and V@GYD, respectively, suggesting that the efficient TM sites exist in Ti@GYD and V@GYD, which makes these catalysts exhibit better HER performance than other TM@GYD composites.

5.3.3 MXene Substrates

In recent years, MXenes, as the novel two-dimensional support materials with a rich variety and adjustable surface chemistry, have attracted tremendous attention. Zhang et al. [85] subjected the $Mo_2TiC_2T_x$ MXene to electrochemical exfoliation treatment in an H_2SO_4 solution with a scan rate of 20 mV s^{-1} between 0 and −0.53 V versus RHE to obtain Mo-vacancy-rich defective $Mo_2TiC_2T_x$ MXene nanosheets ($Mo_2TiC_2T_x$–V_{Mo}) (Figure 5.13a). Then the single-atomic Pt was anchored on the surface of $Mo_2TiC_2T_x$–V_{Mo} via an in-situ electrochemical deposition process to produce the desired catalyst ($Mo_2TiC_2T_x$–Pt_{SA}). In addition, it was proved that the Pt atoms were stabilized in the Mo vacancy by forming three Pt-C covalent bonds with the surrounding carbon, leading to an electron transfer from the Pt to the $Mo_2TiC_2T_x$ support to generate excellent stability toward HER (Figure 5.13b and c). The DFT calculation results indicated that the electronic structure was redistributed by the incorporation of Pt atoms into the lattices of Mxene, leading to a higher occupied density of states near the Fermi level in $Mo_2TiC_2O_2$–Pt_{SA} than that in bare MXene, finally leading to the enhanced catalytic activity (Figure 5.13d and e). Based on first-principles calculations, Cheng et al. [86] proposed that the single-TM atoms (TM=Mn, Fe, Co, Ni, Mo, Ru, Pd, Ag, Ir, and Au) supported on Cr_2CO_2 (TM/Cr_2CO_2) MXene could be used as the potential bifunctional catalysts for HER and OER. They found that Ni/Cr_2CO_2 exhibited the best activity for both HER and OER during all TM/Cr_2CO_2 models, and the predicted results showed that Ni/Cr_2CO_2 could be experimentally synthesized (Figure 5.13f and g). The computational results confirmed that the electrons could transfer from the Ni atom to the surface O atoms of Cr_2CO_2, which promoted the binding strength of Ni with Cr_2CO_2 (Figure 5.13h–k). This powerful bond between Ni atom and Cr_2CO_2 allowed atomic Ni atoms to maintain atomic-level presence on the surface, so as not to aggregate into nanoclusters, which endow the high stability of Ni/Cr_2CO_2 as a SAC catalyst for water splitting.

5.3.4 Carbon Nanotube Substrates

As identified, the atomic-level distribution of Pt on the carbon-support material can greatly increase the number of active centers. However, due to the weak interaction between the carbon substrate and the metal atoms, the stability of the carbon-supported Pt SAC material has been greatly challenged. Tavakkoli et al. [71] prepared a pseudo-atomic-scale Pt that stably existed on single-walled carbon nanotubes

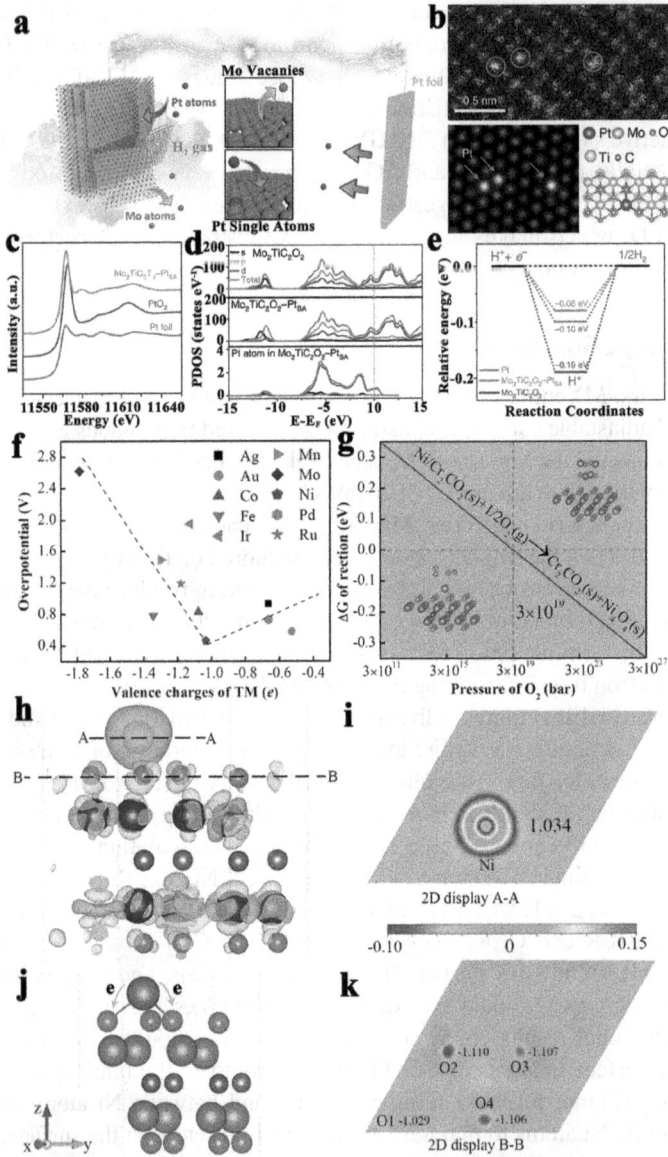

FIGURE 5.13 (a) Schematic illustration of the electrochemical exfoliation process of MXene with immobilized Pt single atoms. (b) HAADF–STEM image of $Mo_2TiC_2T_x$–Pt_{SA}. (c) Normalized XANES spectra at the Pt L_3-edge of Pt foil, PtO_2, and $Mo_2TiC_2T_x$–Pt_{SA}. (d) Calculated PDOS of $Mo_2TiC_2O_2$ and $Mo_2TiC_2O_2$–Pt_{SA}, with aligned Fermi level. (e) Calculated free energy profiles of HER at the equilibrium potential for $Mo_2TiC_2O_2$, $Mo_2TiC_2O_2$–Pt_{SA}, and Pt/C. (Ref. [85] Copyright 2018 Springer Nature.) (f) Relationship of overpotentials with valence charges of TM. (g) Reaction Gibbs free energy of the formation of Ni oxide from Ni active site. Charge density difference of Ni/Cr_2CO_2 surface, (h, j) three-dimensional view, and the schematic of charges transfer of Ni/Cr_2CO_2 system, two-dimensional enlarge view of (l) A–A and (k) B–B planes, respectively. (Ref. [86] Copyright 2019 American Chemical Society.)

by electrodeposition. Pt mainly existed on the surface of carbon nanotubes in the form of single atoms or sub-nano Pt clusters composed of several to dozens of atoms through controlling a short electrodeposition time (Figure 5.14a and b). The DFT calculation results showed that compared to other graphene materials, the sidewalls of SWNT could effectively fix Pt atoms so that they could achieve atomic-level dispersion on the surface of SWNT (Figure 5.14c and d). Chen et al. [70] used the co-decoration of CNTs with N and S to assist the atomic-level distribution of Fe−N$_x$ species on the surface of CNTs as bifunctional OER/ORR catalysts to achieve an excellent bifunctional activity in an alkaline medium. First, the CNT was wrapped with 2,2-bipyridine and Fe salt precursor, and pyrolyzed in a nitrogen atmosphere at 900 °C, and then etched in an acid solution to obtain the final product S, N−Fe/N/C−CNT (Figure 5.14e). Compared with the S-free sample, the introduction of S could promote the atomic-level dispersion of Fe−N species in the final sample. The FT of the Fe K-edge EXAFS results showed that there was only Fe−N shell at about 1.5 Å and no obvious Fe-Fe interaction could be detected (Figure 5.14f), indicating that Fe atoms were atomically dispersed in the as-obtained product. DFT calculations further indicated that the introduction of S could lead to the uneven distribution of charges in the carbon skeleton, which made it easier for the positively charged C atoms to adsorb O species, thereby accelerating the progress of the OER and ORR (Figure 5.14g and h). In addition, Cheng et al. [72] prepared atomically dispersed bimetallic FeNi embedded in N-doped carbon nanotubes via a modified one-pot pyrolysis technique (Figure 5.14i). The obtained electrocatalyst achieved a low potential gap ΔE of 0.81 V to generate an ORR current density of 3 mA·cm^{-2} and an OER current density of 10 mA·cm^{-2} with a lower catalyst loading of 0.2 mg·cm^{-2} (Figures 5.14m and n). The XAS and XPS results indicated the possible existence of bimetallic Ni and Fe atoms bridged by nitrogen, which could facilitate the adsorption and dissociation of the intermediates in the OER process (Figure 5.14j–l). Thus, the synergistic effect between adjacent Ni and Fe atoms in the FeNi bimetallic sites can achieve high activity for OER and ORR compared with single Fe SAs/NCNT or Ni SAs/NCNT sites.

5.3.5 CARBON NANOBALL (NANOCAGE) SUBSTRATES

Liu et al. [68] anchored Pt single atoms on nanosized onion-like carbon (OLC) substrate, which owned a highly curved surface structure to mimic active sites at the corner and edges of particles. They first prepared OLC via a thermal annealing method, followed by an ALD treatment to deposited Pt sites on the surface of substrates at 150 °C (Figure 5.15a and b). The obtained OLC-supported Pt SACs with only 0.27 wt% Pt loading exhibited a high efficiency for HER, which could be comparable to commercial Pt/C catalysts (20 wt% Pt) (Figure 5.15c). DFT calculations suggested that the OLC surface curved like a tip could accumulate electrons around the Pt sites from the substrate, leading to a strong localized electric field to enhance the catalytic activity for HER (Figure 5.15d and e). In addition, Zhang et al. [114] fabricated Pt SACs immobilized on hierarchical N-doped carbon nanocages (named hNCNC) by a simple impregnation-adsorption technique. They first trapped [PtCl6]$^{2-}$ anions into micropore of hCNC, followed by a dechlorinated treatment at 70°C to separate

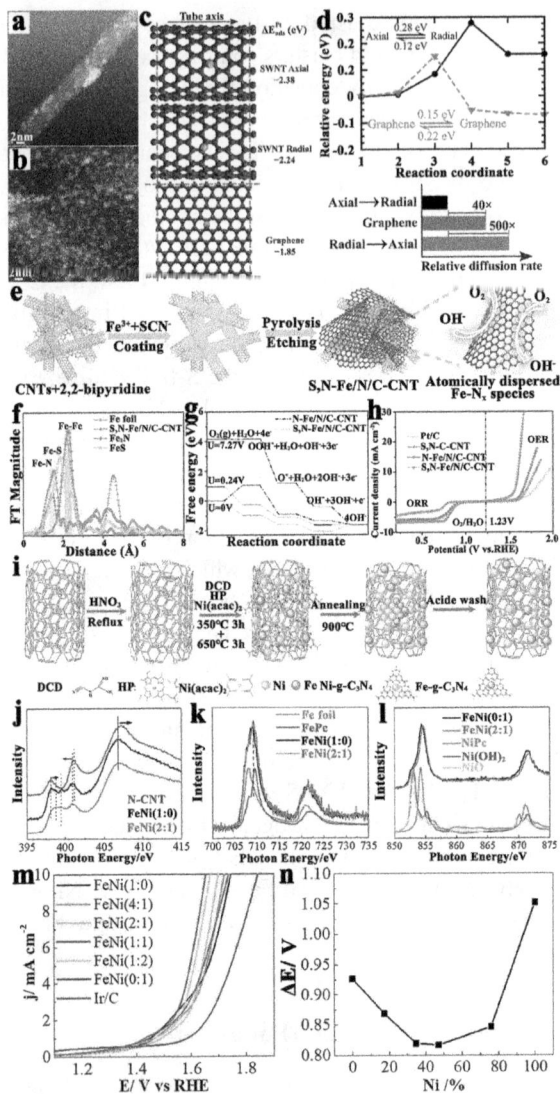

FIGURE 5.14 (a, b) STEM photos of the purified SWNTs after 400 activation cycles. (c) Adsorption of Pt onto (14,0) SWNT and graphene. (d) Comparison of minimum energy profiles for Pt diffusion between two adjacent sites on graphene and SWNT. (Ref. [71] Copyright 2017 American Chemical Society.) (e) Synthetic process for S,N–Fe/N/C–CNT. (b) XRD pattern of S,N–Fe/N/C–CNT. (f) Fourier transforms of the Fe K-edge EXAFS oscillations of Fe foil, reference samples FeS, Fe_2N, and S,N–Fe/N/C–CNT. (g) Energy profiles for the ORR pathway on S,N–Fe/N/C–CNT, and N–Fe/N/C–CNT catalysts under alkaline conditions. (h) The overall polarization curves of all samples in the whole ORR and OER region in a 0.1 M KOH solution. (Ref. [70] Copyright 2019 Wiley-VCH.) (i) Scheme of the synthesis of the bimetallic FeNi catalysts on acid-treated CNTs. N K-edge (j), Fe L-edge (k) and Ni L-edge (l) for FeNi (0:1), FeNi (2:1), NiPc, $Ni(OH)_2$ and NiO. (m) LSV curves of the catalysts for OER. (n) Plots of ΔE to achieve an OER current density of $10\,mA\,cm^{-2}$ and ORR current density of $3\,mA \cdot cm^{-2}$. (Ref. [72] Copyright 2019 Wiley-VCH.)

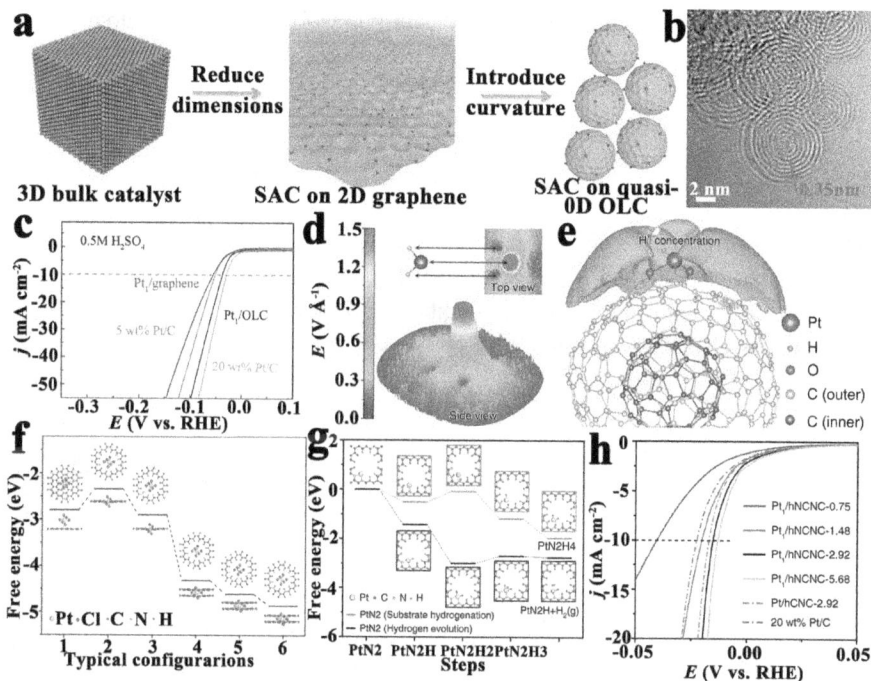

FIGURE 5.15 (a) The schematic illustrates the approach taken to design the Pt₁/OLC catalyst. (b) TEM image of Pt₁/OLC. (c) LSV of Pt₁/OLC in comparison with 5 and 20 wt% commercial Pt/C and Pt₁/graphene (0.33%) in a 0.5 M H_2SO_4 electrolyte. (d) The map of the electric field of the H_2Pt_1/OLC system at the tip-like Pt site with an equilibrium potential. (Ref. [68] Copyright 2019 Springer Nature.) (e) The Gouy–Chapman–Stern model. (f) Six typical configurations of $[PtCl_6]^{2-}$ on different supports and corresponding calculated free energies. Six-Graphitic bi-layer with the micropore decorated by two py-N atoms. (g) Free energy of the hydrogen evolution and support hydrogenation for a Pt atom bonding with two py-N atoms (PtN₂). (h) LSV of Pt₁/hNCNC in 0.5 mol·L⁻¹ H_2SO_4 solution. (Ref. [114] Copyright 2019 Springer Nature.)

trapped Pt single atoms from Pt/hCNC catalyst. DFT indicated that different types of micropores exhibited different free energy of adsorption for $[PtCl6]^{2-}$ anions in the solution (Figure 5.15f). The adsorption free energy of hNCNC (4.6 eV) is higher than that of the graphitic plane (2.8 eV), indicating that the anions are easy to be captured by micropore trapping on hNCNC. The nitrogen anchoring of carbon substrate finally facilitates the preparation of Pt SACs and improves the stabilization of Pt single atoms trapped in hNCNC (Figure 5.15g and h).

5.3.6 N-Doped Carbon Supports

The research of SACs [120–124] indicates that the N-doped porous carbon matrix can be employed to support the isolated single atoms for their merits of high conductivity, good endurance, and confined microenvironment for electrocatalysis. The existence of the N element can serve as anchor points for doped single atoms. More importantly, as traditional nanoparticles or clusters are atomically divided into isolated single atoms anchored on carbon-support materials, the single atoms would exhibit surprising catalytic properties. For example, Lou et al. [73] have relocated the Pt single atoms into the interior of the carbon matrix from the surface of the parent carbon sphere based on a dynamic reaction approach. The adjacent C/N atoms become electrocatalytically active for HER (as shown in Figure 5.16a and b) with the influence of the isolated Pt centers [73]. The confinement effects of the carbon matrix are responsible for high stability of the hybrids along with the intensive bonding between the Pt single atoms and the neighboring C/N atoms. Li et al. [74] achieved the synthesis of Mo SACs using chitosan. The Mo SACs showed highly efficient

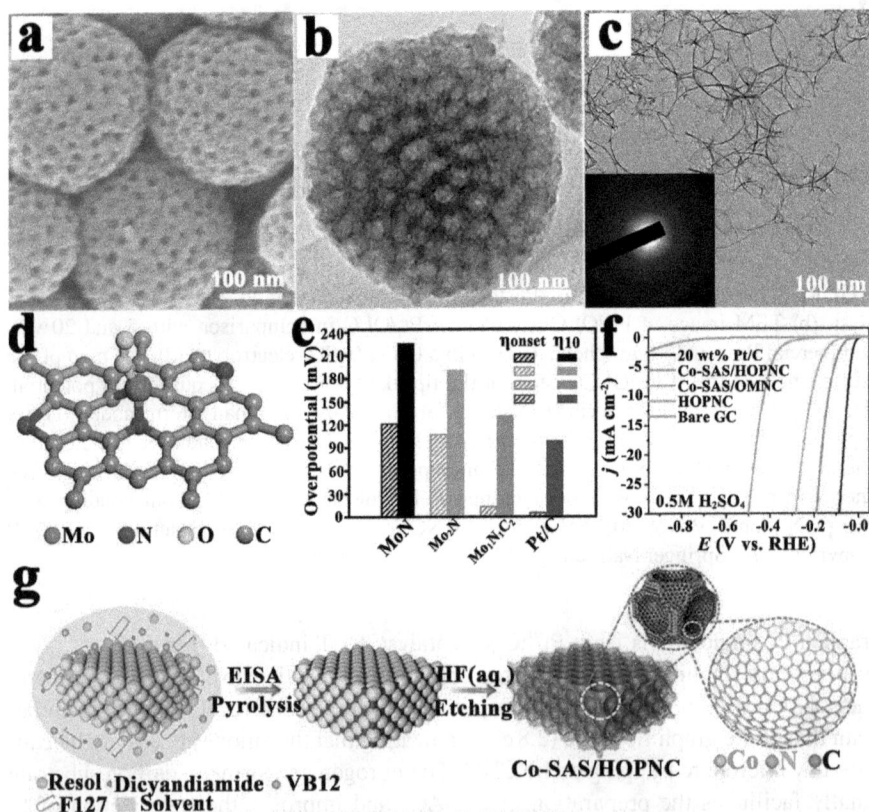

FIGURE 5.16 (a) SEM images and (b) TEM image of Pt@PCM. (Ref. [73] Copyright 2021 American Association for the Advancement of Science.) (c) TEM image of $Mo_1N_1C_2$ (inset: SAED pattern). (d) Atomic structure model of the $Mo_1N_1C_2$. (e) Overpotential for $Mo_1N_1C_2$ compared with Mo_2C, MoN, and 20% Pt/C. (Ref. [74] Copyright 2017 Wiley-VCH Verlag GmbH & Co. KGaA, Weinheim.) (f) HER polarization curves for the Co–SAS/OMNC, Co–SAS/HOPNC, bare glassy carbon (GC) electrode, HOPNC, and 20 wt% Pt/C. (g) Schematic illustration of the preparation of Co–SAS/HOPNC. (Ref. [76] Copyright 2018 PNAS.)

HER activity with a overpotential of $\eta_{10} = 132\,\text{mV}$ at $10\,\text{mA·cm}^{-2}$) and a Tafel slope of $90\,\text{mV·dec}^{-1}$ in $0.1\,\text{M}$ KOH solution, which was much better than those of Mo_2C and MoN (Figure 5.16c–e). Wang et al. [76] used a facile dual-template cooperative pyrolysis approach to prepare Co SACs with Co single atoms buried in N-doped hierarchically ordered porous carbon. They showed competitive bifunctional electrocatalytic performance for ORR and HER (Figure 5.16f and g). The Co–N coordinated active sites were abundant and atomically isolated. The mesopores generated a high surface area to expose more active sites. The hierarchically ordered macro/mesoporous framework shortened the molecule/ion diffusion length in the electrolyte.

5.3.7 MOF DERIVATIVE SUPPORTS

The selection rules of the supporting material for electrocatalysis include high specific surface area and rational structure, which determine the transport of reactants and products and the contact of the reactants with the active sites [125,126]. MOF-derived materials with high surface area, well-engineered porous structure, and multimetal coordination sites have attracted increasing attention and shown promising catalytic activity in SACs research. For example, the zeolitic imidazolate frameworks, ZIFs [121] have been demonstrated to be an ideal host for the construction of catalytic metal sites through pyrolysis without additional substrates [127].

Zhao et al. [82] revealed the dominant activity and selectivity contribution from single-atom electrocatalysis. They showed a coupled evolution of both single-atom cobalt sites and Co@C@N sites in the pyrolysis of Zn/Co bimetallic metal-organic frameworks and further demonstrated that removing the metallic particles through a sulfur-assisted site purification strategy could induce no decline for both ORR and HER activities, indicating the main catalytic contribution was from the atomic cobalt sites. Chen et al. [51] achieved the rational design of a single-W-atom catalyst, with W atoms anchored on an N-doped carbon matrix derived from a MOF (UiO-66-NH$_2$) for HER applications, which exhibited high activity and stability in both acidic and alkaline media. The aggregation of W species was prevented by the uncoordinated amine groups in the UiO-66-NH$_2$. Qu et al. [80] reported a practical method for the synthesis of SACs via direct atoms (Cu atoms) emitting from bulk metals (copper foam) followed by the trapping of nitrogen-rich porous carbon with the aid of ammonia (Figure 5.17a–c). Ammonia hauled out copper atoms and anchored them to the volatilized zinc nodes of the ZIF-8 at high temperatures. The Cu–SAs/N–C SAC showed superior ORR performance and good electrochemical and thermal stability (Figure 5.17d). Yang et al. [81] reported a top-down fabrication method to distribute Ni nanoparticles (NPs) on the surface of defect-containing N-doped carbon (NC) substrate. The NPs turned into single Ni atoms by a thermal diffusion mechanism. The NC substrate could stabilize the reactive sites effectively and achieve fast adsorption of reactants and desorption of products. The Ni single atoms exposed on the surface of the porous NC support could provide fast mass transfer for CO_2 electroreduction with high activity, selectivity, and stability. As shown in Figure 5.17e–g, the transformation of Ni NPs to SAs depends on the defects in NC. The Ni nanoparticles can break surface C–C bonds and drill into the carbon substrate with pores left on the surface.

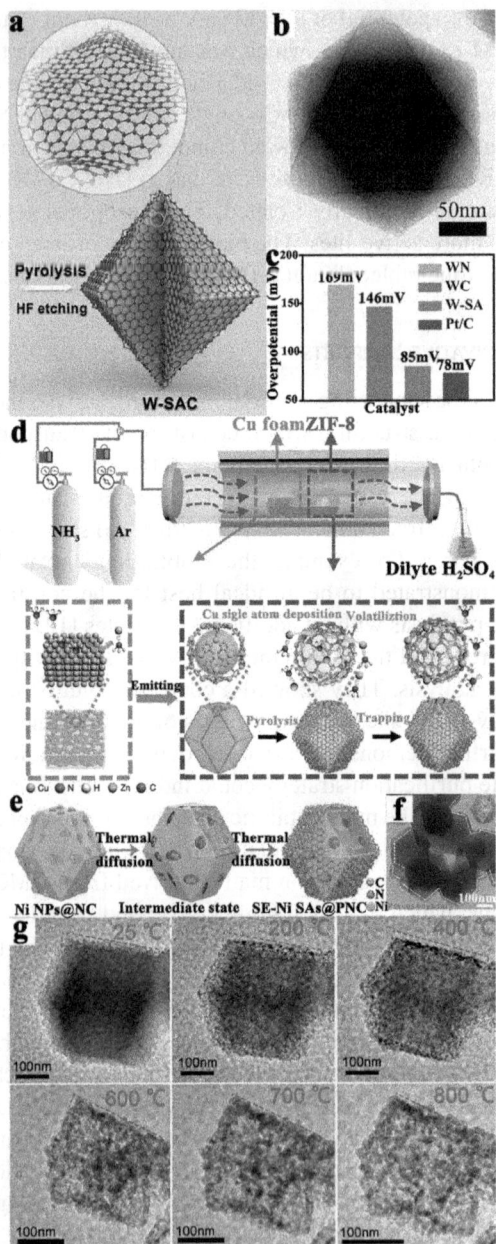

FIGURE 5.17   (a) Illustration of W–SAC. (b) TEM characterizations of the W–SAC. (c) Overpotential for W–SAC compared with WN, WC, and 20% Pt/C. (Ref. [51] Copyright 1999–2021 John Wiley & Sons, Inc.) (d) Fabrication of Cu–SA s/N–C. (Ref. [80] Copyright 2021 Springer Nature.) (e) The transformation of Ni NPs into Ni single atoms. (f) TEM photo surface-enhanced (SE) porous (P) of SE–Ni SAs@PNC. (g) Representative movie TEM images acquired at 25 °C, 200 °C, 400 °C, 600 °C, 700 °C, and 800 °C of Ni NPs@NC pyrolyzed in situ. (Ref. [81] Copyright 1999–2021 John Wiley & Sons.)

5.3.8 OXIDE AND SULFIDE SUBSTRATES

The stability of SACs depends heavily on the binding between the single atoms and the substrates [128]. Metal oxides are typical substrates for SACs, in which FeO_x is often chosen as a substrate for stable SACs due to its abundant defects or vacancies [129]. Studies have shown that there is still a lack of efficient methods to produce SACs and single-cluster catalysts (SCCs) with high loading and purity. Liu et al. [88] tried a strategy for the production of SACs and SCCs anchored on oxide substrates. By modeling Pt, Pd, and Ni SACs anchored on Fe_2O_3, the constituent nanoparticles (NP) of 17 Pt atoms could be leached at $U = 0.82\,V$, as shown in Figure 5.18a. The remaining 3 Pt atoms are dispersed atomically and coordinated to 3 or 4 surface O atoms, which is in the same SAC configuration as in the reported Pt_1/FeO_x SACs. Pd and Ni SAs anchored on the Fe_2O_3 surface present reasonably large U windows in the range of 0.61–0.82 V and 0.01–0.37 V, as indicated by Figure 5.18b and c, respectively. Thus, this method is a general approach suitable for the fabrication of typical substrates with high purity and loading of SACs. It can be applied to any target metal without the aid of any specific chemistry and can be easily controlled to achieve accurate purification and the theoretical maximum of SAC/SCC loading concentration (CL). Liu et al. [89] explored the factors related to the stability of SACs by combining various theoretical approaches. It was found that the stability and reactivity of SACs largely depended on the conditions, such as the reaction temperature, partial pressure, the size of NPs, and the reducibility of substrates. By comparing the chemical potentials, it could be found that Au SAs preferred to adsorb on CeO_2 step sites rather than on Au NPs or step surfaces, regardless of the presence of reactants (Figure 5.18d). Therefore, the dispersion and stability of SACs could be improved by fabricating cerium oxide substrate with a high step concentration. On the reducible substrate, Au SAs could strongly couple to the redox properties of the substrate and transfer charge to the metal cation, resulting in positively charged Au^+ atoms that became very stable upon CO adsorption and the more reducible substrate (Figure 5.18e). Rational design of the atomic structure between single atoms and supports could also improve the intrinsic activity and durability of SACs [130]. Chen et al. [131] prepared an Rh SAC-doped CuO NAs/CF (Rh SAC-CuO NAs/CF) by a facile cation exchange method (Figure 5.18f). HAADF-STEM analysis indicates the individual Rh atoms marked in red circles scattered in the lattice of CuO (Figure 5.18g). As shown in the EXAFS spectrum (Figure 5.18h), a distinct peak contributed by the Rh-O interaction is observed at about 1.5 Å. Rh-Rh peak or Rh-O-Rh peak are both absent in Rh SAC-CuO NAs, proving that the Rh single atoms are dispersed monotonically in CuO NAs. After doping Rh single atoms, some of the Cu atoms in $Cu(OH)_2$ were replaced by Rh^{3+}, which was stabilized by the surrounding O atoms. Rh existed as a single atom in the CuO lattice, after the cation exchange and calcination processes, forming a mixture of Rh single atoms doped into CuO NAs, which effectively reduced the adsorption energy of the reaction intermediates and improved the reaction kinetics of Rh SAC–CuO. Rh SAC–CuO NAs/CF showed an overpotential of 197 mV at 10 mA·cm^{-2} for OER and 44 mV at 10 mA·cm^{-2} for HER, which was close to that of commercial Pt/C, indicating that the performance of the low catalytic activity CuO nanowires

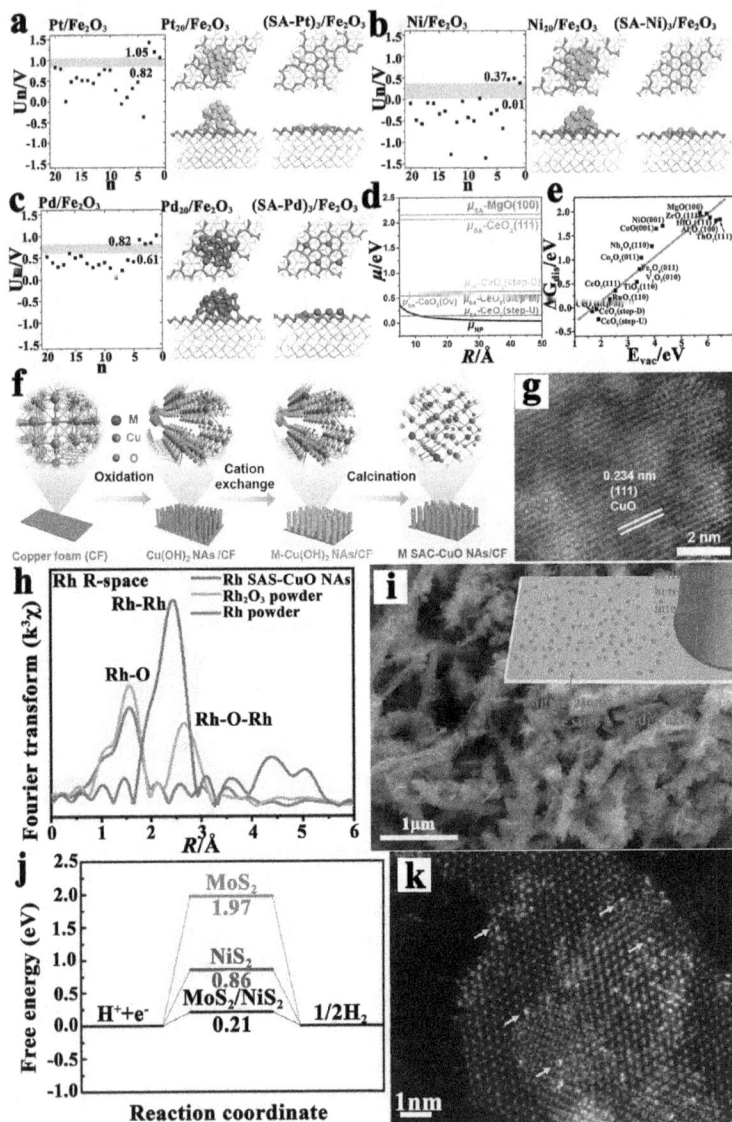

FIGURE 5.18 (a–c) Threshold leaching potentials Un (vs SHE) for Mn/Fe$_2$O$_3$(0001), with M = Pt (a), Pd (b), and Ni (c). (Ref. [88] Copyright 2020 ACS.) (d) Chemical potential and structures of NPs and SAs in a vacuum. Comparison between chemical potential of NPs, μNP(R), and that of SAs, μSA, with respect to the curvature of NPs. (e) ΔG$_{dis}$ versus the support vacancy formation energy. (Ref. [89] Copyright 2017 ACS.) (f) Schematic illustration and corresponding atom structure of products. (g) HAADF-STEM image and (h) the Fourier transformed (FT) k^3-weighted w(k)-function of the EXAFS spectra for the Rh K-edge of Rh SAC–CuO NAs, Rh$_2$O$_3$ powder, and Rh powder. (Ref. [131] Copyright 2020 ACS. (i, j) SEM and HAADF-STEM images of Pt@MoS$_2$/NiS$_2$. The inset of (i) is the model of the ganoderma-like structure and possible active sites. (k)The calculated Gibbs free-energy diagram of HER at different active sites for MoS$_2$/NiS$_2$. (Ref. [133] Copyright 2018 Wiley-VCH.)

could be greatly improved by the Rh single-atom doping, which greatly contributed to the catalytic performance of CuO nanowires.

MoS_2 is considered a potential alternative to Pt-based catalysts. However, the surface of the inert substrate limits its electrocatalytic activity. Therefore, many techniques have been developed to improve the catalytic performance of MoS_2, such as activating the basal plane by single-atom doping of MoS_2. Wang et al. [92] proposed a strategy to decorate Ni single-atom at specific sites. Because the Ni single atoms decorated on MoS_2 nanosheets had good stability in potential cycling, the coordination of Ni single-atom could activate the S-edge of MoS_2 rather than the inert basal plane, resulting in a significant enhancement of HER activity. Since the HER occurred only at the surface [132], the previously reported high-performance Pt-based catalysts could not fully utilize each Pt atom. It was shown that Pt single atoms could increase the intrinsic activity of each active site by tuning the chemical bond energy between H and S/Mo atoms [92]. For example, Guan et al. [133] prepared a ganoderma-like MoS_2/NiS_2 heterogeneous nanostructure with Pt single atoms. By doping Pt single atoms, the maximum atom utilization and excitation of potential catalytic sites could be provided, which significantly improved the catalytic performance [134]. This ganoderma-like heterostructure can effectively confine and disperse the MoS_2 nanosheets to fully expose the edge sites of MoS_2 (Figure 5.18i), providing more opportunities to trap Pt atoms, and the ΔG_H value of ganoderma-like MoS_2/NiS_2 is much lower than that of pure MoS_2 and NiS_2, indicating that the heterogeneous structure can modify the electronic structure to improve the catalytic activity (Figure 5.18j). Due to the good dispersion and the exposed large specific surface area, individual Pt atoms can be easily anchored on the MoS_2 nanosheets with a loading rate of 1.8 wt% (Figure 5.18k). The catalyst shows Pt-like HER catalytic activity with an ultra-low overpotential of 34 mV at a current density of 10 mA·cm^{-2} due to the ganoderma-like structure and the doping of Pt atoms. Luo et al. [135] also demonstrated that the doping of Pd atoms into MoS_2 and the occupation of Pd atoms in the original Mo sites could lead to the creation of sulfur vacancies and the conversion of the poorly conducting 2H phase MoS_2 to the stable 1T phase. Due to the activation of defective sites as well as the 1T basal plane, the heteroatom-doped Pd-MoS_2 exhibited the highest HER performance achieved in acidic solutions, exceeding previous MoS_2-based materials.

5.3.9 OTHER SUBSTRATES

Many studies have shown that the supported Pt nanostructures can usually improve their catalytic activity through favorable local catalyst-platinum interface interaction [136]. Pt-based compounds are considered one of the most effective catalysts for HER under acidic conditions because the Pt–H bond has the proper strength, which is helpful for hydrogen desorption from the catalyst surface [137]. In recent years, great progress has been made in the research of atomic Pt supported on various substrates. For example, Jiang et al. [98] reported a Pt SAC modified nanoporous $Co_{0.85}Se$ (Pt/-np–$Co_{0.85}Se$) as an efficient electrocatalyst for HER. As shown in Figure 5.19a, the loose interatomic distance and spectral lines with different intensities indicate that Pt atoms are in the form of single-atoms, and elemental analysis of the EDX spectrum

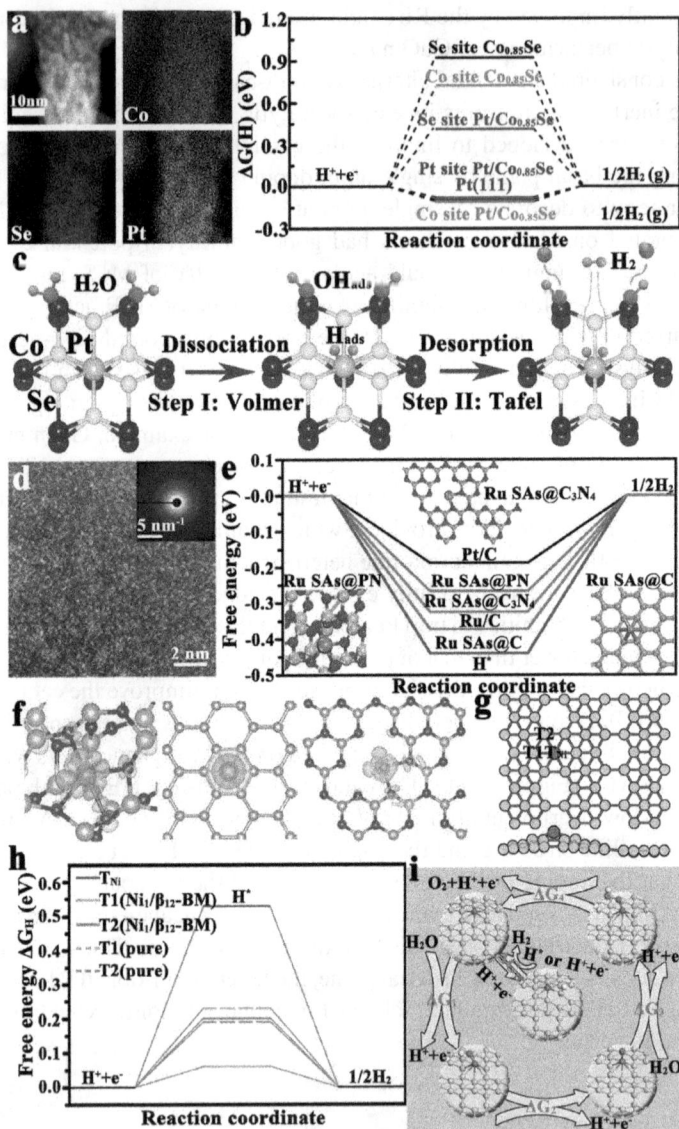

FIGURE 5.19 (a)The STEM-EDX mapping of Pt/np-Co$_{0.85}$Se. (b) HER procedure of Pt/-np-Co$_{0.85}$Se in neutral media. (c) ΔG_H diagrams at different active sites of Co$_{0.85}$Se (004), Pt/Co$_{0.85}$Se (004), and Pt (111). (Ref. [98]Copyright 2019 nature.) (d) Aberration-corrected HAADF-STEM image (inset: SAED pattern) of Ru SAs@PN. (e) The calculated free-energy diagram of the HER at the equilibrium potential for Pt/C, Ru/C, Ru SAs@C, Ru SAs@C$_3$N$_4$, and Ru SAs@PN. (f)The charge density difference of the Ru SAs supported on the phosphorus nitride imide nanotubes (HPN), graphene, and C$_3$N$_4$ matrix. (Ref. [36] Copyright 2018 Wiley-VCH.) (g) T$_1$ and T$_2$ sites on the top of B atoms and T$_{Ni}$ sites on the top of Ni atom in Ni$_1$/β$_{12}$-BM. (h) corresponding ΔG_H as compared with that of T$_1$ and T$_2$ sites on pure β$_{12}$-BM. (i) Elementary reactions of OER and HER and the structures of the adsorbed states for each species, including *H, *OH, *O, and *OOH. (Ref. [141] Copyright 2017 ACS.)

further proves that Co, Se, and Pt are uniformly distributed on the surface of Pt/-np-$Co_{0.85}Se$. XAS study combined with DFT calculation reveals that Pt single atoms can induce intensive charge redistribution on the Pt/np-$Co_{0.85}Se$ interface, which improves the adsorption/desorption behavior of H and promotes the water-splitting process (Figure 5.19b and c) so that the obtained Pt/np-$Co_{0.85}Se$ shows high catalytic performance. In a neutral medium, it has an initial overpotential close to zero with a low Tafel slope of 35 mv·dec^{-1}, which is superior to commercial Pt/C catalysts and other reported transition metal-based compounds. Compared with Pt, Ru metal shows similar hydrogen bond strength to Pt. Yang et al. [36] reported a PN-assisted strategy to obtain Ru single-site catalyst (Ru SAs@PN) for HER. To elucidate the form of the Ru atoms, HAADF-STEM was used to observe the isolated and heavy Ru SAs in the PN substrates (Figure 5.19d). HER performance can be well described by Gibbs free energy (δG_H) in the intermediate state [138]. When δG_H approaches 0 eV, the catalyst exhibits better HER performance. As shown in Figure 5.19h, the δG_H of Ru SAs@PN is 0.27 eV, which is closer to zero than that of Ru/C, Ru SAs@C_3N_4, and Ru SAs@C, leading to the adsorption-desorption behavior of Ru SAs@PN. Ru SAs@c and Ru SAs@PN show completely different performances due to the differently coordinated electronic structures of different substrates (Figure 5.19e). Compared with C_3N_4 and carbon paper, more electrons can be transferred between Ru SAs and HPN nano-tubes, leading to more positive charges in Ru SAs @PN, showing the more significant hydrogen adsorption and promoting HER performance of Ru SAs@PN (Figure 5.19f). For the whole water-splitting process, bifunctional SACs can achieve higher perfor-mance and reduce costs [139,140]. For example, Nair et al. [99] systematically studied the catalytic activity of SACs doped with transition metals for HER and OER and proved that Pt single atoms could provide bifunctional activity for OER and HER. Ling et al. [141] reported the preparation of nanosheet-supported bifunctional SACs, namely β_{12}-boron monolayer supported nickel atom (Ni_1/β_{12}-BM) to realize the whole electrochemical water splitting. The enhanced HER activity of Ni_1/β_{12}-BM was attrib-uted to the cooperative or competitive effect of tensile strain (Figure 5.19g) and the charge transfer caused by Ni adsorption (Figure 5.19h). For OER, there are four basic steps: (i) an H_2O molecule dissociates into an OH group, which is adsorbed on the catalyst surface ($^*OH$); (ii) further dissociation of $^*OH$ into O group ($^*O$); (iii) the $^*O$ reacts with another H_2O molecule and generates an OOH group ($^*OOH$); and (iv) the final product O_2 is formed and then released (Figure 5.19i). Ni_1/β_{12}-BM shows remarkable electrocatalytic performance. The calculated overpotentials of oxygen evolution/hydrogen evolution reaction are only 0.40/0.06 V, showing both good cata-lytic activities for HER and OER.

The remarkable electrocatalytic activity of many SACs in acidic or basic media has been reported, however, HER electrocatalysis in neutral media is less reported. Zhang et al. [142] synthesized PtSA–NT–NF catalyst consisting of single Pt atoms on CoP-based nanotube (NT) arrays supported by nickel foam (NF) (Figure 5.20a), which contains an array of NTs on the NF and has a centimeter-scale size (Figure 5.20b). The atomic-resolution (AR) HAADF images indicate a large number of Pt single atoms are well dispersed on the NTs (Figure 5.20c). Because of the higher Pt atom utiliza-tion and the positive modulation of the SAC electronic structure by the support, the overpotential (η) value of PtSA–NT–NF (24 mV) is only 7 mV higher than that of the

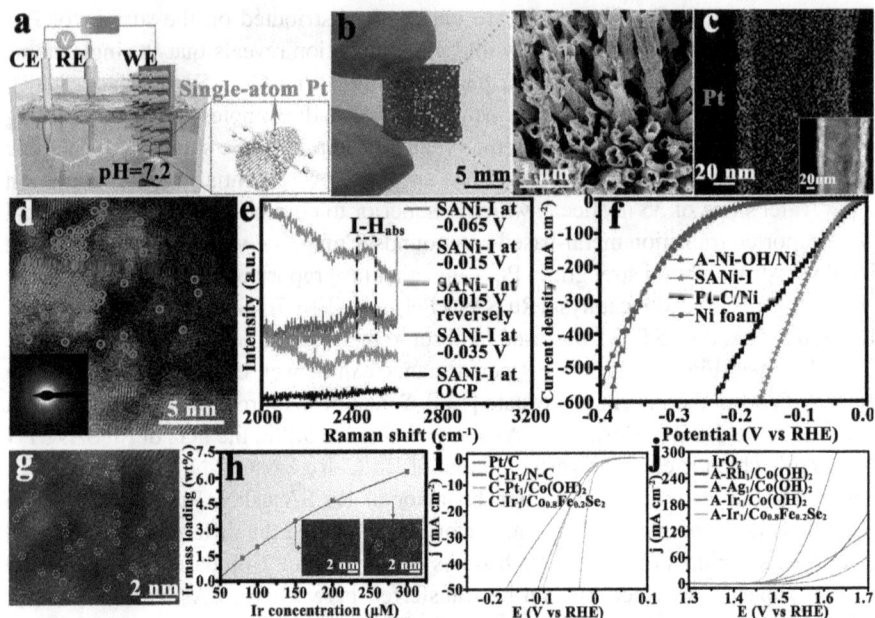

FIGURE 5.20 (a) Synthesis process. (b) Optical and SEM photos of PtSA–NT–NF. (c) Atomic-resolution (AR) HAADF image of the NT. (Ref. [142] Copyright 2017 Wiley-VCH.) (d) Magnified HAADF-STEM image and (e) In-situ Raman spectra of SANi-I; (f) HER performance of SANi–I, A–Ni–OH, Pt–C/Ni, and Ni foam. (Ref. [100] Copyright 2019 Wiley-VCH.) (g) HAADF-STEM images of C–Ir_1/Co(OH)$_2$. (h) Ir mass loadings as a function of Ir concentration in the 1 M KOH electrolyte for cathodic deposition. (i) Polarization curves of SACs for HER and OER. (Ref. [144] Copyright 2019 Wiley-VCH.)

reference Pt/C (17 mV) at $j_{HER} = 10$ mA·cm^{-2}, suggesting that Pt single atoms can contribute significantly to the HER property of PtSA–NT–NF in neutral media. Zhao et al. [100] reported a single-atom nickel iodide (SANi-I) electrocatalyst. HAADF-STEM images of SANi-I confirm that the iodine atoms are atomically dispersed, as shown in Figure 5.20d. In-situ Raman spectroscopy (Figure 5.20e) shows that hydrogen atoms (H$_{ads}$) are adsorbed by individual iodine atoms to form I-H$_{ads}$ intermediates, thus facilitating the HER process and allowing SANi-I to exhibit good electrocatalytic activity and stability in alkaline media (Figure 5.20f).

SAC exhibits excellent catalytic performance due to its maximum atom utilization and unique electronic structure [143]. However, special requirements for anchoring metals or substrates need to be addressed in the synthesis of SACs. Zhang et al. [144] reported an electrochemical deposition method suitable for various metals and substrates for the preparation of SACs. As shown in Figure 5.20g, the Ir single atoms are deposited on the Co$_{0.8}$Fe$_{0.2}$Se$_2$@Ni foam at both cathode and anode and integrated into a two-electrode battery. HAADF-STEM (Figure 5.20g) shows that the Ir single atoms are uniformly distributed on the substrate. As shown in Figure 5.20h, with increasing Ir concentration, the mass loading of Ir species continues to increase. Ir single-atom loading can reach 3.5% when the Ir concentration is increased to 150 μM.

Ir clusters appear under a mass loading of 4.7% when the concentration is further increased to 200 μM. Therefore, for the formation of SAC, there is a mass load upper limit between 3.5% and 4.7% (Figure 5.20h). The electrochemical deposition processes are similar to the nucleation in solution phase synthesis. In this case, the upper mass loading limit of SAC corresponds to the minimum supersaturation level on the substrate. The separately dispersed metal atoms begin to nucleate and form clusters once the mass loading exceeds the minimum supersaturation level. Therefore, the exploration of the minimum supersaturation level is crucial for the formation of SACs. The Ir single-atom shows outstanding catalytic activity on $Co_{0.8}Fe_{0.2}Se_2$ due to the efficient charge transfer of the substrate and the effective metal-substrate interaction, which gives higher HER activity than commercial Pt/C. Compared to the benchmark IrO_2, $A–Ag_1/Co(OH)_2$, $A–Ir_1/Co(OH)_2$, $A–Rh_1/Co(OH)_2$, and $A–Ir_1/Co_{0.8}Fe_{0.2}Se_2$ exhibit higher OER activity (Figure 5.20i).

5.4 TYPICAL RESULTS OF HER AND OER CATALYZED BY SINGLE-ATOM CATALYSTS

The development of single-atom catalysis has become one of the most active areas for its great potential in water electrolysis. Herein we list some typical results for HER and OER in different electrolytes, as shown in Table 5.1. The list emphasizes the overpotential values of different SACs in comparable electrolytes at a similar current, about 10 mA·cm^{-2}. The overpotentials of noble-metal-based ones are quite competitive compared with those of Fe, Co, Ni, W, Cu based catalysts. The development of common transition metal-based SACs needs further research to make up for their intrinsic deficiency. Compared with the low HER overpotential of Pt SACs in H_2SO_4 electrolyte, Ru-based ones show very competitive overpotentials in both H_2SO_4 and KOH electrolytes for HER and OER. Another very obvious phenomenon is that the overpotentials vary with the coordination environments of the single-atom because of the different supports. The phenomenon reflects the decisive role of the coordination environments supplied by the supports.

5.5 SUMMARY

SACs have offered a chance for researchers to investigate the water electrolysis mechanism step by step because of their high activity, selectivity, and fine-tuned coordination environment of the single atoms. The unsaturated coordination environments enable their theoretical efficiency to reach 100% in catalytic reactions. In-depth researches reveal that the supports are as important as the single atoms because their function includes tuning the inherent element property and the coordination environment besides supplying the geometric structure, and determining the loading amount of metal atoms. The tunable metal-support interaction will introduce different electronic structures with discrete energy level distributions, which determine the high selectivity of the catalytic reactions. The strong metal-support coupling also stabilizes the single atoms in various electrolytes.

The challenge of single-atom catalysis includes the development of novel fabrication strategies, the increase of metal loading, the microstructure design of supports,

TABLE 5.1

List of Typical Results of Single-Atom Catalysts in Water Electrolysis

Reactions	Catalysts	Electrolytes	Overpotentials	References
HER	Pt SASs/AG	0.5 M H_2SO_4	0.012 V @ 10 mA·cm^{-2}	[117]
	Pt_1/hNCNC-2.92	0.5 M H_2SO_4	0.015 V @ 10 mA·cm^{-2}	[69]
	Ru SAs@PN	0.5 M H_2SO_4	0.024 V @ 10 mA·cm^{-2}	[36]
	A–CoPt–NC	0.5 M H_2SO_4	0.027 V @ 10 mA·cm^{-2}	[145]
	Pt/p–MWCNTs	0.5 M H_2SO_4	0.044 V @ 10 mA·cm^{-2}	[146]
	Ru@Co–SAs/N–C	0.5 M H_2SO_4	0.057 V @ 10 mA·cm^{-2}	[147]
	Ru–MoS$_2$/CC	0.5 M H_2SO_4	0.061 V @ 10 mA·cm^{-2}	[148]
	Fe/GD	0.5 M H_2SO_4	0.066 V @ 10 mA·cm^{-2}	[56]
	A–Ni@DG	0.5 M H_2SO_4	0.070 V @ 10 mA·cm^{-2}	[57]
	$Mo_2TiC_2T_x$–Pt$_{SA}$	0.5 M H_2SO_4	0.077 V @ 100 mA·cm^{-2}	[85]
	Ni/GD	0.5 M H_2SO_4	0.088 V @ 10 mA·cm^{-2}	[56]
	CoSAs/PTF-600	0.5 M H_2SO_4	0.094 V @ 10 mA·cm^{-2}	[83]
	W–SAC	0.5 M H_2SO_4	0.105 V @ 10 mA·cm^{-2}	[51]
	NiSA–MoS$_2$	0.5 M H_2SO_4	0.110 V @ 10 mA·cm^{-2}	[92]
	Cu@MoS$_2$	0.5 M H_2SO_4	0.131 V @ 10 mA·cm^{-2}	[149]
	RuSA–N–S–$Ti_3C_2T_x$	0.5 M H_2SO_4	0.151 V @ 10 mA·cm^{-2}	[38]
	SACo–N/C	0.5 M H_2SO_4	0.169 V @ 10 mA·cm^{-2}	[150]
	Pt–1T'MoS$_2$	0.5 M H_2SO_4	0.180 V @ 10 mA·cm^{-2}	[151]
	Ru@Co–SAs/N–C	1 M KOH	0.007 V @ 10 mA·cm^{-2}	[147]
	Ru–NC-700	1 M KOH	0.012 V @ 10 mA·cm^{-2}	[102]
	Ru–MoS$_2$/CC	1 M KOH	0.041 V @ 10 mA·cm^{-2}	[148]
	A–CoPt–NC	1 M KOH	0.050 V @ 10 mA·cm^{-2}	[145]
	Ir_1@Co/NC	1 M KOH	0.060 V @ 10 mA·cm^{-2}	[152]
	SANi–PtNWs	1 M KOH	0.070 V @ 11 mA·cm^{-2}	[153]
	NiSA–MoS$_2$	1 M KOH	0.098 V @ 10 mA·cm^{-2}	[92]
	$Mo_1N_1C_2$	0.1 M KOH	0.132 V @ 10 mA·cm^{-2}	[74]
	Co_1/PCN	1 M KOH	0.138 V @ 10 mA·cm^{-2}	[154]
	SACo–N/C	1 M KOH	0.178 V @ 10 mA·cm^{-2}	[150]
	Fe–N$_4$ SAs/NPC	1 M KOH	0.202 V @ 10 mA·cm^{-2}	[155]
	PtSA–NT–NF	1 M PBS	0.024 V @ 10 mA·cm^{-2}	[11]
	Pt/np–$Co_{0.85}$Se	1 M PBS	0.050 V @ 10 mA·cm^{-2}	[98]
	Ru@Co–SAs/N–C	1 M PBS	0.055 V @ 10 mA·cm^{-2}	[147]
	$Mo_2TiC_2T_x$–Pt$_{SA}$	0.5 M PBS	0.061 V @ 10 mA·cm^{-2}	[85]
	Ru–MoS$_2$/CC	1 M PBS	0.114 V @ 10 mA·cm^{-2}	[148]
OER	Ru/CoFe–LDH	1 M KOH	0.198 V @ 10 mA·cm^{-2}	[37]
	Au@Ni$_2$P–350°C	1 M KOH	0.240 V @ 10 mA·cm^{-2}	[156]
	Ir_1@Co/NC	1 M KOH	0.260 V @ 10 mA·cm^{-2}	[152]
	A–Ni@DG	1 M KOH	0.270 V @ 10 mA·cm^{-2}	[57]
	w–Ni(OH)$_2$	1 M KOH	0.273 V @ 10 mA·cm^{-2}	[157]
	Ir@Co	1 M KOH	0.273 V @ 10 mA·cm^{-2}	[158]
	Co–N$_x$/C NRA	0.1 M KOH	0.300 V @ 10 mA·cm^{-2}	[159]
	HCM@Ni–N	1 M KOH	0.304 V @ 10 mA·cm^{-2}	[49]
	Co–Fe–N–C	1 M KOH	0.309 V @ 10 mA·cm^{-2}	[77]

(Continued)

TABLE 5.1 (*Continued*)
List of Typical Results of Single-Atom Catalysts in Water Electrolysis

Reactions	Catalysts	Electrolytes	Overpotentials	References
	SCoNC	0.1 M KOH	0.310 V @ 10 mA·cm^{-2}	[44]
	Ni–NHGF	1 M KOH	0.331 V @ 10 mA·cm^{-2}	[61]
	CoNi–SAs/NC	1 M KOH	0.340 V @ 10 mA·cm^{-2}	[21]
	0.5 wt% Pt/NiO	1 M KOH	0.358 V @ 10 mA·cm^{-2}	[160]
	CoIr-0.2	1 M PBS	0.373 V @ 10 mA·cm^{-2}	[161]
	Co–C$_3$N$_4$/CNT	1 M KOH	0.380 V @ 10 mA·cm^{-2}	[120]
	P–O/FeN$_4$–CNS	0.1 M KOH	0.390 V @ 10 mA·cm^{-2}	[162]
	Fe–N$_4$ SAs/NPC	1 M KOH	0.440 V @ 10 mA·cm^{-2}	[155]
	Au$_1$N$_x$	0.1 M KOH	0.450 V @ 10 mA·cm^{-2}	[163]
	Ru$_1$–Pt$_3$Cu	0.1 M HClO$_4$	0.220 V @ 10 mA·cm^{-2}	[164]
	Ru–N–C	0.5 M H$_2$SO$_4$	0.267 V @ 10 mA·cm^{-2}	[14]
	CoIr-0.2	1 M PBS	0.373 V @ 10 mA·cm^{-2}	[161]

and the catalytic reaction track. The reported metal loading needs further improvement to increase the number of active sites to make the SACs as competitive as the traditional noble metal catalysts. Although many substrates have been explored to supply stable coordination environments for single atoms in various catalytic reactions, the physical or chemical bonding is still challenged by the agglomeration once the loading amount reaches a quite low threshold value. Either novel supports or new stabilizing strategies need further investigation to conquer the formation of nanocluster or nanocrystal. Another challenge is the current data focus on the efficiency and overpotential to describe the catalytic performance of SACs. Accurate and measurable descriptors are still under development for every step in water electrolysis to define the limiting step. Proper descriptors will help researchers find the research direction to accelerate the limiting steps rather than explain the water electrolysis mechanism with general macro parameters.

ACKNOWLEDGMENT

This work is financially supported by the National Natural Science Foundation of China (Grant No. 51572166). W. X. Li acknowledges research supported by the Program for Professor of Special Appointment (Eastern Scholar: TP2014041) at Shanghai Institutions of Higher Learning. We also appreciate the useful discussion with Prof. Mingyuan Zhu and Prof Ying Li and the help from Mingjie Sun, Jiancheng Li, Zulin Sun, Huidong Xu, Yuqi Zhang, Xin Xing, Yao Xu, and Haobo Liu.

REFERENCES

1. B. Lee, J. Heo, S. Kim, C. Sung, C. Moon, S. Moon, H. Lim, Economic feasibility studies of high pressure PEM water electrolysis for distributed H$_2$ refueling stations, *Energy Convers. Manage.*, 162 (**2018**) 139–144.

2. S. Shiva Kumar, V. Himabindu, Hydrogen production by PEM water electrolysis: A review, *Mater. Sci. Energy Technol.*, 2 (**2019**) 442–454.
3. S. van Renssen, The hydrogen solution? *Nat. Climate Change*, 10 (**2020**) 799–801.
4. C. Zhu, Q. Shi, S. Feng, D. Du, Y. Lin, Single-atom catalysts for electrochemical water splitting, *ACS Energy Lett.*, 3 (**2018**) 1713–1721.
5. B. Qiao, A. Wang, X. Yang, L.F. Allard, Z. Jiang, Y. Cui, J. Liu, J. Li, T. Zhang, Single-atom catalysis of CO oxidation using Pt_1/FeO_x, *Nat. Chem.*, 3 (**2011**) 634–641.
6. J. Lin, A. Wang, B. Qiao, X. Liu, X. Yang, X. Wang, J. Liang, J. Li, J. Liu, T. Zhang, Remarkable performance of Ir_1/FeO_x single-atom catalyst in water gas shift reaction, *J. Am. Chem. Soc.*, 135 (**2013**) 15314–15317.
7. J.C. Li, Z. Wei, D. Liu, D. Du, Y. Lin, M. Shao, Dispersive single-atom metals anchored on functionalized nanocarbons for electrochemical reactions, *Top. Curr. Chem. (Cham.)*, 377 (**2019**) 4.
8. J. Liu, Catalysis by supported single metal atoms, *ACS Catal.*, 7 (**2016**) 34–59.
9. C. Zhu, S. Fu, Q. Shi, D. Du, Y. Lin, Single-atom electrocatalysts, *Angew Chem Int Ed Engl*, 56 (**2017**) 13944–13960.
10. A. Wang, J. Li, T. Zhang, Heterogeneous single-atom catalysis, *Nat. Rev. Chem.*, 2 (**2018**) 65–81.
11. L. Zhang, L. Han, H. Liu, X. Liu, J. Luo, Potential-cycling synthesis of single platinum atoms for efficient hydrogen evolution in neutral media, *Angew. Chem. Int. Ed. Engl.*, 56 (**2017**) 13694–13698.
12. N. Cheng, S. Stambula, D. Wang, M.N. Banis, J. Liu, A. Riese, B. Xiao, R. Li, T.K. Sham, L.M. Liu, G.A. Botton, X. Sun, Platinum single-atom and cluster catalysis of the hydrogen evolution reaction, *Nat. Commun.*, 7 (**2016**) 13638.
13. C. Meng, T. Ling, T.Y. Ma, H. Wang, Z. Hu, Y. Zhou, J. Mao, X.W. Du, M. Jaroniec, S.Z. Qiao, Atomically and electronically coupled Pt and CoO hybrid nanocatalysts for enhanced electrocatalytic performance, *Adv. Mater.*, 29 (**2017**) 1604607.
14. L. Cao, Q. Luo, J. Chen, L. Wang, Y. Lin, H. Wang, X. Liu, X. Shen, W. Zhang, W. Liu, Z. Qi, Z. Jiang, J. Yang, T. Yao, Dynamic oxygen adsorption on single-atomic Ruthenium catalyst with high performance for acidic oxygen evolution reaction, *Nat. Commun.*, 10 (**2019**) 4849.
15. C.H. Chen, D. Wu, Z. Li, R. Zhang, C.G. Kuai, X.R. Zhao, C.K. Dong, S.Z. Qiao, H. Liu, X.W. Du, Ruthenium-based single-atom alloy with high electrocatalytic activity for hydrogen evolution, *Adv. Energy Mater.*, 9 (**2019**) 1803913.
16. H. Fei, J. Dong, M.J. Arellano-Jimenez, G. Ye, N. Dong Kim, E.L. Samuel, Z. Peng, Z. Zhu, F. Qin, J. Bao, M.J. Yacaman, P.M. Ajayan, D. Chen, J.M. Tour, Atomic cobalt on nitrogen-doped graphene for hydrogen generation, *Nat. Commun.*, 6 (**2015**) 8668.
17. P. Yin, T. Yao, Y. Wu, L. Zheng, Y. Lin, W. Liu, H. Ju, J. Zhu, X. Hong, Z. Deng, G. Zhou, S. Wei, Y. Li, Single cobalt atoms with precise N-coordination as superior oxygen reduction reaction catalysts, *Angew. Chem. Int. Ed. Engl.*, 55 (**2016**) 10800–10805.
18. L. Zhao, Y. Zhang, L.-B. Huang, X.-Z. Liu, Q.-H. Zhang, C. He, Z.-Y. Wu, L.-J. Zhang, J. Wu, W. Yang, L. Gu, J.-S. Hu, L.-J. Wan, Cascade anchoring strategy for general mass production of high-loading single-atomic metal-nitrogen catalysts, *Nat. Commun.*, 10 (**2019**) 1278.
19. J. Chen, H. Li, C. Fan, Q. Meng, Y. Tang, X. Qiu, G. Fu, T. Ma, Dual single-atomic Ni-N4 and Fe-N4 sites constructing janus hollow graphene for selective oxygen electrocatalysis, *Adv. Mater.*, 32 (**2020**) e2003134.
20. H. Zhang, L. Yu, T. Chen, W. Zhou, X.W.D. Lou, Surface modulation of hierarchical MoS_2 nanosheets by Ni single atoms for enhanced electrocatalytic hydrogen evolution, *Adv. Funct. Mater.*, 28 (**2018**) 1807086.

21. X. Han, X. Ling, D. Yu, D. Xie, L. Li, S. Peng, C. Zhong, N. Zhao, Y. Deng, W. Hu, Atomically dispersed binary Co-Ni sites in nitrogen-doped hollow carbon nanocubes for reversible oxygen reduction and evolution, *Adv. Mater.*, 31 (**2019**) e1905622.

22. D. Ji, L. Fan, L. Li, S. Peng, D. Yu, J. Song, S. Ramakrishna, S. Guo, Atomically transition metals on self-supported porous carbon flake arrays as binder-free air cathode for wearable zinc-air batteries, *Adv. Mater.*, 31 (**2019**) e1808267.

23. K.C. Kwon, J.M. Suh, R.S. Varma, M. Shokouhimehr, H.W. Jang, Electrocatalytic water splitting and CO_2 reduction: Sustainable solutions via single-atom catalysts supported on 2D materials, *Small Methods*, 3 (**2019**) 1800492.

24. T. Sun, L. Xu, D. Wang, Y. Li, Metal organic frameworks derived single atom catalysts for electrocatalytic energy conversion, *Nano Res.*, 12 (**2019**) 2067–2080.

25. J. Zhang, C. Liu, B. Zhang, Insights into single-atom metal–support interactions in electrocatalytic water splitting, *Small Methods*, 3 (**2019**) 1800481.

26. H. Li, H. Zhu, Z. Zhuang, S. Lu, F. Duan, M. Du, Single-atom catalysts for electrochemical clean energy conversion: recent progress and perspectives, *Sustainable EnergyFuels*, 4 (**2020**) 996–1011.

27. W. Liu, H. Zhang, C. Li, X. Wang, J. Liu, X. Zhang, Non-noble metal single-atom catalysts prepared by wet chemical method and their applications in electrochemical water splitting, *J. Energy Chem.*, 47 (**2020**) 333–345.

28. Z. Shi, W. Yang, Y. Gu, T. Liao, Z. Sun, Metal-nitrogen-doped carbon materials as highly efficient catalysts: Progress and rational design, *Adv. Sci.*, 7 (**2020**) 2001069.

29. W. Zang, Z. Kou, S.J. Pennycook, J. Wang, Heterogeneous single atom electrocatalysis, where "singles" are "married", *Adv. Energy Mater.*, 10 (**2020**) 1903181.

30. Q. Zhang, J. Guan, Atomically dispersed catalysts for hydrogen/oxygen evolution reactions and overall water splitting, *J. Power Sources*, 471 (**2020**) 228446.

31. X. Fang, Q. Shang, Y. Wang, L. Jiao, T. Yao, Y. Li, Q. Zhang, Y. Luo, H.L. Jiang, Single Pt atoms confined into a metal-organic framework for efficient photocatalysis, *Adv. Mater.*, 30 (**2018**) 1705112.

32. J. Kim, C.-W. Roh, S.K. Sahoo, S. Yang, J. Bae, J.W. Han, H. Lee, Highly durable platinum single-atom alloy catalyst for electrochemical reactions, *Adv. Energy Mater.*, 8 (**2018**) 1701476.

33. X. Zeng, J. Shui, X. Liu, Q. Liu, Y. Li, J. Shang, L. Zheng, R. Yu, Single-atom to single-atom grafting of Pt_1 onto $Fe-N_4$ center: Pt_1@Fe-N-C multifunctional electrocatalyst with significantly enhanced properties, *Adv. Energy Mater.*, 8 (**2018**) 1701345.

34. J. Zhang, X. Wu, W.C. Cheong, W. Chen, R. Lin, J. Li, L. Zheng, W. Yan, L. Gu, C. Chen, Q. Peng, D. Wang, Y. Li, Cation vacancy stabilization of single-atomic-site Pt_1/$Ni(OH)_x$ catalyst for diboration of alkynes and alkenes, *Nat. Commun.*, 9 (**2018**) 1002.

35. L. Zhuang, Y. Jia, H. Liu, X. Wang, R.K. Hocking, H. Liu, J. Chen, L. Ge, L. Zhang, M. Li, C.L. Dong, Y.C. Huang, S. Shen, D. Yang, Z. Zhu, X. Yao, Defect-induced Pt-Co-Se coordinated sites with highly asymmetrical electronic distribution for boosting oxygen-involving electrocatalysis, *Adv. Mater.*, 31 (**2019**) e1805581.

36. J. Yang, B. Chen, X. Liu, W. Liu, Z. Li, J. Dong, W. Chen, W. Yan, T. Yao, X. Duan, Y. Wu, Y. Li, Efficient and robust hydrogen evolution: Phosphorus nitride imide nanotubes as supports for anchoring single ruthenium sites, *Angew. Chem. Int. Ed. Engl.*, 57 (**2018**) 9495–9500.

37. P. Li, M. Wang, X. Duan, L. Zheng, X. Cheng, Y. Zhang, Y. Kuang, Y. Li, Q. Ma, Z. Feng, W. Liu, X. Sun, Boosting oxygen evolution of single-atomic ruthenium through electronic coupling with cobalt-iron layered double hydroxides, *Nat. Commun.*, 10 (**2019**) 1711.

38. V. Ramalingam, P. Varadhan, H.C. Fu, H. Kim, D. Zhang, S. Chen, L. Song, D. Ma, Y. Wang, H.N. Alshareef, J.H. He, Heteroatom-mediated interactions between ruthenium single atoms and an MXene support for efficient hydrogen evolution, *Adv. Mater.*, 31 (**2019**) e1903841.

39. H. Zhang, W. Zhou, X.F. Lu, T. Chen, X.W. Lou, Implanting isolated Ru atoms into edge-rich carbon matrix for efficient electrocatalytic hydrogen evolution, *Adv. Energy Mater.*, 10 (**2020**) 2000882.
40. A. Zitolo, N. Ranjbar-Sahraie, T. Mineva, J. Li, Q. Jia, S. Stamatin, G.F. Harrington, S.M. Lyth, P. Krtil, S. Mukerjee, E. Fonda, F. Jaouen, Identification of catalytic sites in cobalt-nitrogen-carbon materials for the oxygen reduction reaction, *Nat. Commun.*, 8 (**2017**) 957.
41. M.D. Hossain, Z. Liu, M. Zhuang, X. Yan, G.-L. Xu, C.A. Gadre, A. Tyagi, I.H. Abidi, C.-J. Sun, H. Wong, A. Guda, Y. Hao, X. Pan, K. Amine, Z. Luo, Rational design of graphene-supported single atom catalysts for hydrogen evolution reaction, *Adv. Energy Mater.*, 9 (**2019**) 1803689.
42. K. Qi, X. Cui, L. Gu, S. Yu, X. Fan, M. Luo, S. Xu, N. Li, L. Zheng, Q. Zhang, J. Ma, Y. Gong, F. Lv, K. Wang, H. Huang, W. Zhang, S. Guo, W. Zheng, P. Liu, Single-atom cobalt array bound to distorted 1T MoS_2 with ensemble effect for hydrogen evolution catalysis, *Nat. Commun.*, 10 (**2019**) 5231.
43. Z. Zhang, C. Feng, C. Liu, M. Zuo, L. Qin, X. Yan, Y. Xing, H. Li, R. Si, S. Zhou, J. Zeng, Electrochemical deposition as a universal route for fabricating single-atom catalysts, Nat Commun, 11 (**2020**) 1215.
44. J. Wu, H. Zhou, Q. Li, M. Chen, J. Wan, N. Zhang, L. Xiong, S. Li, B.Y. Xia, G. Feng, M. Liu, L. Huang, Densely populated isolated single Co-N site for efficient oxygen electrocatalysis, *Adv. Energy Mater.*, 9 (**2019**) 1900149.
45. X. Song, S. Chen, L. Guo, Y. Sun, X. Li, X. Cao, Z. Wang, J. Sun, C. Lin, Y. Wang, General dimension-controlled synthesis of hollow carbon embedded with metal singe atoms or core-shell nanoparticles for energy storage applications, *Adv. Energy Mater.*, 8 (**2018**) 1801101.
46. Y. Qu, L. Wang, Z. Li, P. Li, Q. Zhang, Y. Lin, F. Zhou, H. Wang, Z. Yang, Y. Hu, M. Zhu, X. Zhao, X. Han, C. Wang, Q. Xu, L. Gu, J. Luo, L. Zheng, Y. Wu, Ambient synthesis of single-atom catalysts from bulk metal via trapping of atoms by surface dangling bonds, *Adv. Mater.*, 31 (**2019**) e1904496.
47. D.W. Su, J. Ran, Z.W. Zhuang, C. Chen, S.Z. Qiao, Y.D. Li, G.X. Wang, Atomically dispersed Ni in cadmium-zinc sulfide quantum dots for high-performance visible-light photocatalytic hydrogen production, *Sci. Adv.*, 6 (**2020**) eaaz8447.
48. C. Lei, Y. Wang, Y. Hou, P. Liu, J. Yang, T. Zhang, X. Zhuang, M. Chen, B. Yang, L. Lei, C. Yuan, M. Qiu, X. Feng, Efficient alkaline hydrogen evolution on atomically dispersed Ni–N_x species anchored porous carbon with embedded Ni nanoparticles by accelerating water dissociation kinetics, *Energy Environ. Sci.*, 12 (**2019**) 149–156.
49. H. Zhang, Y. Liu, T. Chen, J. Zhang, J. Zhang, X.W.D. Lou, Unveiling the activity origin of electrocatalytic oxygen evolution over isolated Ni atoms supported on a N-doped carbon matrix, *Adv. Mater.*, 31 (**2019**) e1904548.
50. S. Liang, C. Zhu, N. Zhang, S. Zhang, B. Qiao, H. Liu, X. Liu, Z. Liu, X. Song, H. Zhang, C. Hao, Y. Shi, A novel single-atom electrocatalyst Ti_1/rGO for efficient cathodic reduction in hybrid photovoltaics, *Adv. Mater.*, 32 (**2020**) e2000478.
51. W. Chen, J. Pei, C.T. He, J. Wan, H. Ren, Y. Wang, J. Dong, K. Wu, W.C. Cheong, J. Mao, X. Zheng, W. Yan, Z. Zhuang, C. Chen, Q. Peng, D. Wang, Y. Li, Single tungsten atoms supported on MOF-derived N-doped carbon for robust electrochemical hydrogen evolution, *Adv. Mater.*, 30 (**2018**) 1800396.
52. Y. Yang, Y. Qian, H. Li, Z. Zhang, Y. Mu, D. Do, B. Zhou, J. Dong, W. Yan, Y. Qin, L. Fang, R. Feng, J. Zhou, P. Zhang, J. Dong, G. Yu, Y. Liu, X. Zhang, X. Fan, O-coordinated W-Mo dual-atom catalyst for pH-universal electrocatalytic hydrogen evolution, *Sci. Adv.*, 6 (**2020**) eaba6586.
53. L. Wang, X. Duan, X. Liu, J. Gu, R. Si, Y. Qiu, Y. Qiu, D. Shi, F. Chen, X. Sun, J. Lin, J. Sun, Atomically dispersed Mo supported on metallic Co_9S_8 nanoflakes as an advanced

noble-metal-free bifunctional water splitting catalyst working in universal pH conditions, *Adv. Energy Mater.*, 10 (**2019**) 1903137.

54. H. Yan, Y. Lin, H. Wu, W. Zhang, Z. Sun, H. Cheng, W. Liu, C. Wang, J. Li, X. Huang, T. Yao, J. Yang, S. Wei, J. Lu, Bottom-up precise synthesis of stable platinum dimers on graphene, *Nat. Commun.*, 8 (**2017**) 1070.

55. C. Lei, H. Chen, J. Cao, J. Yang, M. Qiu, Y. Xia, C. Yuan, B. Yang, Z. Li, X. Zhang, L. Lei, J. Abbott, Y. Zhong, X. Xia, G. Wu, Q. He, Y. Hou, Fe-N₄ sites embedded into carbon nanofiber integrated with electrochemically exfoliated graphene for oxygen evolution in acidic medium, *Adv. Energy Mater.*, 8 (**2018**) 1801912.

56. Y. Xue, B. Huang, Y. Yi, Y. Guo, Z. Zuo, Y. Li, Z. Jia, H. Liu, Y. Li, Anchoring zero valence single atoms of nickel and iron on graphdiyne for hydrogen evolution, *Nat. Commun.*, 9 (**2018**) 1460.

57. L. Zhang, Y. Jia, G. Gao, X. Yan, N. Chen, J. Chen, M.T. Soo, B. Wood, D. Yang, A. Du, X. Yao, Graphene defects trap atomic Ni species for hydrogen and oxygen evolution reactions, *Chem*, 4 (**2018**) 285–297.

58. X. Gao, Y. Zhou, Y. Tan, S. Liu, Z. Cheng, Z. Shen, Graphyne doped with transition-metal single atoms as effective bifunctional electrocatalysts for water splitting, *Appl. Surf. Sci.*, 492 (**2019**) 8–15.

59. X. Gao, L. Mei, Y. Zhou, Z. Shen, Impact of electron transfer of atomic metals on adjacent graphyne layers on electrochemical water splitting, *Nanoscale*, 12 (**2020**) 7814–7821.

60. S. Lin, H. Xu, Y. Wang, X.C. Zeng, Z. Chen, Directly predicting limiting potentials from easily obtainable physical properties of graphene-supported single-atom electrocatalysts by machine learning, *J. Mater. Chem. A*, 8 (**2020**) 5663–5670.

61. H. Fei, J. Dong, Y. Feng, C.S. Allen, C. Wan, B. Volosskiy, M. Li, Z. Zhao, Y. Wang, H. Sun, P. An, W. Chen, Z. Guo, C. Lee, D. Chen, I. Shakir, M. Liu, T. Hu, Y. Li, A.I. Kirkland, X. Duan, Y. Huang, General synthesis and definitive structural identification of MN₄C₄ single-atom catalysts with tunable electrocatalytic activities, *Nat. Catal.*, 1 (**2018**) 63–72.

62. H. Fei, J. Dong, C. Wan, Z. Zhao, X. Xu, Z. Lin, Y. Wang, H. Liu, K. Zang, J. Luo, S. Zhao, W. Hu, W. Yan, I. Shakir, Y. Huang, X. Duan, Microwave-assisted rapid synthesis of graphene-supported single atomic metals, *Adv. Mater.*, 30 (**2018**) e1802146.

63. Y. Zhou, G. Gao, Y. Li, W. Chu, L.W. Wang, Transition-metal single atoms in nitrogen-doped graphenes as efficient active centers for water splitting: A theoretical study, *Phys. Chem. Chem. Phys.*, 21 (**2019**) 3024–3032.

64. X. Gao, Y. Zhou, S. Liu, Z. Cheng, Y. Tan, Z. Shen, Single cobalt atom anchored on N-doped graphyne for boosting the overall water splitting, *Appl. Surf. Sci.*, 502 (**2020**) 144155.

65. X. Lv, W. Wei, H. Wang, B. Huang, Y. Dai, Holey graphitic carbon nitride (g-CN) supported bifunctional single atom electrocatalysts for highly efficient overall water splitting, *Appl. Catal., B*, 264 (**2020**) 118521.

66. X.P. Yin, H.J. Wang, S.F. Tang, X.L. Lu, M. Shu, R. Si, T.B. Lu, Engineering the coordination environment of single-atom platinum anchored on graphdiyne for optimizing electrocatalytic hydrogen evolution, *Angew. Chem. Int. Ed. Engl.*, 57 (**2018**) 9382–9386.

67. T. He, S.K. Matta, G. Will, A. Du, Transition-metal single atoms anchored on graphdiyne as high-efficiency electrocatalysts for water splitting and oxygen reduction, *Small Methods*, 3 (**2019**) 1800419.

68. D. Liu, X. Li, S. Chen, H. Yan, C. Wang, C. Wu, Y.A. Haleem, S. Duan, J. Lu, B. Ge, P.M. Ajayan, Y. Luo, J. Jiang, L. Song, Atomically dispersed platinum supported on curved carbon supports for efficient electrocatalytic hydrogen evolution, *Nat. Energy*, 4 (**2019**) 512–518.

69. Z. Zhang, Y. Chen, L. Zhou, C. Chen, Z. Han, B. Zhang, Q. Wu, L. Yang, L. Du, Y. Bu, P. Wang, X. Wang, H. Yang, Z. Hu, The simplest construction of single-site catalysts by the synergism of micropore trapping and nitrogen anchoring, *Nat. Commun.*, 10 (**2019**) 1657.

70. P. Chen, T. Zhou, L. Xing, K. Xu, Y. Tong, H. Xie, L. Zhang, W. Yan, W. Chu, C. Wu, Y. Xie, Atomically dispersed iron-nitrogen species as electrocatalysts for bifunctional oxygen evolution and reduction reactions, *Angew. Chem. Int. Ed. Engl.*, 56 (**2017**) 610–614.

71. M. Tavakkoli, N. Holmberg, R. Kronberg, H. Jiang, J. Sainio, E.I. Kauppinen, T. Kallio, K. Laasonen, Electrochemical activation of single-walled carbon nanotubes with pseudo-atomic-scale platinum for the hydrogen evolution reaction, *ACS Catal.*, 7 (**2017**) 3121–3130.

72. Y. Cheng, S. He, J.P. Veder, R. De Marco, S.Z. Yang, S. Ping Jiang, Atomically dispersed bimetallic FeNi catalysts as highly efficient bifunctional catalysts for reversible oxygen evolution and oxygen reduction reactions, *ChemElectroChem*, 6 (**2019**) 3478–3487.

73. H. Zhang, P. An, W. Zhou, B.Y. Guan, P. Zhang, J. Dong, X.W. Lou, Dynamic traction of lattice-confined platinum atoms into mesoporous carbon matrix for hydrogen evolution reaction, *Sci. Adv.*, 4 (**2018**) eaao6657.

74. W. Chen, J. Pei, C.T. He, J. Wan, H. Ren, Y. Zhu, Y. Wang, J. Dong, S. Tian, W.C. Cheong, S. Lu, L. Zheng, X. Zheng, W. Yan, Z. Zhuang, C. Chen, Q. Peng, D. Wang, Y. Li, Rational design of single molybdenum atoms anchored on N-doped carbon for effective hydrogen evolution reaction, *Angew. Chem. Int. Ed. Engl.*, 56 (**2017**) 16086–16090.

75. T. Li, J. Liu, Y. Song, F. Wang, Photochemical solid-phase synthesis of platinum single atoms on nitrogen-doped carbon with high loading as bifunctional catalysts for hydrogen evolution and oxygen reduction reactions, *ACS Catal.*, 8 (**2018**) 8450–8458.

76. T. Sun, S. Zhao, W. Chen, D. Zhai, J. Dong, Y. Wang, S. Zhang, A. Han, L. Gu, R. Yu, X. Wen, H. Ren, L. Xu, C. Chen, Q. Peng, D. Wang, Y. Li, Single-atomic cobalt sites embedded in hierarchically ordered porous nitrogen-doped carbon as a superior bifunctional electrocatalyst, *Proc. Natl. Acad. Sci. U. S. A.*, 115 (**2018**) 12692–12697.

77. L. Bai, C.S. Hsu, D.T.L. Alexander, H.M. Chen, X. Hu, A cobalt-iron double-atom catalyst for the oxygen evolution reaction, *J. Am. Chem. Soc.*, 141 (**2019**) 14190–14199.

78. H. Wei, H. Wu, K. Huang, B. Ge, J. Ma, J. Lang, D. Zu, M. Lei, Y. Yao, W. Guo, H. Wu, Ultralow-temperature photochemical synthesis of atomically dispersed Pt catalysts for the hydrogen evolution reaction, *Chem. Sci.*, 10 (**2019**) 2830–2836.

79. X. Wen, L. Bai, M. Li, J. Guan, Atomically dispersed cobalt- and nitrogen-codoped graphene toward bifunctional catalysis of oxygen reduction and hydrogen evolution reactions, *ACS Sustainable Chem. Eng.*, 7 (**2019**) 9249–9256.

80. Y. Qu, Z. Li, W. Chen, Y. Lin, T. Yuan, Z. Yang, C. Zhao, J. Wang, C. Zhao, X. Wang, F. Zhou, Z. Zhuang, Y. Wu, Y. Li, Direct transformation of bulk copper into copper single sites via emitting and trapping of atoms, *Nat. Catal.*, 1 (**2018**) 781–786.

81. J. Yang, Z. Qiu, C. Zhao, W. Wei, W. Chen, Z. Li, Y. Qu, J. Dong, J. Luo, Z. Li, Y. Wu, In situ thermal atomization to convert supported nickel nanoparticles into surface-bound nickel single-atom catalysts, *Angew. Chem. Int. Ed. Engl.*, 57 (**2018**) 14095–14100.

82. W. Zhao, G. Wan, C. Peng, H. Sheng, J. Wen, H. Chen, Key single-atom electrocatalysis in metal-organic framework (MOF)-derived bifunctional catalysts, *ChemSusChem*, 11 (**2018**) 3473–3479.

83. J.-D. Yi, R. Xu, G.-L. Chai, T. Zhang, K. Zang, B. Nan, H. Lin, Y.-L. Liang, J. Lv, J. Luo, R. Si, Y.-B. Huang, R. Cao, Cobalt single-atoms anchored on porphyrinic triazine-based frameworks as bifunctional electrocatalysts for oxygen reduction and hydrogen evolution reactions, *J. Mater. Chem. A*, 7 (**2019**) 1252–1259.

84. Q. Zuo, T. Liu, C. Chen, Y. Ji, X. Gong, Y. Mai, Y. Zhou, Ultrathin metal-organic framework nanosheets with ultrahigh loading of single Pt atoms for efficient visible-light-driven photocatalytic H$_2$ evolution, *Angew. Chem. Int. Ed. Engl.*, 58 (**2019**) 10198–10203.

85. J. Zhang, Y. Zhao, X. Guo, C. Chen, C.-L. Dong, R.-S. Liu, C.-P. Han, Y. Li, Y. Gogotsi, G. Wang, Single platinum atoms immobilized on an MXene as an efficient catalyst for the hydrogen evolution reaction, *Nat. Catal.*, 1 (**2018**) 985–992.

86. Y. Cheng, J. Dai, Y. Song, Y. Zhang, Nanostructure of Cr_2CO_2 MXene supported single metal atom as an efficient bifunctional electrocatalyst for overall water splitting, *ACS Appl. Energy Mater.*, 2 (**2019**) 6851–6859.

87. H. Xu, T. Liu, S. Bai, L. Li, Y. Zhu, J. Wang, S. Yang, Y. Li, Q. Shao, X. Huang, Cation exchange strategy to single-atom noble-metal doped CuO nanowire arrays with ultralow overpotential for H2O splitting, *Nano Lett.*, 20 (**2020**) 5482–5489.

88. J.-C. Liu, H. Xiao, J. Li, Constructing high-loading single-atom/cluster catalysts via an electrochemical potential window strategy, *J. Am. Chem. Soc.*, 142 (**2020**) 3375–3383.

89. J.C. Liu, Y.G. Wang, J. Li, Toward rational design of oxide-supported single-atom catalysts: Atomic dispersion of gold on ceria, *J. Am. Chem. Soc.*, 139 (**2017**) 6190–6199.

90. Y. Guan, Y. Feng, J. Wan, X. Yang, L. Fang, X. Gu, R. Liu, Z. Huang, J. Li, J. Luo, C. Li, Y. Wang, Ganoderma-like MoS_2/NiS_2 with single platinum atoms doping as an efficient and stable hydrogen evolution reaction catalyst, *Small*, 14 (**2018**) e1800697.

91. Z. Luo, Y. Ouyang, H. Zhang, M. Xiao, J. Ge, Z. Jiang, J. Wang, D. Tang, X. Cao, C. Liu, W. Xing, Chemically activating MoS_2 via spontaneous atomic palladium interfacial doping towards efficient hydrogen evolution, *Nat. Commun.*, 9 (**2018**) 2120.

92. Q. Wang, Z.L. Zhao, S. Dong, D. He, M.J. Lawrence, S. Han, C. Cai, S. Xiang, P. Rodriguez, B. Xiang, Z. Wang, Y. Liang, M. Gu, Design of active nickel single-atom decorated MoS_2 as a pH-universal catalyst for hydrogen evolution reaction, *Nano Energy*, 53 (**2018**) 458–467.

93. X. Xu, H. Xu, D. Cheng, Design of high-performance MoS_2 edge supported single-metal atom bifunctional catalysts for overall water splitting via a simple equation, *Nanoscale*, 11 (**2019**) 20228–20237.

94. C. Ling, L. Shi, Y. Ouyang, X.C. Zeng, J. Wang, Nanosheet supported single-metal atom bifunctional catalyst for overall water splitting, *Nano Lett.*, 17 (**2017**) 5133–5139.

95. A. Mohajeri, N.L. Dashti, Cooperativity in bimetallic SACs: An efficient strategy for designing bifunctional catalysts for overall water splitting, *J. Phys. Chem. C*, 123 (**2019**) 30972–30980.

96. S. Yang, J. Kim, Y.J. Tak, A. Soon, H. Lee, Single-atom catalyst of platinum supported on titanium nitride for selective electrochemical reactions, *Angew. Chem. Int. Ed. Engl.*, 55 (**2016**) 2058–2062.

97. M. Zhou, J.E. Dick, A.J. Bard, Electrodeposition of isolated platinum atoms and clusters on bismuth-characterization and electrocatalysis, *J. Am. Chem. Soc.*, 139 (**2017**) 17677–17682.

98. K. Jiang, B. Liu, M. Luo, S. Ning, M. Peng, Y. Zhao, Y.-R. Lu, T.-S. Chan, F.M.F. de Groot, Y. Tan, Single platinum atoms embedded in nanoporous cobalt selenide as electrocatalyst for accelerating hydrogen evolution reaction, *Nat. Commun.*, 10 (**2019**) 1743.

99. A.S. Nair, R. Ahuja, B. Pathak, Unraveling the single-atom electrocatalytic activity of transition metal-doped phosphorene, *Nanoscale Adv.*, 2 (**2020**) 2410–2421.

100. Y. Zhao, T. Ling, S. Chen, B. Jin, A. Vasileff, Y. Jiao, L. Song, J. Luo, S.-Z. Qiao, Nonmetal single-iodine-atom electrocatalysts for the hydrogen evolution reaction, *Angew. Chem. Int. Ed.*, 58 (**2019**) 12252–12257.

101. J. Zhu, L.S. Hu, P.X. Zhao, L.Y.S. Lee, K.Y. Wong, Recent advances in electrocatalytic hydrogen evolution using nanoparticles, *Chem. Rev.*, 120 (**2020**) 851–918.

102. B. Lu, L. Guo, F. Wu, Y. Peng, J.E. Lu, T.J. Smart, N. Wang, Y.Z. Finfrock, D. Morris, P. Zhang, N. Li, P. Gao, Y. Ping, S. Chen, Ruthenium atomically dispersed in carbon outperforms platinum toward hydrogen evolution in alkaline media, *Nat. Commun.*, 10 (**2019**) 631.

103. Y. Wang, X. Huang, Z. Wei, Recent developments in the use of single-atom catalysts for water splitting, *Chin. J. Catal.*, 42 (**2021**) 1269–1286.

104. W.H. Lee, Y.-J. Ko, J.-Y. Kim, B.K. Min, Y.J. Hwang, H.-S. Oh, Single-atom catalysts for the oxygen evolution reaction: recent developments and future perspectives, *Chem. Commun.*, 56 (**2020**) 12687–12697.

105. N. Daelman, M. Capdevila-Cortada, N. Lopez, Dynamic charge and oxidation state of Pt/CeO$_2$ single-atom catalysts, *Nat. Mater.*, 18 (**2019**) 1215–1221.

106. K.L. Zhou, Z. Wang, C.B. Han, X. Ke, C. Wang, Y. Jin, Q. Zhang, J. Liu, H. Wang, H. Yan, Platinum single-atom catalyst coupled with transition metal/metal oxide heterostructure for accelerating alkaline hydrogen evolution reaction, *Nat. Commun.*, 12 (**2021**) 3783.

107. Y. Qu, L. Wang, Z. Li, P. Li, Q. Zhang, Y. Lin, F. Zhou, H. Wang, Z. Yang, Y. Hu, M. Zhu, X. Zhao, X. Han, C. Wang, Q. Xu, L. Gu, J. Luo, L. Zheng, Y. Wu, Ambient synthesis of single-atom catalysts from bulk metal via trapping of atoms by surface dangling bonds, *Adv. Mater.*, 31 (**2019**) 1904496.

108. J. Chen, H. Li, C. Fan, Q. Meng, Y. Tang, X. Qiu, G. Fu, T. Ma, Dual single-atomic Ni-N4 and Fe-N4 sites constructing janus hollow graphene for selective oxygen electrocatalysis, *Adv. Mater.*, 32 (**2020**) 2003134.

109. X. Han, X. Ling, D. Yu, D. Xie, L. Li, S. Peng, C. Zhong, N. Zhao, Y. Deng, W. Hu, Atomically dispersed binary Co-Ni sites in nitrogen-doped hollow carbon nanocubes for reversible oxygen reduction and evolution, *Adv. Mater.*, 31 (**2019**) 1905622.

110. Y. Lin, K. Sun, S. Liu, X. Chen, Y. Cheng, W.-C. Cheong, Z. Chen, L. Zheng, J. Zhang, X. Li, Y. Pan, C. Chen, Construction of CoP/NiCoP nanotadpoles heterojunction interface for wide pH hydrogen evolution electrocatalysis and supercapacitor, *Adv. Energy Mater.*, 9 (**2019**) 1901213.

111. X. Liu, W. Li, X. Zhao, Y. Liu, C.-W. Nan, L.-Z. Fan, Two birds with one stone: Metal–organic framework derived micro-/nanostructured Ni$_2$P/Ni hybrids embedded in porous carbon for electrocatalysis and energy storage, *Adv. Funct. Mater.*, 29 (**2019**) 1901510.

112. C.C. Yang, S.F. Zai, Y.T. Zhou, L. Du, Q. Jiang, Fe3C-Co nanoparticles encapsulated in a hierarchical structure of N-doped carbon as a multifunctional electrocatalyst for ORR, OER, and HER, *Adv. Funct. Mater.*, 29 (**2019**) 1901949.

113. A. Chen, X. Yu, Y. Zhou, S. Miao, Y. Li, S. Kuld, J. Sehested, J. Liu, T. Aoki, S. Hong, M.F. Camellone, S. Fabris, J. Ning, C. Jin, C. Yang, A. Nefedov, C. Wöll, Y. Wang, W. Shen, Structure of the catalytically active copper–ceria interfacial perimeter, *Nat. Catal.*, 2 (**2019**) 334–341.

114. Z. Zhang, Y. Chen, L. Zhou, C. Chen, Z. Han, B. Zhang, Q. Wu, L. Yang, L. Du, Y. Bu, The simplest construction of single-site catalysts by the synergism of micropore trapping and nitrogen anchoring, *Nat. Commun.*, 10 (**2019**) 1–7.

115. G. Han, Y. Zheng, X. Zhang, Z. Wang, Y. Gong, C. Du, M.N. Banis, Y.-M. Yiu, T.-K. Sham, L. Gu, Y. Sun, Y. Wang, J. Wang, Y. Gao, G. Yin, X. Sun, High loading single-atom Cu dispersed on graphene for efficient oxygen reduction reaction, *Nano Energy*, 66 (**2019**) 104088.

116. H.Y. Zhuo, X. Zhang, J.X. Liang, Q. Yu, H. Xiao, J. Li, Theoretical understandings of graphene-based metal single-atom catalysts: Stability and catalytic performance, *Chem. Rev.*, 120 (**2020**) 12315–12341.

117. S. Ye, F. Luo, Q. Zhang, P. Zhang, T. Xu, Q. Wang, D. He, L. Guo, Y. Zhang, C. He, X. Ouyang, M. Gu, J. Liu, X. Sun, Highly stable single Pt atomic sites anchored on aniline-stacked graphene for hydrogen evolution reaction, *Energy Environ. Sci.*, 12 (**2019**) 1000–1007.

118. S. Tang, X. Zhou, T. Liu, S. Zhang, T. Yang, Y. Luo, E. Sharman, J. Jiang, Single nickel atom supported on hybridized graphene–boron nitride nanosheet as a highly active bifunctional electrocatalyst for hydrogen and oxygen evolution reactions, *J. Mater. Chem. A*, 7 (**2019**) 26261–26265.

119. S. Wang, E. Iyyamperumal, A. Roy, Y. Xue, D. Yu, L. Dai, Vertically aligned BCN nanotubes as efficient metal-free electrocatalysts for the oxygen reduction reaction: A synergetic effect by co-doping with boron and nitrogen, *Angew. Chem. Int. Ed. Engl.*, 50 (**2011**) 11756–11760.

120. Z. Yao, Y. Jiao, Y. Zhu, Q. Cai, A. Vasileff, H.L. Lu, Y. Han, Y. Chen, S.Z. Qiao, Molecule-level g-C_3N_4 coordinated transition metals as a new class of electrocatalysts for oxygen electrode reactions, *J. Am. Chem. Soc.*, 139 (**2017**) 3336.

121. L. Fan, P.F. Liu, X. Yan, L. Gu, Z.Z. Yang, H.G. Yang, S. Qiu, X. Yao, Atomically isolated nickel species anchored on graphitized carbon for efficient hydrogen evolution electrocatalysis, *Nat. Commun.*, 7 (**2016**) 10667.

122. H. Zhang, G. Liu, L. Shi, J. Ye, Single atom catalysts: Emerging multifunctional materials in heterogeneous catalysis, *Adv. Energy Mater.*, 8 (**2018**) 1701343.

123. Y. Chen, S. Ji, C. Chen, Q. Peng, D. Wang, Y. Li, Single-atom catalysts: Synthetic strategies and electrochemical applications, *Joule*, 2 (**2018**) 1242–1264.

124. P. Liu, Y. Zhao, R. Qin, S. Mo, G. Chen, L. Gu, D.M. Chevrier, P. Zhang, Q. Guo, D. Zang, Photochemical route for synthesizing atomically dispersed palladium catalysts, *Science*, 352 (**2016**) 797–801.

125. J. Zhang, Z. Zhao, Z. Xia, L. Dai, A metal-free bifunctional electrocatalyst for oxygen reduction and oxygen evolution reactions, *Nat. Nanotechnol.*, 10 (**2015**) 444–452.

126. T. Sun, N. Shan, L. Xu, J. Wang, J. Chen, A.A. Zakhidov, R.H. Baughman, General synthesis of 3D ordered macro-/mesoporous materials by templating mesoporous silica confined in opals, *Chem. Mater.*, 30 (**2018**) 1617–1624.

127. Q.L. Zhu, W. Xia, L.R. Zheng, R. Zou, Z. Liu, Q. Xu, Atomically dispersed Fe/N-doped hierarchical carbon architectures derived from a metal–organic framework composite for extremely efficient electrocatalysis, *ACS Energy Lett.*, 2 (**2017**) 504–511.

128. P. Liu, Y. Zhao, R. Qin, S. Mo, G. Chen, L. Gu, D.M. Chevrier, P. Zhang, Q. Guo, D.J.S. Zang, Photochemical route for synthesizing atomically dispersed palladium catalysts, *Science*, 352 (**2016**) 797–801.

129. R. Lang, W. Xi, J.C. Liu, Y.T. Cui, T. Li, A.F. Lee, F. Chen, Y. Chen, L. Li, L. Li, Non defect-stabilized thermally stable single-atom catalyst, *Nat. Commun.*, 10 (**2019**) 234.

130. Y. Chen, S. Ji, C. Chen, Q. Peng, D. Wang, Y.J.J. Li, Single-atom catalysts: Synthetic strategies and electrochemical applications, *Nano-Micro Lett.*, 2 (**2018**) 1242–1264.

131. H. Xu, T. Liu, S. Bai, L. Li, Y. Zhu, J. Wang, S. Yang, Y. Li, Q. Shao, X. Huang, Cation exchange strategy to single-atom noble-metal doped CuO nanowire arrays with ultralow overpotential for H2O splitting, *Nano Lett.*, 20 (**2020**) 5482–5489.

132. P.X. Liu, Y. Zhao, R.X. Qin, S.G. Mo, G.X. Chen, L. Gu, D.M. Chevrier, P. Zhang, Q. Guo, D.D. Zang, B.H. Wu, G. Fu, N.F. Zheng, Photochemical route for synthesizing atomically dispersed palladium catalysts, *Science*, 352 (**2016**) 797–801.

133. Y. Guan, Y. Feng, J. Wan, X. Yang, L. Fang, X. Gu, R. Liu, Z. Huang, J. Li, J. Luo, C. Li, Y. Wang, Ganoderma-like MoS_2/NiS_2 with single platinum atoms doping as an efficient and stable hydrogen evolution reaction catalyst, *Small*, 14 (**2018**) 1800697.

134. B.T. Qiao, A.Q. Wang, X.F. Yang, L.F. Allard, Z. Jiang, Y.T. Cui, J.Y. Liu, J. Li, T. Zhang, Single-atom catalysis of CO oxidation using Pt_1/FeO_x, *Nat. Chem.*, 3 (**2011**) 634–641.

135. Z. Luo, Y. Ouyang, H. Zhang, M. Xiao, J. Ge, Z. Jiang, J. Wang, D. Tang, X. Cao, C. Liu, W. Xing, Chemically activating MoS_2 via spontaneous atomic palladium interfacial doping towards efficient hydrogen evolution, *Nat. Commun.*, 9 (**2018**) 2120.

136. H. Yin, S. Zhao, K. Zhao, A. Muqsit, H. Tang, L. Chang, H. Zhao, Y. Gao, Z. Tang, Ultrathin platinum nanowires grown on single-layered nickel hydroxide with high hydrogen evolution activity, *Nat. Commun.*, 6 (**2015**) 6430.

137. E.J. Popczun, J.R. McKone, C.G. Read, A.J. Biacchi, A.M. Wiltrout, N.S. Lewis, R.E. Schaak, Nanostructured nickel phosphide as an electrocatalyst for the hydrogen evolution reaction, *J. Am. Chem. Soc.*, 135 (**2013**) 9267–9270.

138. J.K. Norskov, T. Bligaard, A. Logadottir, J.R. Kitchin, J.G. Chen, S. Pandelov, J.K. Norskov, Trends in the exchange current for hydrogen evolution, *J. Electrochem. Soc.*, 152 (**2005**) J23–J26.

139. H. Wang, H.-W. Lee, Y. Deng, Z. Lu, P.-C. Hsu, Y. Liu, D. Lin, Y. Cui, Bifunctional non-noble metal oxide nanoparticle electrocatalysts through lithium-induced conversion for overall water splitting, *Nat. Commun.*, 6 (**2015**) 7261.

140. Y. Wu, G.-D. Li, Y. Liu, L. Yang, X. Lian, T. Asefa, X. Zou, Overall water splitting catalyzed efficiently by an ultrathin nanosheet-built, hollow Ni_3S_2-based electrocatalyst, *Adv. Funct. Mater.*, 26 (**2016**) 4839–4847.

141. C. Ling, L. Shi, Y. Ouyang, X.C. Zeng, J. Wang, Nanosheet supported single-metal atom bifunctional catalyst for overall water splitting, *Nano Lett.*, 17 (**2017**) 5133–5139.

142. L. Zhang, L. Han, H. Liu, X. Liu, J. Luo, Potential-cycling synthesis of single platinum atoms for efficient hydrogen evolution in neutral media, *Angew. Chem. Int. Ed.*, 56 (**2017**) 13694–13698.

143. Y. Zhao, T. Ling, S. Chen, B. Jin, S. Qiao, Non-metal single-iodine-atom electrocatalysts for the hydrogen evolution reaction, *Angew. Chem.*, 131 (**2019**) 12380–12385.

144. Z. Zhang, C. Feng, C. Liu, M. Zuo, L. Qin, X. Yan, Y. Xing, H. Li, R. Si, S. Zhou, J. Zeng, Electrochemical deposition as a universal route for fabricating single-atom catalysts, *Nat. Commun.*, 11 (**2020**) 1215.

145. L. Zhang, Y. Jia, H. Liu, L. Zhuang, X. Yan, C. Lang, X. Wang, D. Yang, K. Huang, S. Feng, X. Yao, Charge polarization from atomic metals on adjacent graphitic layers for enhancing the hydrogen evolution reaction, *Angew. Chem. Int. Ed.*, 58 (**2019**) 9404–9408.

146. J. Ji, Y. Zhang, L. Tang, C. Liu, X. Gao, M. Sun, J. Zheng, M. Ling, C. Liang, Z. Lin, Platinum single-atom and cluster anchored on functionalized MWCNTs with ultrahigh mass efficiency for electrocatalytic hydrogen evolution, *Nano Energy*, 63 (**2019**) 103849.

147. S. Yuan, Z. Pu, H. Zhou, J. Yu, I.S. Amiinu, J. Zhu, Q. Liang, J. Yang, D. He, Z. Hu, G. Van Tendeloo, S. Mu, A universal synthesis strategy for single atom dispersed cobalt/metal clusters heterostructure boosting hydrogen evolution catalysis at all pH values, *Nano Energy*, 59 (**2019**) 472–480.

148. D. Wang, Q. Li, C. Han, Z. Xing, X. Yang, Single-atom ruthenium based catalyst for enhanced hydrogen evolution, *Appl. Catal., B*, 249 (**2019**) 91–97.

149. L. Ji, P. Yan, C. Zhu, C. Ma, W. Wu, C. Wei, Y. Shen, S. Chu, J. Wang, Y. Du, J. Chen, X. Yang, Q. Xu, One-pot synthesis of porous 1T-phase MoS2 integrated with single-atom Cu doping for enhancing electrocatalytic hydrogen evolution reaction, *Appl. Catal., B*, 251 (**2019**) 87–93.

150. Y. Wang, L. Chen, Z. Mao, L. Peng, R. Xiang, X. Tang, J. Deng, Z. Wei, Q. Liao, Controlled synthesis of single cobalt atom catalysts via a facile one-pot pyrolysis for efficient oxygen reduction and hydrogen evolution reactions, *Sci. Bull.*, 64 (**2019**) 1095–1102.

151. C. Wu, D. Li, S. Ding, Z.U. Rehman, Q. Liu, S. Chen, B. Zhang, L. Song, Monoatomic platinum-anchored metallic MoS_2: Correlation between surface dopant and hydrogen evolution, *J. Phys. Chem. Lett.*, 10 (**2019**) 6081–6087.

152. W.-H. Lai, L.-F. Zhang, W.-B. Hua, S. Indris, Z.-C. Yan, Z. Hu, B. Zhang, Y. Liu, L. Wang, M. Liu, R. Liu, Y.-X. Wang, J.-Z. Wang, Z. Hu, H.-K. Liu, S.-L. Chou, S.-X. Dou, General π-electron-assisted strategy for Ir, Pt, Ru, Pd, Fe, Ni single-atom electrocatalysts with bifunctional active sites for highly efficient water splitting, *Angew. Chem. Int. Ed.*, 58 (**2019**) 11868–11873.

153. M. Li, K. Duanmu, C. Wan, T. Cheng, L. Zhang, S. Dai, W. Chen, Z. Zhao, P. Li, H. Fei, Y. Zhu, R. Yu, J. Luo, K. Zang, Z. Lin, M. Ding, J. Huang, H. Sun, J. Guo, X. Pan, W.A. Goddard, P. Sautet, Y. Huang, X. Duan, Single-atom tailoring of platinum nanocatalysts for high-performance multifunctional electrocatalysis, *Nat. Catal.*, 2 (**2019**) 495–503.

154. L. Cao, Q. Luo, W. Liu, Y. Lin, X. Liu, Y. Cao, W. Zhang, Y. Wu, J. Yang, T. Yao, S. Wei, Identification of single-atom active sites in carbon-based cobalt catalysts during electrocatalytic hydrogen evolution, *Nat. Catal.*, 2 (**2019**) 134–141.

155. Y. Pan, S. Liu, K. Sun, X. Chen, B. Wang, K. Wu, X. Cao, W.-C. Cheong, R. Shen, A. Han, Z. Chen, L. Zheng, J. Luo, Y. Lin, Y. Liu, D. Wang, Q. Peng, Q. Zhang, C. Chen, Y. Li, A bimetallic Zn/Fe polyphthalocyanine-derived single-atom Fe-N$_4$ catalytic site: A superior trifunctional catalyst for overall water splitting and Zn–air batteries, *Angew. Chem. Int. Ed.*, 57 (**2018**) 8614–8618.

156. C. Cai, S. Han, Q. Wang, M. Gu, Direct observation of Yolk–Shell transforming to gold single atoms and clusters with superior oxygen evolution reaction efficiency, *ACS Nano*, 13 (**2019**) 8865–8871.

157. J. Yan, L. Kong, Y. Ji, J. White, Y. Li, J. Zhang, P. An, S. Liu, S.-T. Lee, T. Ma, Single atom tungsten doped ultrathin α-Ni(OH)$_2$ for enhanced electrocatalytic water oxidation, *Nat. Commun.*, 10 (**2019**) 2149.

158. D.D. Babu, Y. Huang, G. Anandhababu, X. Wang, R. Si, M. Wu, Q. Li, Y. Wang, J. Yao, Atomic iridium@cobalt nanosheets for dinuclear tandem water oxidation, *J. Mater. Chem. A*, 7 (**2019**) 8376–8383.

159. I.S. Amiinu, X. Liu, Z. Pu, W. Li, Q. Li, J. Zhang, H. Tang, H. Zhang, S. Mu, From 3D ZIF nanocrystals to Co-N$_x$/C nanorod array electrocatalysts for ORR, OER, and Zn–air batteries, *Adv. Funct. Mater.*, 28 (**2018**) 1704638.

160. C. Lin, Y. Zhao, H. Zhang, S. Xie, Y.-F. Li, X. Li, Z. Jiang, Z.-P. Liu, Accelerated active phase transformation of NiO powered by Pt single atoms for enhanced oxygen evolution reaction, *Chem. Sci.*, 9 (**2018**) 6803–6812.

161. Y. Zhang, C. Wu, H. Jiang, Y. Lin, H. Liu, Q. He, S. Chen, T. Duan, L. Song, Atomic iridium incorporated in cobalt hydroxide for efficient oxygen evolution catalysis in neutral electrolyte, *Adv. Mater.*, 30 (**2018**) 1707522.

162. H. Sun, S. Liu, M. Wang, T. Qian, J. Xiong, C. Yan, Updating the intrinsic activity of a single-atom site with a P–O bond for a rechargeable Zn–air battery, *ACS Appl. Mat. Interfaces*, 11 (**2019**) 33054–33061.

163. L. Liu, H. Su, F. Tang, X. Zhao, Q. Liu, Confined organometallic Au$_1$N$_x$ single-site as an efficient bifunctional oxygen electrocatalyst, *Nano Energy*, 46 (**2018**) 110–116.

164. Y. Yao, S. Hu, W. Chen, Z.-Q. Huang, W. Wei, T. Yao, R. Liu, K. Zang, X. Wang, G. Wu, W. Yuan, T. Yuan, B. Zhu, W. Liu, Z. Li, D. He, Z. Xue, Y. Wang, X. Zheng, J. Dong, C.-R. Chang, Y. Chen, X. Hong, J. Luo, S. Wei, W.-X. Li, P. Strasser, Y. Wu, Y. Li, Engineering the electronic structure of single atom Ru sites via compressive strain boosts acidic water oxidation electrocatalysis, *Nat. Catal.*, 2 (**2019**) 304–313.

6 Zn-Air Battery Application of Atomically Dispersed Metallic Materials

Mingjie Wu, Gaixia Zhang, Hariprasad Ranganathan, and Shuhui Sun
Institut National de la Recherche Scientifique (INRS)

CONTENTS

6.1 INTRODUCTION

The development of low-cost oxygen electrocatalysts to efficiently catalyze either oxygen reduction reaction (ORR) or oxygen evolution reaction (OER) or both is one of the key bottlenecks in the advanced electrochemical energy storage and conversion technologies, including metal-air batteries, fuel cells, and water splitting.[1,2]

DOI: 10.1201/9781003153436-6

Among various metal-air battery technologies, the rechargeable Zn-air batteries (ZABs) based on the reversible formation and decomposition of $Zn(OH)_4^{2-}$ possessing a high theoretical specific energy density (1,086 Wh kg^{-1}).[3] Compared with the other metal-air batteries (metal = Li/Na/K/Mg/Al/Fe), ZABs are attracting more attention due to their high energy density, low-cost, aqueous electrolyte system, easy processing, safety, slow self-discharge and high reversibility.[1] Accordingly, massive research efforts have been dedicated to enhance the slow ORR and OER kinetics and improve the round-trip efficiency of ZABs by developing efficient heterogeneous electrocatalysts, such as supported transition-metal-based nanoparticle catalysts (e.g., oxides, sulfides, phosphides, nitrides, alloys).[4]

Among all heterogeneous electrocatalysts, metal single-atom catalysts (SACs) comprising isolated metal atoms anchored on appropriate supports, such as heteroatom (N, S, P)-doped carbon and oxide supports, have received wide interest.[5] The resulting uniform active centers consisting of metal sites and coordinating atoms are attractive for their utmost atom-utilization efficiency, high selectivity, and electrocatalytic activity.[6] The concept of "single-atom catalysts" was first reported for the single Pt atoms supported on iron oxide to study the CO oxidation in 2011.[7] The near 100% atom-utilization efficiency contributes to the minimal cost and maximal mass activity, which is particularly important for the noble metal catalysts.[8–10] Moreover, the different catalytic activities and behaviors of these metal SACs highly depend on the coordination configurations of single-atom sites (i.e., type and quantity of center metal and coordinating atoms, type of supports, and environmental atoms or groups surrounding active centers).[11,12] The distinctive coordinated environments induce unique and strong electronic metal-support interactions (EMSI).[13] Consequently, the facilitated electron transfer modulates the electronic structure of metal atoms, which can greatly improve the intrinsic activities. In addition, by virtue of the identical geometric structure of active centers, density functional theory (DFT) calculations shed light on the relationship between activity and electronic structure and promote the understanding of the catalytic mechanisms at the atomic level.

Thanks to the firstly reported single-atom Pt catalyst, various SACs have been successfully explored in the field of electrochemistry. The rapid progress in this field produced many new synthetic approaches and made great advances in the design and identification of efficient active sites. However, the efficiency and stability of SACs as the bifunctional ORR and OER catalysts for rechargeable ZABs are yet to fulfill the requirement of commercialization. Therefore, it is of significance to further explore efficient catalysts with high electrochemical stability to obtain high round-trip efficiency and long cycle life of the ZABs. To date, most reviews focus on the preparation methods for SACs, such as defect engineering, space confinement, coordination design, and so on.[14–17] The progress and challenges of the electrochemical applications of SACs in the bifunctional air electrode of ZABs have not been systematically summarized. Therefore, the recent progress of SAC electrocatalysis for the application of ZABs is urgent and necessary. Accordingly, in this review, we specifically and systematically elucidate the EMSI of SACs with a focus on the impact of the atomic coordinating structure and the substrates on the oxygen electrocatalytic activity and stability in ZABs. Finally, a perspective is also proposed to highlight the challenges and strategies of SACs in liquid or solid-state ZABs.

6.2 CONFIGURATIONS OF ZABs

Zinc is the most common anode material in primary metal-air batteries, which have been the predominant energy source for hearing aids. Rechargeable zinc-air batteries for EVs were heavily investigated between ca. 1975 and 2000. There are two types of rechargeable liquid ZABs proposed: mechanically and electrically rechargeable forms.[18] The difference between them is that the oxygen electrochemical processes (oxygen reduction/evolution reaction, ORR/OER) happen at two separated air electrodes (tri-electrode configuration) or one single air electrode (two-electrode configuration), respectively. Although the tri-electrode configuration provides a higher battery cycling durability and much more loose requirements of catalysts preparation than the two-electrode configuration, it unavoidably pays the price of increasing the volume and weight of the batteries that end up with reduced volumetric energy as well as power density. Currently, the electrically rechargeable ZABs with a flowing electrolyte design greatly improve the durability of the zinc electrode. The basic structure of a rechargeable flow battery is schematically shown in Figure 6.1a, which is typically composed of four main components: a zinc anode, membrane separator, electrolyte and bifunctional air cathode. In addition, the solid-state ZABs are also proposed. Up to date, there are two main configurations for the flexible solid-state ZABs: sandwich type and cable type.[19,20] The basic structure of a solid-state ZAB is similar to liquid ZABs which is consisting of a zinc electrode, a solid-state electrolyte and a porous air electrode with active materials. The air electrode consists of a substrate and a catalyst layer providing the accommodation for both the ORR and OER (Figure 6.1b). The bifunctional activity of the air electrode significantly determines the power performance of rechargeable ZABs. For the alkaline solid-state ZABs, the electrolyte membrane can conduct hydroxide ions and prevent the oxidized zinc ions from migration to the air electrode. The importance of electrolyte membranes for solid-state ZABs has been underestimated compared to electrode materials for a

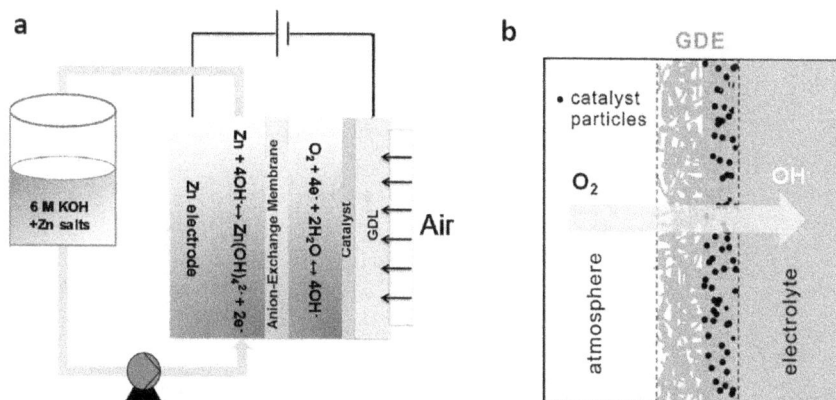

FIGURE 6.1 (a) Scheme of the liquid Zn-air flow-cell. (b) The structure of a gas diffusion electrode in contact with the liquid electrolyte. (Reproduced with permission from Ref.[21] Copyright 2014, The Royal Society of Chemistry.)

long time. Both kinetics and stability of the solid-state ZABs are seriously limited by the electrolyte membrane.

6.3 REACTION MECHANISM OF SACs IN ZABs

During discharge, the zinc electrode is oxidized to Zn^{2+} and then liberates the electrons to the air electrode through the external circuit. At the same time, atmospheric oxygen diffuses into the porous air electrode and is reduced through the ORR. Based on the different electrolyte systems, ZABs with alkaline and neutral electrolyte, respectively, may show different electron transfer processes for oxygen electrocatalysis. It has been reported that the ORR and OER reactions at the air electrode can proceed via a 4e$^-$ transfer process or 2e$^-$ transfer process, which depends on both the oxygen electrocatalysts and the electrolyte systems.[22,23] In the case of the 2e$^-$ pathway in the neutral electrolyte, an aprotic 2e$^-$ ORR process ($O_2 + 2e^- \rightarrow O_2^{2-}$) in ZABs using aqueous $Zn(OTf)_2$ electrolyte was reported, which involves the reversible ZnO_2 formation and decomposition ($Zn + O_2 \leftrightarrow ZnO_2$).[24] This reversible reaction is achieved successfully by mostly depending on the construction of a localized H_2O-poor and zinc ion (Zn^{2+})-rich environment in the inner Helmholtz layer (IHL). The zinc-O_2/zinc peroxide (ZnO_2) chemistry significantly extends the reversibility of ZABs. By contrast, ORR can also proceed via a serial $2 \times 2e^-$ pathway in alkaline electrolyte accompanied by the formation of peroxide species ($O_2 + H_2O + 2e^- \rightarrow HO_2^- + OH^-$) and the further reduction of peroxide ($HO_2^- + H_2O + 2e^- \rightarrow 3OH^-$).[25] The associative conventional mechanism for the 2e$^-$ and 4e$^-$ oxygen reduction in the alkaline electrolyte is shown below (* refers to a surface site):

2e$^-$ process:

$$O_2 + H_2O + e^- \rightarrow OOH^* + OH^- \qquad (6.1)$$

$$OOH^* + H_2O + e^- \rightarrow H_2O_2 + OH^- \qquad (6.2)$$

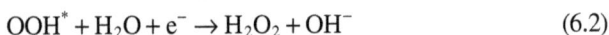

4e$^-$ process (O* mechanism):

$$O_2 + H_2O + e^- \rightarrow OOH^* + OH^- \qquad (6.3)$$

$$OOH^* + e^- \rightarrow O^* + OH^- \qquad (6.4)$$

$$O* + H_2O + e^- \rightarrow OH^* + OH^- \qquad (6.5)$$

$$OH^* + e^- \rightarrow OH^- \qquad (6.6)$$

For practical alkaline ZABs applications, electrocatalysts that promote oxygen electrocatalysis via a direct four-electron reduction pathway are highly pursued. For the transition metal oxide, the abundant −OH groups on the surface prohibit the direct adsorption of O_2 molecules and the ORR undergoes outer sphere electron transfer (OSET) which favors the 2e$^-$ pathway (Figure 6.2a and b).[26] The ORR pathway and

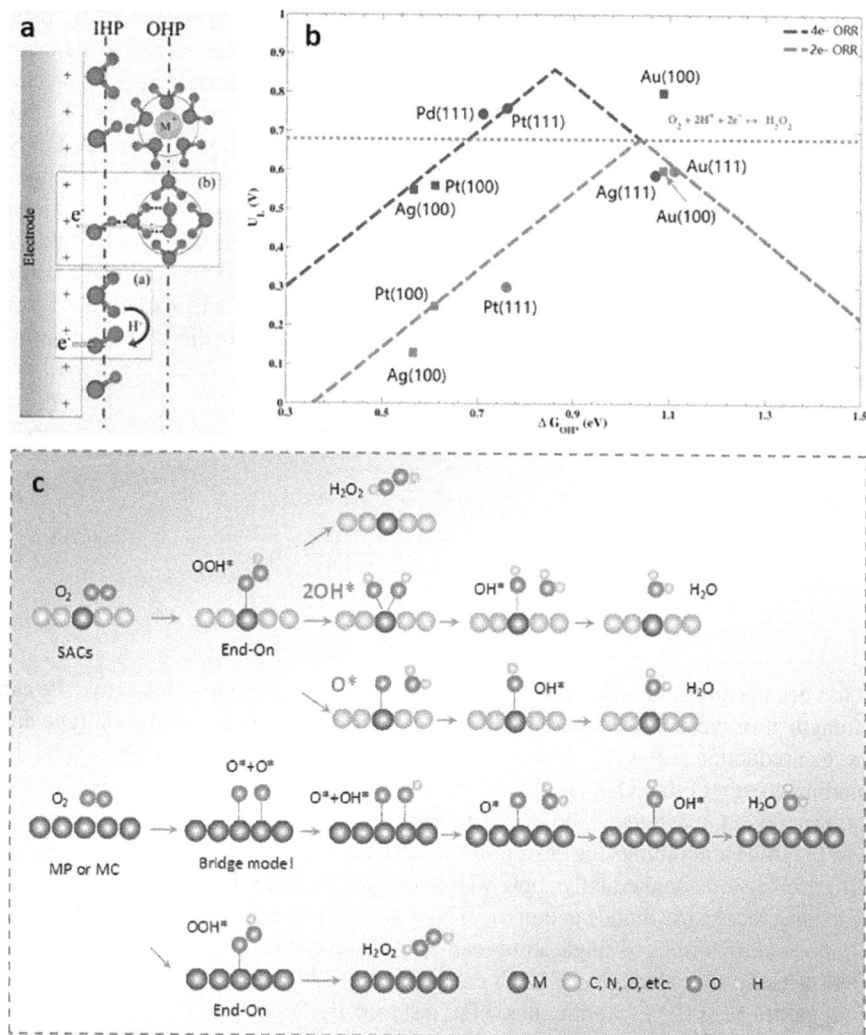

FIGURE 6.2 (a) Schematic illustration of the inner- and outer-sphere (IHP and OHP) electron transfer processes. (b) Volcano plots for the 2e⁻ and 4e⁻ of ORR. (Reproduced with permission from Ref.[34] Copyright 2012, American Chemical Society.) (c) 2e⁻ and 4e⁻ pathways of ORR on SACs and metal nanoparticles or metal clusters.)

activity can be controlled by regulating oxygen adsorption configurations and oxygen intermediates binding energy involving outer-sphere electron transfer (ISET) processes. In an ORR process, the activation of O_2 molecules adsorbed on metal sites as the first step typically proceeds via three conventional adsorption types such as side-on, end-on and bridge adsorption pathways.[27] The adsorption type of O_2 molecules is critical in determining the oxygen reduction pathway. As shown in Figure 6.2c, the bridge adsorption with each O atom coordinated to the metal atom site on the metal particle (MP) or metal cluster (MC) catalysts tends to proceed by the direct 4e⁻ pathway, whereas the end-on coordination with only one O atom coordinated to the metal atom sites favors the 2e⁻ pathway.[28] In addition to the conventional O* ORR mechanism on the SACs ($O_2 \rightarrow OOH^* \rightarrow O^* \rightarrow OH^* \rightarrow H_2O$), the unconventional 2OH* ORR mechanism ($O_2 \rightarrow OOH^* \rightarrow 2OH^* \rightarrow OH^* \rightarrow H_2O$) is also proposed.[29] For the Fe–N₄ catalysts, previous DFT studies based on conventional adsorption mechanisms abnormally underestimated the half-wave potential value for ORR. This unconventional 2OH* mechanism is more suitable to predict the half-wave potential values of Fe–N₄ catalysts. Compared with the well-known scaling relation $\Delta G(O^*) = 2\Delta G(OH^*)$, a new scaling relation is identified for the 2OH* mechanism: $\Delta G(2OH^*) - \Delta G(O^*) + 1.5\,eV$ on single M–N₄ active sites. The 2OH* mechanism in alkaline media can be composed of four steps (Figure 6.2a):[29]

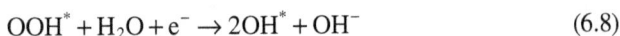

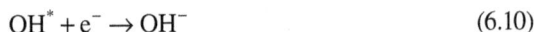

$$O_2 + H_2O + e^- \rightarrow OOH^* + OH^- \tag{6.7}$$

$$OOH^* + H_2O + e^- \rightarrow 2OH^* + OH^- \tag{6.8}$$

$$2OH^* + e^- \rightarrow OH^* + OH^- \tag{6.9}$$

$$OH^* + e^- \rightarrow OH^- \tag{6.10}$$

It has been demonstrated that the geometric structure modification of the active site can strengthen or weaken the adsorption of O_2, which changes their adsorption type and oxygen reduction pathway.[30] In addition, based on the thermodynamics principle, the binding energy of the *OOH intermediate is vital in determining the selectivity of the oxygen reduction pathway. The strong binding energy of *OOH favors the decomposition of *OOH and inhibits the desorption of *OOH to form H_2O_2.[31] This design concept discussed above could equally apply well to design SACs for different electrocatalytic reactions. Due to the unique geometrical configuration of SACs, the adsorption type of O_2 molecules on isolated single atoms can only be the end-on adsorption. However, the ORR activity and selectivity of SACs can be controlled by regulating the coordination configurations of single-atomic sites.[32] For instance, the 2e⁻ pathway of Fe SACs can be shifted toward the 4e⁻ pathway by replacing the neighboring O with N coordination.[33] The aforementioned discussion shows that the preparation of SACs with precise and uniform active centers is significant for efficient oxygen electrocatalysis.

6.4   CHALLENGES OF SACs IN ZABs

Compared with metal nanoparticles or bulk catalysts, the specific activity of SACs with low coordination and unsaturated configuration can be significantly enhanced. However, the surface energy of single metal atoms is much higher than metal nanoparticles and the individual metal atoms tend to aggregate into clusters or particles (Figure 6.3).[35,36] Therefore, the synthesis of SACs with high loading of metal is a big challenge.[37,38] Also, the EMSI by coordinating bonds with the substrate stabilizes the individual metal atoms, which allows the manipulation of their electrocatalytic activity and selectivity.[39] However, most SACs will generate a wide

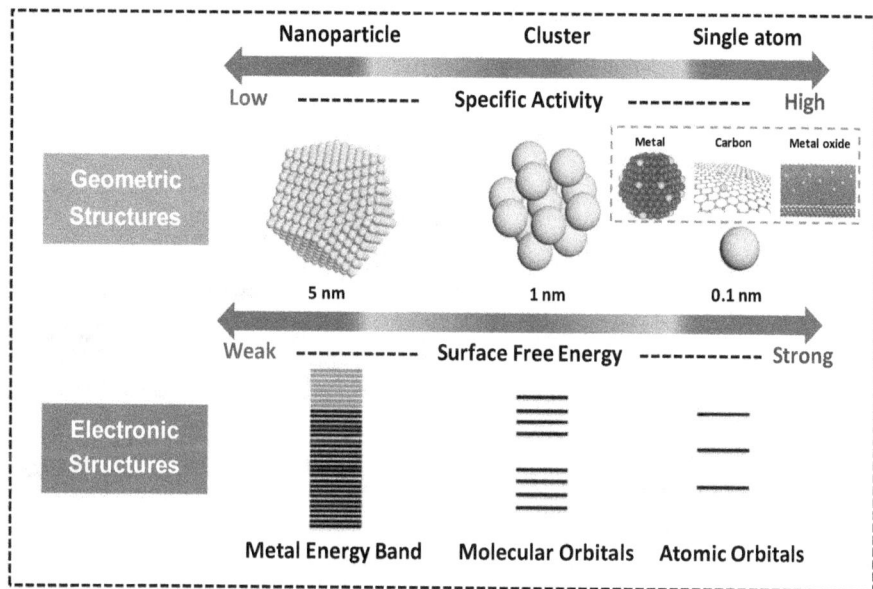

FIGURE 6.3 Geometric structures, electronic structures, surface free energy, and specific activity of single atom, clusters, and nanoparticles.[35]

range of coordination structures during preparation. The precise modulation of the atomic configurations of the active center still represents the most innovative line of research. In addition, the ideal substrate (e.g., metal, metal oxide, metal carbide, and carbon) not only stabilizes the individual metal atoms but also guarantees high electrochemical surface area during the oxygen electrocatalysis process.[40–43] For the carbon-based materials yielded by the pyrolysis method, serious carbon corrosion of amorphous carbon, especially under high oxidative conditions (OER process), will result in significant charge- and mass transport resistances, due to the changes in the carbon lattice structures and morphologies.[44,45] In addition, the kinetic losses may also result from the dissolution of active metal sites and the destruction of the atomic configurations of the active center. Currently, for the carbon-based SACs, a high degree of graphitization of carbon substrates is required to enhance the corrosion resistance of the carbon, thus improving SACs stability in rechargeable ZABs.[46,47]

6.5 SYNTHESIS AND CHARACTERIZATION METHODS OF SACs

6.5.1 Synthesis of SACs

The supporting materials for SACs have been extended from metal (hydro)oxides (e.g., FeO_x, CoO_x, NiO, CuO, ZnO, Al_2O_3, CeO_2, TiO_2 and $Co(OH)_2$, etc.) to heteroatom-doped (e.g., N, O, S, P, etc.) porous carbon, metal-organic frameworks (MOFs), covalent organic frameworks (COFs) and two-dimensional materials (e.g., graphene and MXene), etc.[39,42,48,49] As discussed above, the precise synthesis

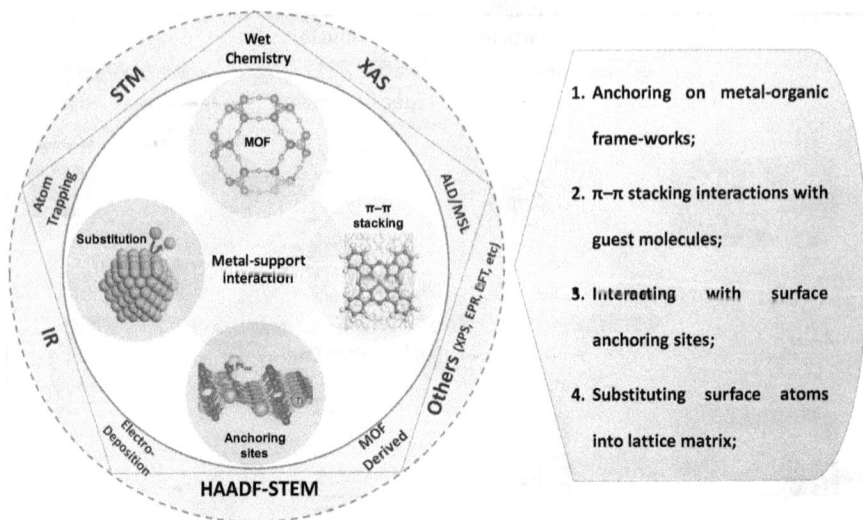

FIGURE 6.4 Summary of synthesis and characterization methods of SACs and the single atom-support interaction in different forms.

of well-defined SACs with specific active center is the key to designing efficient catalysts. To stop metal atom aggregation and obtain specific and highly dispersed single atoms, many effective synthetic approaches have been reported (Figure 6.4).[14] For example, mass-selected soft landing (MSL) and atomic layer deposition (ALD) are two powerful techniques that have been used in preparing SACs.[50,51] However, these two methods both demand the use of high-cost equipment while the low yield limits their large-scale manufacturing. Currently, wet-chemical synthetic methods, including co-precipitation and impregnation methods, have been widely employed because of their easy operation and large-scale production.[52,53] In this approach, the precursors with single-atom metal species are anchored on the supports through the coordination between the ligands of the precursor and the surface groups of the supports. In this case, the anchoring sites with various surface groups or defects of the support play an active role in stabilizing single-atom metal species.[54] Besides, during the post-treatment processes, to remove useless or even poisonous ligands, the EMSI is crucially important in avoiding the aggregation of single atoms on the support surface.[53] In addition to the above-mentioned methods, some ingenious synthetic methods have also been developed, such as atom trapping, electrochemical deposition, ball milling, MOF, and MOF-derived synthesis methods.[55–57] For the SACs as oxygen electrocatalysts in ZABs, those based on various carbon supports have attracted specific interest due to their high surface area and excellent conductivity. Particularly, pyrolysis or carbonization of Zn-containing MOFs under a protected atmosphere (i.e., N_2, NH_3, Ar, or H_2) has been widely developed in the preparation of SACs.[58,59] Optimization of the precursors and controlling the synthetic procedures play a key role in modulating the geometric structure of SACs. The introduction of Zn metal nodes can efficiently change the spatial distribution of the transition metal atoms (i.e.,

Fe, Co, Ni, Cu, etc.), and during the pyrolysis process, Zn can be easily evaporated due to its low boiling point to form SACs.[60] Despite the great progress that has been achieved in the synthesis of SACs, it is still a challenge to further increase their activity without sacrificing their stability.[61] The wide working potential range from ≈ 0.6 V during discharge to ≈ 2.0 V during charging for the ZABs makes strict demands on the electrochemical stability of SACs under the wide potential window.[62]

6.5.2 Characterizations of SACs

It is incontrovertible that the isolated single atoms act as the real active centers of SACs for the electrochemical reaction. Therefore, the determination of their spatial distribution and coordination configuration is crucial for the rational and efficient design of the SACs. Advanced characterization techniques can reveal the specific geometric structures and chemical composition of metal active centers.[54] The clear coordination configuration combining theoretical calculations allows for in-depth understanding of the catalytic mechanisms at the atomic level.[63] So far, some advanced atomic resolution characterization techniques have been developed, such as scanning tunneling microscopy (STM), high-angle annular dark-field imaging scanning tunneling microscopy (HAADF-STEM) and infrared (IR) spectroscopy, Mössbauer spectroscopy and X-ray absorption spectroscopy (XAS) including extended X-ray absorption fine structure (EXAFS) and X-ray absorption near-edge structure spectroscopies (XANES) (Figure 6.4).[64] HAADF-STEM can directly image SACs with high spatial resolutions to determine the size and distribution of individual metal atoms and provide their local structural information. For instance, the structure of Fe–N–C centers can be demonstrated by HAADF-STEM images and electron energy loss spectroscopy (EELS) by revealing the coordination atom of the single metal centers to indicate the formation of Fe–N_x bonds.[65] Additionally, XAS has become a valuable tool in the investigation of SACs due to its intense and tunable X-ray beams. XAS with high flux X-rays is appropriate for the investigation of bulk materials and reveals the information of the overall structure of SACs rather than just local or surface.[66] STM is also an excellent technique to image the catalyst surfaces at the atomic level.[67] However, it has a high requirement against the samples and demands extremely clean and stable surfaces. Thus, this characterization technique is best suited for samples with flat surfaces and single composition. IR spectroscopy with CO is also widely used to distinguish single metal atoms on various supports based on the peak position of CO_{ad}. For example, the IR spectra of CO adsorbed onto Pt on supports (e.g., TiO_2, FeO_x, CeO_2, etc.) can distinguish the single Pt atom and Pt metal.[68,69] Furthermore, some in-situ/operando characterization techniques derived from the above-mentioned techniques are also developed to study the activity, stability, and reaction mechanisms of SACs by identifying changes in spectra during catalytic reactions.[70]

6.6 ELECTRONIC METAL-SUPPORT INTERACTION IN SACs

Since the concept of "electronic factor" was first proposed by G. M. Schwab in the 1930s, electrons transferring between the metal and the support have begun to reveal the impacts of electronic interaction on the catalytic activity of the supported catalyst.[71]

Subsequently, the concept of "strong metal-support interaction" (SMSI), as a key factor affecting catalytic activity, selectivity and stability, was brought forward by S. J. Tauster in 1978.[72] For the deep mechanism of SMSI, electron transfer between the metal and the support has been generally accepted. To give a more detailed description of the enhanced properties of supported catalysts than the term of SMSI, C. T. Campbell proposed the concept of "EMSI" in 2012 which reflects the fact that the associated charge transfer due to the strong chemical bonding between the metal and the support contributes to better catalytic effects.[73,74] Meanwhile, the successful putting forward and development of SACs provide an ideal platform for the study of EMSI than metal particles or clusters.[7] The EMSI derived from the chemical bonding between metal single-atom and support is always uniform. The detailed information about their electronic state can be obtained by both advanced techniques and theoretical calculations, which significantly promotes the understanding of EMSI. The EMSI not only plays a vital role in stabilizing the single metal atoms but also significantly affects the electronic structure of the SACs,[75,76] which is strongly determined by the support type. The active centers with superior adsorption and activation toward molecular reactants enhance SACs' catalytic activity. A deep understanding of EMSI is vital for the rational design of SACs with high loading, excellent stability, and superior catalytic activity. In the case of the oxygen electrocatalysis, the single atoms can be stabilized on various supports to obtain high-performance SACs through (i) spatial confinement by MOFs, (ii) π-π stacking interactions between guest molecules and host materials, (iii) interacting with surface anchoring sites and (iv) substituting surface atoms into lattice matrix (Figure 6.4).[77–79] Spatial confinement to achieve the separation and encapsulation of suitable mononuclear metal precursors (e.g., Zeolites, MOFs, and COFs) is an effective way to prevent the agglomeration of single atoms.[80] Among these, MOFs have been widely employed as ideal supports to prepare SACs in recent years due to their unique structural diversity, atomic-level structural uniformity, adjustable chemical functionality and uniform porous structure.[81] MOFs as organic-inorganic hybrid crystalline porous materials are constructed through the periodic coordination of metal ions and organic ligands. The organic linkers in MOFs, having a variety of configurations, determine the final structural diversity.[82] Also, the formation of π-π stacking or electrostatic interactions between the guest molecules (e.g., transition metal phthalocyanines and macrocyclic compounds) and the carbon-based host materials (e.g., CNTs and graphene) is another way to design SACs.[83,84] It is a facile and effective strategy to achieve large-scale production of SACs. Moreover, the M–N$_4$ units, the most popular active sites in the transition metal phthalocyanines, can be attached to carbon-support surfaces directly by the strong noncovalent π-π interactions.[85,86] For the interaction with the surface anchoring sites, the specific defect sites on the surface of supports play a vital role in directly stabilizing single atoms. For instance, the doped heteroatoms (e.g., N, S, O, P, etc.) in the carbon materials could serve as anchoring sites to directly stabilize metal single atoms. Besides, the intrinsic coordinatively unsaturated surface atoms, oxygen functional groups and surface vacancies are also employed as anchoring sites to bond metal single atoms.[87] On the surface of the oxide support, the excess oxygen ions, metal vacancies, and oxygen vacancies may coordinate with transition metal atoms through M–O(OH) (metal-oxo/hydroxyl group) interactions.[35,88] Other than single-atom stabilization by surface anchoring sites, the single atoms can be directly introduced into the lattice matrix of the support by substitution. For instance, single-atom alloys (SAAs) are one of the most typical representatives that have generated significant interest.[89] Specifically, the single atoms dispersed on the surface lattice matrix of another metal show high metal utilization efficiency and thermodynamic stability.

6.7 ACTIVE METAL CENTER MODULATION TO OPTIMIZE THE EMSI

As discussed above, the different EMSI in the coordination configurations of single-atom sites significantly determines the catalytic activity, selectivity, and stability of oxygen electrocatalysts.[90,91] In particular, atomically dispersed transition metal-nitrogen moieties supported on carbon (M–N–Cs), a unique class of SACs, have attracted extensive research interest due to their high electrical conductivity and activity.[92] Take the typical M–N–Cs, for instance, the single metal atoms tended to form planar $M–N_4$ moieties by directly coordinating with four pyrrolic or pyridinic N atoms and then the formed well-defined $M–N_4$ sites are anchored within a carbon skeleton to form M–N–C active centers.[93] The single metal sites as the absorption sites strongly interact with oxygen and oxygenic intermediates (O_2^*, O^*, OOH^*, and OH^*) to proceed to electrocatalytic processes. Furthermore, the carbon skeleton with a large surface area and electrical conductivity plays a vital role in affording efficient electron and mass transport. In recent years, M–N–Cs have been extensively investigated as electrocatalysts for the ORR and OER. However, the activity and stability of M–N–Cs in the rechargeable ZABs are hitherto unsatisfactory, largely because most M–N–Cs were generally prepared through pyrolysis of nitrogen-containing molecular or polymeric precursors and the as-synthesized M–N–Cs contain a lot of amorphous carbon and limited well-defined $M–N_4$ sites. While the physicochemical characteristics (e.g., degree of graphitization, carbon structure, nitrogen-doping type, and surface area) of the synthesized M–N–Cs highly determine their catalytic activity and stability.[94] These limitations represent the key challenges to achieve higher electrocatalytic performance and promote their practical applications in rechargeable ZABs. To date, massive efforts have been devoted to preparing M–N–Cs featuring diverse types of metal centers, coordination shells, and carbon skeletons, including fabricating dual-metal sites, neighboring heteroatoms, and coordination number regulations.[95] Their numerous empty d-orbitals of transition metal atoms could result in various types of coordination structures, as shown in Figure 6.6. To reveal the relationship between local atomic structure and the intrinsic electrocatalytic performance, five factors (including center metal atoms, coordinated atoms, environmental atoms, guest groups, and carbon support) are proposed based on the structure and performance of M–N–Cs. These factors involve two main aspects: (i) the structure of the active center and (ii) the geometry configuration of the carbon support (Figure 6.5a).[96] The transition metal centers of typical carbon-based SACs (e.g., Fe, Co, Mn, Ni, Cu, etc.) coordinate with their neighboring heteroatoms (e.g., C, N, O, S, P, B, etc.) in the same plane. The center metal atoms acting as the actual adsorption sites directly affect the intrinsic catalytic activity of SACs. For instance, the ORR performance of M–N–Cs with Mn, Fe, Co, Ni, and Cu has been investigated, and their intrinsic ORR

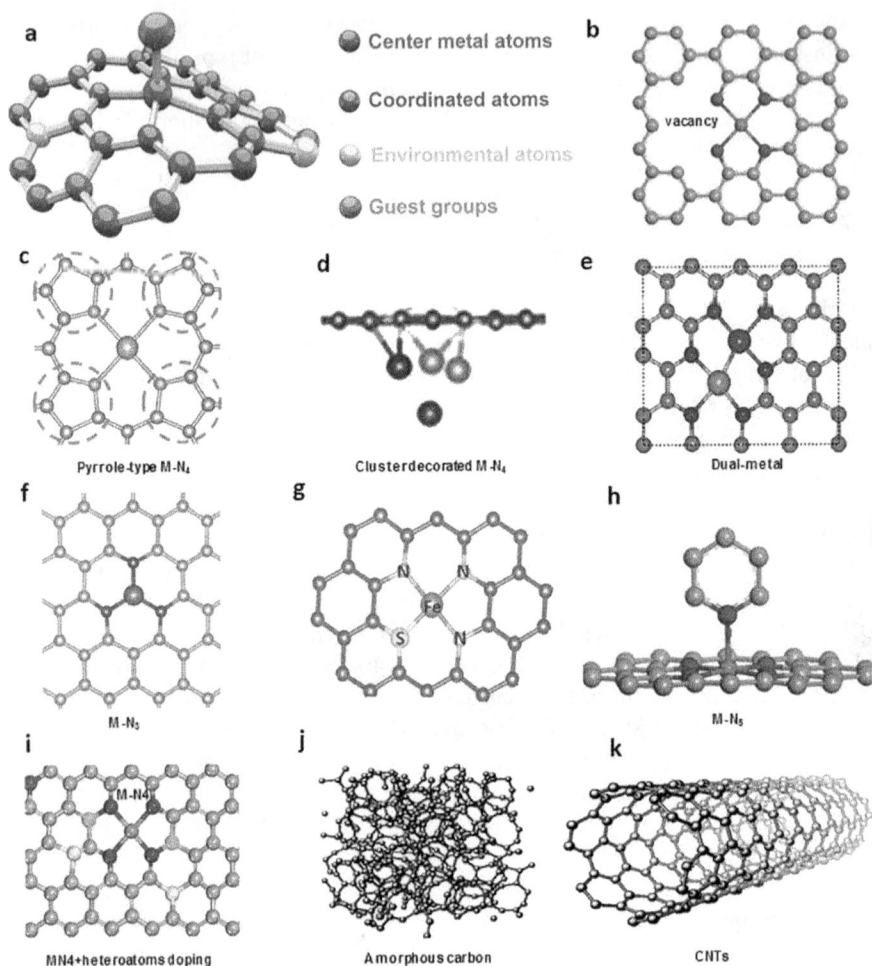

FIGURE 6.5 (a) Geometric structure of carbon-based SACs that consists of center metal atoms, coordinated atoms, environmental atoms, and guest groups. (Reproduced with permission from Ref.[15] Copyright 2021, Wiley-VCH.) Some typical geometric structures of M–N–Cs, including (b) M–N₃C₇ structure with micropores,[25] (c) pyrrole-type FeN₄ structures,[93] (d) cluster decorated MN₄,[110] (e) dual-metal atoms structure,[100,101] (f) metal-N₃ structure,[111,112] (g) FeN₃S structure,[103] (h) metal-N₅ structure,[25] (i) metal-N₄ sites with heteroatoms doping,[113] (j) amorphous carbon, and (k) carbon nanotubes (CNTs). (Reproduced with permission from Ref.[114] Copyright 2019, The Royal Society of Chemistry.)

activity follows the following order: Fe > Co > Cu > Mn > Ni.[97] The OER performance of M–N–Cs with Fe, Co, Ni has also been studied with the following activity trend: Ni > Co > Fe.[98] Additionally, for the M–N₄ structure, the types of N atoms coordinating with C atoms are also varied (Figure 6.5b and c). Pyrrole- and pyridine-type FeN₄ structures are two of the typical representatives reported mostly.[99] The cluster decorated M–N₄ and incorporation of dual transition metal single atoms into the M–N–C structure has been reported (Figure 6.5d and e).[100,101] Besides, heteroatoms (e.g., B, P, S, etc.) with different atomic radii and electronegativity can partially replace the nitrogen atom of the M–N_4 sites (Figure 6.5g).[102,103] Meanwhile, these heteroatoms can also be doped in the carbon skeleton as the environmental atoms of M–N_4, which generate long-range interactions between the heteroatoms doped and metal centers (Figure 6.5i).[104,105] In addition, some potential guest groups, such as small molecules and inorganic particles, can be further grafted on the center metal atoms due to their unsaturated coordination (Figure 6.5h).[106,107] The changes in the coordination structures directly influence the electronic structure of the active centers and regulate their intrinsic activity. Furthermore, carbon-support materials for ORR and OER should possess high porosity, high conductivity, large surface area and high electrochemical stability to achieve maximal activity. The porous structure not only facilitates mass transport but also increases electrochemically available active sites. Moreover, to ensure the efficient electron and mass transport during cycling of the rechargeable ZABs, the strong carbon corrosion resistance of the substrate is particularly important. Recently it has been found that anchoring M–N_4 active sites on some high-graphitization-degree carbon (e.g., CNTs, graphene) effectively improves their stability (Figure 6.5f and g).[108,109]

6.8 DESIGN STRATEGIES OF SACs WITH STRONG EMSI FOR ZABs

6.8.1 Pyrolysis and Pyrolysis-Free Synthetic Approaches

For the rechargeable ZABs, the development of low-cost electrocatalysts for efficient ORR and OER is essential. Accordingly, the pyrolysis of metal-containing complexes such as MOFs, or organic polymers is the most straightforward way to fabricate SACs for ORR and OER which has been widely reported.[115] Careful selection of the precursors and precise control of synthesis parameters play a major role in achieving the desired single-atom sites. For instance, Zhang et al. reported a polymerization-pyrolysis strategy to synthesize atomically dispersed Fe–N_4 catalysts (Figure 6.6a). The highly dispersed single-atom metal sites can be achieved from the proposed predesigned bimetallic Zn/Fe polyphthalocyanine by a low-temperature solvent-free solid-phase synthesis. The ZAB with the obtained catalyst shows a high stability at 2.0 mA cm^{-2} without obvious decay even after 108 cycles.[116] Although high-temperature pyrolysis is a facile method to prepare SACs, unpredictable structural changes in the support and the formed inorganic materials affect their activity and stability due to the attacks of oxygen-free radicals. To solve this problem, pyrolysis-free synthetic approaches toward M–N_x coordinated SACs have been proposed. However, it remains a great challenge to precisely control active centers with reversible oxygen reactions. MOF-based materials, particularly the ZIFs, have become one of the most popular precursors for the synthesis of ORR catalysts. Recently, Chen et al. developed pristine Co–Zn heterometallic ZIFs without pyrolysis as the air electrode of ZABs. The results show that the generation of unoccupied 3d-orbitals at metal sites by a competing coordination strategy effectively facilitates oxygen electrocatalysis. The ZAB employed with the as-prepared ZIFs catalysts

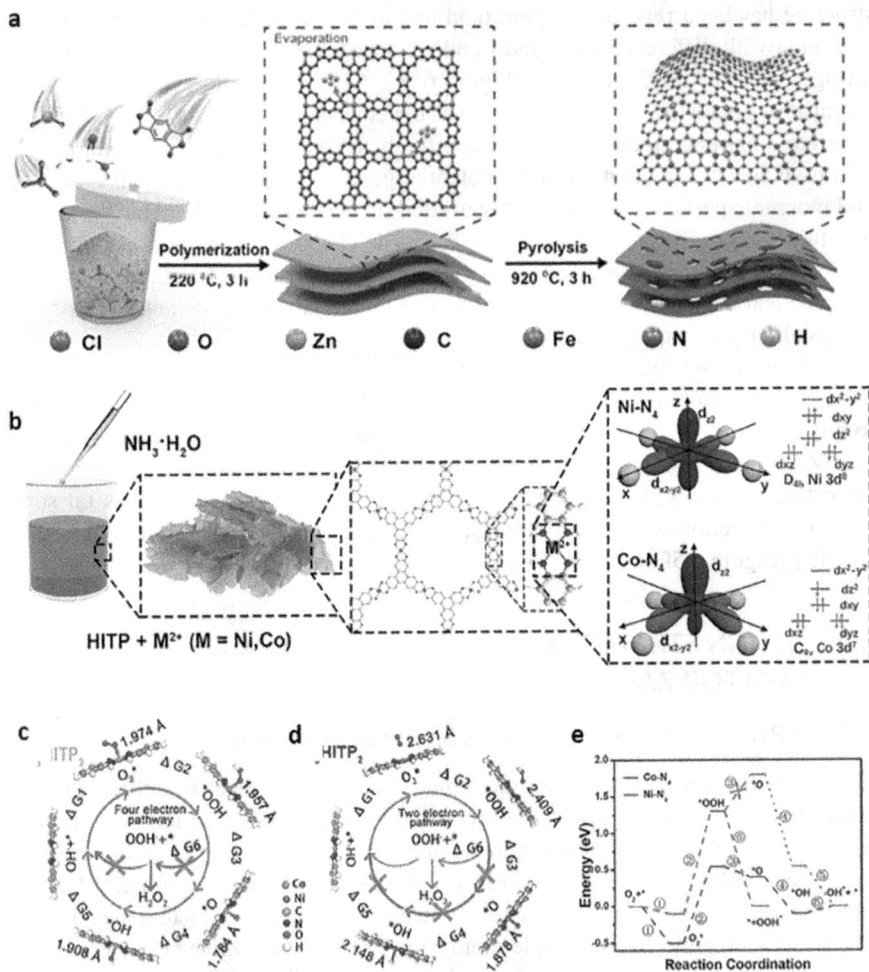

FIGURE 6.6 (a) Synthesis procedure of Fe–N$_4$ SAs/NPC. (Reproduced with permission from Ref.[116] Copyright 2018, Wiley-VCH.) (b) Schematic illustration of the synthesis procedure of the M$_3$HITP$_2$. Catalytic cycle and bond lengths of ORR intermediates on (c) Co$_3$HITP$_2$ and (d) Ni$_3$HITP$_2$. (e) Free energy diagrams of ORR on Ni–N$_4$ and Co–N$_4$. (Reproduced with permission from Ref.[117] Copyright 2020, Wiley-VCH.)

shows enhanced energy efficiency and excellent cyclability of 1,250 h at 15 mA cm^{-2}. To demonstrate the role of metal centers in the MOF on ORR performance, Peng et al. developed conductive coordination polymers by a simple ammonia-assisted procedure (Figure 6.6b).[117]

Compared to Ni coordination polymers (Ni$_3$HITP$_2$), the unpaired 3d electrons in Co coordination polymers (Co$_3$HITP$_2$) with reduced electric conductivity contribute to higher electrocatalytic activities in both ORR and OER. DFT is then employed to unveil the origin of activity and the result indicated a transition from the two-electron pathway on Ni_3HITP_2 to the four-electron pathway on Co_3HITP_2 (Figure 6.6c–e). ZAB employed with Co_3HITP_2 at $5 \, mA \, cm^{-2}$ operated stably for over 80 h.

6.8.2 Intermolecular Interactions between Phthalocyanine-Based Materials and Supports

In addition to the direct modulation of the MOF-based materials, metal phthalocyanine and some phthalocyanine-based layered two-dimensional conjugated MOF assembled with the carbon supports via intermolecular interactions are also reported to synthesize SACs.[118] For instance, Peng et al. developed a fully π-conjugated iron phthalocyanine (FePc)-rich covalent organic framework (COF) (Figure 6.7a).[119] Different from the pyrolysis procedure of randomly creating single-atom sites, the obtained COF with pre-assembled Fe–N–C centers is directly riveted onto the graphene support. The electron localization function (ELF) analysis shows only van der Waals interactions. By comparing the charge density differences, the Fe–C electron pathway was confirmed by the fact that the graphene electrons were attracted to the N-coordinated Fe sites (Figure 6.7b). The as-synthesized SAC shows increased ORR catalytic performance (Figure 6.7c) and the ZAB based on the mixture of SAC and IrO_2 exhibited a long-life cycle of over 300h. Feng et al. developed a copper phthalocyanine-based 2D conjugated MOF supported on the CNTs as an air electrode of ZABs.[120] The obtained 2D conjugated MOF with square-planar cobalt bis (dihydroxy) complexes ($Co-O_4$) as linkages ($PcCu-O_8-Co$) shows a high crystalline structure (Figure 6.7d). In-situ Raman spectro-electrochemistry and DFT verified the Co centers as the catalytic sites (Figure 6.7e).

6.8.3 A Dual-Atom Centered Site or Cluster/Nanoparticle Decorated M–N–C Site

As depicted in Figure 6.2a, O_2 molecules are preferably adsorbed onto SACs via end-on models whereas, in the bridge model, O_2 molecules tend to adsorb on an MC. Compared with the end-on adsorption, the bridge adsorption is more favorable toward the $4e^-$ ORR pathway.[28] Given this, dual-atom-centered sites have been proposed to be an effective route to further enhance their catalytic activity.[100,121,122] Among these dual-atom centered site structures, the incorporation of dual transition metal single atoms into the M–N–C structure has been well studied, which could result in the synergistic effects between the two metallic atoms due to the charge redistribution and d-band center shifts.[123–125] For instance, the incorporation of Fe–Co dual-metal single atoms into the M–N–C structure to form N-coordinated dual-atom centered sites contribute to superior ORR activity (Figure 6.8a).[126] The change of the electronic structure and the geometry configuration of the active sites would stretch the O–O bond length and weaken the bonding. Aside from the $M-N_4$ structure with dual-metal atoms, the combination of $M-N_4$ sites with only one metal atom would also be conducive to enhance the electrocatalytic performance.[127,128] Fu et al. prepared an atomic-dispersed

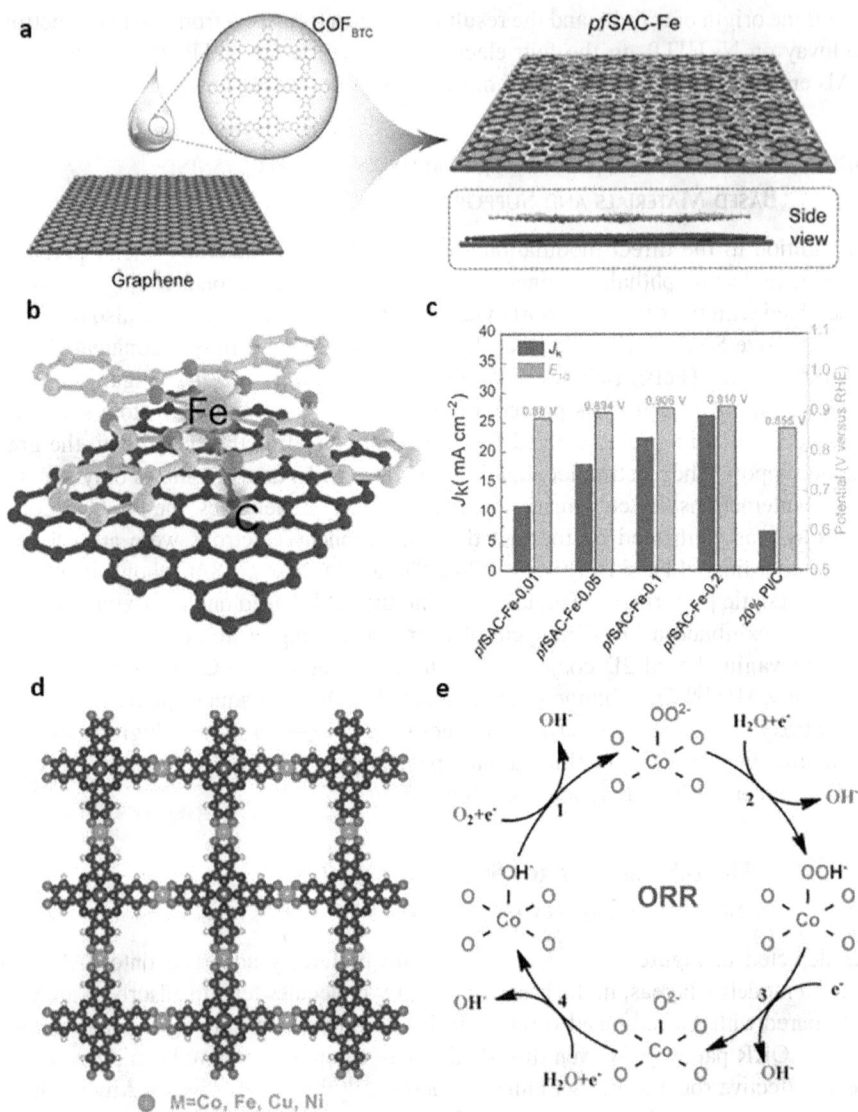

FIGURE 6.7 (a) Synthesis process of the FePc-rich COF. (b) Simulated structures based on the Fe K-edge EXAFS result. (c) Comparison of kinetic current density and half-wave potentials. (Reproduced with permission from Ref.[119] Copyright 2019 American Association for the Advancement of Science.) (d) Schematic structure of PcCu-O$_8$-M (M=Co, Fe, Ni, Cu). (e) Proposed reaction mechanism. (Reproduced with permission from Ref.[120] Copyright 2019, Wiley-VCH.)

Cu–N$_4$ and Zn–N$_4$ on the N-doped carbon support (Cu/Zn–NC).[129] The corresponding DFT calculation indicates that the Cu–N$_4$ with end-on adsorption acts as the main active center for ORR. The increase of the electronegativity of the Cu center due to the electron transfer from the Zn atom to the d-orbit of the Cu atom is contributed to

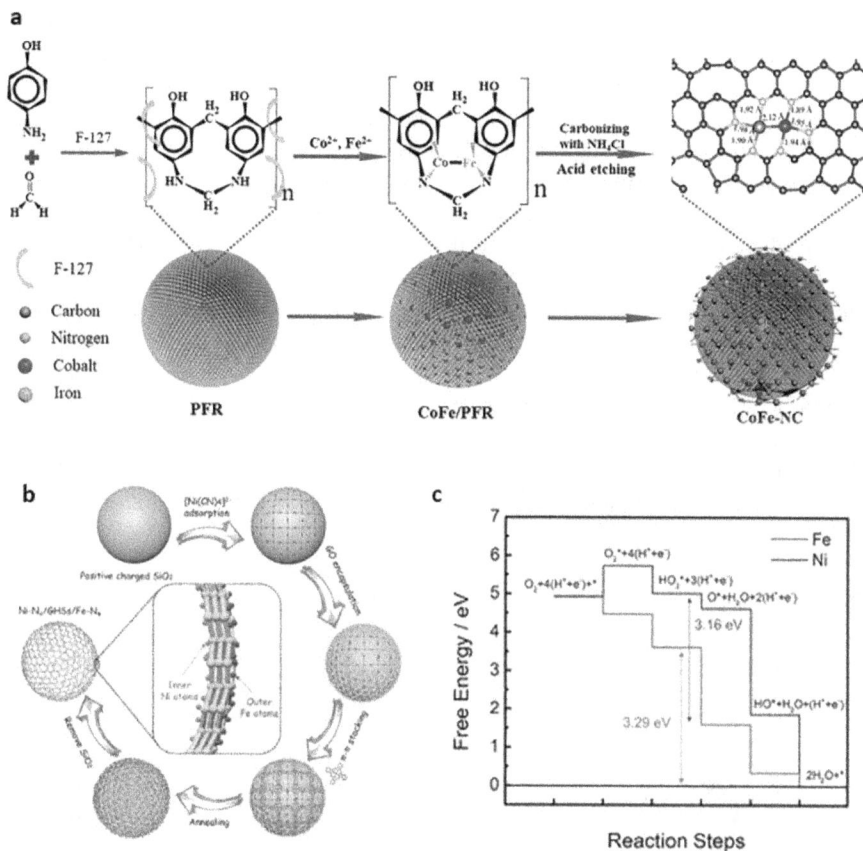

FIGURE 6.8 (a) Synthesis procedure of CoFe–NC. (b) HAADF-STEM image of CoFe–NC. (c) LSV curves in O_2-saturated 0.1 M KOH. (Reproduced with permission from Ref.[132] Copyright 2021, Royal Society of Chemistry.) (d) Synthesis procedure of SACs with dual Ni–N_4 and Fe–N_4 sites. (e) ORR and (f) OER polarization curves in O_2-saturated 0.1 m KOH. (g) The DFT-optimized adsorption configurations of reaction adsorbates. (h) Free energy diagrams of oxygen electrocatalytic reactions ($U = 0$ V). (i) PDOS of Fe–N_4 and Ni–N_4 single-atom sites. (Reproduced with permission from Ref.[131] Copyright 2020, Royal Society of Chemistry.)

the absorption of intermediates and the breaking of O–O bonds. Additionally, certain metal SACs can deliver excellent electrolytic performance for certain reactions. Most SACs with a single catalytic function, focus on either the ORR or the OER process, which cannot fulfill the requirements of the rechargeable ZABs. Since the ORR and OER activities of SACs with different transition-metal centers follow a different order, a Janus catalyst with a selective combination of single-atom is expected to endow excellent bifunctional property.[130] For instance, Ma et al. developed a step-by-step self-assembly strategy to allow NiN_4 and FeN_4 sites on the inner and outer walls of graphene hollow nanospheres, respectively (Figure 6.8b).[131] Their DFT calculations reveal that the outer Fe–N_4 dominantly contributes to efficient ORR and the inner Ni–N_4 clusters are responsible for the excellent OER (Figure 6.8c).

Recently, the synergistic effect of single-atom centered site and cluster/nanoparticle is also demonstrated to improve ORR activity.[110,133,134] For instance, Wang et al. reported an efficient and durable ORR catalyst for rechargeable ZABs that consists of atomically dispersed Co single atoms (Co–SAs) and small Co nanoparticles (Co–SNPs). Their DFT calculations reveal that the rate-determining step (RDS) for the Co–N_4 system intermediate (*OH + e⁻ → OH⁻, * represents catalytic active sites) limits the whole ORR reaction on the Co–N_4 site (Figure 6.9a and b). In the case of Co–N_4@Co_{12} and Co–N_4@$Co_{2layers}$, the lower energy barriers of the RDS (*OOH + e⁻ → *O + OH⁻) for the Co–N_4@Co_{12} indicate a more efficient ORR catalytic activity than the Co–N_4 site. The d band of the active Co atoms in Co–N_4@Co_{12} and Co–N_4@$Co_{2layers}$ shows obvious shifts of Co 3d states to lower energy levels compared to Co–N_4 (Figure 6.9c). The interaction between Co–N_4 moieties and Co_{12} clusters with optimized surface adsorption capacity of intermediates is in favor of a higher intrinsic

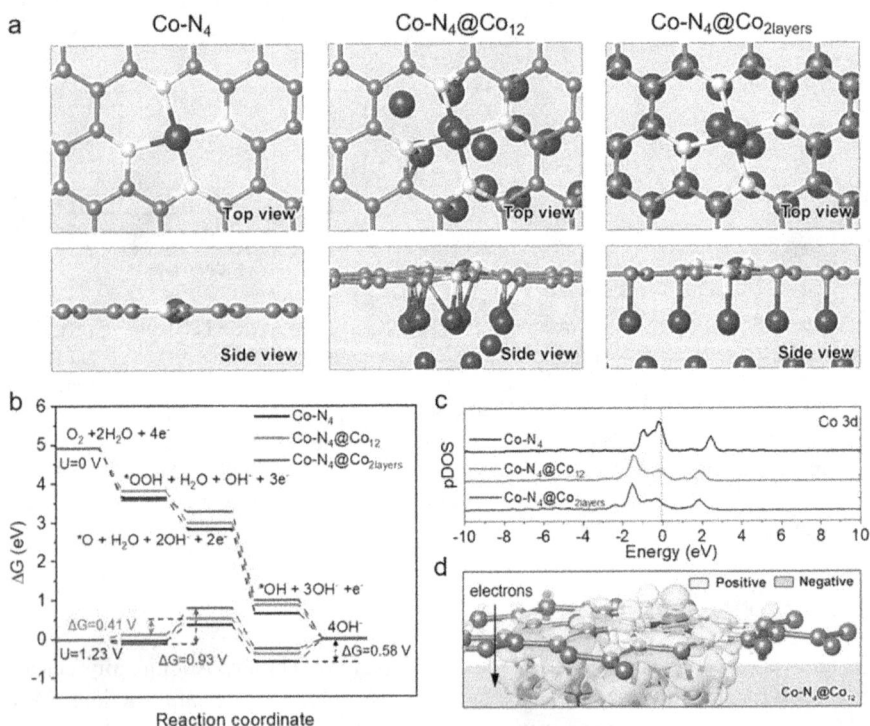

FIGURE 6.9 (a) Top views (upper panels) and side views (bottom panels) of the optimized atomic structures of Co–N_4, Co–N_4@Co_{12}, and Co–N_4@$Co_{2layers}$. (b) ORR free energy diagrams at $U = 0$, and 1.23 V along the 4-electrons pathway. (c) Calculated projected density of states (pDOS) of Co d band. (d) The charge density difference for Co–N_4@Co_{12}. (Reproduced with permission from Ref.[133] Copyright 2021, Wiley-VCH.)

ORR catalytic activity compared to the Co–N$_4$ site. However, in the case of Co–N$_4$@ Co$_{2layers}$, the too-high oxidation state of the Co site would lead to an excessive weak adsorption capacity of intermediates, which in turn impedes the ORR (Figure 6.9d). Recently, Sun et al. synthesized and constructed a FeCo–N–C catalyst containing highly active nanoparticle and M–N$_4$ composite sites (M/FeCo–SAs–N–C). The corresponding DFT reveals that there is also a strong interaction between M–NPs and FeN$_4$ sites, which can activate the O–O bond, thus facilitating a direct 4e$^-$ process.

6.8.4 Secondary Heteroatom Dopant Incorporation

Although the M–N–Cs through coordination with four N atoms have a great advantage due to their appealing properties, the strong electronegativity of the N atom is likely to undesirably increase the free energy for adsorption of reaction intermediates on the metal center. The Sabatier principle shows that the best catalysts should bond O$_2$ atoms and intermediates with optimal strength. The too strong interaction is not beneficial for the desorption of products. To further optimize the local coordination environment of the metal center, introducing secondary heteroatom dopants with relatively weak electronegativity (S and P) as the environmental atoms or coordination atoms has been studied. For instance, Feng et al. developed atomically dispersed nickel coordinated with nitrogen and sulfur species in porous carbon nanosheets for efficient OER (1.51 V at 10 mA cm^{-2} and a Tafel slope of 45 mV dec^{-1}).[135] Additionally, a single-atom copper catalyst with S atoms as the environmental atoms of Cu–N$_4$ exhibited enhanced ORR performance.[136] Recently, Chen et al. synthesized a single-atom Co catalyst doped with S (CoSA/N, S–HCS) for efficient ORR and OER (Figure 6.10a and b).[137] Compared with the sample without S doping, CoSA/N, S–HCS exhibited superior ORR and OER catalytic activity in terms of half-wave potential (0.85 V) and an overpotential (306 mV at the current density of 10 mA cm^{-2}) (Figure 6.10c). Besides, Müllen et al. developed Fe–N–Cs with S and F-co-doping as an efficient ORR catalyst with half-wave potentials of 0.91 in alkaline solution (Figure 6.10d).[138] Their DFT calculations suggested that the OH* reduction was the RDS for FeN$_4$ active sites and the ORR process on the Fe–SA–NSFC was thermodynamically favored. The co-doping of S and F significantly lowers the free energy for OH* reduction on the FeN$_4$ active sites (Figure 6.10e).

6.8.5 Substrate Selectivity and Optimization

Defects on the surface of carbon (such as vacancy or topological defects) can act as "traps" to anchor single-atom metal, stabilize metal centers and restrict undesirable agglomeration. Taking advantage of the electronic redistribution and coordination environment of the stable M–N binding, atomically dispersed M–N–C centers on the defect "holes" tend to obtain more suitable surface adsorption capacity of intermediates. Therefore, the local carbon structure surrounding FeN$_4$ moiety plays a key role in the final catalytic activity. By tailoring the coordination environment of nitrogen to get different FeN$_4$ structures, mainly including bulk and zigzag-edge-hosting, FeN$_4$ sites with different structures have been identified as an effective way to adjust the adsorption free energies of intermediates during the ORR/OER process.[139] The

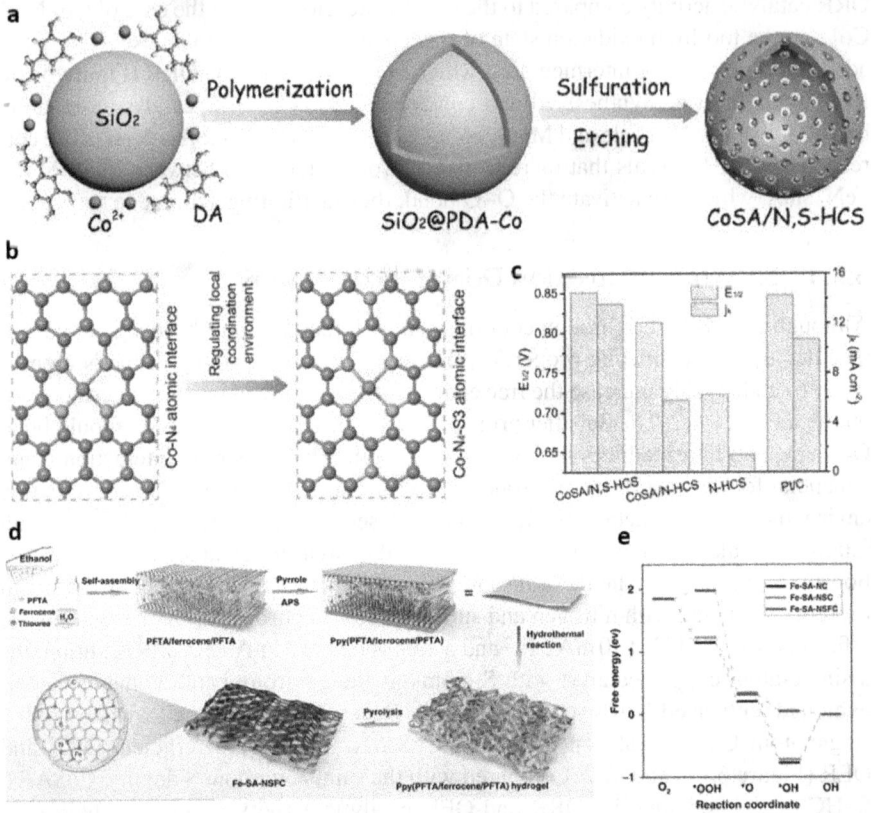

FIGURE 6.10 (a) Schematic illustration of the synthetic procedure of CoSA/N, S–HCS. (b) The atomic interface model of CoSA/N, S–HCS. (c) Comparison of half-wave potential ($E_{1/2}$) and kinetic current density (j_k). (Reproduced with permission from Ref.[137] Copyright 2020, Royal Society of Chemistry.) (d) Schematic illustration of the synthetic process for the Fe–SA–NSFC. (e) Free energy diagram for the catalysts. (Reproduced with permission from Ref.[138] Copyright 2020, Springer Nature.)

corresponding results suggest that defective zigzag-edge FeN_4 configuration may act as a highly active site. For instance, the micropores in the M-N-Cs result in the formation of more FeN_4 sites located at the edge, which exhibits superior ORR performance compared with FeN_4 sites in a plane. Meanwhile, the graphitizing degree of carbon support plays a key role in determining electrochemical stability, especially in the harsh OER process. Highly graphitic carbon nanotubes and graphene with strong oxidation resistance have been identified as the ideal supports of the single atoms. However, integrating the more favorable FeN_4 sites into a highly graphitic carbon substrate to achieve bifunctionality while maintaining high catalytic activity is a significant challenge. Recently, Chen et al. developed a self-sacrificed template approach to integrate edge-enriched FeN_4 sites in the highly graphitic carbon nanosheets.[140] The Fe clusters formed during the synthesis process catalyze the growth of graphitic carbon and induce the preferential anchoring of FeN_4 surrounding porous structure. The as-synthesized catalyst shows superior catalytic activity and stability for the ORR (half-wave potential of 0.89 V) and OER (an overpotential of 370 mV at 10 mA cm^{-2}). The assembled rechargeable ZABs exhibited a high energy efficiency and enhanced cycling stability for over 240 cycles. Liu et al. developed a cobalt single-atom anchored on nitrogen-doped graphene-sheet@tube (CoSAs-NGST) to study the morphology effects on the surficial electric structure (Figure 6.11a).[141] The hybrid structure with a bamboo-like graphene tube and sheet not only enhances the dispersity of single atoms but also induces defect state evolution.

Their DFT modeling demonstrates that the coupling effect of $Co-N_4$-tube and $Co-N_4$-sheet contributes to enhanced ORR and OER activities (Figure 6.11b–d). The excellent bifunctional catalytic performance of CoSAs–NGST exhibits enhanced

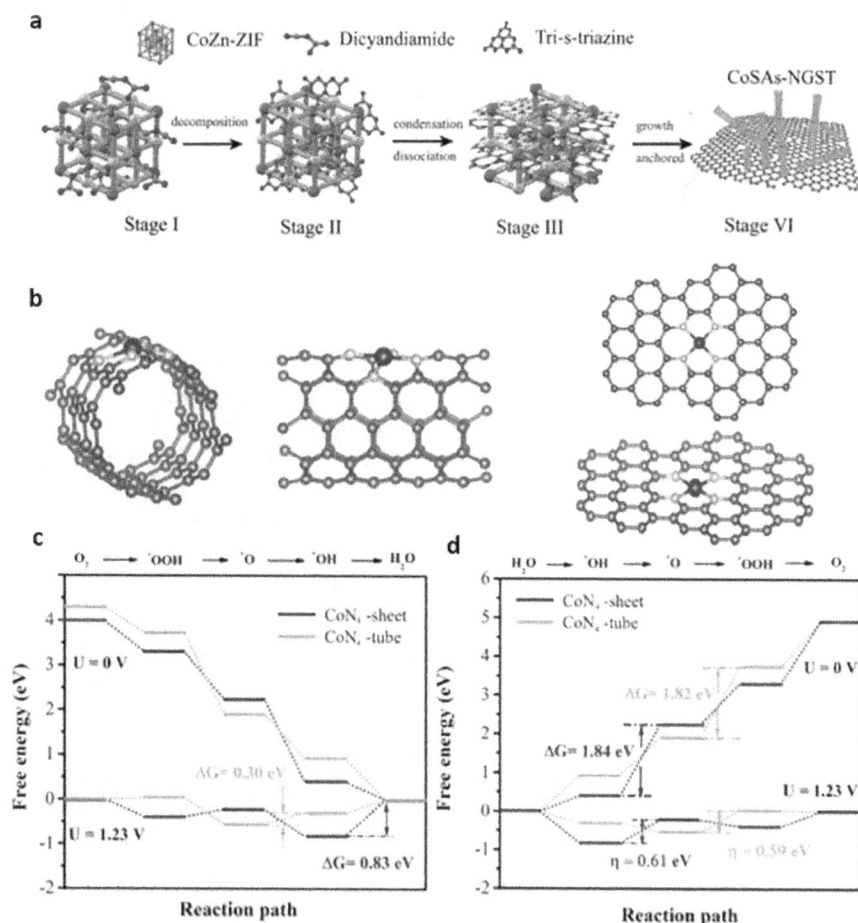

FIGURE 6.11 (a) Schematic illustration of the synthesis procedure of CoSAs-NGST. (b) DFT modeling of CoN_4 on a tube and a sheet. Free energy diagram of (c) ORR and (d) OER process on two models. (Reproduced with permission from Ref.[141] Copyright 2020, Wiley-VCH.)

stability in ZABs and shows a small voltage gap of 0.93 V at 5 mA cm^{-2}. So far, most of the SACs with M–N–C moieties are based on carbonaceous substrates (amorphous carbon, graphene, carbon nanotubes, carbon nitrides, etc.), which usually show improvements in ORR. However, the carbonaceous catalysts for the OER process still suffer from insufficient activity and durability due to the slow oxidization of the carbonaceous matrix under high potential (>1.8 V).[142] Metal carbides with hybridization between the d-orbitals of transition metals and p-orbitals of carbon has also been well studied as a promising substrate.[143] Most recently, Wang et al. developed a durable and conductive tungsten carbide (WC$_x$) support where the atomic FeNi catalytic sites are weakly bonded with the surface W and C atoms (Figure 6.12a). They believe that the formation of heteroatom bonds of the metal atoms, such as metal-N$_x$–C$_y$, could have unfavorably influenced the catalytic activity. Stabilizing catalytic metal atoms on the surface of WC$_x$ support without the aid of strong heteroatom coordination is expected to further improve catalytic activity and durability. They found that these atomically dispersed FeNi atoms are supported on the surface of WC$_x$ by metal-metal interactions (Figure 6.12b and c). Moreover, they used the DFT calculations to elucidate the origin of activity and the impacts of the surface oxidation on the OER performance. The results show that WC$_x$–FeNi exhibits the lowest OER overpotential (0.16 V) compared to WC$_x$–Ni and WC$_x$–Fe catalysts (Figure 6.12c and d). Moreover, the synergistic effect between Fe and Ni in the O-bridged FeNi moieties formed during the OER further improves OER catalytic activity (Figure 6.12f and g).

6.9   CONCLUSION AND OUTLOOK

SACs with maximum atom-utilization efficiency and tunable active sites offer major advantages in electrocatalysis for bifunctional ORR and OER. The unsaturated environment of metal single atoms enables suitable adsorption sites of reactant molecules while the EMSI greatly enhances their intrinsic activity by regulating the electronic structure. As the metal materials downsize to the atomic scale, the induced strong EMSI can determine the rate of reaction. Due to the significantly different effects of various EMSI on the electrocatalysis activity, stability, and selectivity, it is necessary to deeply explore the mechanism with the help of advanced ex/in-situ characterization techniques and theoretical calculations. SACs, bridging the fields of homogeneous and heterogeneous catalysis, provide well-defined model active centers which allow a deep understanding of the catalytic reactions at the atomic level. The powerful theoretical calculations combined with the experimental characterization techniques enable us to study the EMSI directly and obtain much detailed electronic information on the reaction. In this review, we summarized recent advances of the SACs as the bifunctional oxygen electrocatalysis and their application in rechargeable ZABs. Despite the huge achievements that have been made in the study of synthesis and mechanism of SACs, the progress of SACs in the bifunctional air electrode of ZABs is not enough. By systematically discussing the challenges and reaction mechanisms of SACs for oxygen electrocatalysis, we review the importance and design strategies of EMSI for enhanced oxygen electrocatalysis. Several major challenges and subjects of SACs for bifunctional oxygen electrocatalysis should be given more attention:

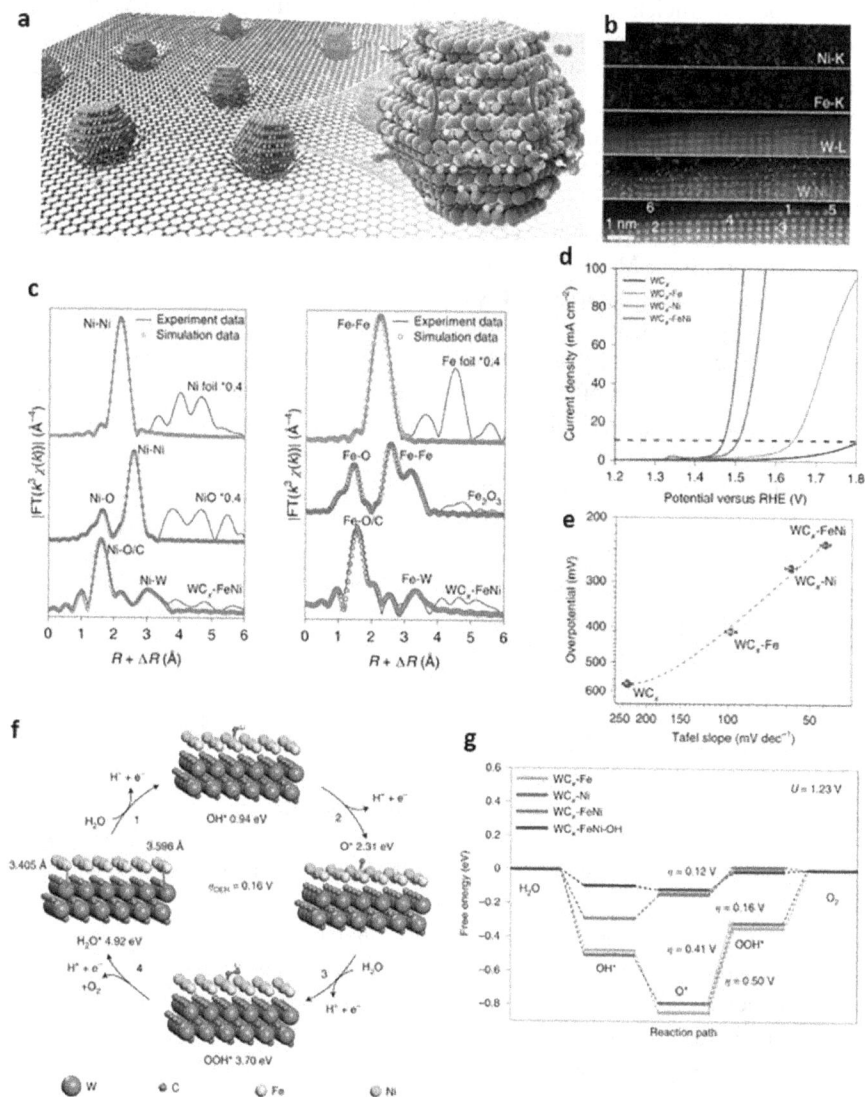

FIGURE 6.12 (a) Schematic illustration of Fe and Ni atoms stabilized on WC_x nanocrystallites. (b) STEM spectrum imaging of the edge of the nanocrystallite. (c) k^3-weighted FT spectra of the Ni K-edge and the Fe K-edge. (d) The OER polarization curves of different catalysts. (e) The OER data analysis of overpotentials at the current density of $10\,mA\,cm^{-2}$ and Tafel slopes. (f) The reaction paths on WC_x–FeNi and the corresponding free energies. (g) The reaction paths of OER at a set potential of 1.23 V. (Reproduced with permission from[144] Copyright 2021, Springer Nature.)

1. To obtain efficient ORR and OER activity, the synthesis of SACs with high loading of single metal atoms is crucial. Increasing the metal loading of SACs is effective to enhance the catalytic performance. However, the high surface free energy of single atoms leads to the migration and agglomeration of metal atoms to form nanoparticles. More attention should be paid to the development of new approaches for fabricating SACs with high metal loading.

2. The air electrode of the rechargeable ZABs demands more types of support materials for the M N–C's Carbon corrosion of the carbon-based materials, especially under the harsh OER process, results in significant charge- and mass transport resistances. Single atoms anchored on the carbon substrates, with a high degree of graphitization, play a key role in enhancing their stability. The exploration of new approaches for improving the degree of graphitization without sacrificing the metal loading is challenging. Therefore, developing new advanced support materials with high conductivity and corrosion resistance for supporting the high loading of single atoms is highly desired.

3. Modulation of the MOF-based materials or assembly of the metal phthalocyanine seems to be a promising route to facilitate oxygen electrocatalysis and maintain high stability. However, it remains a great challenge to control active centers, precisely with reversible oxygen reactions. More efforts should be devoted to exploring new ligands and synthesis methods to regulate their electronic states and structural properties.

4. For the synthesis of SACs, structural modifications (including center metal atoms, coordinated atoms, environmental atoms, guest groups, and carbon support) would lead to the manipulation of the electronic structures of the active centers. The precise modulation of the atomic configurations of the active center plays an important role in determining the reaction dynamics and eventual electrocatalytic activity and stability. The coexistence of multi-metal sites or multi-heteroatom dopants in SACs provides more opportunities for bifunctional oxygen electrocatalysis.

5. The design of SACs for bifunctional oxygen electrocatalysis is not limited to the known M–N–C sites. To prepare SAs on different supports ((hydro)oxides, sulfides, selenides, alloy, etc.), the generation of various defects, such as oxygen vacancy, metal vacancy, lattice distortion, and strain change, will affect the microscopic coordination environment. The related explorations of the effects on the bifunctional oxygen electrocatalysis are significant but rarely studied.

6. The understanding of the effect of EMSI of SACs on oxygen electrocatalysis under reaction conditions is still limited, due to the difficulty in detection of the intermediates adsorbed on active sites. Although some in-situ or operando characterization techniques have been developed, it is still a challenging task to accurately probe the transient state occurring on the active sites. Therefore, advanced techniques with faster detection should be developed to study the effect of EMSI on the elementary reactions of oxygen electrocatalysis.

REFERENCES

1. M. Wu, G. Zhang, L. Du, D. Yang, H. Yang and S. Sun, *Small Methods*, 2020, **5**, 2000868.
2. Y. Li, Q. Li, H. Wang, L. Zhang, D. P. Wilkinson and J. Zhang, *Electrochem. Energy Rev.*, 2019, **2**, 518–538.
3. L. Peng, L. Shang, T. Zhang and G. I. N. Waterhouse, *Adv. Energy Mater.*, 2020, **10**, 2003018.
4. M. Wu, G. Zhang, M. Wu, J. Prakash and S. Sun, *Energy Storage Mater.*, 2019, **21**, 253–286.
5. Y. Chen, S. Ji, C. Chen, Q. Peng, D. Wang and Y. Li, *Joule*, 2018, **2**, 1242–1264.
6. J. Wang, Z. Huang, L. Lu, Q. Jia, L. Huang, S. Chang, M. Zhang, Z. Zhang, S. Li, D. He, W. Wu, S. Zhang, N. Toshima and H. Zhang, *Green Chem.*, 2020, **22**, 1269–1274.
7. Y. Wang, F. Chu, J. Zeng, Q. Wang, T. Naren, Y. Li, Y. Cheng, Y. Lei and F. Wu, *ACS Nano*, 2021, **15**, 210–239.
8. S. Sun, G. Zhang, N. Gauquelin, N. Chen, J. Zhou, S. Yang, W. Chen, X. Meng, D. Geng, M. N. Banis, R. Li, S. Ye, S. Knights, G. A. Botton, T.-K. Sham and X. Sun, *Sci. Rep.*, 2013, **3**.
9. M. Li, K. Duanmu, C. Wan, T. Cheng, L. Zhang, S. Dai, W. Chen, Z. Zhao, P. Li, H. Fei, Y. Zhu, R. Yu, J. Luo, K. Zang, Z. Lin, M. Ding, J. Huang, H. Sun, J. Guo, X. Pan, W. A. Goddard, P. Sautet, Y. Huang and X. Duan, *Nat. Catal.*, 2019, **2**, 495–503.
10. B. Wu, T. Lin, R. Yang, M. Huang, H. Zhang, J. Li, F. Sun, F. Song, Z. Jiang, L. Zhong and Y. Sun, *Green Chem.*, 2021, **23**, 4753–4761.
11. A. Han, B. Wang, A. Kumar, Y. Qin, J. Jin, X. Wang, C. Yang, B. Dong, Y. Jia, J. Liu and X. Sun, *Small Methods*, 2019, **3**, 1800471.
12. D. Gao, T. Liu, G. Wang and X. Bao, *ACS Energy Lett.*, 2021, **6**, 713–727.
13. J. Zhang, H. Yang and B. Liu, *Adv. Energy Mater.*, 2020, **11**, 2002473.
14. N. Cheng, L. Zhang, K. Doyle-Davis and X. Sun, *Electrochem. Energy Rev.*, 2019, **2**, 539–573.
15. C. X. Zhao, B. Q. Li, J. N. Liu and Q. Zhang, *Angew. Chem. Int. Ed. Engl.*, 2021, **60**, 4448–4463.
16. Y. Zhang, L. Guo, L. Tao, Y. Lu and S. Wang, *Small Methods*, 2018, **3**, 1800406.
17. F. Dong, M. Wu, G. Zhang, X. Liu, D. Rawach, A. C. Tavares and S. Sun, *Chem. Asian J.*, 2020, **15**, 3737–3751.
18. L. Li, C. Liu, G. He, D. Fan and A. Manthiram, *Energy Environ. Sci.*, 2015, **8**, 3274–3282.
19. J. Pan, Y. Y. Xu, H. Yang, Z. Dong, H. Liu and B. Y. Xia, *Adv. Sci.*, 2018, **5**, 1700691.
20. X. Cai, L. Lai, J. Lin and Z. Shen, *Mater. Horiz.*, 2017, **4**, 945–976.
21. Y. Li and H. Dai, *Chem. Soc. Rev.*, 2014, **43**, 5257–5275.
22. T. Sun, S. Mitchell, J. Li, P. Lyu, X. Wu, J. Perez-Ramirez and J. Lu, *Adv. Mater.*, 2021, **33**, e2003075.
23. X. Liu, X. Fan, B. Liu, J. Ding, Y. Deng, X. Han, C. Zhong and W. Hu, *Adv. Mater.*, 2021, e2006461. DOI: 10.1002/adma.202006461.
24. R. F. Service, *Science*, 2021, **372**, 890–891.
25. C. Wan, X. Duan and Y. Huang, *Adv. Energy Mater.*, 2020, **10**, 1903815.
26. N. Ramaswamy, U. Tylus, Q. Jia and S. Mukerjee, *J. Am. Chem. Soc.*, 2013, **135**, 15443–15449.
27. S. Takase, Y. Aoto, D. Ikeda, H. Wakita and Y. Shimizu, *Electrocatalysis*, 2019, **10**, 653–662.
28. Z. Teng, Q. Zhang, H. Yang, K. Kato, W. Yang, Y.-R. Lu, S. Liu, C. Wang, A. Yamakata, C. Su, B. Liu and T. Ohno, *Nat. Catal.*, 2021, **4**, 374–384.
29. L. Zhong and S. Li, *ACS Catal.*, 2020, **10**, 4313–4318.

30. X. Zhang, Y. Xia, C. Xia and H. Wang, *Trends Chem.*, 2020, **2**, 942–953.
31. Y. Jiang, P. Ni, C. Chen, Y. Lu, P. Yang, B. Kong, A. Fisher and X. Wang, *Adv. Energy Mater.*, 2018, **8**, 1801909.
32. Y. Wang, G. I. N. Waterhouse, L. Shang and T. Zhang, *Adv. Energy Mater.*, 2020, **11**, 2003323.
33. K. Jiang, S. Back, A. J. Akey, C. Xia, Y. Hu, W. Liang, D. Schaak, E. Stavitski, J. K. Norskov, S. Siahrostami and H. Wang, *Nat. Commun.*, 2019, **10**, 3997.
34. V. Viswanathan, H. A. Hansen, J. Rossmeisl and J. K. Norskov, *J. Phys. Chem. Lett.*, 2012, **3**, 2948–2951.
35. L. Liu and A. Corma, *Chem. Rev.*, 2018, **118**, 4981–5079.
36. X. F. Yang, A. Wang, B. Qiao, J. Li, J. Liu and T. Zhang, *Acc. Chem Res.*, 2013, **46**, 1740–1748.
37. L. Jiao, R. Zhang, G. Wan, W. Yang, X. Wan, H. Zhou, J. Shui, S. H. Yu and H. L. Jiang, *Nat. Commun.*, 2020, **11**, 2831.
38. J. Wu, L. Xiong, B. Zhao, M. Liu and L. Huang, *Small Methods*, 2019, **4**, 1900540.
39. J. Li, Q. Guan, H. Wu, W. Liu, Y. Lin, Z. Sun, X. Ye, X. Zheng, H. Pan, J. Zhu, S. Chen, W. Zhang, S. Wei and J. Lu, *J. Am. Chem. Soc.*, 2019, **141**, 14515–14519.
40. S. K. Sahoo, Y. Ye, S. Lee, J. Park, H. Lee, J. Lee and J. W. Han, *ACS Energy Lett.*, 2018, **4**, 126–132.
41. S. Li, B. Chen, Y. Wang, M.-Y. Ye, P. A. van Aken, C. Cheng and A. Thomas, *Nat. Mater.*, 2021. DOI: 10.1038/s41563-021-01006-2.
42. R. Lang, X. Du, Y. Huang, X. Jiang, Q. Zhang, Y. Guo, K. Liu, B. Qiao, A. Wang and T. Zhang, *Chem. Rev.*, 2020, **120**, 11986–12043.
43. S. Zhao, G. Chen, G. Zhou, L. C. Yin, J. P. Veder, B. Johannessen, M. Saunders, S. Z. Yang, R. De Marco, C. Liu and S. P. Jiang, *Adv. Funct. Mater.*, 2019, **30**, 1906157.
44. J. Li, M. Chen, D. A. Cullen, S. Hwang, M. Wang, B. Li, K. Liu, S. Karakalos, M. Lucero, H. Zhang, C. Lei, H. Xu, G. E. Sterbinsky, Z. Feng, D. Su, K. L. More, G. Wang, Z. Wang and G. Wu, *Nat. Catal.*, 2018, **1**, 935–945.
45. D. Yang, H. Tan, X. Rui and Y. Yu, *Electrochem. Energy Rev.*, 2019, **2**, 395–427.
46. D. Yang, D. Chen, Y. Jiang, E. H. Ang, Y. Feng, X. Rui and Y. Yu, *Carbon Energy*, 2020, **3**, 50–65.
47. M. Wu, G. Zhang, Y. Hu, J. Wang, T. Sun, T. Regier, J. Qiao and S. Sun, *Carbon Energy*, 2020, **3**, 176–187.
48. L. Jiao and H.-L. Jiang, *Chem*, 2019, **5**, 786–804.
49. A. Wang, J. Li and T. Zhang, *Nat. Rev. Chem.*, 2018, **2**, 65–81.
50. L. Zhang, M. N. Banis and X. Sun, *Natl. Sci. Rev.*, 2018, **5**, 628–630.
51. K. C. Kwon, J. M. Suh, R. S. Varma, M. Shokouhimehr and H. W. Jang, *Small Methods*, 2019, **3**, 1800492.
52. X. I. Pereira-Hernandez, A. DeLaRiva, V. Muravev, D. Kunwar, H. Xiong, B. Sudduth, M. Engelhard, L. Kovarik, E. J. M. Hensen, Y. Wang and A. K. Datye, *Nat. Commun.*, 2019, **10**, 1358.
53. K. Liu, X. Zhao, G. Ren, T. Yang, Y. Ren, A. F. Lee, Y. Su, X. Pan, J. Zhang, Z. Chen, J. Yang, X. Liu, T. Zhou, W. Xi, J. Luo, C. Zeng, H. Matsumoto, W. Liu, Q. Jiang, K. Wilson, A. Wang, B. Qiao, W. Li and T. Zhang, *Nat. Commun.*, 2020, **11**, 1263.
54. D. Liu, Q. He, S. Ding and L. Song, *Adv. Energy Mater.*, 2020, **10**, 2001482.
55. Z. Zhang, C. Feng, C. Liu, M. Zuo, L. Qin, X. Yan, Y. Xing, H. Li, R. Si, S. Zhou and J. Zeng, *Nat. Commun.*, 2020, **11**, 1215.
56. B. W. Zhang, Y. X. Wang, S. L. Chou, H. K. Liu and S. X. Dou, *Small Methods*, 2019, **3**, 1800497.
57. Y. Qu, Z. Li, W. Chen, Y. Lin, T. Yuan, Z. Yang, C. Zhao, J. Wang, C. Zhao, X. Wang, F. Zhou, Z. Zhuang, Y. Wu and Y. Li, *Nat. Catal.*, 2018, **1**, 781–786.

58. J. Xi, H. S. Jung, Y. Xu, F. Xiao, J. W. Bae and S. Wang, *Adv. Funct. Mater.*, 2021, **31**, 2008318.
59. G. Wan, P. Yu, H. Chen, J. Wen, C. J. Sun, H. Zhou, N. Zhang, Q. Li, W. Zhao, B. Xie, T. Li and J. Shi, *Small*, 2018, **14**, e1704319.
60. P. Yin, T. Yao, Y. Wu, L. Zheng, Y. Lin, W. Liu, H. Ju, J. Zhu, X. Hong, Z. Deng, G. Zhou, S. Wei and Y. Li, *Angew. Chem. Int. Ed. Engl.*, 2016, **55**, 10800–10805.
61. M. Wu, G. Zhang, N. Chen, Y. Hu, T. Regier, D. Rawach and S. Sun, *ACS Energy Lett.*, 2021, 1153–1161. DOI: 10.1021/acsenergylett.1c00037.
62. M. Wu, Q. Wei, G. Zhang, J. Qiao, M. Wu, J. Zhang, Q. Gong and S. Sun, *Adv. Energy Mater.*, 2018, **8**, 1801836.
63. N. Daelman, M. Capdevila-Cortada and N. Lopez, *Nat. Mater.*, 2019, **18**, 1215–1221.
64. L. Wang, L. Huang, F. Liang, S. Liu, Y. Wang and H. Zhang, *Chinese J. Catal.*, 2017, **38**, 1528–1539.
65. L. Zhao, Y. Zhang, L. B. Huang, X. Z. Liu, Q. H. Zhang, C. He, Z. Y. Wu, L. J. Zhang, J. Wu, W. Yang, L. Gu, J. S. Hu and L. J. Wan, *Nat. Commun.*, 2019, **10**, 1278.
66. L. Fang, S. Seifert, R. E. Winans and T. Li, *Small Methods*, 2021, **5**, 2001194.
67. Y. Shang, X. Xu, B. Gao, S. Wang and X. Duan, *Chem. Soc. Rev.*, 2021, **50**, 5281–5322.
68. S. Hoang, S. Guo, A. J. Binder, W. Tang, S. Wang, J. J. Liu, H. Tran, X. Lu, Y. Wang, Y. Ding, E. A. Kyriakidou, J. Yang, T. J. Toops, T. R. Pauly, R. Ramprasad and P. X. Gao, *Nat. Commun.*, 2020, **11**, 1062.
69. F. Maurer, J. Jelic, J. Wang, A. Gänzler, P. Dolcet, C. Wöll, Y. Wang, F. Studt, M. Casapu and J.-D. Grunwaldt, *Nat. Catal.*, 2020, **3**, 824–833.
70. J. Li and J. Gong, *Energy Environ. Sci.*, 2020, **13**, 3748–3779.
71. J. Yang, W. Li, D. Wang and Y. Li, *Adv. Mater.*, 2020, **32**, e2003300.
72. S. Tauster, *J. Catal.*, 1978, **55**, 29–35.
73. C. T. Campbell, *Nat. Chem.*, 2012, **4**, 597–598.
74. S. Zhang, Z. Xia, T. Ni, Z. Zhang, Y. Ma and Y. Qu, *J. Catal.*, 2018, **359**, 101–111.
75. Y. Shi, Z. R. Ma, Y. Y. Xiao, Y. C. Yin, W. M. Huang, Z. C. Huang, Y. Z. Zheng, F. Y. Mu, R. Huang, G. Y. Shi, Y. Y. Sun, X. H. Xia and W. Chen, *Nat. Commun.*, 2021, **12**, 3021.
76. J. Li, L. Zhang, K. Doyle-Davis, R. Li and X. Sun, *Carbon Energy*, 2020, **2**, 488–520.
77. C. Gao, J. Low, R. Long, T. Kong, J. Zhu and Y. Xiong, *Chem. Rev.*, 2020, **120**, 12175–12216.
78. H. Zhang, G. Liu, L. Shi and J. Ye, *Adv. Energy Mater.*, 2018, **8**, 1701343.
79. M. T. Greiner, T. E. Jones, S. Beeg, L. Zwiener, M. Scherzer, F. Girgsdies, S. Piccinin, M. Armbruster, A. Knop-Gericke and R. Schlogl, *Nat. Chem.*, 2018, **10**, 1008–1015.
80. Z. Song, L. Zhang, K. Doyle-Davis, X. Fu, J. L. Luo and X. Sun, *Adv. Energy Mater.*, 2020, **10**, 2001561.
81. Y. Zheng and S.-Z. Qiao, *Natl. Sci. Rev.*, 2018, **5**, 626–627.
82. S. Li, Y. Gao, N. Li, L. Ge, X. Bu and P. Feng, *Energy Environ. Sci.*, 2021, **14**, 1897–1927.
83. W. Wan, C. A. Triana, J. Lan, J. Li, C. S. Allen, Y. Zhao, M. Iannuzzi and G. R. Patzke, *ACS Nano*, 2020, **14**, 13279–13293.
84. B. Q. Li, C. X. Zhao, S. Chen, J. N. Liu, X. Chen, L. Song and Q. Zhang, *Adv. Mater.*, 2019, **31**, e1900592.
85. X. Zhang, Y. Wang, M. Gu, M. Wang, Z. Zhang, W. Pan, Z. Jiang, H. Zheng, M. Lucero, H. Wang, G. E. Sterbinsky, Q. Ma, Y.-G. Wang, Z. Feng, J. Li, H. Dai and Y. Liang, *Nat. Energy*, 2020, **5**, 684–692.
86. X. Zhang, Y. Feng, S. Tang and W. Feng, *Carbon*, 2010, **48**, 211–216.
87. Y. Cheng, S. Yang, S. P. Jiang and S. Wang, *Small Methods*, 2019, **3**, 1800440.
88. R. Gao, J. Wang, Z.-F. Huang, R. Zhang, W. Wang, L. Pan, J. Zhang, W. Zhu, X. Zhang, C. Shi, J. Lim and J.-J. Zou, *Nat. Energy*, 2021. DOI: 10.1038/s41560-021-00826-5.

89. G. Giannakakis, M. Flytzani-Stephanopoulos and E. C. H. Sykes, *Acc. Chem. Res.*, 2019, **52**, 237–247.
90. J. Zhang, C. Liu and B. Zhang, *Small Methods*, 2019, **3**, 1800481.
91. T. W. van Deelen, C. Hernández Mejía and K. P. de Jong, *Nat. Catal.*, 2019, **2**, 955–970.
92. H. Y. Zhuo, X. Zhang, J. X. Liang, Q. Yu, H. Xiao and J. Li, *Chem. Rev.*, 2020, **120**, 12315–12341.
93. N. Zhang, T. Zhou, M. Chen, H. Feng, R. Yuan, C. A. Zhong, W. Yan, Y. Tian, X. Wu, W. Chu, C. Wu and Y. Xie, *Energy Environ. Sci.*, 2020, **13**, 111–118.
94. D. Lyu, Y. B. Mollamahale, S. Huang, P. Zhu, X. Zhang, Y. Du, S. Wang, M. Qing, Z. Q. Tian and P. K. Shen, *J. Catal.*, 2018, **368**, 279–290.
95. B. Lu, Q. Liu and S. Chen, *ACS Catal.*, 2020, **10**, 7584–7618.
96. C. Zhu, Q. Shi, S. Feng, D. Du and Y. Lin, *ACS Energy Lett.*, 2018, **3**, 1713–1721.
97. H. Peng, F. Liu, X. Liu, S. Liao, C. You, X. Tian, H. Nan, F. Luo, H. Song, Z. Fu and P. Huang, *ACS Catal.*, 2014, **4**, 3797–3805.
98. H. Fei, J. Dong, Y. Feng, C. S. Allen, C. Wan, B. Volosskiy, M. Li, Z. Zhao, Y. Wang, H. Sun, P. An, W. Chen, Z. Guo, C. Lee, D. Chen, I. Shakir, M. Liu, T. Hu, Y. Li, A. I. Kirkland, X. Duan and Y. Huang, *Nat. Catal.*, 2018, **1**, 63–72.
99. L. Yang, D. Cheng, H. Xu, X. Zeng, X. Wan, J. Shui, Z. Xiang and D. Cao, *Proc. Natl. Acad. Sci.*, 2018, **115**, 6626–6631.
100. J. Wang, Z. Huang, W. Liu, C. Chang, H. Tang, Z. Li, W. Chen, C. Jia, T. Yao, S. Wei, Y. Wu and Y. Li, *J. Am. Chem. Soc.*, 2017, **139**, 17281–17284.
101. F. Wang, W. Xie, L. Yang, D. Xie and S. Lin, *J. Catal.*, 2021, **396**, 215–223.
102. Y. Chen, R. Gao, S. Ji, H. Li, K. Tang, P. Jiang, H. Hu, Z. Zhang, H. Hao, Q. Qu, X. Liang, W. Chen, J. Dong, D. Wang and Y. Li, *Angew. Chem. Int. Ed. Engl.*, 2021, **60**, 3212–3221.
103. M. Wang, W. Yang, X. Li, Y. Xu, L. Zheng, C. Su and B. Liu, *ACS Energy Lett.*, 2021, **6**, 379–386.
104. M. Wang, H. Ji, S. Liu, H. Sun, J. Liu, C. Yan and T. Qian, *Chem. Eng. J.*, 2020, **393**, 124702.
105. J. Zhang, Y. Zhao, C. Chen, Y. C. Huang, C. L. Dong, C. J. Chen, R. S. Liu, C. Wang, K. Yan, Y. Li and G. Wang, *J. Am. Chem. Soc.*, 2019, **141**, 20118–20126.
106. X. Zeng, J. Shui, X. Liu, Q. Liu, Y. Li, J. Shang, L. Zheng and R. Yu, *Adv. Energy Mater.*, 2018, **8**, 1701345.
107. D. Qi, Y. Liu, M. Hu, X. Peng, Y. Qiu, S. Zhang, W. Liu, H. Li, G. Hu, L. Zhuo, Y. Qin, J. He, G. Qi, J. Sun, J. Luo and X. Liu, *Small*, 2020, **16**, e2004855.
108. M. Tavakkoli, E. Flahaut, P. Peljo, J. Sainio, F. Davodi, E. V. Lobiak, K. Mustonen and E. I. Kauppinen, *ACS Catal.*, 2020, **10**, 4647–4658.
109. L. Yang, L. Shi, D. Wang, Y. Lv and D. Cao, *Nano Energy*, 2018, **50**, 691–698.
110. S. H. Yin, J. Yang, Y. Han, G. Li, L. Y. Wan, Y. H. Chen, C. Chen, X. M. Qu, Y. X. Jiang and S. G. Sun, *Angew. Chem.*, 2020, **132**, 22160–22163.
111. Q. K. Li, X. F. Li, G. Zhang and J. Jiang, *J. Am. Chem. Soc.*, 2018, **140**, 15149–15152.
112. J. Feng, L. Zheng, C. Jiang, Z. Chen, L. Liu, S. Zeng, L. Bai, S. Zhang and X. Zhang, *Green Chem.*, 2021, DOI: 10.1039/d1gc01914g.
113. Y. Chen, S. Ji, S. Zhao, W. Chen, J. Dong, W. C. Cheong, R. Shen, X. Wen, L. Zheng, A. I. Rykov, S. Cai, H. Tang, Z. Zhuang, C. Chen, Q. Peng, D. Wang and Y. Li, *Nat. Commun.*, 2018, **9**, 5422.
114. G. Qin, K. R. Hao, Q. B. Yan, M. Hu and G. Su, *Nanoscale*, 2019, **11**, 5798–5806.
115. D. Liu, J. C. Li, S. Ding, Z. Lyu, S. Feng, H. Tian, C. Huyan, M. Xu, T. Li, D. Du, P. Liu, M. Shao and Y. Lin, *Small Methods*, 2020, **4**, 1900827.
116. Y. Pan, S. Liu, K. Sun, X. Chen, B. Wang, K. Wu, X. Cao, W. C. Cheong, R. Shen, A. Han, Z. Chen, L. Zheng, J. Luo, Y. Lin, Y. Liu, D. Wang, Q. Peng, Q. Zhang, C. Chen and Y. Li, *Angew. Chem. Int. Ed. Engl.*, 2018, **57**, 8614–8618.

117. Y. Lian, W. Yang, C. Zhang, H. Sun, Z. Deng, W. Xu, L. Song, Z. Ouyang, Z. Wang, J. Guo and Y. Peng, *Angew. Chem. Int. Ed. Engl.*, 2020, **59**, 286–294.
118. Z. Xu, W. Deng and X. Wang, *Electrochem. Energy Rev.*, 2021, **4**, 269–335.
119. P. Peng, L. Shi, F. Huo, C. Mi, X. Wu, S. Zhang and Z. Xiang, *Sci. Adv.*, 2019, **5**, eaaw2322.
120. H. Zhong, K. H. Ly, M. Wang, Y. Krupskaya, X. Han, J. Zhang, J. Zhang, V. Kataev, B. Büchner, I. M. Weidinger, S. Kaskel, P. Liu, M. Chen, R. Dong and X. Feng, *Angew. Chem.*, 2019, **131**, 10787–10792.
121. Y. Wang, Z. Li, P. Zhang, Y. Pan, Y. Zhang, Q. Cai, S. R. P. Silva, J. Liu, G. Zhang, X. Sun and Z. Yan, *Nano Energy*, 2021, **87**, 106147.
122. M. Xiao, H. Zhang, Y. Chen, J. Zhu, L. Gao, Z. Jin, J. Ge, Z. Jiang, S. Chen, C. Liu and W. Xing, *Nano Energy*, 2018, **46**, 396–403.
123. L. Cao, Y. Shao, H. Pan and Z. Lu, *J. Phys. Chem. C*, 2020, **124**, 11301–11307.
124. J. Xu, S. Lai, D. Qi, M. Hu, X. Peng, Y. Liu, W. Liu, G. Hu, H. Xu, F. Li, C. Li, J. He, L. Zhuo, J. Sun, Y. Qiu, S. Zhang, J. Luo and X. Liu, *Nano Res.*, 2020, **14**, 1374–1381.
125. T. He, A. R. Puente Santiago and A. Du, *J. Catal.*, 2020, **388**, 77–83.
126. J. Wang, W. Liu, G. Luo, Z. Li, C. Zhao, H. Zhang, M. Zhu, Q. Xu, X. Wang, C. Zhao, Y. Qu, Z. Yang, T. Yao, Y. Li, Y. Lin, Y. Wu and Y. Li, *Energy Environ. Sci.*, 2018, **11**, 3375–3379.
127. D. Zhang, W. Chen, Z. Li, Y. Chen, L. Zheng, Y. Gong, Q. Li, R. Shen, Y. Han, W. C. Cheong, L. Gu and Y. Li, *Chem. Commun.*, 2018, **54**, 4274–4277.
128. Q. Shi, Y. He, X. Bai, M. Wang, D. A. Cullen, M. Lucero, X. Zhao, K. L. More, H. Zhou, Z. Feng, Y. Liu and G. Wu, *Energy Environ. Sci.*, 2020, **13**, 3544–3555.
129. M. Tong, F. Sun, Y. Xie, Y. Wang, Y. Yang, C. Tian, L. Wang and H. Fu, *Angew. Chem. Int. Ed. Engl.*, 2021. DOI: 10.1002/anie.202102053.
130. K. H. Roh, D. C. Martin and J. Lahann, *Nat. Mater.*, 2005, **4**, 759–763.
131. J. Chen, H. Li, C. Fan, Q. Meng, Y. Tang, X. Qiu, G. Fu and T. Ma, *Adv. Mater.*, 2020, **32**, e2003134.
132. K. Wang, J. Liu, Z. Tang, L. Li, Z. Wang, M. Zubair, F. Ciucci, L. Thomsen, J. Wright and N. M. Bedford, *J. Mater. Chem. A*, 2021. DOI: 10.1039/d1ta02925h.
133. Z. Wang, C. Zhu, H. Tan, J. Liu, L. Xu, Y. Zhang, Y. Liu, X. Zou, Z. Liu and X. Lu, *Adv. Funct. Mater.*, 2021. DOI: 10.1002/adfm.202104735, 2104735.
134. G. Chen, T. Wang, P. Liu, Z. Liao, H. Zhong, G. Wang, P. Zhang, M. Yu, E. Zschech, M. Chen, J. Zhang and X. Feng, *Energy Environ. Sci.*, 2020, **13**, 2849–2855.
135. Y. Hou, M. Qiu, M. G. Kim, P. Liu, G. Nam, T. Zhang, X. Zhuang, B. Yang, J. Cho, M. Chen, C. Yuan, L. Lei and X. Feng, *Nat. Commun.*, 2019, **10**, 1392.
136. Z. Jiang, W. Sun, H. Shang, W. Chen, T. Sun, H. Li, J. Dong, J. Zhou, Z. Li, Y. Wang, R. Cao, R. Sarangi, Z. Yang, D. Wang, J. Zhang and Y. Li, *Energy Environ. Sci.*, 2019, **12**, 3508–3514.
137. Z. Zhang, X. Zhao, S. Xi, L. Zhang, Z. Chen, Z. Zeng, M. Huang, H. Yang, B. Liu, S. J. Pennycook and P. Chen, *Adv. Energy Mater.*, 2020, **10**, 2002896.
138. Y. Zhou, X. Tao, G. Chen, R. Lu, D. Wang, M. X. Chen, E. Jin, J. Yang, H. W. Liang, Y. Zhao, X. Feng, A. Narita and K. Mullen, *Nat. Commun.*, 2020, **11**, 5892.
139. K. Liu, G. Wu and G. Wang, *J. Phys. Chem. C*, 2017, **121**, 11319–11324.
140. M. Xiao, Z. Xing, Z. Jin, C. Liu, J. Ge, J. Zhu, Y. Wang, X. Zhao and Z. Chen, *Adv. Mater.*, 2020, **32**, e2004900.
141. J. Ban, X. Wen, H. Xu, Z. Wang, X. Liu, G. Cao, G. Shao and J. Hu, *Adv. Funct. Mater.*, 2021, **31**, 2010472.
142. Q. Shi, C. Zhu, D. Du and Y. Lin, *Chem. Soc. Rev.*, 2019, **48**, 3181–3192.
143. S. Back and Y. Jung, *ACS Energy Lett.*, 2017, **2**, 969–975.
144. S. Li, B. Chen, Y. Wang, M. Y. Ye, P. A. van Aken, C. Cheng and A. Thomas, *Nat. Mater.*, 2021. DOI: 10.1038/s41563-021-01006-2.

7 CO$_2$ Electroreduction Applications of Atomically Dispersed Metallic Materials

Leiduan Hao and Zhenyu Sun
Beijing University of Chemical Technology

CONTENTS

7.1   INTRODUCTION

Due to anthropogenic activities, the concentration of CO$_2$ in the atmosphere has been increasing since industrial revolution, which causes global warming and detrimental effect, threatening the sustainable living and development of human society [1]. Reducing CO$_2$ level in the atmosphere is an urgent issue worldwide. Electrochemical CO$_2$ reduction (ECR) is recognized as a promising alternative because it can store electrical energy from renewable sources in the form of chemical energy through the conversion of CO$_2$ to value-added chemicals [2–6]. Additionally, ECR is generally operated under mild conditions with water as economical hydrogen source [7,8]. However, it is far from the practical application of ECR considering the immature of the technology, mainly limited by large overpotentials, low product selectivity and current density, high cost of electrocatalysts, stability issue, and so on [9–11]. To improve the energy efficiency and viability of ECR, it is important to lower the operational overpotential, suppress the side reactions (e.g., hydrogen evolution reaction (HER)), and maintain large current density, where the development of robust electrocatalysts is a critical aspect.

DOI: 10.1201/9781003153436-7

Single-atom catalysts (SACs) featuring isolated metal center coordinated with non-metal atoms have emerged as the frontier of catalysis and received massive attention in various catalytic processes [12–18]. SACs can integrate the advantages of both homogeneous and heterogeneous catalysts to exhibit high intrinsic catalytic activity as well as maximus atom utilization [19]. In addition, the structure of SACs offers an ideal model for the exploration of the catalytic mechanism, which is meaningful for fundamental understanding and future design of efficient catalysts. Recently, the application of SACs for ECR has achieved significant improvements in aspects of catalyst design and structure-activity relationship study [20–25]. This chapter focused on the advances of SACs in ECR. Various metal single atoms and their ECR performance regarding product selectivity, catalytic activity, and mechanism are demonstrated.

7.2 SACs FOR ECR

7.2.1 SINGLE NI ATOM CATALYSTS

The report published by Strasser et al. was considered to be the first work that applied heterogeneous catalyst with single-metal sites in the ECR reactions [26]. Metal-containing N-doped porous carbon black (M–N–C) was used as a catalyst for CO_2 electroreduction to produce CO and hydrocarbon products. It was found that the CO selectivity was mostly dependent on the nitrogen moieties rather than the metal ions, which include Fe, Mn, and FeMn. Instead, the metal ions played a critical role in determining the product from further reduction of CO. In the case of Ni single metal sites anchored on N-doped carbon materials, the catalytic active centers and selectivity for ECR were explored by Kamiya et al. [27]. Ni–N–Gr was synthesized through the heat treatment of Ni complex and graphene oxides. The X-ray photoelectron spectroscopy (XPS) result indicated that the valence state of Ni in Ni–N–Gr was Ni^{2+}. The Fourier transform extended X-ray absorption fine structure (EXAFS) spectra for Ni–N–Gr and Ni metal revealed the presence of Ni–N coordination structure and no Ni–Ni was observed. Ni–N–Gr displayed a Faradaic efficiency (FE) of more than 90% for CO production from ECR at −0.7 to −0.9 V vs. reversible hydrogen electrode (RHE). Control experiments using Cu–N–Gr and N–Gr as catalysts suggested that both Ni and N were responsible for the performance of Ni–N–Gr. Therefore, different from the previous report with Fe-, Mn- and FeMn–N–C, the Ni–N could act as the catalytic active centers in Ni–N–Gr for ECR to produce CO.

Li et al. fabricated Ni single atoms-containing N-doped porous carbon (Ni SAs/-N–C) by using a ZIF-assisted strategy (Figure 7.1a) [28]. ZIF-8 was used as a precursor to adsorb Ni^{2+} ions. After pyrolysis at 1,000°C, the Zn nodes evaporated and the ZIF-8 was transformed into N-doped porous carbon, during which the Ni^{2+} ions were stabilized by the N-rich defects through ionic exchange with Zn^{2+} ions. The Ni SAs can be clearly observed from the high-angle annular dark-field scanning transmission electron microscopy (HAADF-STEM) images of the obtained material (Figure 7.1b and c). The valence state of the Ni species was revealed to be between 0 and 2 from the XPS analysis, which was in accordance with the X-ray absorption near-edge structure (XANES) results. The EXAFS spectra and the corresponding fitting curves

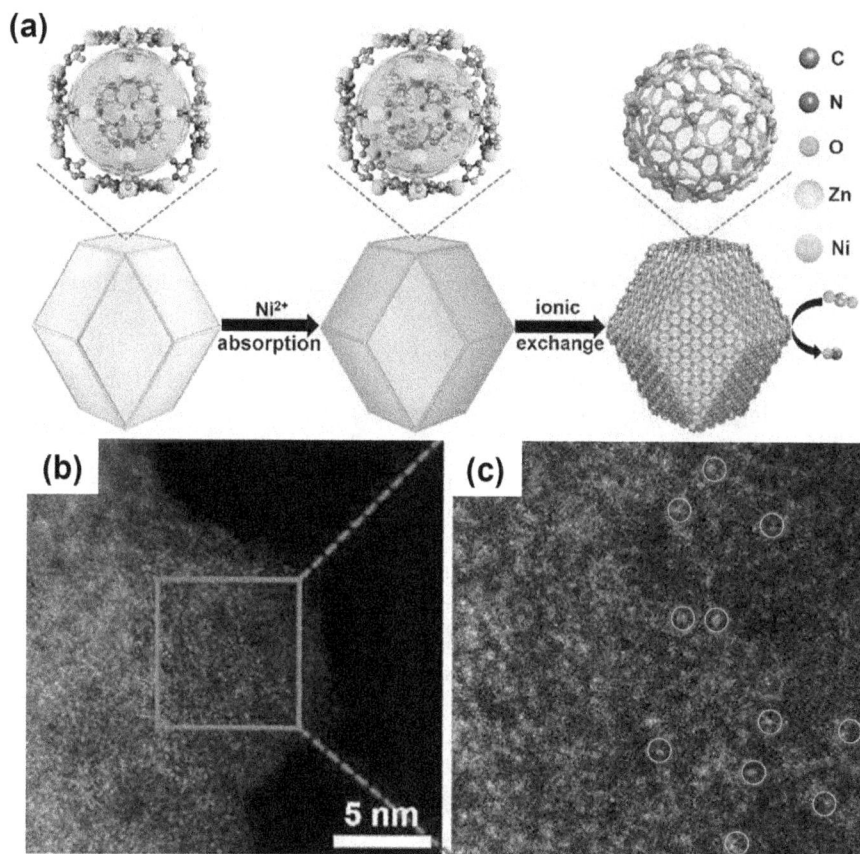

FIGURE 7.1 (a) Scheme of the formation of Ni SAs/N–C. (b, c) Magnified HAADF-STEM images of Ni SAs/N–C. The Ni single atoms are marked with circles.

indicate the dominant presence of Ni–N coordination with a Ni–N$_3$ architecture. This Ni SAs/N–C exhibited a FE of 71.9% for CO at −0.9 V vs. RHE, and the partial current density of CO at −1.0 V could reach 7.37 mA cm^{-2} with a turnover frequency of 5,273 h^{-1}. This Ni SAs/N–C could also outperform the Ni nanoparticles and the commercial Ni foam as catalysts for ECR, which could be attributed to the abundant surface active sites and faster electron transfer for the formation of CO$_2$$^{·-}$ intermediate. However, the FE for CO was depressed over Ni SAs/N–C due to the simultaneously enhanced HER activity.

In a later work by Wu et al., exclusive Ni–N$_4$ sites were constructed by adopting a topo-chemical transformation strategy [29]. With a carbon-coated layer, the Ni-doped g–C$_3$N$_4$, in which the Ni was coordinated by two O atoms and four N atoms, was transformed into Ni–N$_4$–C after pyrolysis, during which the oxygen bond was removed and exclusive Ni–N$_4$ structure was formed. As a comparison, without the carbon layer, Ni particles (Ni@N–C) were also obtained. In ECR, Ni–N$_4$–C achieved a FE of 99% for CO at −0.81 V vs. RHE, much higher than that of Ni@N–C (65%).

Besides the higher electrochemical active surface area (ECSA) and faster charge transfer of Ni–N$_4$–C, density functional theory (DFT) calculation results indicated that the Ni–N$_4$ sites could lower the formation energy of COOH*, thus assisting the formation of CO and showing high activity in ECR.

ECR is a promising approach to convert CO$_2$ into value-added chemicals, during which the massive preparation of electrocatalysts with high efficiency and stability is critical yet challenging. Wu et al. developed a simple method that involved solid-state diffusion between the N-doped carbon phase and the Ni foil to fabricate single-atom Ni catalysts [30]. As shown in Figure 7.2a, through solid diffusion, Ni nanoparticles from Ni foil diffuse to the N–C soil, which is formed from the heat treatment of the melamine layer on the Ni foil. These Ni nanoparticles can act as a catalyst for the growth of nitrogen-doped carbon nanotubes (N–CNTs) on the N–C soil, resulting in freshly prepared carbon papers (F–CPs) after peeling off from the Ni foil. With the subsequent acid-leaching, hierarchical carbon papers (H–CPs) containing Ni single atoms in its structure were obtained. As shown in Figure 7.2b, X-ray diffraction (XRD) patterns of the materials indicate that the peak assigned to the Ni nanoparticles is decreased in H–CPs compared with that in F–CPs, suggesting the partial removal of Ni nanoparticles after acid treatment. The same phenomenon can be observed in the EXAFS spectra that the Ni–Ni bonding is decreased (Figure 7.2c). Instead, a new Ni–N bonding appears in H–CPs. Cyclic voltammetry (CV) curves further prove that the Ni species in H–CPs are mainly Ni single atoms coordinated with N (Figure 7.2d). In addition, H–CPs possessed a higher BET surface area (143.97 m^{2}g^{-1} vs. 113.21 m^{2}g^{-1}) and CO$_2$ adsorption capacity (7.64 cm^{3}g^{-1} vs. 3.85 cm^{3}g^{-1}) than the F–CPs due to the dissolution of Ni nanoparticles. As a result, in ECR reaction, H–CPs exhibited a FE over 90.8% for CO in a wide range of working

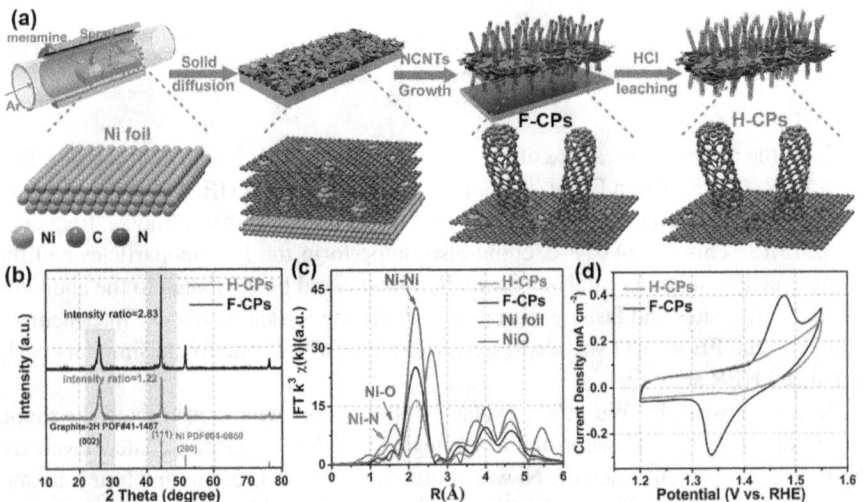

FIGURE 7.2 (a) Scheme of the formation of H–CPs. (b) XRD patterns of H–CPs and F–CPs. (c) k^3-weighted χ(k) function of the EXAFS spectra. (d) CV curves in 0.1 M KOH electrolyte at a scan rate of 5 mV s^{-1}.

potentials from −0.7 to −1.2 V and a specific CO current density of 60.11 mA cm^{-2} at −1.2 V vs. RHE, which was seven times higher than that of F–CPs. A mechanism study demonstrated that H–CPs containing Ni single atoms were more favorable for the desorption of the CO* intermediate, while the binding strength of CO* to Ni nanoparticles was too strong. H–CPs displayed great potential for practical application since their preparation could be programmable and scalable, and they had long-term stability for CO FE and current density over 40 h at −1.0 V vs. RHE.

In another report, Wang et al. synthesized Ni SACs using activated carbon black (CB) as the support [31]. This method enabled gram-scale SAC production in a one-batch synthesis, during which Ni^{2+} was adsorbed on the surface of activated CB followed by heat treatment with urea as the nitrogen source. The obtained Ni–NCB displayed high activity for ECR to produce CO. In a traditional H-cell, the FE of CO reached 99% at −0.681 V vs. RHE. Considering that high product selectivity and large current density are important issues for the scaling up of ECR toward practical applications, the authors operated the ECR using an anion membrane electrode assembly (MEA) to further boost the performance of Ni–NCB. MEA could prevent the electrocatalysts from contacting the aqueous electrolyte, thus suppressing the hydrogen evolution side reaction [32,33]. Meanwhile, fuel cell technology is also beneficial to CO$_2$ diffusion, which overcomes the CO$_2$ solubility issue in aqueous electrolyte used in H-cell [34,35]. As a result, a current density above 100 mA cm^{-2} and near 100% selectivity of CO were achieved on the MEA. Moreover, the Ni–NCB electrocatalyst could be integrated into a 10 × 10 cm^2 modular cell while still maintaining a large current density and high FE for CO. This work provided a promising route for approaching practical expectations.

Other than tuning the electronic structure of the SACs, designing the morphology of the supportive materials can also benefit the ECR performance [36]. A highly porous support with high surface area and CO$_2$ adsorption capacity will increase the CO$_2$ concentration around the electrocatalysts, which may lead to a large current density [37]. Nam et al. prepared three-dimensional (3D) porous Ni and N-doped carbon for ECR to produce CO [38]. The nitrogen-coordinated single-atom Ni acted as the catalytic active sites for ECR while the 3D structure of the electrocatalyst brought high surface area favoring the accessibility of CO$_2$. Combining with the boosting effect of MEA, a FE of 99% for CO and a high current density over 300 mA cm^{-2} were achieved, which reached the suggested rate (200–400 mA cm^{-2}) for profitable ECR [39].

The electrolyte plays an important role in promoting the performance of ECR since it can interact with CO$_2$ and the reaction intermediate as well as the product, leading to different activity and selectivity [40,41]. Ionic liquids (ILs) have been used as efficient electrolytes for ECR, in which ILs not only possessed high CO$_2$ solubility but also activated CO$_2$ as a co-catalyst [42,43]. However, there are also drawbacks of ILs, such as the high cost and slow mass transfer due to the high viscosity [44]. Taking advantage of the benefits of ILs as electrolyte for ECR and in the meantime avoiding the shortcomings, Zhao et al. designed a strategy to confine ILs into the structure of single-atom Ni–N catalyst [45]. As shown in Figure 7.3a, an IL of 1-butyl-3-methylimidazolium hexafluorophosphate ([BMIM][PF$_6$]) is impregnated into the nanosized channels and pores of Ni–N structure under vacuum through capillarity forces. The resulted Ni–N@ ILs maintain the shape of the original Ni–N containing single-atom Ni sites. With the

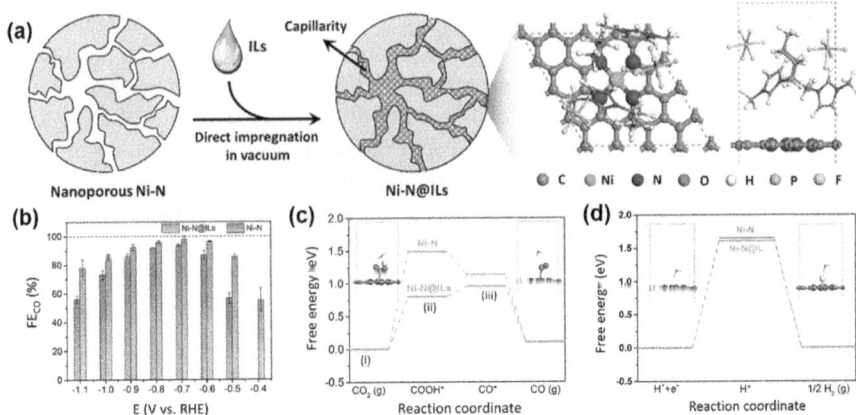

FIGURE 7.3 (a) Schematic illustration of the nanoconfined ILs strategy. (b) CO FE in the potential range of −0.4 to −1.1 V vs. RHE. (c) The calculated free energy diagrams for CO_2-to-CO conversion on Ni–N and Ni–N@ILs. (d) The calculated free energy diagrams for HER on Ni–N and Ni–N@ILs.

synergistic effect between Ni–N and the introduced [BMIM][PF_6], the onsite potential for ECR to CO was decreased by 0.07 V compared Ni–N@ILs and Ni–N. In addition, the FE for CO over Ni–N@ILs is higher than that of Ni–N in the whole potential range in Figure 7.3b. Ni–N@ILs exhibit a CO FE of 98% at −0.7 V vs. RHE. At −1.0 V vs. RHE, the current density of CO reaches 66.1 mA cm^{-2}, which is benefited from the high solubility of CO_2 in [BMIM][PF_6]. The composite could remain in its structure for over 50 h of electrolysis with both stable CO FE and current density. Mechanism study in Figure 7.3c and d display that after confining [BMIM][PF_6], the energy barrier of the rate-determining step of ECR from CO_2 to COOH* is decreased from 1.49 to 0.80 eV without affecting the H* adsorption, thus suppressing the competitive HER and resulting in high CO FE.

The above work provides a strategy for tuning the catalytic performance of SACs through the confinement of ILs into the pristine catalyst structure and the creation of a solid/liquid interface, which is promising for optimizing the electrocatalytic reaction environment [46]. Doping of heteroatom is another approach for manipulating the electronic structure of single-atom sites to achieve efficient ECR. Zhu et al. adopted a one-step polymer-assisted pyrolysis strategy to synthesize an F, N-doped carbon anchored single-atom Ni catalyst (Ni–SAs@FNC) [47]. All the precursors including glucose, melamine, Ni salt, and polytetrafluoroethylene (PTFE) were pyrolyzed to obtain this Ni–SAs@FNC. The fluorine-containing PTFE acted as the sacrificial reagent to assist the dispersion of Ni for the formation of Ni single atoms as well as the source of F dopant around the Ni sites. Ni–SAs@FNC exhibited a FE of 95% for CO in a wide potential range from −0.67 to −0.97 V vs. RHE. Mechanism study through in-situ attenuated total reflection-infrared spectroscopy (ATR-IR) revealed that the rate-limiting step for the ECR to CO was the formation of *COOH intermediate. DFT calculations demonstrated that the electron configuration of the active

Ni–N$_4$ sites was modulated by F doping, which resulted in a lower energy barrier for CO$_2$ activation and favored the generation of *COOH intermediate.

Despite the progresses that have been made using various single Ni atom catalysts, it is still challenging to realize the practical application of ECR for CO production. Quantum mechanics study predicted that the highest current density for Ni–N$_4$ single-atom sites could reach 700 mA cm^{-2} [48], which is much higher than the results accomplished by present research. With the efforts devoted to the development of efficient catalysts together with advanced electrolysis technology, it is promising to realize profitable production of CO from ECR using Ni SACs.

7.2.2 SINGLE FE ATOM CATALYSTS

Iron-based catalysts are attractive and promising owing to the non-toxicity, low cost, and abundance on earth. Fe–N–C materials have been predicted and reported to be active for ECR [26,49]. Strasser et al. proved that single-atom Fe catalyst was active for the ECR to yield CO, although the FE (60%) was relatively low [26]. Fontecave et al. prepared a series of Fe–N–C materials with different amounts of Fe through the pyrolysis of ferrous acetate, phenanthroline, and ZIF-8(Zn) [50]. Structural characterization demonstrated that these Fe–N–C materials contained both Fe single atoms and nanoparticles with different ratios. The structure-selectivity relationship for ECR was studied, which revealed that FeN$_4$ was the active species for CO production and Fe nanoparticles mainly led to the HER. Moreover, the product ratio (CO/H$_2$) could be controlled by the ratio of FeN$_4$ single-atom sites and Fe nanoparticles. The material with sole FeN$_4$ species exhibited a FE of 91% for CO at a low overpotential of 190 mV. In a later work of Tour et al. [51], the relationship between the structure of the Fe catalysts and the ECR performance was further explored. Through annealing, a mixture of graphene oxide (GO) and FeCl$_3$, single-atom Fe on nitrogen-doped graphene (Fe/NG) was obtained. The ECR performance of the materials prepared under different annealing temperatures was studied and it turned out that a maximum FE of about 80% for CO was achieved at −0.6 V vs. RHE over Fe/NG-750 with an annealing temperature of 750°C and a Fe loading of 1.25%. As shown in Figure 7.4a, the catalysts with lower or higher Fe amount (0.5Fe/NG-750 or 2Fe/NG-750) and without Fe or N doping (NG-750 or Fe/G-750) are less active for ECR to CO. As shown in Figure 7.4b, EXAFS study demonstrates that higher annealing temperature or Fe loading can result in more Fe–Fe coordination and the formation of Fe nanoparticles. XANES spectra further confirm that the catalytic active sites possess the geometry of pyridinic Fe–N$_4$ (Figure 7.4c and d). The relative amount of Fe–Fe bonds from Fe nanoparticles and Fe–N bonds from Fe–N$_4$ were calculated subsequently, thus proving the critical role of the single-atom Fe–N$_4$ sites in boosting the performance of ECR to CO.

Despite the above discovery that the Fe–N$_4$ single sites are active for the production of CO in ECR, Wang et al. designed a route for the synthesis of single-atom Fe catalysts containing Fe–N$_5$ sites [52]. As shown in Figure 7.5a, by pyrolysis of different starting materials, catalysts containing Fe nanoparticles or FeN$_4$ sites can be obtained. When pyrolysis of hemin, melamine and graphene, instead of forming FeN$_4$ sites, an additional axial ligand can be generated and coordinated with FeN$_4$

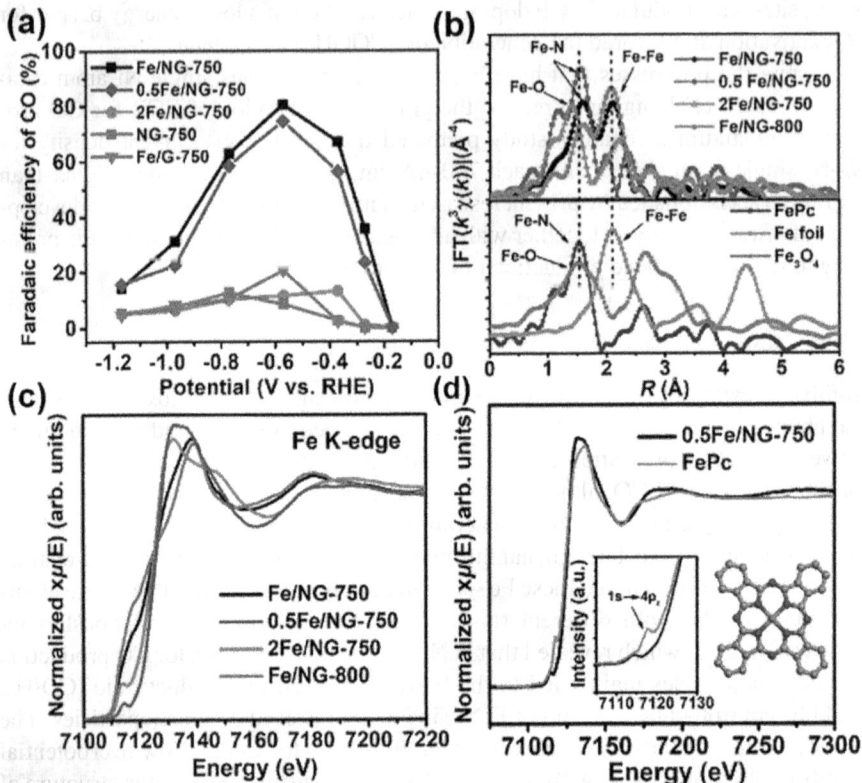

FIGURE 7.4   (a) FE of CO for ECR. (b) Fourier transform magnitudes of the experimental Fe K-edge EXAFS spectra. (c, d) The normalized Fe K-edge XANES spectra.

to form the FeN_5 sites. The free energy profile in Figure 7.5b reveals that for the ECR to CO, the key step from CO_2 to *COOH is much easier over FeN_5 sites than over FeN_4 sites. In addition, FeN_4 sites could adsorb the CO strongly, resulting in a higher desorption energy and a low CO selectivity. While FeN_5 sites possessed a weak binding strength to CO, which facilitates the desorption of CO thus bringing in high CO selectivity. Therefore, the catalyst with FeN_5 sites displayed a high FE of over 97% for CO at a low overpotential of 0.35 V. Further theoretical study indicated that the pyrrolic N ligand in FeN_5 was responsible for the rapid CO desorption and high selectivity since it reduced the electron density of Fe 3d orbitals and the Fe–CO π back-donation.

It is known that the electronic properties of the SACs can be affected by the surrounding environment [53–56]. Li et al. reported the design of catalyst support to modulate the electronic structure of Fe–N_4 sites [57]. As shown in Figure 7.6, the Fe–N_4 sites hosted by bulk graphene suffer from the poor desorption of CO from the active sites. While the creation of pore edged and the anchoring of Fe–N_4 sites on the resulted porous graphene (Fe–N–G–p) could assist the desorption of CO. Using hydrogen peroxide as the etching reagent, Fe–N–G–p was prepared and applied as

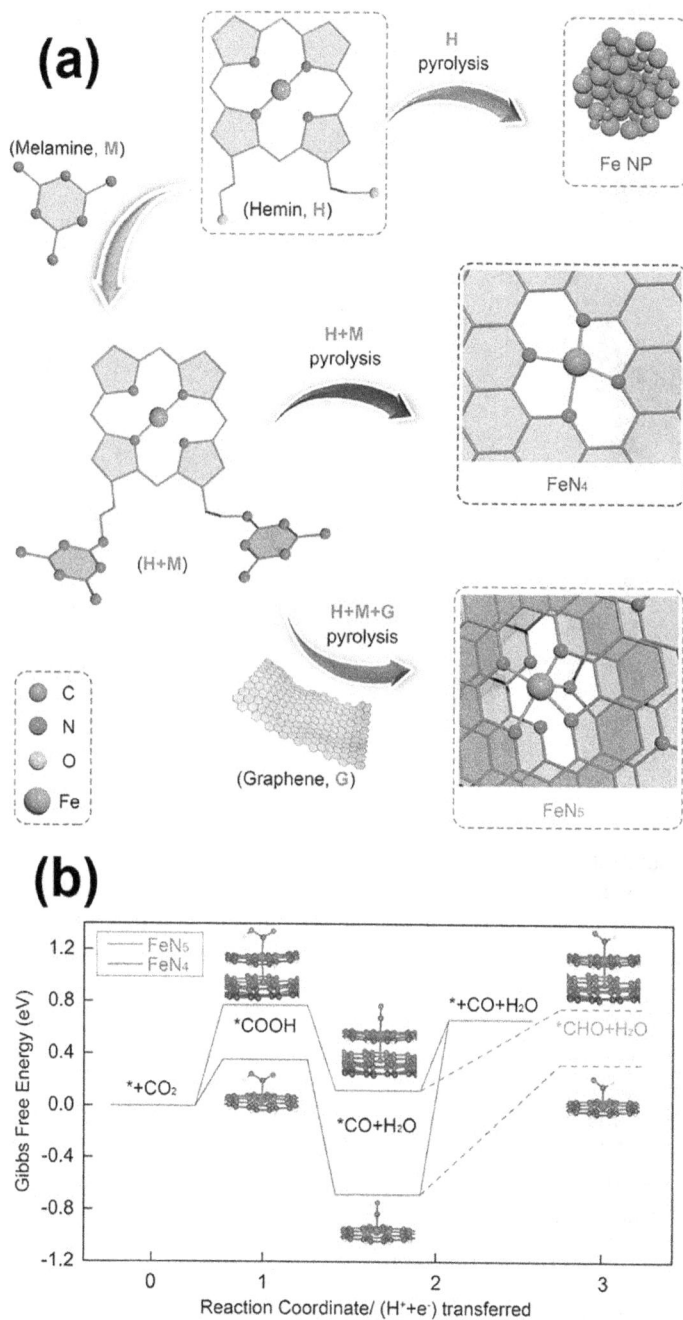

FIGURE 7.5 (a) Schematic synthesis of Fe nanoparticles, FeN$_4$, and FeN$_5$ catalysts. (b) Free-energy profile with the optimized intermediates for the electroreduction of CO$_2$ to CO. The asterisk (*) denotes the free adsorption site, and *M (M=COOH, CO, CHO) indicates the adsorbed chemical species.

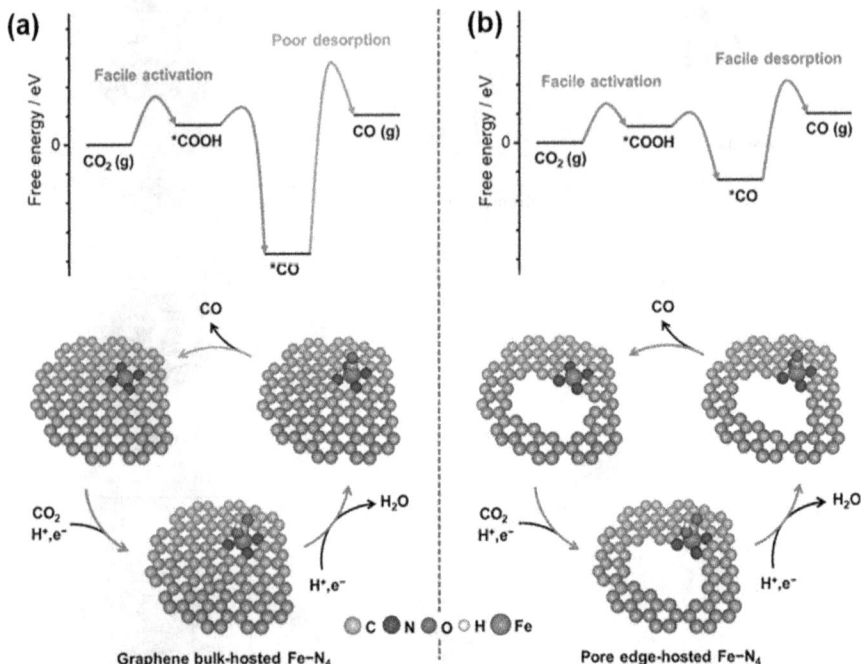

FIGURE 7.6 Schematic showing the electrocatalytic CO_2 reduction behaviors on (a) pore-deficient graphene bulk-supported Fe–N$_4$ and (b) pore-rich graphene-supported Fe–N$_4$.

the electrocatalyst in ECR. At -0.58 V vs. RHE, a FE of 94% for CO was achieved with a turnover frequency of $1,630\,h^{-1}$, which was over three times higher than the pore-free graphene-supported counterpart. Mechanism exploration suggested that the introduction of pore edges could increase the length and decrease the strength between Fe and CO, thus weakening the adsorption of CO on the FeN$_4$ sites and facilitating CO desorption process.

Bearing the same concept in mind, Li group reported another work that highlighted the importance of support architecture on the performance of SAC for ECR [58]. As shown in Figure 7.7a, commercial CNT is partially transformed to graphene nanoribbon (GNR) and the resulted CNT@GNR is used as a support for a single Fe atom catalyst. Through tuning the amount of $KMnO_4$, Fe–N/CNT@GNR with different surface areas and mass transport properties can be fabricated, as shown in Figure 7.7b. The Fe single atoms can be observed clearly in the HAADF-STEM images of Fe–N/CNT@GNR-2 (Figure 7.7c and d). As shown in Figure 7.7e, the ECR performance of the materials treated with different amounts of $KMnO_4$ is studied and it turns out that Fe–N/CNT@GNR-2 can give the best activity for CO production benefiting from its high electrochemically active surface area and sufficient mass transport. At -0.76 V vs. RHE, Fe–N/CNT@GNR-2 exhibits a CO FE of 96% with a partial current density of $22.6\,mA\,cm^{-2}$. DFT calculations demonstrated that the hierarchically porous structure of the support provided basal plane and edge sites

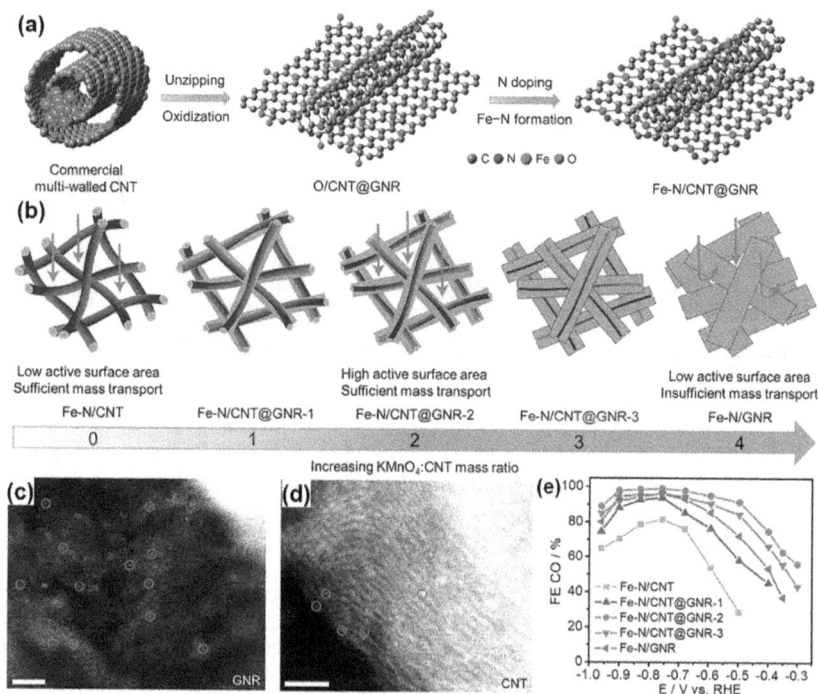

FIGURE 7.7 (a) Schematic illustration of transforming multiwalled CNTs into Fe-N/CNT@GNR. (b) Structural evolution from CNTs to CNT@GNR to GNR by adjusting $KMnO_4$:CNT mass ratios. (c, d) HAADF-STEM images. (e) CO FE over various catalysts.

for anchoring Fe–N_4 centers, which favored the activation of CO_2 and suppressed the hydrogen evolution side reaction.

Although Fe SACs displayed good performance for ECR to CO production, the potential range afforded high FE was narrow due to the competing HER [59,60]. In a very recent report, Chen et al. developed a strategy that could selectively remove the sites in the Fe–N–C catalyst for H_2 production and only leave the CO-production sites [61]. DFT calculations revealed that the pyridinic and pyrrolic N atoms favored the proton activation, while the graphitic N was beneficial to the CO_2 activation (Figure 7.8a). By adding 5% of H_2 during the pyrolysis process for the catalyst preparation, the "protophilic sites" were successfully eliminated with only the graphitic N dominated in the resulted material. Compared with the material pyrolyzed only in N_2 atmosphere and the ones without Fe, H_2–FeN_4/C exhibits a much higher FE for CO, as shown in Figure 7.8b. At −0.6 V vs. RHE, the FE of CO reaches 97% and the FE remains above 90% over a wide potential range from −0.3 to −0.8 V vs. RHE (Figure 7.8b). Besides, the performance of H_2–FeN_4/C is stable during 24 h of electrolysis at −0.6 V vs. RHE (Figure 7.8c). Further mechanistic study demonstrated that the formation of atomic interface between the graphitic N and the Fe single-atom sites could assist the adsorption of CO_2 and water molecule, which boosted the CO_2 activation and the generation of *COOH intermediate.

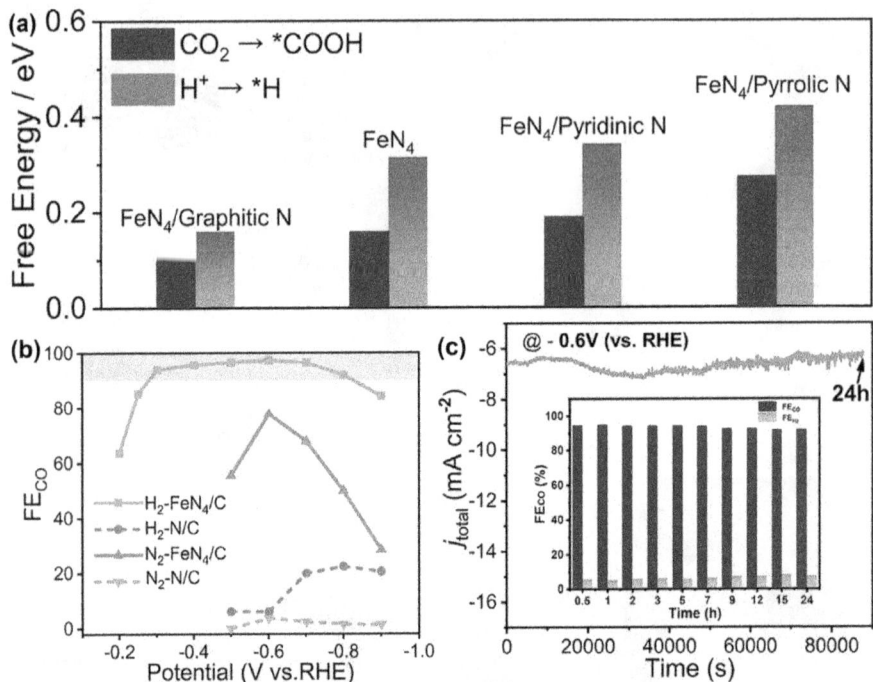

FIGURE 7.8   (a) Comparison of free energies for the steps from CO_2 to *COOH and from H^+ to *H. (b) CO FE of different catalysts. (c) Stability test.

7.2.3   SINGLE CO ATOM CATALYSTS

Chen et al. synthesized single-atom Co supported on polymer-derived hollow N-doped porous carbon spheres (HNPCSs) [62]. As shown in Figure 7.9a, the HNPCSs support is prepared through a pyrolysis and etching process. The Co–N_5/HNPCSs catalyst is subsequently obtained through the coordination between Co from the cobalt phthalocyanine (CoPc) and N. The Co single atoms can be observed from the HAADF-STEM images (Figure 7.9b and c). Compared with the CoPc, Co–N_5/HNPCSs display a high FE for CO production over a broad range of potential. Especially, a FE of 99.4% for CO can be achieved at −0.79 V vs. RHE, which is more than 15 times higher than that of the CoPc (Figure 7.9d). The experimental and theoretical results suggest that the single-atom Co–N_5 sites are critical for the activation of CO_2 to form the COOH* intermediate and the desorption of CO.

    The effect of coordination number on the performance of Co SACs for ECR was studied by Wu et al [63]. The catalysts were prepared by pyrolysis of bimetallic Co/Zn–ZIF, during which the Zn could be evaporated and the Co single atoms were generated with the formation of nitrogen-doped porous carbon. The coordination number of Co was regulated by the pyrolysis temperature. At a temperature of 800°C, 900°C, and 1,000°C, single Co atom catalysts with N coordination numbers of 2, 3, and 4 were synthesized, respectively. The ECR performance of Co–N_2, Co–N_3, and

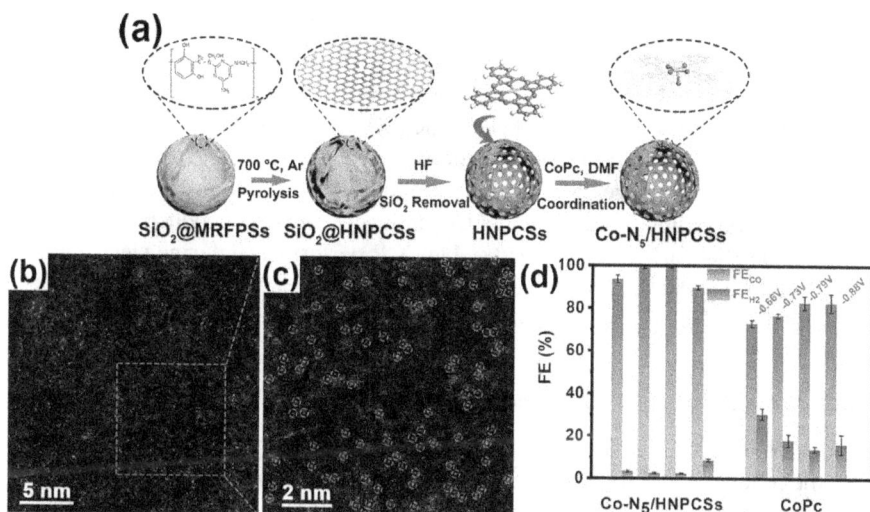

FIGURE 7.9 (a) Schematic illustration of the synthesis of Co–N$_5$/HNPCSs. (b, c) HAADF-STEM images of Co–N$_5$/HNPCSs. (d) CO and H$_2$ FE of Co–N$_5$/HNPCSs and CoPc.

Co–N$_4$ was evaluated, among which Co–N$_2$ exhibited the best selectivity and activity for CO formation. At an overpotential of 0.52 V, the FE for CO was 94% with a current density of 18.1 mA cm^{-2}. Moreover, a high turnover frequency of 18,200 h^{-1} was achieved. DFT calculations indicated that a lower coordination number could benefit the CO$_2$ activation for the formation of CO$_2^-$ intermediate, thus leading to high ECR activity.

In addition to the coordination number, the coordination environment of Co also affects the ECR performance by tuning the binding strength of CO$_2$ and its subsequent activation [64]. The above conclusions provide guidance for future design of efficient Co SACs for ECR.

7.2.4 SINGLE ZN ATOM CATALYSTS

Liu et al. reported the fabrication of single Zn atom catalyst anchored on N-doped graphene (Zn–N–G) [65]. The effect of pyrolysis temperature was studied and it turned out that the material pyrolyzed at 800°C (Zn–N–G-800) possessed the best performance for ECR in producing CO. A FE of 90.8% for CO was achieved at −0.5 V vs. RHE, which corresponded to an overpotential of 0.39 V. However, the CO partial current density was low at this potential and the largest current density was 11.2 mA cm^{-2} at −0.8 V vs. RHE. Mechanism exploration demonstrated that Zn–N$_x$ sites were the catalytic active species, which could facilitate the formation of COOH* intermediate as well as the desorption of CO, thus yielding high activity for ECR to CO production. In another report by Xu et al. [66], Zn–N$_4$ was disclosed as the catalytic active site for ECR to CO, and the FE for CO was improved to 95% at −0.43 V vs. RHE.

In a recent work by Xin et al. [67], Zn single atoms supported on microporous N-doped carbon (SA–Zn/MNC) were fabricated. Different from the above-reported results, the Zn–N$_4$ sites in SA–Zn/MNC were active in producing CH$_4$ from ECR. At −1.8 V vs. the saturated calomel electrode (SCE), a FE of 85% with a CH$_4$ partial current density of 31.8 mA cm^{-2} was achieved. Mechanistic study implied that the O atom in the *OCHO intermediate could bond to Zn single atom more easily than the C atom, thus blocking the formation of CO and facilitating the production of CH$_4$. In addition, the SA–Zn/MNC displayed stable ECR performance during electrolysis of 35 h, which was benefited from the MNC support with rapid charge transport, easy access to the active sites as well as high stability.

7.2.5 SINGLE CU ATOM CATALYSTS

Besides Zn, single Cu atom catalysts could also afford CH$_4$ from ECR [68]. In addition, the product could be tuned between C$_2$H$_2$ and CH$_4$ based on the Cu concentration in the materials, which was realized by regulating the pyrolysis temperature during the preparation process. Specifically, the concentration of Cu was decreased with increasing pyrolysis temperature, and a higher content of Cu favored the formation of C$_2$H$_2$ while a lower content of Cu yielded CH$_4$ as the main product. DFT calculations revealed that the material with higher content of Cu contained more neighboring Cu–N$_2$ sites, which favored the C$_2$H$_2$ generation with a low free energy for C–C coupling. The isolated Cu–N$_4$, the neighboring Cu–N$_4$, and the isolated Cu–N$_2$ preferred the formation of methane.

Cu-based electrocatalysts are well recognized for ECR to produce hydrocarbons and alcohols [69,70]. Unlike other single metal atom catalysts for CO production, Cu single atoms were reported for the generation of methanol [71]. Isolated Cu decorated through-hole carbon nanofibers (CuSAs/TCNFs) were fabricated with high mechanical strength and used directly as the electrode for ECR. The morphology of the CuSAs/TCNFs is shown in Figure 7.10. The through-hole structure enables easy access to the Cu single atoms and smooth mass transport. As a result, a FE of 44% for methanol was achieved at −0.9 V vs. RHE. Moreover, nearly pure C1 products (methanol and CO) were produced with a current density of 93 mA cm^{-2}. Mechanism study demonstrated that the binding energy between Cu single atoms and the *CO intermediate was relatively high, which led to further reduction to methanol instead of CO. Although Cu SACs are inclined to yield hydrocarbon products, by tuning the coordination of the active sites, Deng et al. discovered that Cu SAs/NC containing Cu–N$_4$ sites exhibited high activity for ECR to CO [72]. DFT calculations suggested that Cu–N$_4$ sites could stabilize the *COOH intermediate and easily release *CO, leading to efficient CO production. Cu SAs/NC displayed a FE of 92% for CO at −0.9 V vs. RHE.

7.2.6 OTHER SINGLE METAL ATOM CATALYSTS

Other than the above-mentioned metals, other single-atom metals have also been reported for ECR. For example, Sun et al. synthesized single-atom Sb catalyst through a simple pyrolysis method [73]. Compared with other Sb catalysts such as Sb

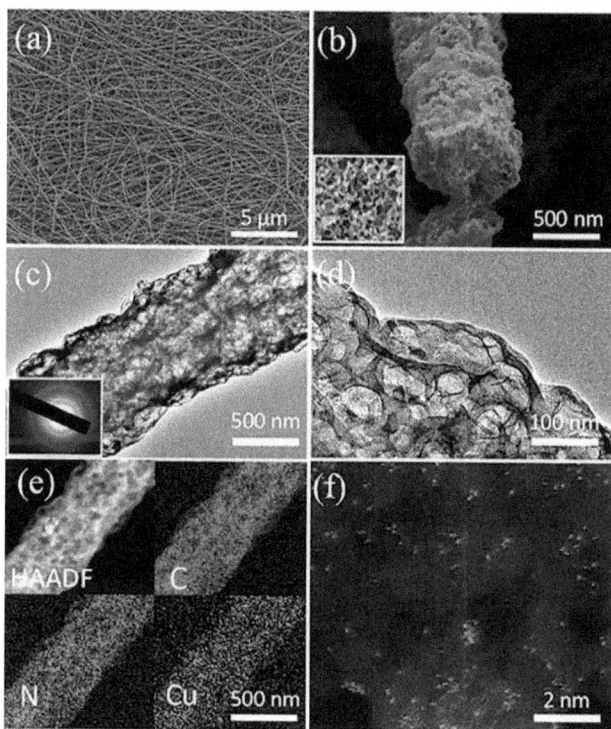

FIGURE 7.10 The morphology of the CuSAs/TCNFs. (a, b) SEM images. (c, d) TEM images. (e) EDX mapping of a single CuSAs/TCNFs nanofiber. (f) HAADF-STEM images.

nanoparticles and bulk Sb, single-atom Sb sites exhibited much higher efficiency for ECR to CO production with a turnover frequency of 16,500 h^{-1} at -0.9 V vs. RHE. The performance of the single Sb atom catalyst was attributed to its suppression effect for the hydrogen evolution side reaction.

Pd-based catalysts have shown to be active for CO production from ECR [74,75]. Due to the high price of Pd, researchers tried to reduce the loading of Pd while remaining its high activity. Chen et al. synthesized a nitrogen-doped carbon-supported Pd single-atom catalyst (Pd–NC) and used it as an electrocatalyst for ECR [76]. Compared with Pd nanoparticle counterpart, Pd–NC exhibited a much higher mass activity (373.0 vs. 28.5 mA mg$^{-1}_{Pd}$) for CO at -0.8 V vs. RHE. Mechanistic study showed that different from the Pd nanoparticles that generally involved the formation of palladium hydride as active species [77,78], Pd–N$_4$ sites were recognized as the active center for CO generation. In addition, *HOCO was determined as the key intermediate and Pd–N$_4$ sites could assist this catalytic pathway as well as the desorption of CO.

Zhang et al. reported the preparation of Bi SAC and its application in ECR to CO [79]. The formation process of the Bi SAC was shown in Figure 7.11. Starting from the Bi–MOF, the in-situ environmental transmission electron microscopy (ETEM) images demonstrate the evolution of Bi from nanoparticles to single atoms with the assistance of NH$_3$ from the decomposition of dicyandiamide (DCD). The resulted Bi

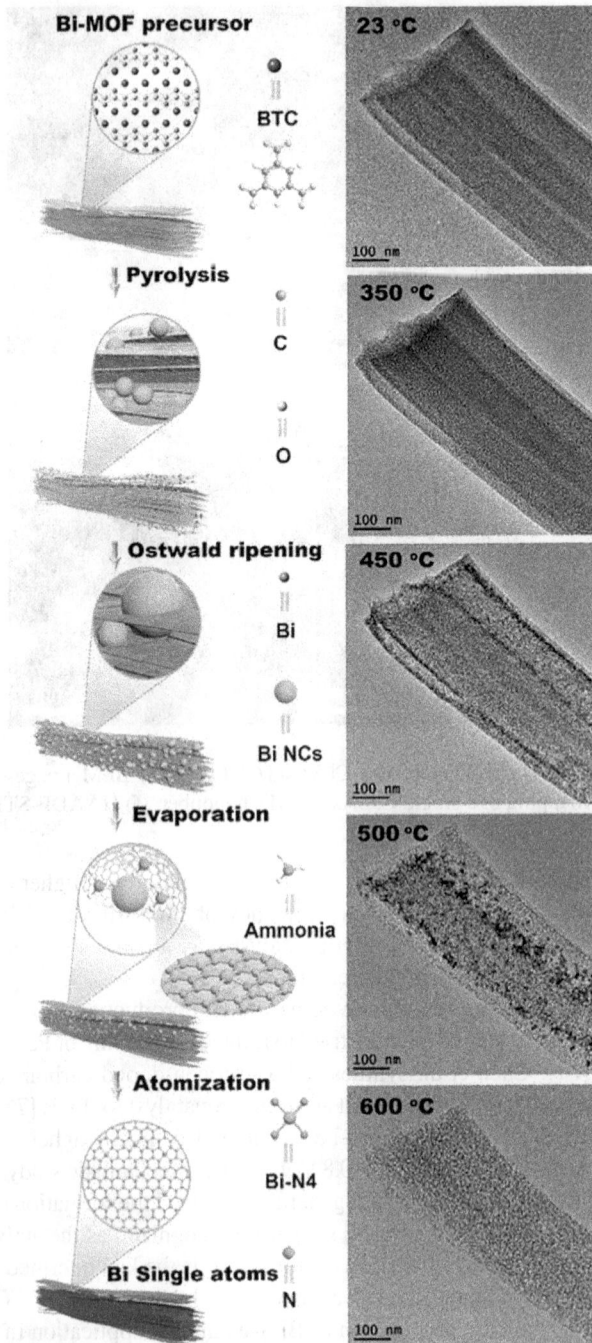

FIGURE 7.11 Scheme of the transformation from Bi–MOF to single Bi atoms and corresponding representative TEM images of Bi–MOF pyrolyzed at different temperatures with the assistance of DCD in situ.

SAs/NC exhibited a FE of 97% for CO at −0.5 V vs. RHE with a current density of 5.1 mA cm⁻² and the CO partial current density of 3.9 mA cm⁻², which was 7.5 times higher than that of the Bi nanoparticles. DFT calculations revealed that the BiN₄ sites possessed lower Gibbs free energy for the formation of *COOH intermediate, thus showing better performance for ECR to CO.

In-based electrocatalysts have been reported for the ECR to produce formate [80–82]. However, the efficiency was unsatisfactory considering the high overpotential and the low FE. Wang et al. designed single atom In catalyst with the assistance of ZIF-8 as the precursor of N-doped carbon [83]. The obtained In-SAs/NC was applied as the electrocatalyst for ECR and it afforded a FE of 96% for formate at a low applied potential of −0.65 V vs. RHE. Additionally, a turnover frequency of 12,500 h⁻¹ was achieved at −0.95 V vs. RHE. DFT calculations revealed that the formation of HCOO* on the single In atom sites was energetically more favorable than COOH*, thus leading to the formate production instead of CO. Compared with In nanoparticles, the In$^{\delta+}$-N₄ sites from In-SAs/NC possessed lower free energy for HCOO* generation, which was responsible for the high efficiency of In-SAs/NC for ECR to formate.

Sn-based electrocatalysts are known for their activity in converting CO_2 to formate [84–87]. For single-atom Sn catalysts, the coordination environment has a great effect on the selectivity of ECR, which has been unraveled by Zhang et al. through constructing Sn single atoms with different coordinated structures, that is the Sn–C_2O_2F and the Sn–N_4 [88]. Instead of affording formate, the Sn–C_2O_2F sites delivered CO as a product from ECR with a FE higher than 90.0% while formate was the main product over Sn–N_4 sites. DFT calculations suggested that the Sn–C_2O_2F structure preferred to adsorb *COOH intermediate rather than *H formation, thus leading to CO generation and suppressing hydrogen and formate production.

It was reported that thermal treatment together with the interaction between the metal species and the support can transform the metal nanoparticles into metal single-atom atoms [89–91]. Recently, Ag single atoms supported on MnO_2 were prepared through thermal treatment and the surface reconstruction of MnO_2 [92]. The obtained Ag₁/ MnO_2 exhibited high efficiency for ECR to CO with a FE of 95.7% at −0.85 V vs. RHE. Theoretical study revealed that Ag₁ was the active site for ECR. The high efficiency was attributed to the high electronic density of Ag₁, which facilitated the adsorption and activation of CO_2.

7.3 SUMMARY

Over the past several years, SACs have shown promising applications for ECR. The performance of CO production over non-noble metal SACs is comparable to noble metals. Especially, in some cases, the selectivity and current density for CO have reached the threshold for profitable ECR. However, many issues remain as significant challenges. For example, there is not a universal strategy for massive and economical fabrication of SACs, which needs to be considered for the large-scale application of ECR. Most of the reported SACs afford CO from ECR with several reports yielding other C1 products (e.g., formate, methane, and methanol). It is promising yet challenging to design SACs with unique electronic structures for the formation of multi-carbon products. Although much progress has been made, it is still difficult to determine the genuine catalytic active sites for ECR. Future efforts should be devoted to the combination of in-situ characterizations and theoretical studies for further understanding of the evolution of the single-atom sites and the catalytic mechanism.

REFERENCES

1. M. He, Y. Sun, B. Han, *Angew. Chem. Int. Ed.*, 2013, 52, 9620–9633.
2. O. S. Bushuyev, P. De Luna, C. T. Dinh, I. Tao, G. Saur, J. van de Lagemaat, S. O. Kelley, E. H. Sargent, *Joule*, 2018, 2, 825–832.
3. Y. Yang, Y. Zhang, J.-S. Hu, L.-J. Wan, *Acta Phys.-Chim. Sin*, 2020, 36, 1906085.
4. Y.-H. Wang, W.-J. Jiang, W. Yao, Z.-L. Liu, Z. Liu, Y. Yang, L.-Z. Gao, *Rare Metals*, 2021, 40, 2327–2353.
5. J. Chen, T. Wang, Z. Li, B. Yang, Q. Zhang, L. Lei, P. Feng, Y. Hou, *Nano Res.*, 2021, 14, 3188–3207.
6. Y. Zhu, X. Cui, H. Liu, Z. Guo, Y. Dang, Z. Fan, Z. Zhang, W. Hu, *Nano Res.*, 2021, 14, 4471–4486.
7. S. Zhang, X.-T. Gao, P.-F. Hou, T.-R. Zhang, P. Kang, *Rare Metals*, 2021, 40, 3117–3124.
8. N. Corbin, J. Zeng, K. Williams, K. Manthiram, *Nano Res.*, 2019, 12, 2093–2125.
9. F. Li, D. R. MacFarlane, J. Zhang, *Nanoscale*, 2018, 10, 6235–6260.
10. Y. Wang, P. Han, X. Lv, L. Zhang, G. Zheng, *Joule*, 2018, 2, 2551–2582.
11. F. Pan, Y. Yang, *Energy Environ. Sci.*, 2020, 13, 2275–2309.
12. Y. Peng, B. Lu, S. Chen, *Adv. Mater.*, 2018, 30, 1801995.
13. B. Qiao, A. Wang, X. Yang, L. F. Allard, Z. Jiang, Y. Cui, J. Liu, J. Li, T. Zhang, *Nat. Chem.*, 2011, 3, 634–641.
14. G. Ding, L. Hao, H. Xu, L. Wang, J. Chen, T. Li, X. Tu, Q. Zhang, *Commun. Chem.*, 2020, 3, 43.
15. L. Hao, A. Auni, G. Ding, X. Li, H. Xu, T. Li, Q. Zhang, *RSC Adv.*, 2021, 11, 25348–25353.
16. X. Su, X.-F. Yang, Y. Huang, B. Liu, T. Zhang, *Acc. Chem. Res.*, 2019, 52, 656–664.
17. Z. Song, L. Zhang, K. Doyle-Davis, X. Fu, J.-L. Luo, X. Sun, *Adv. Energy Mater.*, 2020, 10, 2001561.
18. R. Liu, H.-L. Fei, G.-L. Ye, *Tungsten*, 2020, 2, 147–161.
19. Z.-X. Wei, Y.-T. Zhu, J.-Y. Liu, Z.-C. Zhang, W.-P. Hu, H. Xu, Y.-Z. Feng, J.-M. Ma, *Rare Metals*, 2021, 40, 767–789.
20. Y. Cheng, S. Yang, S. P. Jiang, S. Wang, *Small Methods*, 2019, 3, 1800440.
21. M. Li, H. Wang, W. Luo, P. C. Sherrell, J. Chen, J. Yang, *Adv. Mater.*, 2020, 32, 2001848.
22. T. N. Nguyen, M. Salehi, Q. V. Le, A. Seifitokaldani, C. T. Dinh, *ACS Catal.*, 2020, 10, 10068–10095.
23. D. K. Yadav, D. K. Singh, V. Ganesan, *Curr. Opin. Electrochem.*, 2020, 22, 87–93.
24. C. Xu, A. Vasileff, Y. Zheng, S.-Z. Qiao, *Adv. Mater. Interfaces*, 2021, 12, 6800–6819.
25. J. Zhang, W. Cai, F. X. Hu, H. Yang, B. Liu, *Chem. Sci.*, 2021, 12, 6800–6819.
26. A. S. Varela, N. Ranjbar Sahraie, J. Steinberg, W. Ju, H.-S. Oh, P. Strasser, *Angew. Chem. Int. Ed.*, 2015, 54, 10758–10762.
27. P. Su, K. Iwase, S. Nakanishi, K. Hashimoto, K. Kamiya, *Small*, 2016, 12, 6083–6089.
28. C. Zhao, X. Dai, T. Yao, W. Chen, X. Wang, J. Wang, J. Yang, S. Wei, Y. Wu, Y. Li, *J. Am. Chem. Soc.*, 2017, 139, 8078–8081.
29. X. Li, W. Bi, M. Chen, Y. Sun, H. Ju, W. Yan, J. Zhu, X. Wu, W. Chu, C. Wu, Y. Xie, *J. Am. Chem. Soc.*, 2017, 139, 14889–14892.
30. C. Zhao, Y. Wang, Z. Li, W. Chen, Q. Xu, D. He, D. Xi, Q. Zhang, T. Yuan, Y. Qu, J. Yang, F. Zhou, Z. Yang, X. Wang, J. Wang, J. Luo, Y. Li, H. Duan, Y. Wu, Y. Li, *Joule*, 2019, 3, 584–594.

31. T. Zheng, K. Jiang, N. Ta, Y. Hu, J. Zeng, J. Liu, H. Wang, *Joule*, 2019, 3, 265–278.
32. G. K. S. Prakash, F. A. Viva, G. A. Olah, *J. Power Sour.*, 2013, 223, 68–73.
33. R. J. Lim, M. Xie, M. A. Sk, J.-M. Lee, A. Fisher, X. Wang, K. H. Lim, *Catal. Today*, 2014, 233, 169–180.
34. D. M. Weekes, D. A. Salvatore, A. Reyes, A. Huang, C. P. Berlinguette, *Acc. Chem. Res.*, 2018, 51, 910–918.
35. M. Mushtaq, X.-W. Guo, J.-P. Bi, Z.-X. Wang, H.-J. Yu, *Rare Metals*, 2018, 37, 520–526.
36. G. Centi, *SmartMat*, 2020, 1, e1005.
37. S. Ma, P. Su, W. Huang, S. P. Jiang, S. Bai, J. Liu, *ChemCatChem*, 2019, 11, 6092–6098.
38. H.-Y. Jeong, M. Balamurugan, V. S. K. Choutipalli, E.-S. Jeong, V. Subramanian, U. Sim, K. T. Nam, *J. Mater. Chem. A*, 2019, 7, 10651–10661.
39. M. Jouny, W. Luc, F. Jiao, *Ind. Eng. Chem. Res.*, 2018, 57, 2165–2177.
40. D. Gao, R. M. Arán-Ais, H. S. Jeon, B. Roldan Cuenya, *Nat. Catal.*, 2019, 2, 198–210.
41. R. M. Arán-Ais, D. Gao, B. Roldan Cuenya, *Acc. Chem. Res.*, 2018, 51, 2906–2917.
42. B. A. Rosen, A. Salehi-Khojin, M. R. Thorson, Z. Wei, D. T. Whipple, P. Kenis, R. I. Masel, *Science*, 2011, 334, 643–644.
43. D. Yang, Q. Zhu, B. Han, *Innovation*, 2020, 1, 100016.
44. J. F. Wishart, *Energy Environ. Sci.*, 2009, 2, 956–961.
45. W. Ren, X. Tan, X. Chen, G. Zhang, K. Zhao, W. Yang, C. Jia, Y. Zhao, S. C. Smith, C. Zhao, *ACS Catal.*, 2020, 10, 13171–13178.
46. S. Ringe, E. L. Clark, J. Resasco, A. Walton, B. Seger, A. T. Bell, K. Chan, *Energy Environ. Sci.*, 2019, 12, 3001–3014.
47. S.-G. Han, D.-D. Ma, S.-H. Zhou, K. Zhang, W.-B. Wei, Y. Du, X.-T. Wu, Q. Xu, R. Zou, Q.-L. Zhu, *Appl. Catal., B*, 2021, 283, 119591.
48. M. D. Hossain, Y. Huang, T. H. Yu, W. A. Goddard Iii, Z. Luo, *Nat. Commun.*, 2020, 11, 2256.
49. V. Tripkovic, M. Vanin, M. Karamad, M. E. Björketun, K. W. Jacobsen, K. S. Thygesen, J. Rossmeisl, *J. Phys. Chem. C*, 2013, 117, 9187–9195.
50. T. N. Huan, N. Ranjbar, G. Rousse, M. Sougrati, A. Zitolo, V. Mougel, M. Fontecave, *ACS Catal.*, 2017, 7, 1520–1525.
51. C. Zhang, S. Yang, J. Wu, M. Liu, S. Yazdi, M. Ren, J. Sha, J. Zhong, K. Nie, A. S. Jalilov, Z. Li, H. Li, B. I. Yakobson, Q. Wu, E. Ringe, H. Xu, P. M. Ajayan, J. M. Tour, *Adv. Energy Mater.*, 2018, 8, 1703487.
52. H. Zhang, J. Li, S. Xi, Y. Du, X. Hai, J. Wang, H. Xu, G. Wu, J. Zhang, J. Lu, J. Wang, *Angew. Chem. Int. Ed.*, 2019, 58, 14871–14876.
53. D. Liu, X. Li, S. Chen, H. Yan, C. Wang, C. Wu, Y. A. Haleem, S. Duan, J. Lu, B. Ge, P. M. Ajayan, Y. Luo, Y. Jiang, L. Song, *Nat. Energy*, 2019, 4, 512–518.
54. J. Zhang, Y. Zhao, C. Chen, Y.-C. Huang, C.-L. Dong, C.-J. Chen, R.-S. Liu, C. Wang, K. Yan, Y. Li, G. Wang, *J. Am. Chem. Soc.*, 2019, 141, 20118–20126.
55. A. Wang, J. Li, T. Zhang, *Nat. Rev. Chem.*, 2018, 2, 65–81.
56. C.-H. Yang, F. Nosheen, Z.-C. Zhang, *Rare Metals*, 2021, 40, 1412–1430.
57. F. Pan, B. Li, E. Sarnello, Y. Fei, X. Feng, Y. Gang, X. Xiang, L. Fang, T. Li, Y. H. Hu, G. Wang, Y. Li, *ACS Catal.*, 2020, 10, 10803–10811.
58. F. Pan, B. Li, E. Sarnello, Y. Fei, Y. Gang, X. Xiang, Z. Du, P. Zhang, G. Wang, H. T. Nguyen, T. Li, Y. H. Hu, H.-C. Zhou, Y. Li, *ACS Nano*, 2020, 14, 5506–5516.
59. J. Gu, C.-S. Hsu, L. Bai, H. M. Chen, X. Hu, *Science*, 2019, 364, 1091–1094.
60. M. Liu, Y. Pang, B. Zhang, P. De Luna, O. Voznyy, J. Xu, X. Zheng, C. T. Dinh, F. Fan, C. Cao, F. P. G. de Arquer, T. S. Safaei, A. Mepham, A. Klinkova, E. Kumacheva, T. Filleter, D. Sinton, S. O. Kelley, E. H. Sargent, *Nature*, 2016, 537, 382–386.
61. C. Liu, Y. Wu, K. Sun, J. Fang, A. Huang, Y. Pan, W.-C. Cheong, Z. Zhuang, Z. Zhuang, Q. Yuan, H. L. Xin, C. Zhang, J. Zhang, H. Xiao, C. Chen, Y. Li, *Chem*, 2021, 7, 1297–1307.

62. Y. Pan, R. Lin, Y. Chen, S. Liu, W. Zhu, X. Cao, W. Chen, K. Wu, W.-C. Cheong, Y. Wang, L. Zheng, J. Luo, Y. Lin, Y. Liu, C. Liu, J. Li, Q. Lu, X. Chen, D. Wang, Q. Peng, C. Chen, Y. Li, *J. Am. Chem. Soc.*, 2018, 140, 4218–4221.

63. X. Wang, Z. Chen, X. Zhao, T. Yao, W. Chen, R. You, C. Zhao, G. Wu, J. Wang, W. Huang, J. Yang, X. Hong, S. Wei, Y. Wu, Y. Li, *Angew. Chem. Int. Ed.*, 2018, 57, 1944–1948.

64. Z. Geng, Y. Cao, W. Chen, X. Kong, Y. Liu, T. Yao, Y. Lin, *Appl. Catal. B*, 2019, 240, 234–240.

65. Z. Chen, K. Mou, S. Yao, L. Liu, *ChemSusChem*, 2018, 11, 2944–2952.

66. F. Yang, P. Song, X. Liu, B. Mei, W. Xing, Z. Jiang, L. Gu, W. Xu, *Angew. Chem. Int. Ed.*, 2018, 57, 12303–12307.

67. L. Han, S. Song, M. Liu, S. Yao, Z. Liang, H. Cheng, Z. Ren, W. Liu, R. Lin, G. Qi, X. Liu, Q. Wu, J. Luo, H. L. Xin, *J. Am. Chem. Soc.*, 2020, 142, 12563–12567.

68. A. Guan, Z. Chen, Y. Quan, C. Peng, Z. Wang, T.-K. Sham, C. Yang, Y. Ji, L. Qian, X. Xu, G. Zheng, *ACS Energy Lett.*, 2020, 5, 1044–1053.

69. G. Zhu, Y. Li, H. Zhu, H. Su, S. H. Chan, Q. Sun, *Nano Res.*, 2017, 10, 1641–1650.

70. X. Zhang, Y. Zhang, F. Li, C. D. Easton, A. M. Bond, J. Zhang, *Nano Res.*, 2018, 11, 3678–3690.

71. H. Yang, Y. Wu, G. Li, Q. Lin, Q. Hu, Q. Zhang, J. Liu, C. He, *J. Am. Chem. Soc.*, 2019, 141, 12717–12723.

72. F. Yang, X. Mao, M. Ma, C. Jiang, P. Zhang, J. Wang, Q. Deng, Z. Zeng, S. Deng, *Carbon*, 2020, 168, 528–535.

73. M. Jia, S. Hong, T.-S. Wu, X. Li, Y.-L. Soo, Z. Sun, *Chem. Commun.*, 2019, 55, 12024–12027.

74. Y. Zhou, N. Han, Y. Li, *Acta Phys.-Chim. Sin*, 2020, 36, 2001041.

75. H. Huang, H. Jia, Z. Liu, P. Gao, J. Zhao, Z. Luo, J. Yang, J. Zeng, *Angew. Chem. Int. Ed.*, 2017, 56, 3594–3598.

76. Q. He, J. H. Lee, D. Liu, Y. Liu, Z. Lin, Z. Xie, S. Hwang, S. Kattel, L. Song, J. G. Chen, *Adv. Funct. Mater.*, 2020, 30, 2000407.

77. W. Sheng, S. Kattel, S. Yao, B. Yan, Z. Liang, C. J. Hawxhurst, Q. Wu, J. G. Chen, *Energy Environ. Sci.*, 2017, 10, 1180–1185.

78. Q. Yi, F. Niu, *Rare Metals*, 2011, 30, 332.

79. E. Zhang, T. Wang, K. Yu, J. Liu, W. Chen, A. Li, H. Rong, R. Lin, S. Ji, X. Zheng, Y. Wang, L. Zheng, C. Chen, D. Wang, J. Zhang, Y. Li, *J. Am. Chem. Soc.*, 2019, 141, 16569–16573.

80. A. Zhang, Y. Liang, H. Li, X. Zhao, Y. Chen, B. Zhang, W. Zhu, J. Zeng, *Nano Lett.*, 2019, 19, 6547–6553.

81. J. Zhang, R. Yin, Q. Shao, T. Zhu, X. Huang, *Angew. Chem. Int. Ed.*, 2019, 58, 5609–5613.

82. W. Ma, S. Xie, X.-G. Zhang, F. Sun, J. Kang, Z. Jiang, Q. Zhang, D.-Y. Wu, Y. Wang, *Nat. Commun.*, 2019, 10, 892.

83. H. Shang, T. Wang, J. Pei, Z. Jiang, D. Zhou, Y. Wang, H. Li, J. Dong, Z. Zhuang, W. Chen, D. Wang, J. Zhang, Y. Li, *Angew. Chem. Int. Ed.*, 2020, 59, 22465–22469.

84. B. Kumar, V. Atla, J. P. Brian, S. Kumari, T. Q. Nguyen, M. Sunkara, J. M. Spurgeon, *Angew. Chem. Int. Ed.*, 2017, 56, 3645–3649.

85. S. Zhang, P. Kang, T. J. Meyer, *J. Am. Chem. Soc.*, 2014, 136, 1734–1737.

86. Y. Zhao, J. Liang, C. Wang, J. Ma, G. G. Wallace, *Adv. Energy Mater.*, 2018, 8, 1702524.

87. J. Wu, X. Bai, Z. Ren, S. Du, Z. Song, L. Zhao, B. Liu, G. Wang, H. Fu, *Nano Res.*, 2021, 14, 1053–1060.

88. W. Ni, Y. Gao, Y. Lin, C. Ma, X. Guo, S. Wang, S. Zhang, *ACS Catal.*, 2021, 5212–5221.

89. Q. Fan, P. Hou, C. Choi, T.-S. Wu, S. Hong, F. Li, Y.-L. Soo, P. Kang, Y. Jung, Z. Sun, *Adv. Energy Mater.*, 2020, 10, 1903068.

90. Z. Li, R. Wu, L. Zhao, P. Li, X. Wei, J. Wang, J. S. Chen, T. Zhang, *Nano Res.*, 2021, 14, 3795–3809.
91. J.-J. Wang, X.-P. Li, B.-F. Cui, Z. Zhang, X.-F. Hu, J. Ding, Y.-D. Deng, X.-P. Han, W.-B. Hu, *Rare Metals*, 2021, 40, 3019–3037.
92. N. Zhang, X. Zhang, L. Tao, P. Jiang, C. Ye, R. Lin, Z. Huang, A. Li, D. Pang, H. Yan, Y. Wang, P. Xu, S. An, Q. Zhang, L. Liu, S. Du, X. Han, D. Wang, Y. Li, *Angew. Chem. Int. Ed.*, 2021, 60, 6170–6176.

8 Electrochemical Nitrogen Reduction Application of Atomically Dispersed Metallic Materials

Revanasiddappa Manjunatha,
Shu-Qi Deng, Ejikeme Raphael Ezeigwe,
Wei Yan, and Jiujun Zhang
Shanghai University

Li Dong
Shanghai University
Zhaoqing Leoch Battery Technology Co. Ltd

CONTENTS

DOI: 10.1201/9781003153436-8

8.1 INTRODUCTION

The inevitable and continuous growth of the human population has spawned a persistently increasing demand for energy, the bulk of which has been generated from fossil fuels. However, in light of the planet's constantly decreasing fossil fuel reserves and the environmental pollution caused by the consumption of fossil fuels, intense efforts have been invested in exploring alternative energy production technologies based on clear and sustainable energy sources such as solar, wind, waterfall, etc., which are environmentally friendly and energetically satisfactory.[1] However, the electrical energy generated from such sustainable sources is intermittent and needs to be smoothed and stored for practical applications. For electricity energy storage and conversion, water electrolysis to produce hydrogen has been recognized as a feasible, reliable, efficient, and practical option. Unfortunately, hydrogen gas cannot be liquidized by any pressure at ambient temperature, which makes a difficult for its storage and transportation. Therefore, hydrogen storage has been identified as a challenge for its application as a fuel. There are many ways to store hydrogen including converting it into other chemicals such as ammonia, methanol, borohydride, etc., liquidation at extremely low temperature such as below $-253°C$, compression at high pressure, and physical adsorption in some solid inorganic chemicals and liquid organic chemicals. As a hydrogen carrier, ammonia (NH_3) shows promise to become one of the best substitutes for pure hydrogen because of its relatively high hydrogen content of ~17.65 wt%. Indeed, as a totally carbon-free fuel (i.e., zero emission of "green-house" gasses) that can be easily liquefied at 10 atm and ambient temperature, ammonia is a much safer source of energy than pure hydrogen (ammonia has a very narrow lower-upper flammability limit range from 16% to 25% v/v in comparison with 4%–75% v/v for hydrogen).[2,3] As of now, about 80% of the produced ammonia is being used to synthesize fertilizers for agriculture purposes and the remaining 20% is utilized in various applications such as dry industries, pharmaceuticals, and explosives.[4]

Currently, the most common process for industrial-scale ammonia production is the Haber–Bosch process. Based on the catalytic synthesis of ammonia from pure nitrogen and hydrogen at high pressures (150–300 atm) and temperatures (400°C–500°C), the Haber–Bosch process uses iron and ruthenium (the latter is used in the Kellogg Advanced Ammonia Process) as catalysts.[5] The reaction of ammonia synthesis is entirely controlled by the equilibrium, which can be shifted under high pressure in the direction of product formation in accordance with *Le Chatelier's* principle. Although the thermochemistry suggests that the reactor should be externally cooled, the high activation energy barrier for the dissociation of nitrogen molecules ($E_{(N≡N)} = 941$ kJ mol^{-1}) dictates the need for high operating temperatures.[6] Most of the hydrogen generated in the Haber–Bosch procedure is produced by using methane steam reformation, the only products of which are hydrogen and CO_2. This places the Haber–Bosch method of ammonia synthesis at the top of the list of industrial fabrication processes directly responsible for the emission of greenhouse gases.[7] Therefore, novel electrocatalysts are required for the efficient and "green" electrosynthesis of ammonia directly from nitrogen (or air as a nitrogen source) and water (i.e., a source of protons).

Electrochemical nitrogen reduction reaction (NRR) to NH_3 using water and nitrogen at ambient conditions is the most sustainable and CO_2-free approach. However, there are two major bottleneck issues associated with NRR in aqueous solutions.[8,9] From a thermodynamic perspective, the breaking of the most stable $N\equiv N$ bond requires a high reduction potential that causes parasitic hydrogen evolution reaction (HER), as a result, extremely low Faradaic efficiency under ambient conditions; from a kinetic point of view, NRR in aqueous conditions has a low probability of occurrence for the endergonic charge-transfer steps involve solvent reorganization, resulting in sluggish reaction kinetics.[10] Therefore, the scientific community must explore new catalysts that can promote NRR at low overpotential and suppress the competing HER to achieve a maximum NH_3 formation rate. The reaction kinetic could be improved by lowering the energy barrier of the rate-determining step. Thus, various electrocatalysts ranging from noble metals to metal-free materials have been employed for NRR.[11,12] Although theoretical and experimental investigations have demonstrated that noble metal-based materials could catalyze NRR, their scarce availability, and high-cost limit widespread applications.[13]

Generally, electrocatalysts' activity and selectivity mainly depend on the number of active sites and intrinsic activity of each of its sites.[14] Reduction of the overall particle size of electrocatalyst can increase the number of active sites by exposing a much larger surface area and the intrinsic activity by varying the electronic structure and surface configuration for the quantum size effect.[15] Although the solution sounds very simple, practically, it is difficult to reduce the particle size of the electrocatalyst because of small particles' high surface energy. As a result, they can easily agglomerate and lead to instability of the catalyst.[15] Fortunately, continuous efforts from the scientific community have resulted in rapid development in this research area.[16,17] The synthesis of monodispersed nanocrystals with uniform size unravels the relation between catalytic activity and its size.[18] Besides, the controllable preparation of nanocrystals with well-defined specific planes can help understand the effect of nanocrystal planes toward the catalytic performance and the role of shape-dependent catalysis.[19] Apart from size and shape-dependent catalysis, in bimetallic or multi-metallic, and intermetallic nanocrystals, the catalytic performance depends on synergistic effect, strain effect and ligand effects as well. Therefore, a thorough understanding of all these effects could pave the way to the rational design of highly active next-generation nanocatalysts.[20]

With the rapid progress of nanotechnologies, the development of nanoscience and the advancement of characterization techniques have enabled further reduction of nanocrystal into atomically isolated metal atoms embedded in suitable supporting materials. This one-of-a-kind new catalyst is popularly known as a single atomic site heterogeneous catalyst or single-atom catalyst (SAC). Although in 2003, Fu et al.[21] and in 2006, Bashyam and Zelenay[22] proposed convincing evidence that catalytic activity was not due to the metal nanoparticles instead of ionic or oxidized metal species.

Exactly after five years, by employing advanced experimental and characterization techniques, for the first time, Qiao et al.[23] synthesized a SAC that consisted of isolated platinum (Pt) single atoms anchored onto the surface of iron oxide nano-crystallites. In just a decade, single-atom catalysis presented a new paradigm in heterogeneous catalysis and became a highly transversal field of contemporary chemical research.[24] SACs have several advantages such as the maximum utilization efficiency of metal (almost 100%), excellent quantum size effect, unsaturated coordination environment, strong metal-support interaction, and unique electronic structure compared to nanoparticle counterparts, as shown in Figure 8.1A.[25] Besides, to avoid agglomeration into particles, SACs possess strong interaction or considerable charge transfer between single metal atoms and coordination species of solid support. As a result, SACs obtain a unique electronic structure carrying some charges, distinguishable from conventional metal nanoparticles.[20] It is worth mentioning that SACs not only have inherent advantages of a homogeneous catalyst such as highly uniform active sites and tunable coordination environment but also has virtues of heterogeneous catalyst like excellent recyclability and ease of separation from the reaction mixture, thus, which can patch up the existing gap between homogeneous and heterogeneous catalysts, as illustrated in Figure 8.1B.[25] Owing to all mentioned merits of SACs, since the last decade, this area of research has evolved significantly, as shown in Figure 8.1C.[24]

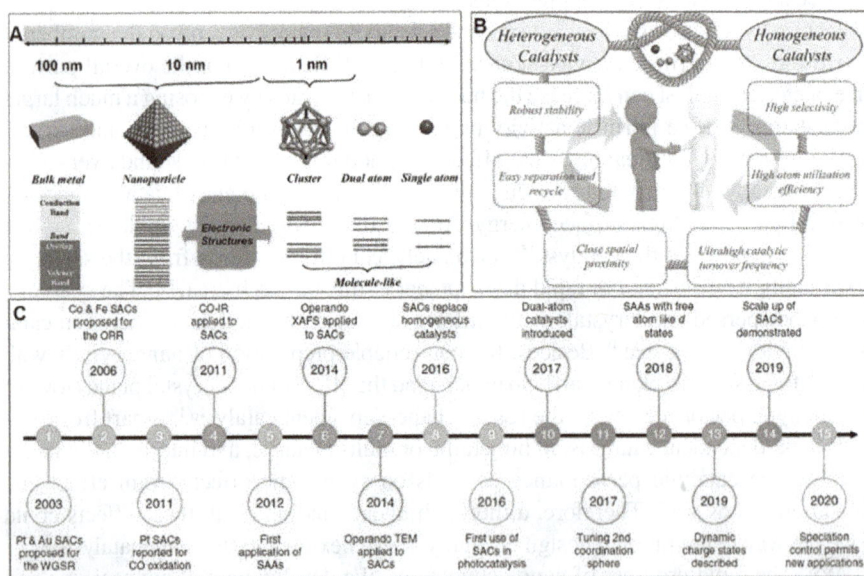

FIGURE 8.1 Electronic structures of bulk metal, nanoparticles, and single-atom (A). SAC possesses advantages from both homogeneous and heterogeneous catalysts (B). (Adopted from Ref.[25]) Progress in single-atom catalysis. ORR, oxygen reduction reaction; WGSR, water-gas shift reaction; SAA, single-atom alloy; TEM, transmission electron microscopy; XAFS, X-ray absorption fine structure; CO-IR, infrared spectroscopy of adsorbed carbon monoxide (C). (Reproduced from Ref.[24])

In this chapter, the recent advances in wet-chemistry SAC synthesis are summarized systematically, and the detail how to realize the atomic dispersions of metal precursors on the various supports and stabilize the as-prepared single atoms against migration and agglomeration are also discussed, which are key points in SACs synthesis. Inspired by the unique electronic and structural properties of SACs, a large number of research groups have focused on the improvement of electrochemical NRR by employing SACs. Due to the rapid development in this research area, several research articles have been published in the last few years. Therefore, this chapter covers difficulties involved in NRR in ambient conditions and analyzes how to mitigate those hurdles using SACs. Furthermore, this chapter also includes the preliminary theoretical screening of SACs for their potential applications in NRR. Finally, the main challenges and future opportunities for the rapid development of SACs in electrocatalytic NRR applications are discussed.

8.2 SYNTHETIC STRATEGIES FOR SACs

As discussed in the introduction section, owing to the high surface energy, SACs always tend to aggregate into nanoclusters or nanoparticles to attain stability. Therefore, preserving the atomic dispersion of metal species after the preparation of SACs is quite challenging. In this regard, atomic layer deposition[26] and mass-selected soft-landing[27] methods have attracted researchers' attention. However, low product yield and the requirement of expensive instruments limit their applications in large-scale productions of SACs.[27] Wet-chemistry synthetic routes have gained popularity because they do not require specialized equipment and are feasible for large-scale manufacturing.[28] Through wet-chemistry synthesis, researchers have employed various strategies to prepare SACs such as impregnation and coprecipitation, coordination design, spatial confinement, and defect engineering. All these different approaches have important and common goals in the synthesis of SACs, which are (i) uniform atomic dispersion all over the support, (ii) strong interaction between isolated single atoms and atoms of supporting material, (iii) enhancement of electron and mass transports as well as better active-site accessibility, and (iv) improved catalytic activity and durability.[29]

8.2.1 Impregnation and Coprecipitation Strategies

Selection of suitable supports with a high specific area and appropriate quantity of metal loading are key parameters involved in SACs preparation when using impregnation and coprecipitation strategies. On carbon-based supporting materials, geometrically isolated single atoms are generally anchored to heteroatoms such as nitrogen, phosphorous, and boron, through covalent tethering, and π–π interactions. However, in the case of metal-oxide solid supports, defects serve as the anchoring sites. For instance, Yang et al.[30] utilized available nitrogen vacant sites in titanium nitride (TiN) to anchor Pt single atoms using the incipient wetness impregnation method. As observed, to obtain Pt single atoms on TiN support (Pt/TiN), the atomic concentration of Pt played a crucial role. When Pt amount was increased from 0.35 to 2 wt%, single atoms started agglomerating into nanoparticles. Interestingly, Pt/TiN catalyst showed better catalytic activity for hydrogen peroxide (H_2O_2) production,

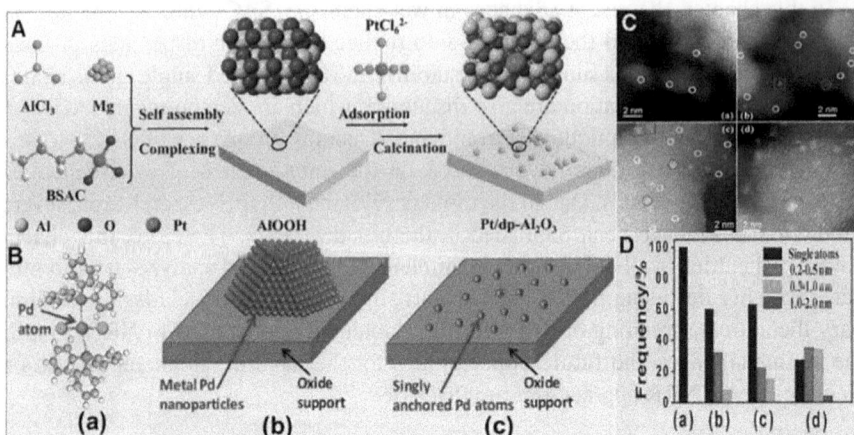

FIGURE 8.2 Schematic representation of the synthesis of Pt/dp–Al$_2$O$_3$ (A). (Reproduced from Ref.[31]) Schematics of a structural model of Pd-based homogenous molecular catalyst Pd(PPh$_3$)$_2$Cl$_2$ (a), Pd nanoparticle catalyst supported on metal oxide (b), and Pd SAC supported on metal oxide (Pd$_1$/TiO$_2$) (c) (B). (Reproduced from Ref.[32]) High-angle annular dark-field scanning transmission electron microscopy (HAADF-STEM) (C) and the frequencies of the observed size scope of Ir/FeO$_x$ with the Ir loadings from 0.01 to 2.40 wt% (a) Ir$_1$/FeO$_x$; (b) Ir/FeO$_x$-0.22; (c) Ir/FeO$_x$-0.32; (d) Ir/FeO$_x$-2.40 (D) (Reproduced from Ref.[36]).

unlike nanoparticle counterpart that was good at four electrons oxygen reduction to water. Shang et al.[31] deliberately prepared defect-rich porous alumina (Al$_2$O$_3$) from crystalline AlOOH sheets using the solvothermal method. They exploited defects, high surface area, and porous structure of Al$_2$O$_3$ to decorate Pt SAC by impregnation, as shown in Figure 8.2A. Lattice oxygen of metal-oxide support could also serve to tether precious metal-single atoms. For example, Zhang et al.[32] synthesized palladium (Pd) single atoms supported on TiO$_2$. Density functional theory (DFT) of TiO$_2$ (101) surface, X-ray absorption near-edge structure (XANES), and extended X-ray absorption fine structure (EXAFS) experiments revealed, on average, that each Pd atom could bond with about four oxygen atoms of TiO$_2$ surface.

In typical organic C–C coupling reactions, Pd complex-based homogenous catalyst such as Pd(PPh$_3$)$_2$Cl$_2$ (Figure 8.2B(a)) has been widely used. As elaborately discussed in the introduction, this type of homogeneous catalyst suffers from separation and recycling. Pd nanoparticles decorated on metal-oxide support (Figure 8.2B(b)) could tackle these limitations; however, multiple metal atoms with different coordination and chemical environment of Pd nanoparticles could hamper the selectivity of the reaction. Pd single atoms anchored on nonmetallic support such as TiO$_2$ (Figure 8.2B(c)) are expected to have a minimal number of choices of binding sites for reactants or intermediates, leading to enhanced selectivity of the organic reaction. Carbon-based supporting materials can provide better electron conductivity, large surface area, and stability under corrosive conditions compared with low electron conductive metal-oxide counterparts. Choi et al.[33] prepared atomically dispersed Pt atoms on sulfur-doped zeolite templated carbon. DFT calculations, XANES, and

EXAFS experiments suggested that Pt single atoms could be anchored to the sulfur content of the carbon in the form of Pt–S_4. High sulfur content (17 wt%) of the as-prepared carbon enabled the Pt atom loading to 5 wt%. As a result, the as-prepared catalyst exhibited enhanced electrochemical H_2O_2 production with 96% selectivity in an acidic medium. Cao et al.[34] have chosen nitrogen-doped carbon for atomically dispersed Ru_1–N_4 on nitrogen-carbon support. Due to its unique structure, single-atom Ru–N–C displayed exceptional catalytic activity toward oxygen evolution reaction (OER) in an acidic medium. Besides, the catalyst did not deactivate or decompose after a continuous 30- h operation. Instead of oxidizing aqueous solutions, Sun et al.[35] used a low-boiling-point, low-polarity solvent such as acetone for atomically dispersed noble metal species on activated carbon by the impregnation method. Furthermore, SACs synthesis using organic solvent not only opens up the involvement of a wide choice of metal precursors such as organometallic salts but also reduces catalyst drying temperature due to their low boiling points.

Generally, using the coprecipitation method, metal-oxide support, and noble metal-single atoms can be prepared simultaneously by controlling pH, solution temperature, and noble metal loading. For example, Lin et al.[36] synthesized iridium (Ir) single atoms on iron oxide substrate (Ir_1/FeO_x) at an optimized 80°C and the controlled pH of the solution at exactly 8. The obtained coprecipitated product was dried at 80 °C followed by a reduction in 10% H_2/He at 300 °C for 0.5 h. It was observed that when Ir loading was increased from 0.01 to 0.22 wt.% along with 60% of single atoms, about 40% of Ir clusters were formed. When the loading was further increased, the cluster size was increased as well, as shown in Figure 8.3C and D. In a few cases, metal-oxide support was prepared by the coprecipitation method, and subsequently metal (noble) single atoms were loaded on support through an electrostatic adsorption process.[37] This process depended entirely on the metal-oxide surface charge and pH of the solution.[20]

The surface charge of metal oxide can be modified into either positive or negative charges by controlling the solution pH. At the point of zero charges (PZC), the metal oxide will have hydroxyl groups, which do not react with metal ions. For instance, surface hydroxyl groups of metal oxide can be converted into positively charged cations by reducing solution pH below the PZC. The negatively charged noble metal anions readily adsorb onto the metal-oxide cations.[37] On the contrary, metal-oxide surface charge can be converted into natively charged anions by increasing the pH above the PZC. Thus, positively charged noble metal cations can be loaded using electrostatic adsorption.[38] In conclusion, impregnation and coprecipitation strategies are a good choice for noble SAC preparation on various metal-oxide supports.

8.2.2 Spatial Confinement Strategy

In this spatial confinement strategy, single metal atoms are spatially confined into molecular cages to subdue migration. This technique involves two steps. The first step is to prepare highly spatial distributed and atomically dispersed metal species by porous materials such as zeolites, metal-organic frameworks (MOFs), and covalent organic frameworks (COFs); the second step is to pyrolyze the prepared porous materials at high temperatures to eliminate organic framework. During pyrolysis, SACs are formed and stabilized by the skeletons of molecular cages.[28]

Zeolites, a type of molecular cages, are aluminosilicate-based highly uniform microporous materials made of SiO_4 and AlO_4 tetrahedral crystalline networks. Several tetrahedrally coordinated elements can connect through bridging oxygen atoms to form rings, cavities, and channels of the zeolite framework that can host various single metal cations.[20] For example, Qiu et al.[39] prepared pure siliceous MFI-type zeolite silicalite-1 (S-1), in which ruthenium (Ru) amine complex was encapsulated. The vacuum decomposition of the resultant intermediate at 450 °C led to the formation of Ru single atom stabilized and supported on S-1 zeolite oxygen atoms (Ru SA/S-1), as shown in Figure 8.3A. Remarkably, this Ru SA/S-1 catalyst exhibited enhanced catalytic NRR compared with the Ru nanoparticle counterpart. Sun et al.[40] employed the same S-1 zeolite to encapsulate rhodium (Rh) single atoms. Like the previous report, in this case, Rh single atoms were within five-membered rings and stabilized by oxygen atoms of the zeolite framework. The resultant catalyst exhibited excellent hydrogen generation rates from ammonia borane hydrolysis. Besides, this catalyst also showed better catalytic performance in shape-selective tandem hydrogenation of various nitroarenes. Although zeolite-based SACs showed high catalytic activity and chemoselectivity in some reactions, the microporosity and small cavities could hamper mass transport. Therefore, new approaches are needed to fix this limitation and further improve the performance of zeolite-based SACs.[20]

MOFs are a new emerging class of porous crystalline materials, also known as porous coordination polymers (PCPs), which have been extensively used as supports in the synthesis of SACs because of their high surface areas, well-defined order,

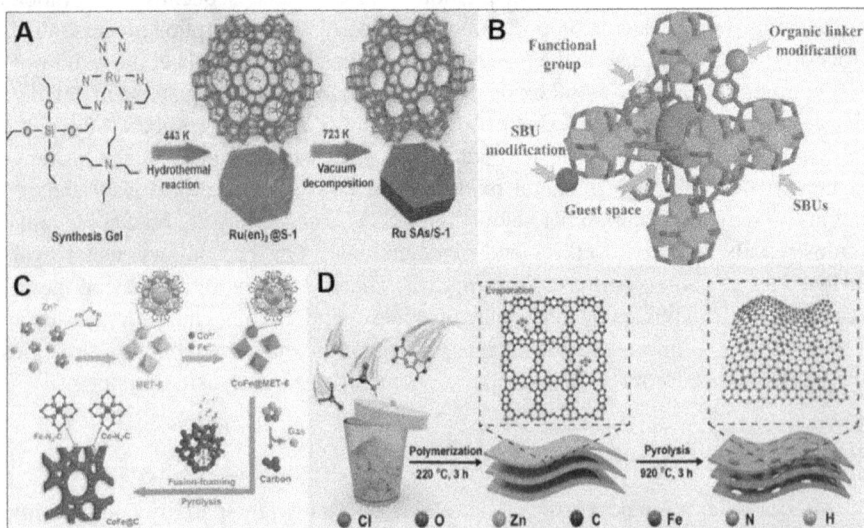

FIGURE 8.3 Schematic representation of the synthesis of Ru SAs/S-1 (A). (Reproduced from Ref.[39]) Various possibilities for the structural modifications of MOF (B). (Reproduced from Ref.[25]) Schematic representation of CoFe@C synthetic route using the fusion-foaming thermal transformation (C). (Reproduced from Ref.[45]) Polymerization-pyrolysis-evaporation strategy for Fe–N_4 SAs/NPC synthesis (D). (Reproduced from Ref.[48])

and flexible fine-tuneability.[41] MOFs can be synthesized using inorganic nodes (also known as secondary building units (SBUs)) and organic ligands such as carboxylate, nitrogen-donor groups, sulfonate, or phosphonate. It is worth mentioning that compared to other porous materials, a plethora of structural modifications of MOFs are its best advantage, as shown in Figure 8.3B.[25] The large pore size of MOF provides enough space for the mononuclear metal spices accommodation, and nitrogen-containing ligands offer nitrogen-doped carbon, which serve as single-atom anchoring sites. The best example is zeolitic imidazolate frameworks (ZIFs), in particular ZIF-8. It consists of a pore size of 3.4 Å, and a nanocavity diameter is ca. 11.6 Å, which is large enough to encapsulate numerous mononuclear metal precursors.[29] Using ZIF-8, Wang et al.[42] prepared iron single atoms anchored on nitrogen-rich carbon nanocages and used to accelerate polysulfide redox conversion for lithium–sulfur batteries. The coordination environment of MOF-derived metal-nitrogen-carbon (M–N–C) catalysts can be modified according to catalytic requirements. For example, Gong et al.[43] prepared a series of single-atom nickel catalysts (Ni_{SA}–N_x–C) with different N coordination numbers by temperature-controlled MOF pyrolysis. This Ni_{SA}–N_2–C, with the lowest N coordination number, exhibited a superior catalytic activity toward CO_2 reduction reaction to Ni_{SA}–N_3–C and Ni_{SA}–N_4–C catalysts.

Addition of dopant such as zinc can play a crucial role in the spatial isolation of mononuclear metal species in the ZIF framework. For instance, Han et al.[44] illustrated a catalyst derived from MOF without zinc dopant, which could give the N-doped carbon embedded Co nanoparticles (CoNPs–N–C). However, Co size was reduced to atomic clusters and further reduced to CoSA after the optimized addition of dopant. Dual-atom metal doped with heteroatom carbon-based catalysts can also be prepared using MOFs. The additional single-atom presence with different chemical environments can provide more catalytic sites than SACs counterparts.[25] However, in developing a facile and scalable dual-atom metal catalyst, synthesis is quite challenging. Zhao et al.[45] prepared atomically dispersed Co and Fe dual atoms embedded in submillimeter-scaled carbon networks with hierarchical porosity from the fusion-foaming thermal transformation of energetic MOFs, as shown n Figure 8.3C. The large carbon network with hierarchical pores could not only provide support to dual atoms but also be conducive to mass transport, electron transfer, and electrolyte penetration. Besides, the strong interaction of dual atoms with the supporting matrix led to superior ORR activity compared to SACs.

Compared to other strategies, the preparation of SACs using MOFs has several advantages such as uniform or precise metal doping with ease, and porosity of support could significantly contribute to catalytic performance. However, expensive ligands, limited self-life of MOFs, and complicated procedures are limitations and need to be addressed.

COFs are porous crystalline organic materials prepared from molecular building blocks by linking light elements such as boron, nitrogen, carbon, and oxygen through covalent bonds in a periodic manner.[46] These frameworks can provide a suitable space for anchoring mononuclear metal species into well-ordered porous structures. For instance, nitrogen atoms of the triazine framework were used to covalently anchor PtSAs.[47] This catalyst showed better ORR activity in an acidic medium with high methanol tolerance. Methanol oxidation mechanism confirmed that for the oxidation, at least three contiguous Pt sites were required. Thus, methanol oxidation on Pt single atomic sites embedded in the triazine framework was forbidden. Zhang et al.[46] prepared uniform molybdenum (Mo) single atomic sites by confining coordination between molybdenyl acetylacetonate and benzoyl salicylal hydrazone ligand. The obtained MoSAs embedded by two-dimensional COF provided the required micro-environment for the selective organic molecules' oxidation reaction.

8.2.3 COORDINATION SITE ATTACHMENT STRATEGY

Materials that contain heteroatoms with lone pairs of electrons such as N, P, O, and S can be used as the coordination sites to anchor metal single-atoms through the strong coordination interaction. These coordination sites act as "paws" to absorb or bind metal precursors or single atoms to avoid migration or agglomeration.[28] Various nitrogen-containing polymers have been employed for SACs synthesis. Pan et al.[48] used polymerization-pyrolysis-evaporation (PPE) strategy (as shown in Figure 8.3D) for the synthesis of atomically dispersed iron on nitrogen-doped porous carbon (NPE). This PPE strategy had two steps. In the first step, bimetallic Zn/Fe polyphthalocyanine (ZnFe-BPPc) conjugated polymer networks with homogeneous distribution of Fe and Zn by a low-temperature solvent-free solid-phase synthesis, and in subsequent step, ZnFe-BPPc was pyrolyzed at 920°C for 3 h. At high temperature, Zn atoms were evaporated and the formation of Fe–Fe bonds was prevented. Besides, PPc could be converted into N-doped porous carbon, and it simultaneously served as Fe anchoring sites. It was observed that the optimized mole ratio of Zn/Fe was crucial in obtaining FeSAs/NPC. In the absence of Zn, Fe–Fe bonds aggregated and formed nanoparticles. Wu et al.[49] prepared FeSAs supported on N and S co-doped hierarchical (Fe$_1$/N, S–PC) porous carbon using the coordination polymer method in the presence of o-phenylenediamine, ammonium persulfate, and potassium ferricyanide. Besides, SiO$_2$ nanoparticles were used in the synthesis as a sacrificial template to get microporous high surface area carbon. Additional S heteroatoms into the carbon structures not only helped stabilize FeSAs but also weakened the O–O bonding for ORR. Thus, the prepared catalyst showed higher activity than the S absent counterpart (Fe$_1$/N–PC).

Carbon-based materials such as carbon nanotubes (CNTs), graphene, and amorphous carbon have been employed as the supporting materials to anchor metal-single atoms due to their excellent electrical, thermal, and anti-corrosion properties. In particular, heteroatoms (N, S, P, B, etc.) doping into the carbon skeleton would induce charge redistribution and provide more anchoring sites.[20] CNTs could be loaded with Ni single atoms (NiSAs) at a relatively high-atomic rate (0.35%–0.88%) through the coordination of doped nitrogen.[50] XPS analysis of N-doped CNTs confirmed that the high loading of NiSAs was due to the high percentage of pyridinic N (47.93%). Pd SAC supported on N-doped graphene (Pd$_1$/N-graphene) was prepared through a freeze-drying-assisted method using dicyanamide, glucose, and Na$_2$PdCl$_4$.[51] The EXAFS analysis revealed that the strong coordination of Pd^{+2} by nitrogen atoms on the N-graphene support could result in atomically dispersed Pd atoms. It was observed that the free-drying step played a significant role in Pd$_1$/N-graphene formation. On contrary, heating of reactants and then pyrolysis could lead to Pd NPs on N-graphene. This Pd_1/N-graphene catalyst exhibited excellent durability for the photothermal hydrogenation of acetylene to ethylene at 125°C due to strong local coordination of Pd atoms by N atoms, which subdued Pd migration and aggregation. Graphene and CNTs hybrid were used as support for dual single atoms (Co, Mo) synthesis.[52] The presence of multiple active sites of the catalyst could have ORR and OER bifunctional activities. Apart from heteroatoms-doped CNTs and graphene, the first-principle theoretical calculations exhibited that transitional metal (Sc, Cu, Mo, Ru, Rh, Pd, Re, Ir, and Pt) atoms could be decorated on the monolayer $g-C_3N_4$ and use as efficient single-atom electrocatalysts for NH_3 production.[53] In particular, Pt atoms anchored on $g-C_3N_4$ could show robust stability, good electrical conductivity, and excellent NRR activity at room temperature. Both DFT and experimental studies suggested that graphdiyne could also be used as the support for anchoring Fe atoms through the formation of covalent Fe–C bonds.[54]

As identified, the preparation of SACs using a coordination strategy is dependent on the following major points.[28]

i. Uniform coordination sites on the support are quite essential to obtain unvarying single-atoms loading;
ii. the appropriate metal precursors should be selected based on their ability to interact with coordination sites;
iii. the reaction conditions should be precisely controlled, such as pH and solvent type to get uniform and reproducible SACs.

8.2.4 DEFECT ENGINEERING STRATEGY

The controlled construction of defects on supports to anchor single atoms is one of the popular techniques for SAC synthesis. Using this strategy, various defects, such as surface disorder, dislocation, heterogeneity, and vacancy, can be synthesized and effectively serve as the anchoring sites and restrain the migration of metal atoms to avoid migration and aggregation.[55] It is well known that strong metal-support interactions (SMSI) play a trivial role in the performance of SACs.[56] Metal oxides are known for SMSI and are thus considered ideal supports to anchor single atoms. Besides, electronic defects can be prepared with ease on their surface. For instance, gold (Au) single atoms could be effectively supported on the defective TiO_2 nanosheets.[55] When TiO_2 is heated at 200 °C under a reduced atmosphere, oxygen vacancy could give high surface defects. Au single atom was filled through Ti–Au–Ti structure formation, as shown in Figure 8.4A Both experimental and DFT studies proved that the defects in TiO_2 nanosheets not only provided good stability to Au single atoms but also promoted catalytic activity by reducing the energy barrier of the reaction. Dvořák et al.[57] adopted photoelectron spectroscopy, scanning tunneling microscopy, and DFT calculations to prove that Pt single atoms on CeO_2 could be stabilized by most common step edges surface defects. They adjusted step density on the CeO_2 support experimentally for maximizing the load of monodispersed Pt^{2+} ions.

Apart from the reducible oxides, γ-Al_2O_3 is one of the most commonly used catalyst supports in particle applications. Electronic defects are not present in the γ-Al_2O_3 in that case, Kwak et al.[58] reported that Al^{3+} centers of γ-Al_2O_3 [i.e., pentacoordinate

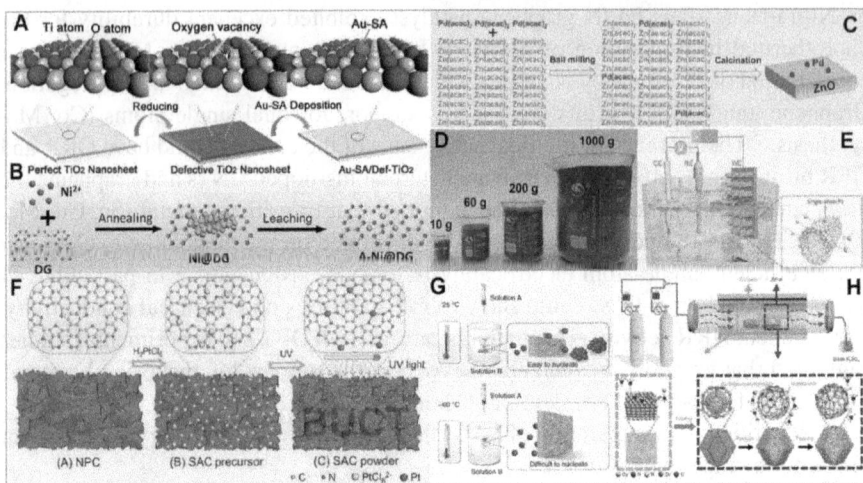

FIGURE 8.4 Schematic representation of Au–SA/TiO$_2$ synthesis (A). (Reproduced from Ref.[55]) Fabrication illustration of A–Ni@DG catalyst (B). (Reproduced from Ref.[62]) Schematic representation of Pd$_1$/ZnO synthesis (C). Digital image of different scale-up batches of Pd$_1$/ZnO catalyst (D). (Reproduced from Ref.[67]) Graphical representation of Pt SACs on CoP-based nano-tube (NT) arrays supported by aNifoam (NF) preparation (E). (Reproduced from Ref.[73]) Schematic illustration of the formation of Pt$_1$/NPC catalyst (F): nitrogen-doped porous carbon (NPC) substrate (a), PtCl$_6^{2-}$ ions adsorbed on the NPC (b), and Pt single atoms embedded on the NPC (c). (Reproduced from Ref.[76]) Schematics of the ultra-low-temperature solution reduction process. At ultra-low temperature, the nucleation of reduced metal atoms can be significantly reduced due to kinetic and thermodynamic processes (top figure). Contrary, at room temperature rapid nuclei formation takes place (bottom figure) (G). (Reproduced from Ref.[79]) Schematic representation of Cu–SA s/N–C synthesis (H). (Reproduced from Ref.[83])

Al^{3+} (Al$_{Penta}^{3+}$)] could serve as the Pt single atoms anchoring sites.[58] They observed that the atomic dispersion of Pt strongly was dependent on the number of available Al$_{Penta}^{3+}$ active sites. When Pt loading exceeded the number of Al$_{Penta}^{3+}$ active sites, raft-structured two-dimensional platinum oxide could be formed. Tang et al.[59] adopted an evaporation-induced self-assembly method to prepare Al$_{Penta}^{3+}$ rich γ-Al$_2$O$_3$. A high quantity of Ru single atoms (1 wt%) could be uniformly dispersed on the as-prepared defect-rich support. Strong interaction between Ru single atoms and Al$_{Penta}^{3+}$ sites was observed, which led to special geometric and electronic features of Ru species.

Despite various efforts put on metal-oxide-supported SACs, defective graphene matrices have been of particular interest because of their large surface area, high electron conductivity, chemical stability, and abundance of defects with or without heteroatoms for potential metal-support coordination.[60] In defective-free graphene, only edges can provide active sites because of their unsaturated coordination, resulting in low metal atoms loading. To solve this issue, Jia et al.[61] developed a defective-rich graphene (DG) with high density of defects using a facile nitrogen removal procedure, which provided a large number of active sites through the strong charge transfer between 2π antibonding state of the carbon atoms and the single metal atoms for better stability. The as-prepared DG was used to trap high loading of nickel (Ni) single atoms (NiSAs) (1.24 wt%) via pyrolysis and leaching steps, as shown in Figure 8.4B.[62] X-ray adsorption characterization and DFT calculations revealed that the variety of defects in graphene could induce different local electronic density of states of NiSAs, resulting in high specific electrocatalytic OER performance. Furthermore, NiSAs trapped on stone-wales graphene defect were found to be active for HER. Wu et al.[63] took Au as a model material and carried out systematic molecular dynamics simulations to study the atomic-scale mechanisms associated with the transformation of a metal nanoparticle into an array of stable SAs on defective carbon surface at high temperature. They found that the thermodynamically stable metal single-atom formation was dependent on the interaction between density/type of defects on the carbon surface, the cohesion energy between metal atoms, and the metal-carbon-binding energy.

In general, chemically exfoliated single-layered metal chalcogenides such as MoS_2, WS_2 contain a large concentration of sulfur vacancies on the basal plane. These sulfur vacancies can anchor metal-single atoms with high affinity. Liu et al.[64] prepared isolated Co atoms dispersed on sulfur vacancies of MoS_2 monolayers using thiourea-based Co species. Experimental studies and DFT calculation revealed that the strong covalent attachment of Co atoms to monolayer MoS_2 via sulfur vacancies could significantly improve the number of Co–S–Mo interfacial sites on the basal plane. Compared with traditionally synthesized $CoMOS_2$ catalyst, the catalyst of Co single atoms dispersed on monolayer MoS_2 showed superior selectivity, stability, and catalytic activity, which was due to the presence of a large number of Co–S–Mo interfacial sites.

Metal nitrides have high resistance to corrosion and acid attack, and superb electrical conducting qualities, and are suitable for supporting single atoms to form SACs. The best example is TiN. For example, Zhang et al.[65] utilized first-principles DFT calculations to show that Pt could be easily stabilized on the nitrogen vacancies of TiN to form single atoms rather than clusters. Yang et al.[30] also experimentally proved this point by preparing Pt single atoms decorated TiN.

Compared with other SAC synthetic strategies, the defect engineering process takes advantage of defects to trap metal precursors and stabilize the trapped metal atoms against sintering through an improved charge-transfer mechanism between defective sites and single atoms. Besides, by controlling the concentration of defects, metal loading can be fine-tuned.[28] Even though there are still many challenges to fully explore the SACs, it is believed that the newly established defect sites-stabilized single atoms mechanism will be helpful for designing and fabricating SACs for practical applications.[56]

8.2.5 Miscellaneous Strategies

8.2.5.1 Ball Milling/Mechanochemical Method

Recently, the ball-milling method has received considerable attention for SAC synthesis because of its unique advantages such as easy scale-up procedure, minimal/no solvent, and wide application range.[66] He et al.[67] reported the synthesis of PdSAs supported on ZnO (Pd$_1$/ZnO) using ball-milling of palladium acetylacetonate [Pd (acac)$_2$] and zinc acetylacetonate [Zn(acac)$_2$] salts for 10 h followed by calcination at

400 °C for 2 h in air. The key point in this synthesis was the weight ratios of $Zn(acac)_2$ and $Pd(acac)_2$ (400:1). After balling process, the average distance between $Pd(acac)_2$ was increased to a considerable extent, as shown in Figure 8.4C, thereby resisting the aggregation of Pd species during the calcination step. Benefited from the feasibility of mass production with a ball-milling approach, they could increase the synthesis scales of Pd_1/ZnO from 10 to 60, 200, and 1,000 g (Figure 8.4D) with phase uniformity. Similarly, Gan et al.[68] adopted this approach for the preparation of AuSAs on CeO_2 support (Au_1/CeO_2). They observed that four different scales of Au_1/CeO_2 (10, 50, 100, and 1,000 g batches) synthesis could give identical catalyst structure as well as catalytic performance.

High-energy ball milling has been employed to cut and reconstruct the chemical bonds of materials or molecules with required energy input. For instance, Deng et al. prepared a single Fe site confined in a graphene matrix by ball milling iron phthalocyanine (FePc) and graphene nanosheets (GNs) at 450 rotations per minute for 20 h. During high-energy ball milling, the outside macrocyclic structure of FePc could be broken, the residual isolated FeN_4 centers could interact with the GNs at the defected site, thereby forming FeN_4 centers embedded graphene matrix. Cui et al.[70] employed the same strategy for synthesizing various transition single metal (Mn, Fe, Co, Ni, or Cu) atoms doped GNs using metal phthalocyanine precursors. All these work suggest that the ball milling/mechanochemical method could be ideal for large-scale SAC synthesis.

8.2.5.2 Electrochemical Deposition Strategy

Single metal atoms can be easily electrodeposited on suitable support using dissolved metal ions in the electrolyte solution. The growth of the single metal atoms on the catalyst support strongly depends on the slow diffusion rate of the metal ions, which can be controlled by varying the concentration of the metal ion precursor. Besides, it's easy to manipulate the size and density of distributions by optimizing the electrodeposition time and metal ions concentration.[71] Thus, this method has been used as a universal route for fabricating a wide range of metal atoms (more than 30 different single atoms) on various supports.[72] Interestingly, the same metal single atoms electrodeposited by cathodic potentials exhibited contrasting electronic states with the anodic electrodeposited counterpart. As a result, the single-atom deposited on the same support showed different electrocatalytic activities for the electrochemical reactions. For example, cathodic deposited iridium (Ir) single atoms on $Co_{0.8}Fe_{0.2}Se_2$ showed better HER activity, and anodic deposited counterpart exhibited better OER due to the distinct electronic states. As mentioned above, when Ir metal ion concentration was increased up to 150 μM, IrSAs could be formed. When it was further raised to 200 μM, Ir clusters were formed at a mass loading of 4.7%. The number of scan cycles and the scan rate also had a severe impact on the size of Ir. Zhang et al.[73] developed a potential-cycling electrodeposition method with a three-electrode cell containing phosphate buffer saline to synthesize a catalyst with a large area of PtSAs on CoP-based nanotube arrays supported by a Ni foam, as shown in Figure 8.4E. The as-prepared binder-free catalyst showed an ultralow overpotential of 24 mV for HER, only 7 mV larger than that of the commercial Pt/C catalyst. Shi et al.[74] first prepared a single layer of Cu atoms on chemically exfoliated MoS_2 (CuSA/ce-MoS_2) by using a Cu underpotential deposition. Then, PtSAs were introduced into the CuSA/ce-MoS$_2$ system through the galvanic exchange of Cu atoms by Pt(II). The open-circuit potential measurement confirmed the galvanic replacement of Cu to Pt. This kind of site-specific electrodeposition enables the formation of energetically favorable metal-support bonds, followed by metallic bonding formation because of the sequential automatic termination. Overall, mild conditions, facile procedure, room-temperature synthesis, and precision are advantages of the electrodeposition method.

8.2.5.3 Photochemical Method

The photochemical method is quite suitable for precious metal single-atoms synthesis. The key features of this method are the involvement of electronically excited states and the adsorption of photons.[75] Utilizing ultraviolet (UV) light, precious metal ions can be directly reduced on various supports without further requirement of physical or chemical post-treatments. For example, Li et. al.[76] prepared isolated Pt single atoms on nitrogen-doped carbon by simple physical mixing of H$_2$PtCl$_6$ solution with high specific area carbon followed by UV-light treatment, as shown in Figure 8.4F. The presence of abundant nitrogen and oxygen heteroatoms, plenty of micro- and meso-pores, and carbon defects were propitious to the chemical adsorption and anchorage of PtCl$_6^{2-}$ ions. After adsorption and drying, a simple UV-light treatment was sufficient for the formation of isolated Pt single atoms on the nitrogen-doped carbon (Pt$_1$/NPC). Besides, high loading (3.8 wt%) of Pt single atoms was also achieved. The as-prepared catalyst exhibited ORR and HER bifunctional activity in both acidic and basic solutions. The remarkable electrocatalytic activity of Pt$_1$/NPC was attributed to the unique electronic structures of Pt–N$_4$ coordination sites. DFT studies also supported the presence of stable Pt–N$_4$ coordination sites. Simple procedures and mild synthetic conditions are the main advantages of the photochemical method. Several articles have been reported based on a room-temperature synthesis of atomically dispersed Pd catalyst on TiO$_2$ nanosheets (Pd$_1$/TiO$_2$) via photochemical strategy.[77,78] The formation of stable Pd–O bonds was the trivial step in these procedures. It was achieved by UV light-induced ethylene glycol (organic ligand) radicals that facilitated the removal of Cl$^-$ (of Pd salt) ions. Heterolytic dissociation of H$_2$ was observed at homogeneous catalysts. However, Pd$_1$/TiO$_2$ also exhibited the same phenomenon because of no Pd–Pd pair availability. Thus, Pd$_1$/TiO$_2$ catalyst presented excellent C=C and C=O hydrogenation activities, 9 and 55 folds better than the commercial Pd counterpart.

8.2.5.4 Low-Temperature Reduction Method

Recently, many ingenious methods have been adopted for SACs synthesis. Low-temperature reduction of metal species into single atoms on the support is one of the versatile methods. In this process, metal species containing liquid solutions are frozen at ultra-low temperature to restrict the random diffusion and migration of metal precursors. With the assistance of chemical or photochemical reduction methods, SACs can be prepared on the support at ultra-low temperature by suppressing rapid metal nuclei formation. For example, Huang et al.[79] prepared atomically dispersed Co by using a water/ethanol mixed solvent with a low freezing point, liquid-phase reduction of Co precursor with hydrazine hydrate at −60 °C. At this low temperature, atomically isolated Co-complex could be formed due to the extremely suppressive nucleation. In contrast, if the same reaction was carried out at room temperature by conventional solution-phase reduction, Co clusters or nanoparticles could develop because of the uncontrollable nucleation and growth, as explained in Figure 8.4G. Therefore, instead of organic reducing agents, UV light can be used as a reducing tool to anchor precious metal single atoms on carbon-based/or metal-oxide-based supports at an ultralow temperature.[80–82] Temperature is the crucial parameter in this method. By decreasing the reaction temperature, the energy barrier and rate of reduced metal atoms to nuclei can be effectively regulated.[79]

8.2.5.5 Atom-Trapping Method

The atom-trapping method is a fairly new strategy to synthesize SACs. In this technique, bulk materials can be transformed directly into single atoms by trapping them on various substrates at high temperatures. For example, Cu single atoms could be decorated on nitrogen-doped carbon using Cu foam as a metal source at 900 °C with the assistance of NH_3.[83] As shown in Figure 8.4H, Cu foam and ZIF-8 powder are kept separately on a porcelain boat. First, the porcelain boat is heated in a tubular furnace at 900 °C in Ar atmosphere to obtain nitrogen-doped carbon. Due to the inert nature of Ar gas, nothing can happen to Cu foam. However, volatile $Cu(NH_3)_x$ species can be formed in a subsequent step in NH_3 gas flow. The as-formed $Cu(NH_3)_x$ species are transported and trapped by the defects on the nitrogen-rich carbon support, the isolated Cu sites are resulted. Likewise, when Pt NPs supported Al_2O_3 were heated at 800 °C in air, volatile PtO_2 can be transferred to the polyhedral CeO_2 and trapped as the isolated single atoms.[84]

The highlight of high-temperature synthesis is that only the single atoms occupy the most stable binding sites. As a result, a sinter-resistant, stable, atomically dispersed catalyst can result. Qu et al.[85] employed a facile dangling bond trapping strategy to prepare SACs under ambient conditions from easily accessible bulk metals such as Fe, Co, Ni, and Cu. When the dangling oxygen groups on graphene oxide support touched the metal foam, electrons were received from metal, forming stable M–O bonds. Using sonication, metal atoms attached to graphene oxide were pulled out from metal foam to obtain isolated metal atoms embedded in the graphene oxide.

8.3 APPLICATION OF SACs IN ELECTROCHEMICAL NRR

As discussed in the introduction section, presently, large-scale NH_3 is produced industrially using the Haber–Bosch process. The electrochemical NH_3 synthesis using air and water at ambient conditions is a feasible alternative to the Haber–Bosch process. However, developing a durable, economical, and energy-efficient electrocatalyst is a challenging task. The unique electronic structure of SACs, because of their strong metal-support interaction, quantum effect, and unsaturated coordination environment, has attained considerable attention as forefront candidates for NRR.

8.3.1 RECENT THEORETICAL ADVANCES OF SACs FOR NRR APPLICATION

Recently, with the improvement of various databases and exponential development in supercomputers, DFT computations and high-throughput screening based on algorithms have paved a way for rapid exploration of suitable single-atom electrocatalysts for NRR. Electrocatalytic NRR activity and stability of SACs strongly depend on interactions between single atoms and support because there is a trade-off between diffusion and aggregation of metal single atoms.[86] Nitrogen-doped carbon materials, two-dimensional metal carbides (MXenes), boron nitride (BN), MoS_2, and several other supports have been employed to anchor metal-single atoms to evaluate their stability and electrocatalytic NRR activity.

8.3.1.1 Theoretical Evaluation of Metal Single Atoms Stabilized on N-Doped Carbon Materials

A literature review of theoretical studies based on SACs for NRR has shown that metal single atoms have been extensively coordinated on various types of N-doped carbon-based materials (such as N-doped graphene, N-doped graphyne, and $g–C_3N_4$) due to their linear scaling relation regulating ability. For instance, Liu et al.[87] presented the NRR activity of 20 different transition metal (TM) centers supported on three different types of N-doped carbon. This study demonstrated that the coordination environment provided by different types of N-doped carbon supports could regulate the linear scaling relations for adsorption of intermediates and thus indirectly affect the NRR activity. The nitrogen adatom adsorption energy could be effectively used to establish intrinsic activity trends of catalysts. Guo et al.[88] systematically explored the catalytic activity and selectivity of carbon-based iron-nitrogen (Fe@N_x, $x = 0–4$) and proved that both were strongly dependent on the coordination environment of the Fe single atom. Among seven different coordinated sites of Fe (as shown in Figure 8.5A), the Fe single atom with two N atoms located at opposite coordination (Fe@$N_{2\text{-opp}}$) exhibits the highest activity and selectivity. Compared to the other six coordinated Fe environments, Fe@$N_{2\text{-opp}}$ showed the least −0.63 V limiting potential exhibits its conducive NRR nature. Chemisorption of divalent nitrogen on the catalyst surface to realize inert N≡N activation was a prerequisite quality of an efficient NRR catalyst. Ling et al.[89] screened single metal atoms (Cu, Pd, Pt, and Mo) supported on N-doped carbon for NRR by testing their N_2 adsorption ability. Mo single atoms supported on N-doped carbon (Mo$_1$–N$_1$C$_2$) illustrated superior side-on and end-on N_2 adsorption compared to the previously stated metal atoms. The first principle suggested that Mo$_1$–N$_1$C$_2$ could catalyze NRR via an enzymatic pathway with an ultralow overpotential of 0.24 V. Besides, only 0.47 eV uphill free-energy sufficient to the removal of produced NH_3 on the catalyst surface.

Among carbon-based materials, N-doped graphene is a widely used model support for SACs. Divacancy is the most common defect present in two-dimensional (2D) graphene sheets. Thus, it can be employed to anchor TM single atoms. Recently, Riyaz and Goel[90] employed nitrogen-passivated di-vacant graphene as support to anchor Cr, Mn, Fe, Mo, and Fe central TM atoms. Preliminary screening experiments suggested that Cr metal atom dispersed on nitrogen-passivated di-vacant graphene catalyst showed a supereminent interaction with N_2 with a binding energy of −0.65 eV.

The detailed theoretical studies have demonstrated that facile N_2 adsorption and activation of the N≡N bond take place on the catalyst due to its high spin density. Inspired by the nitrogenase enzyme's bimetallic FeMo cofactor (Figure 8.5B(i)), He

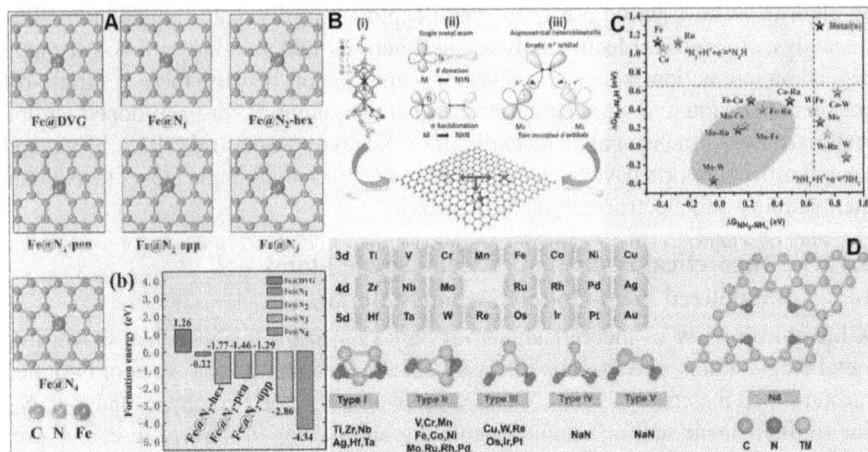

FIGURE 8.5 The geometric structures of Fe single-atom are ingrained in double vacancy graphene (DVG) and N-doped double vacancy graphene (Fe@N$_x$, $x = 0$–4). The calculated formation energy of Fe@N$_x$ is present at the right bottom (A). (Reproduced from Ref.[88]). Nitrogenase enzyme's FeMo cofactor structure (i), Schematic representation of N$_2$ bonding to single (ii) and dual-site TMs (iii) (B). The calculated Gibbs free energy changes for the first and last hydrogenation steps of NRR on various TM atoms. The oval area represents the most promising NRR candidates (C). (Reproduced from Ref.[91]). The lowest energy configurations of triple-transition-metal clusters on N-doped graphene (N$_4$) (D). (Reproduced from Ref.[95])

et al.[91] adopted an asymmetrical dual-metal dimer catalyst for NRR. They selected the random combination of two non-precious metal atoms such as Fe, Co, Mo, W, and Ru for DFT studies. Surprisingly, Mo–Ru, Mo–Co, Mo–W, Mo–Fe, and Fe–Ru dimers exhibited low onset potentials of 0.17, 0.27, 0.28, 0.36, and 0.39 V, respectively. Furthermore, they also calculated the Gibbs free energy changes for the first and the last hydrogenation steps on both single and dual-metal dimer catalysts, as illustrated in Figure 8.5C. The heterobimetallic catalytic centers exhibit lower Gibbs free energy changes (green highlighted area) than single-atom counterparts. The superior catalytic activity is mainly due to the electron donation from the asymmetric metal atoms to the terminal side-on N$_2$ molecule (Figure 8.5B(iii)), which significantly polarizes and weakens the N≡N bond. However, the activation of N≡N bond activation through δ donation or π back donation of SAC (Figure 8.5C(ii)) is still a limited process. Besides, the synergistic effect of the dual-metal dimers is also conducive to moderating the binding energies of the pivotal intermediates. Likewise, several reports have demonstrated that dual-atom embedded N-doped graphene is more beneficial to catalytic NRR than single atom counterparts.[92–94]

Single-cluster catalysts are also extended family members of SACs, gaining popularity in heterogeneous catalysis. They possess well-defined low-nuclearity active centers such as trimer. Zheng et al.[95] tested a series of 3d, 4d, and 5d (Figure 8.5D) transition-metal-atom clusters embedded in N-doped graphene (TM$_3$–N$_4$) for NRR. As per theoretical analysis, Mn$_3$–N$_4$, Fe$_3$–N$_4$, Co$_3$–N$_4$, and Mo$_3$–N$_4$ non-precious catalysts could show better NRR activity. The Co$_3$–N$_4$ catalyst outperformed with a limiting potential of -0.41 V via the enzymatic mechanism. It was worth mentioning that when compared to the CoN_4 single-atom system, the Co_3–N_4 exhibited enhanced NRR performance owing to its unique atomic structure that resulted in robust π back donation to activate the $N\equiv N$ bond.

Generally, high-throughput screening of NRR electrocatalysts requires several calculations in space. It requires not only a long time to screen NRR catalysts but is also expensive. To mitigate this issue, deep neural network (DNN) was used as an alternative to predict better catalysts among TM embedded boron(B)-doped SACs.[96] This model could effectively reduce the time of computation by eliminating non-efficient catalysts from screening. This model was used to screen 26 TM single atoms decorated on B-doped graphene. Among them, Cr single-atom embedded B-graphene showed better NRR catalyst with an overpotential of 0.13 V.

Graphyne is another popular carbon-based support prepared by inserting the acetylenic group ($-C\equiv C-$) between two bonded carbon atoms in graphene. The coexistence of the sp and sp^2 hybrid networks in graphyne endows outstanding properties such as good chemical stability, large surface area, uniformly distributed pores, and superb electronic conductivity.[97] Thus, heteroatoms-doped/unmodified graphyne has been employed as a support to anchor SACs for NRR application. Zhai et al.[98] evaluated the potential of the transitional-metal single-atom-embedded (TM: Sc, Cr, Mn, Fe, Co, Ni, Cu, Zn, Ru, Rh, Pd, and Ag) graphyne monolayer as an NRR electrocatalyst by DFT studies.[98] They screened catalysts by keeping the following three points to ascertain better NRR catalysts: (i) recycling stability (which could be determined by measuring binding energy (E_b) between the TM atoms and graphyne); (ii) adsorption energy of N_2 molecule on the catalyst; and (iii) first protonation step (this step widely is regarded as a potential determining step for NRR). Computational results showed that Mo-embedded graphyne exhibited negative E_b, better N_2 adsorption energy, enlarged N–N bond lengths, and lower change in Gibbs energy for the first hydrogenation step compared to other TM-based catalysts. Mo-embedded graphyne not only exhibited a lower onset NRR potential of -0.33 V but was also capable of effectively restraining the HER. Single tungsten (W) atom supported N-doped graphyne was also reported as a high-performance NRR catalyst under ambient conditions.[99] The catalyst could activate adsorbed N_2 molecules through both end-on and side-on configurations, as shown in Figure 8.6A. However, end-on adsorption configuration is preferred more due to a low onset potential of 0.29 V. Band structure calculation revealed the robust metallic nature of the catalyst, which was conducive to electron transfer efficiency during nitrogen reduction. The density of states (DOS) and change in Gibbs free energy calculations of N_2 on double-atom TM embedded graphyne illustrated the enhanced electron donation, back-donation, and N_2 adsorption ability, suggesting that double-atom TM counterpart was also prominent in N_2 reduction catalysts.[100] The metal-metal bonding in these types of catalysts can offer additional stability for loaded metals. For example, Chen et al.[101] showed that a triple atom TM with ultra-high loading (up to 35.8 wt%) could show extraordinary N_2 reduction activity due to the additional active sites, which provides more electrons for N_2, making facile NRR.

Graphitic carbon nitride (g-C_3N_4) is also a popular carbon support that consists of a fragmented graphene structure connected by tertiary amines. g-C_3N_4 comprises evenly distributed holes, which offer abundant and uniform nitrogen coordinators

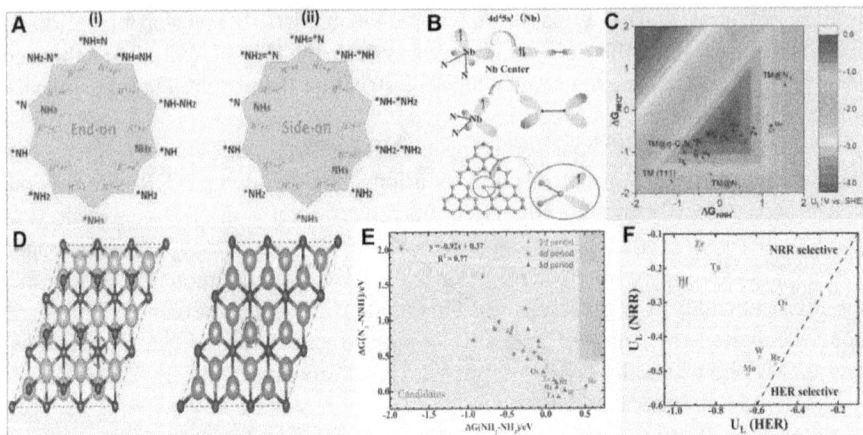

FIGURE 8.6 The schematic representation of end-on (i) and side-on (ii) configurations on the W@N-doped graphyne (A). (Reproduced from Ref.[99]) Electronic diagram illustrating the interaction between N_2 and Nb atom anchored to the support (B). (Reproduced from Ref.[103]). Color-filled contour plot of the limiting potential as a function of Gibbs free energy of *NNH and *NH_2 (C). (Reproduced from Ref.[87]) Optimized models of $TiCO_2$ and Mo_2CO_2. TMSA anchoring sites are highlighted in green dotted circles (D). (Reproduced from Ref.[108]) Screened NRR candidates supported on $Mo_2TiC_2O_2$ substrate (E). Calculated limiting potentials for HER and NRR on the prescreened candidates (F). (Reproduced from Ref.[110])

with rich lone pairs of electrons to anchor metal ions in the ligands.[102] Theoretical calculations suggested that the TM single atoms like vanadium (V), niobium (Nb), and tantalum (Ta) anchored on g-C_3N_4 substrate could exhibit superior NRR activity than graphene support. Nb single-atom on g-C_3N_4 with more negative valence could provide structural advantages for hosting empty d-orbitals for strong N_2 and N_2H adsorption along with additional single d-electrons to further enhance back-donation to activate the N≡N bond, as shown in Figure 8.6B[103]. Liu et al.[87] adopted activity trends, electronic origins, and design strategies to screen TM SACs anchored on N-doped carbon supports. With the help of established scaling relations for adsorption energies of intermediates, they proved that the intrinsic activity was related to the nitrogen adatom adsorption strength. Besides, the coordination environments provided by different N-doped carbons (TM@N_4, TM@N_3, and TM@ g-C_3N_4) could regulate the linear scaling as well and thus indirectly affect the NRR activity. The color contour plots (Figure 8.6C) provide a comprehensive understanding of the effect of both metal centers and supports on the NRR activity. It also offers some guidelines for improving catalytic activity. Cao et al.[104] also explained the importance of N-doped carbon support by analyzing NRR activities of RuSA anchored on C_2N, g-C_3N_4, and γ-graphene ligands. All three supports provided stable anchoring sites for RuSA, g-C_3N_4, and γ-graphene ligands were found effective because of their low HER activity, which was beneficial to NH_3 formation rather than H_2 under applied potential. The molecular dynamics simulation studies revealed that nitrogen vacancies present in the g-C_3N_4 could effectively serve TM (Ti, V, Co, Ni, Zr, Mo,

Ru, and Pt) atoms anchoring sites.[105] Among all TM, Ti@g-C$_3$N$_4$ was the most promising N$_2$ reduction candidate due to its low thermodynamic reaction energy of 0.28 eV with a small energy barrier of 0.57 eV. Unlike g-C$_3$N$_4$, the electrical conductivity of C$_2$N was high because of its continuous network structure. Besides, it showed a high structural stability with the highest eigenfrequency of 46.98 THz, which was close to the graphene (47.97).[106] Thus, C$_2$N was employed to support single and double boron atoms. Both B@C$_2$N and B$_2$@C$_2$N followed an enzymatic pathway for NRR. However, B$_2$@C$_2$N was a more efficient electrocatalyst with an extremely low overpotential of 0.19 eV compared to B@C$_2$N (0.29 eV).

8.3.1.2 Theoretical Evaluation of Metal Single Atoms Stabilized on MXenes, BN, MoS$_2$, and Other Non-Conventional Supports

MXenes are 2D TM carbides or nitrides which are first reported in 2011.[107] They have a general chemical formula of M$_{n+1}$X$_n$T, where M is a TM, X is C and/or N, and T is O/OH/F, (n = 1–3). Owing to the high electrical conductivity, large surface area, and tunable surface composition, MXenes have been explored as electrocatalysts for various reactions and single-atoms support as well.[107] To explore the electrocatalytic properties of MXenes, Huang et al.[108] investigated the potential of Ru and Mo single atoms anchored on the Ti$_2$CO$_2$ and Mo$_2$CO$_2$ MXenes monolayers as NRR catalysts by first-principles computations. The optimized models of both Ti$_2$CO$_2$ and Mo$_2$CO$_2$ are shown in Figure 8.6D. Due to the high conductivity of Mo$_2$CO$_2$, Ru and Mo single atoms decorated Mo$_2$CO$_2$ (Ru@Mo$_2$CO$_2$ and Mo@Mo$_2$CO$_2$) exhibited superior activity than that of Ti$_2$CO$_2$ counterpart. Among Ru@Mo$_2$CO$_2$ and Mo@Mo$_2$CO$_2$, the latter showed robust NRR because of its higher activity and selectivity. The reaction could proceed via a distal or hybrid mechanism with the low overpotential of 0.16 V. The slight modification of surface from Ti$_2$CO$_2$ to Ti$_3$C$_2$O$_2$ could entirely change the interaction between TM single atoms and substrate. Ti$_3$C$_2$O$_2$ was used as a ligand to test a group of TMs (Sc, Ti, V, Cr, Mn, Fe, Co, Ni, Cu, Zn, Mo, Ru, Rh, Pd, Ag, Cd, and Au).[109] The reaction pathways were analyzed by calculating Gibb's free energy and the overpotentials of Fe, Co, Ru, and Rh supported on Ti$_3$C$_2$O$_2$. They found out that along with the first hydrogenation step (N$_2$ to *NNH), the last hydrogenation step (*NH$_2$ to NH$_3$) could also be considered to be PDS. Li et al.[110] screened 3d, 4d, and 5d TMs for NRR activity by employing Mo$_2$TiC$_2$O$_2$ (M$_3$C$_2$O$_2$ type) MXene as the support. The preliminary screening was carried out based on the free energy changes of the first and the last hydrogenation steps. The selected candidates are shown in Figure 8.6E (elements present in the green region) are further scrutinized. By using distal and alternating routes, they further evaluated seven screened catalysts. The key parameters to be considered in examining the kinetic activity of the catalyst were the adsorption energy of N$_2$ and the free energy change of desorption of NH$_3$. The highest energy barrier, such as PDS was found to determine the thermodynamic feasibility of the electrocatalytic process. Among seven screened catalysts, zirconium single-atom embedded Mo$_2$TiC$_2$O$_2$ exhibited the lowest PDS barrier, corresponding to a limiting potential of –0.15 V. The Faradic efficiency (FE) was found to greatly depend on its immunity to HER. This catalyst showed a lower limiting potential for NRR than HER compared to prescreened elements shown in Figure 8.6F. Twenty-eight different TM single atoms were also anchored on the TM nitride type Ti_2NO_2 MXene to examine NRR performance.[111] Based on the scaling relationship between the single N atom binding strength and onset potential, NiSA showed an excellent NRR performance with an onset potential as low as -0.13 V. The Nb_2CN_2 MXene belonging to the TM carbonitride category was also used to support 26 different TM single atoms.[112] The authors considered all three reaction pathways, distal, alternative, and enzymatic mechanisms, to estimate the NRR performance of TM-Nb_2CN_2 catalysts. The distal and alternative routes matched with the end-on adsorption of N_2 on TM-Nb_2CN_2, while the enzymatic mechanism followed the side-on configuration. The detailed studies on the Mn Nb_2CN_2 were given emphasis because of its lowest energy barrier for the protonation of *N_2 to *NNH, compared to other catalysts. Theoretical calculations suggested that this Mn-Nb_2CN_2 catalyst exhibited superior NRR activity and selectivity with the low overpotential of 0.51 V and. Also better kinetic and thermal stability.

Hexagonal boron nitride (h-BN), also known as white graphene, possesses many inherent advantages, such as high chemical stability and excellent oxidation resistance. Especially, h-BN nanosheets can sustain much higher thermal stability (up to 1,000 K) than graphene (below 600 K).[113] However, due to the large bandgap, h-BN suffers from poor electrical conductivity. Both theoretical and experimental attempts have been made to enhance h-BN conductivity. Although a few theoretical studies showed that defective h-BN could be used to anchor metal single-atoms,[113–115] it would be much more catalytically beneficial if h-BN electrical conductivity could be improved before using as a substrate. The simple and effective method is the hybridization of excellent conducting materials such as graphene with h-BN by taking advantage of their similar crystal structures. The hybrid h-BN/graphene (BCN) heterostructures have already been synthesized experimentally.[116] A few theoretical studies have been reported based on BCN substrate to anchor SACs. For instance, Huang et al.[117] evaluated an NRR activity of Mo single-atom decorated on a defective BCN monolayer (Mo@BCN). Among B, N, and C vacancies, the B vacancy showed the most suitable sites for tethering Mo single-atom. After Mo immobilization, BCN electronic property was transformed from semiconductor to metallic that favored charge transport to active sites to drive the NRR. It is worth noting that Mo anchored pristine BN (Mo@BCN-0) shows a stronger H adsorption than N_2, as shown in Gibbs free energy comparison of N_2 and H diagram (Figure 8.7A). In contrast, the N_2 adsorption can be significantly increased with increasing C concentration to $n=5$ (-0.65 eV) than H counterpart (-0.32 eV), suggesting a suppressed HER. The Mo@BCN presented stable and excellent NRR activity with 0.42 V overpotential, followed by an enzymatic pathway. Likewise, Chen et al.[118] also studied NRR activity of Mo single-atom supported on two BC_2N monolayers having different defects. In their work, the effect of defects on NRR activity was studied. For example, the type-A BC_2N monolayer with B and C vacancies exhibited NRR performance via a distal pathway with an overpotential of 0.441 V. However, the type B BC_2N monolayer with C–C vacancies followed an alternating mechanism with an overpotential of 0.516 V. These results suggested that the former BC_2N monolayer had a better NRR activity than the later catalyst. A new study demonstrated that the asymmetric ligands of N_2B_2 embedded in defective BN could serve as a better support than mono-symmetric ligands.[119] The NRR performance of nine different TM single atoms (SAs) immobilized on the

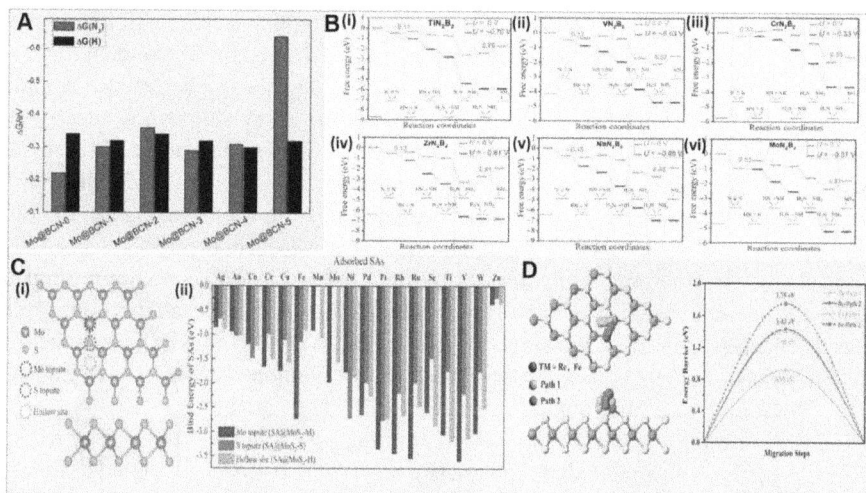

FIGURE 8.7 The comparison between the Gibbs free energies for N_2 and H adsorption on different models. first and second pillars denote $\Delta G(N_2)$ and $\Delta G(H)$, respectively (A). (Reproduced from Ref.[117]) Gibbs energy diagrams for enzymatic pathway for NRR on TiN_2B_2 (i), VN_2B_2 (ii), CrN_2B_2 (iii), ZrN_2B_2 (iv), NbN_2B_2 (v), and MoN_2B_2 (vi) at different applied potentials. Zero-point means free energy of $N_{2(g)}+(6H^++e^-)$ (B). (Reproduced from Ref.[119]) The top, side views (i), and binding energies transitional metal single atoms on $4\times4\times1$ MoS_2 supercell (ii) (C). (Reproduced from Ref.[123]) Top and side views of the migration paths for Re and Fe atoms on MoS_2 surface and their migration energy barrier profiles (D). (Reproduced from Ref.[124])

N_2B_2 nanosheet was investigated. Interestingly, the theoretical calculations revealed that the PDS of NRR on N_2B_2 support followed the last hydrogenation step except for Cr and Mo single atoms, as shown in Figure 8.7B(i–vi). The free energy barriers of PDS were found to be in the order of 0.33 eV (CrN_2B_2) < 0.53 eV (VN_2B_2) < 0.57 eV (MoN_2B_2) < 0.70 eV (TiN_2B_2) < 0.81 eV (ZrN_2B_2) < 0.85 eV (NbN_2B_2) < 0.92 eV (WN_2B_2) < 1.26 eV (HfN_2B_2) = 1.26 eV (TaN_2B_2). It is worth mentioning that the higher spin state of metal single atom can lead to deviation in the scaling relation between key N-containing intermediates. Thus, due to their high spin state, CrN_2B_2, VN_2B_2, and MoN_2B_2 can give higher activity and selectivity to NRR than the other six electrocatalysts.

Among 2D materials, molybdenum disulfide (MoS_2) is one of the highly studied substrates because of its chemical stability and catalytic activity. MoS_2 exhibits a typical 2D structure similar to graphene, composed of three atomic layers stacked together and bonded through van der Waals interactions.[120] Defect-rich MoS_2 itself can catalyze NRR under mild conditions.[121] Thus, TM SAs decorated on defective MoS_2-based catalysts can show superior NRR activity than defective-rich MoS_2. For example, Guo et al.[122] illustrated by DFT calculations that TM SA decorated on defective MoS_2 nanosheet exhibited an enhanced NRR activity than pristine MoS_2. Based on simulation results, they suggested that the defective MoS_2 anchored with Sc, Ti, Cu, Hf, or Zr was suitable for NRR. Besides, they also proposed an empirical descriptor to predict the properties of SACs, which could correlate with the d-band center and N-binding energy. Using first-principle high-throughput screening of several TMs, Yang et al.[123] showed again that MoS_2 monolayer was a potential candidate to anchor TM SAs for NRR. Prior to the NRR activity, they studied the adsorption sites of metal SAs on the $4 \times 4 \times 1$ MoS_2 supercell. As shown in Figure 8.7C(i), metal SAs can adsorb on any top sites of Mo, S, or hollow positions. Figure 8.7C(ii) shows the binding energies required for the earlier-mentioned top sites. Among all, the Mo site was the most stable adsorption configuration. By hydrogenation of PDS and N–N bond elongation strength calculations, $Mo@MoS_2$–M and $W@MoS_2$–M were singled out as the potential candidates. However, further simulations indicated that Gibbs free energy change required for the ammonia desorption $(\Delta G(NH_3^*$ to $NH_3))$ was superior on $Mo@MoS_2$–M $(0.9\,eV)$ to $W@MoS_2$–M $(1.2\,eV)$. Therefore, $Mo@MoS_2$–M was the better NRR electrocatalyst among other tested TM single atoms. The NRR on $Mo@MoS_2$–M preferred the distal route with a limiting potential of $-0.44\,V$. The recent DFT calculations illustrated that rhenium (Re) supported on MoS_2 exhibited a slightly improved limiting potential $(-0.43\,V)$ for NRR compared to the previous work.[124] From a stability point of view, the binding energy of Re on MoS_2 was $-3.36\,eV$, which was higher than Fe single atom decorated MoS_2.[125] Besides, $Re@MoS_2$ sustained high thermal stability up to 500 K, showing a superior thermodynamic strength. The catalyst kinetic stability was also tested by calculating the binding energies required for two different migration paths. The high binding energies as given in Figure 8.7D suggest the satisfactory kinetic stability of the catalyst. Zhang et al.[126] used natural charge population and the projected crystal orbital Hamilton population analysis to evaluate the NRR performance of duel iron atom anchored MoS_2 (Fe_2/MoS_2). NRR on Fe_2/MoS_2 followed the enzymatic pathways, which required a small overpotential of 0.21 V and a low N_2 adsorption energy of $-0.65\,eV$. On the other hand, both single and triatomic Fe anchored MoS_2 followed enzymatic mechanisms with higher overpotentials and adsorption energies. Therefore, the NRR activity on Fe_2/MoS_2 was better than single or triatomic counterparts.

Apart from the above-discussed conventional supports, researchers have also been employed some unique substrates to anchor SACs. For instance, the exohedral metalloborospherene $(B_{40}M)$ cage is a unique ligand and can be used to decorate 4th period (Sc to Cu) metal SAs, owing to its distinctive hole-like hexagonal and heptagonal rings on the sphere and boron inherent electron-deficient nature.[127] As shown in Figure 8.8A, metal SAs can be placed on both centers of 6- and 7-membered rings. However, metal SAs prefer heptagonal rings over hexagonal counterparts due to the strong metal-cage interactions and the feasible metal radii for doping. The suitable metal SA can be identified by adopting the following points: (i) the catalyst should be capable of absorbing N_2 and weakening the N≡N bond for further hydrogenation; (ii) a low overpotential accompanied proton-coupled electron transfer process; and (iii) high selectivity (NRR over HER). Based on the above conditions, DFT simulations suggested tungsten(W) SA could be the most feasible candidate for NRR. Therefore, the detailed studies on W SA anchored $B_{40}M$ suggested the enzymatic pathway was most favorable with a limiting potential of 0.56 V. Boron could also be used to prepare stable two-dimensional metal diborides, which possessed layered structure and could provide a suitable environment for SACs.[128] In this regard, ScB_2, TiB_2, VB_2,

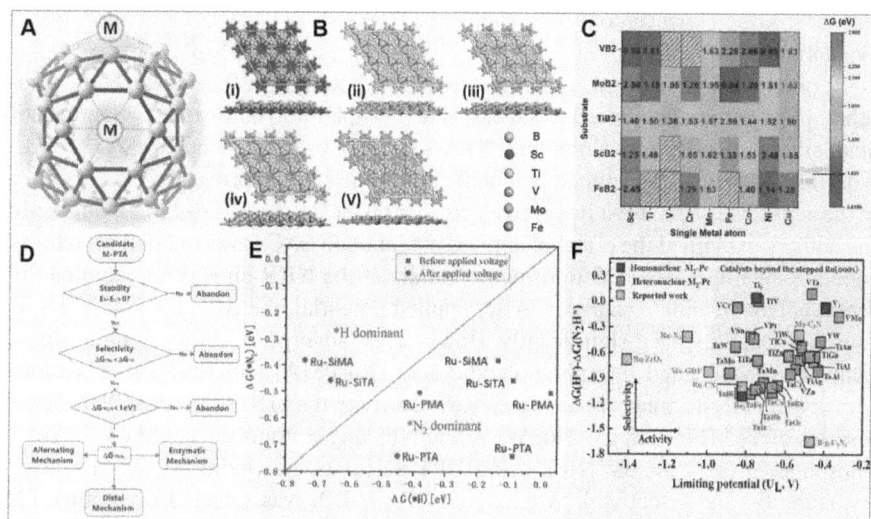

FIGURE 8.8 Stabilized structure of B_{40} and its hexagonal (center region), heptagonal (top region) rings as metal(M) binding sites (A). (Reproduced from Ref.[127]) Optimized 2D structures of (i) ScB_2, (ii) TiB_2, (iii) VB_2, (iv) MoB_2, and (v) FeB_2 (B). (Reproduced from Ref.[128]) Change of free energy in the first step of N_2 hydrogenation reaction on different TM SAs supported on 2D B-defective metal diborides. The maximum free energy change for NRR on the Ru (0001) stepped surface was provided as a reference (C). (Reproduced from Ref.[128]) The screening criteria for the NRR on M-PTA (D). (Reproduced from Ref.[129]) Calculated $\Delta G(^*N_2)$ and $\Delta G(^*H)$ on different polyoxometalate-based electrocatalysts before and after applying voltage. The dashed line indicates $\Delta G(^*N_2) = \Delta G(^*H)$ (E). (Reproduced from Ref.[130]) Limiting potential (U_L) vs. $\Delta G(H^*) - \Delta G(N_2H^*)$ on various catalysts (F). (Reproduced from Ref.[133].)

MoB_2, and FeB_2 could be some of the metal diborides, which might be employed as substrates to screen suitable TM SAs for NRR. Based on their optimized structure (Figure 8.8B), the metal atoms are located in the middle and lower part of the boron ring. Before simulating complete NRR routes on the SACs, the authors searched for the most promising NRR candidate by calculating ΔG of NNH* formation, as shown in Figure 8.8C. Based on the change in the free energy of the first step of hydrogenation, Ti SAs anchored on VB_2 (Ti@VB_2) were found to be the ideal catalyst. The detailed DFT calculations revealed that Ti@VB_2 had an NRR overpotential of -0.61 V, lower than the Ru(0001) stepped surface. Besides, the former catalyst had better NRR selectivity than HER.

Phosphotungstic acid (PTA) belongs to the polyoxometalate family, and it has been used to support and screen 20 different d-block metal atoms for NRR by theoretical calculations.[129] Figure 8.8D illustrates the screening criteria adopted to select the better catalyst. The stability of SAC on the support is considered a prerequisite for NRR. E_b and E_c refer to the thermal stability of the SAC in the bulk metal and PTA. If $(E_{b-}E_c) > 0$, then the SAC is considered stable on the PTA support. Among 20 different TMs, Mo, Tc, and Ru have shown high NRR activity with high selectivity. Since Ru is a noble metal and Tc is a radioactive element, both were skipped for further studies. However, only Mo-supported PTA (Mo-PTA) was chosen for further evaluation. The detailed examination on Mo-supported PTA (Mo-PTA) revealed that NRR proceeded via a distal pathway with an overpotential of 0.26 V. Recently, some other polyoxometalate-based catalysts like phosphomolybdic acid, silicotungstic acid, silicomolybdic acid along with PTA were also used to anchor RuSA.[130] Under working conditions, the applied potential could be used to improve proton adsorption at the active sites because it involved electron transfer ($H^+ + e^- + ^* \rightarrow H^*$). In this study, the authors explained the effect of applied potential on SACs toward NRR selectivity. Figure 8.8E illustrates PDS limiting potentials of the NRR on RuSA decorated different polyoxometalate catalysts. After applied potential, the ΔG values of H^* for all four SACs are changed significantly. However, N_2 adsorption ability remains dominant on RuSA-loaded phosphomolybdic acid (Ru-PMA) and Ru-PTA. Therefore, these two catalysts are prone to high NRR selectivity than HER. Liu et al.[131] designed a series of TM (Fe, Mn, Cr, Mo, W, V, and Nb) single atoms anchored on the black phosphorene (BP) for the NRR using DFT calculations. Owing to superior activity, stability, and selectivity, W-anchored-BP (W-BP) was selected over other TM SACs. The better performance of W-BP was attributed to the WP_3 active sites, which served as the electron adaptors to activate N_2 by donating electrons and regulating the charge transfer between BP and the reaction intermediates.

Molybdenum sulfide selenide (MoSSe) is a new Janus type 2D material in which Mo is sandwiched between S and Se layers, similar to MoS_2. A series of TM SAs were decorated on S or Se vacancies of MoSSe monolayer to explore the feasibility of the resultant SAC for NRR.[132] As per theoretical calculations, the MoSSe monolayer with Mo single-atom embedded at S-vacancy showed an excellent NRR activity with a low onset potential of −0.49 V. Guo et al.[133] synthesized biatom catalysts (BACs) supported on 2D expanded phthalocyanine (Pc) for NRR application. Both homonuclear (M_2-Pc) and heteronuclear (MM'-Pc) BACs were considered for screening. More than 900 different combinations of TMs could form, based on N_2H^* adsorption energy as the activity descriptor for screening criteria. As a result, the number of promising catalysts was reduced to <100 from 900. Large-scale DFT computations revealed that 3 homonuclear and 28 heteronuclear BACs could show excellent NRR activity. Using free energy difference of H^* and N_2H^* as selectivity descriptor, the screening was pinned down to five systems, including Ti_2-Pc, V_2-Pc, TiV-Pc, VCr-Pc, and VTa-Pc, as shown in Figure 8.8F.

8.3.2 Recent Experimental Advances of SACs for NRR Application

Presently, the main issues associated with electrochemical NRR are the high overpotential, low NH_3 yield, and low FE. Various attempts have been made using noble metals (such as Ru, Pt, Pd, Ru, and Au),[11] transitional metals (such as Co, Fe, Ni, Mo, etc.),[134] and metal-free (such as N, P, S heteroatoms-doped carbon-based) materials,[12] and their hybrids catalysts[4] to tackle above limitations. Compared to various electrocatalysts, SACs have distinct advantages such as the homogeneity of the catalytically active sites, the low-coordination environment of metal atoms, and maximum metal utilization efficiency. In this section SACs employed for NRR will be discussed.

8.3.2.1   Noble Metal-Based SACs

Due to the maximum metal utilization efficiency and a modest quantity of metal precursor sufficient for the SACs preparation, various noble metals have been employed for electrochemical NRR applications. Among all noble metal-based SACs, Ru-based catalysts have gained special attention because of their facile N_2 adsorption ability.[135] In 2018, Geng et al.[136] reported Ru SAs anchored on N-doped carbon (Ru SAs/N–C) for NRR. They prepared the catalyst via pyrolysis of Ru-contained derivatives of ZIF-8, as shown in Figure 8.9A. They found that Ru SAs/N–C can outperform in terms of partial current density, rate of NH_3 formation, and FE compared to Ru nanoparticle cousin (Ru NPs/N–C), as shown in Figure 8.9A(ii–iv), respectively. Besides, they calculated the turnover frequency (TOF) numbers of Ru SAs/N–C and Ru NPs/N–C. The TOF number of Ru SAs/N–C was found to be $376\,h^{-1}$, more than ten times larger than that of Ru NPs/N–C ($35\,h^{-1}$) at $-0.2\,V$ vs. RHE.

NRR in an acidic medium generally suffers from undesired HER. To avoid this, glitch alkaline electrolytes can be used as an alternative medium. Yu et al.[137] synthesized Ru SAs/g-C_3N_4 catalyst and used it for NRR in an alkaline ($0.5\,M$ NaOH) environment. The as-prepared catalyst was supported on copper foam to mitigate the semiconducting issues of g-C_3N_4. The DFT studies revealed that the superior NRR activity of Ru SAs/g-C_3N_4 catalyst was due to its stronger N_2 adsorption and the reduced H poising at the active sites. Another simple way to reduce HER is the incorporation of HER-resistant material. For example, Tao et al.[138] integrated zirconium oxide (ZrO_2) along with Ru SAs supported on N-doped carbon (Ru@ZrO_2/NC) to reduce HER. The addition of ZrO_2 not only suppressed HER but also played an important role in stabilizing Ru SAs. The oxygen vacancies of ZrO_2 could bind Ru SAs strongly and significantly reduce the mobility of the anchored Ru SAs. At a constant potential of $-0.21\,V$, the catalyst displayed a stable behavior for more than $60\,h$. Six consecutive N_2 reduction cycles, rates of NH_3 formation, and FE were found to be constant, suggesting the excellent stability of the catalyst.

Theoretical calculations have revealed that MXenes with metal atoms (Mo atoms in particular) composing the terminal surface can show an excellent N_2 adsorption capability and promote NRR under ambient conditions.[139] Peng et al.[140] prepared an ultra-low loaded Ru SAs on Mo_2CT_x MXene nanosheet (SA Ru/Mo_2CT_x) composite and used for NRR at ambient conditions. This catalyst exhibited the highest rate of NH_3 formation ($1.9 \times 10^{-10}\,mol\ cm^{-2}s^{-1}$) and FE (25.7%) at $-0.3\,V$ vs RHE compared to commercial Ru/C ($9.5 \times 10^{-11}\,mol\ cm^{-2}s^{-1}$, 12.7%) and pristine Mo_2CT_x ($5.1 \times 10^{-11}\,mol\ cm^{-2}s^{-1}$, 7.7%). Operando X-ray absorption spectroscopy and DFT studies disclosed that the strong electron back-donation capacity could lead to the superior activity of the catalyst. The powerful electron back-donation property could not only promote N_2 adsorption and activation but also lower the thermodynamic energy barrier of the first hydrogenation step.

The electrochemical cell design also plays a major role in NRR performance. By engineering SACs and electrochemical cells, one can simultaneously regulate the chemical kinetics of NRR.[141] According to Henry's law, the dissolved N_2 concentration in water is in proportion to the N_2 partial pressure (Figure 8.9A(i)). Therefore, NRR is carried out at a high pressure of 55 atm using Rh SA anchored on graphdiyne

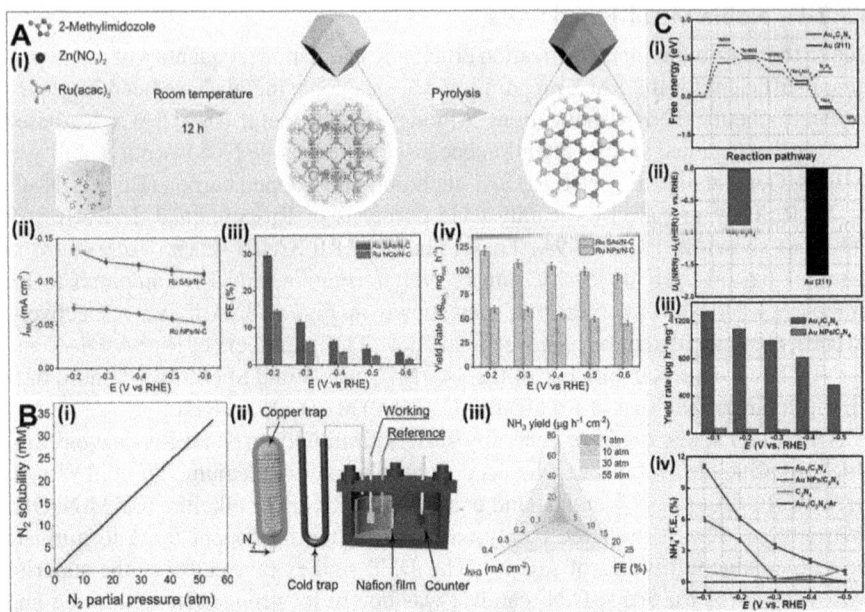

FIGURE 8.9 (i) Schematic representation of a synthetic procedure for Ru SAs/N–C, (ii) current density for NH_3 production, (iii) FE, and (iv) rate of NH_3 formation at different potentials on Ru SAs/N-C (A). (Reproduced from Ref.[136]) (i) A plot of N_2 solubility vs. the partial pressure, (ii) schematic illustration of pressurized electrochemical setup, and (iii) comparison of the NH_3 yield, FE, and j_{NH3} for Rh SA/GDY vs. the exerted N_2 pressures (B). (Reproduced from Ref.[141]) (i) Free energy diagram of NRR on Au_1/C_3N_4 and Au (211) at zero potential, (ii) difference in limiting potentials for NRR and HER, (iii) FE, and (iv) rate of NH_3 formation at different potentials on Au_1/C_3N_4 (C). (Reproduced from Ref.[143])

(Rh SA/GDY) as the catalyst, as shown in Figure 8.9B(ii). The rate of NH_3 formation, FE, and NH_3 partial current (j_{NH3}) are increased with increasing applied pressure (Figure 8.9B(iii)). Besides, all three parameters are increased 7.3-, 4.9-, and 9.2-folds higher than those obtained under ambient conditions.

Au is the second most used noble metal catalyst after Ru due to its affinity, electrochemical N_2 reducing ability, and relatively low hydrogen generation tendency. Au SAs supported on N-doped porous carbons (AuSAs-NDPCs) can act like a frustrated Lewis pair and activate N_2 molecules by electronic polarization.[142] At ambient conditions, at an applied voltage of −0.2 V vs. RHE, the prepared catalyst showed an excellent NH_3 formation rate of 3.8×10^{-11} mol $cm^{-2}s^{-1}$ with 12.3% of FE. The enhanced electrocatalytic activity was due to the synergistic effect between Au SAs and highly open porous structures with a high surface area of NDPCs. Wang et al.[143] theoretically and experimentally comprehended electrocatalytic activities of Au SAs and Au nanoparticles (AuNPs) decorated on C_3N_4 toward NRR. As shown in Figure 8.9D(i), both catalysts follow alternating mechanisms. However, Au SAs coated on C_3N_4 (Au_1/C_3N_4) exhibit a lower ΔG of 1.33 eV than 2.01 eV on Au (111). Besides, the former catalyst shows better selectivity than the latter counterpart

(Figure 8.9D(ii)). Experimental studies revealed that the rate of NH_3 formation on Au_1/C_3N_4 is more than 22 folds higher than $AuNPs/C_3N_4$ along with the improved FE (Figure 8.9D(iii and iv)). Sahoo et al.[144] performed comparative studies of Fe- and Au-based SAs for NRR application. Theoretically, Fe SAs anchored on N-doped carbon ($Fe–C_2N$) could exhibit a lower onset potential of $-0.7\,V$ compared to the -1.6 Von Au SA ($Au–C_2N$) counterpart for PDS. However, experimental studies illustrated that the rate of NH_3 formation and FE were better at $Au–C_2N$ than $Fe–C_2N$. This anomaly was due to the competitive HER and higher desorption energy of NH_3 on $Fe–C_2N$ catalyst. Table 8.1 summarizes those noble metal-based SACs electrocatalysts employed for NRR.

8.3.2.2   Non-Noble Metal-Based SACs

Despite the excellent NRR activity of noble metal-based SACs, their high cost and low abundance are two major limitations, which have encouraged researchers to search for more sustainable NRR catalysts. The non-noble metal-based SACs are a better choice, especially to realize industrial-scale NH_3 production. Among non-noble metal-based SACs, Fe-based catalysts have been extensively employed for NRR. Various synthetic routes are used to prepare Fe SAC on N-doped carbon[145-149] and MoS_2[125] substrates. For instance, Wang et al.[145] prepared Fe SAs anchored on N-doped carbon ($Fe_{SA}–N–C$) using pyrrole, ferrous chloride, and sodium chloride at $600°C$ under Ar flow. Under ambient conditions, the synthesized $Fe_{SA}–N–C$ exhibited an excellent rate of NH_3 formation of $1.2 \times 10^{-10}\,mol\ cm^{-2}s^{-1}$ with 56.5% of FE. In-situ XANES experiments were carried out under operating conditions to confirm Fe single atoms serving as NRR active sites. The high NRR activity of the $Fe_{SA}–N–C$ catalyst was explained based on a hot-atom mechanism. According to this mechanism, the $Fe_{SA}–N–C$ structure could effectively attract the access of N_2 molecules with a small energy barrier of $2.38\,kJ\ mol^{-1}$. A low binding Gibbs free energy of $-0.28\,eV$ was sufficient for N_2 adsorption. Besides, such an exothermic process could provide energy for the subsequent hydrogenation step. Zhang et al.[146] selected a Fe-decorated porphyritic MOF precursor to synthesize Fe SAs implanted in N-doped carbon ($Fe_1–N–C$) via a mixed ligand strategy. For comparison purposes, Co and Ni SACs were also prepared using the above strategy. However, $Fe_1–N–C$ showed a better rate of NH_3 formation of $1.5 \times 10^{-11}\,mol\ cm^{-2}s^{-1}$ with 4.5% of FE, superior to Co and Ni single-atom counterparts. The Fe SAC was also prepared by pyrolysis of bimetallic (Fe/Zn) ZIF-8.[147] In this work, the prepared catalyst was used for NRR in 0.1 M phosphate buffer saline (PBS) at room temperature. The electrocatalytic activity of the isolated single Fe atomic sites anchored onto N-doped carbon (ISAS-Fe/NC) was initially evaluated using CV studies in Ar/N_2 saturated 0.1 M PBS, as given in Figure 8.10A(i). Clearly, ISAS-Fe/NC catalyst shows enhanced reduction current density in N_2 saturated PBS than Ar saturated solution, suggesting a better NRR catalytic activity. Besides, the undesired byproduct hydrazine was not detected after electrolysis, suggesting high selectivity of this catalyst. This work also employed a[15]N_2 isotopic experiment to verify NH_3 formation from NRR. The electroreduction was carried out at $-0.4\,V$ vs. RHE by feeding $^{15}N_2$ isotopic gas. After the electrolysis, the obtained product was analyzed using[1]H NMR. As shown in Figure 8.10A(ii), the doublet pattern with a coupling constant of $72\,Hz$ (which

TABLE 8.1

Summary of Noble Metal-Based SACs Electrocatalysts and Their Experimental Details for NRR.

SAC Type	Catalyst	Electrolyte	NH$_3$ Determination Method	Rate of NH$_3$ Formation (mol cm^{-2} s^{-1})	FE (%)	References
Noble metal-based SACs	Ru SAs/N-C	50 mM H$_2$SO$_4$	Indophenol	2.0×10^{-9}* @ -0.2 V vs. RHE	26.9	Geng et al.[136]
	RuSAs/g-C$_3$N$_4$	0.5 M NaOH	Nessler's reagent and ASE[a]	3.7×10^{-10}* @ 0.05 V vs. RHE	8.3	Yu et al.[137]
	Ru@ZrO$_2$/NC	0.1 M HCl	Indophenol	5.9×10^{-8}* @ -0.21 V vs. RHE	21	Tao et al.[138]
	SA Ru/Mo$_2$CT$_x$	0.5 M K$_2$SO$_4$	Indophenol and ^{1}H NMR[b]	1.9×10^{-10}@ -0.3 V vs. RHE	25.7	Peng et al.[140]
	Rh SA/GDY	5 mM H$_2$SO$_4$ and 0.5 M K$_2$SO$_4$ + 10 mM ascorbic acid	Ion chromatography and NMR[b]	1.2×10^{-9}† @ -0.2 V vs. RHE	20.3	Zou et al.[141]
	AuSAs-NDPCs	0.1 M HCl	Indophenol	3.8×10^{-11}@ -0.2 V vs. RHE	12.3	Qin and Heil[142]
	Au$_1$/C$_3$N$_4$	5 mM H$_2$SO$_4$	Indophenol	2.1×10^{-8}* @ -0.1 V vs. RHE	11.1	Wang et al.[143]
	Au-C$_2$N	0.1 M HCl	Indophenol	2.1×10^{-11}@ -0.2 V vs. RHE	10.1	Kühne et al.[144]

* Rate of NH$_3$ formation presented in mol mg$_{cat}^{-1}$ s^{-1}.

a Ammonium ion-selective electrode.

b ^{1}H nuclear magnetic resonance spectroscopy.

† NRR carried out at 55 atm pressure.

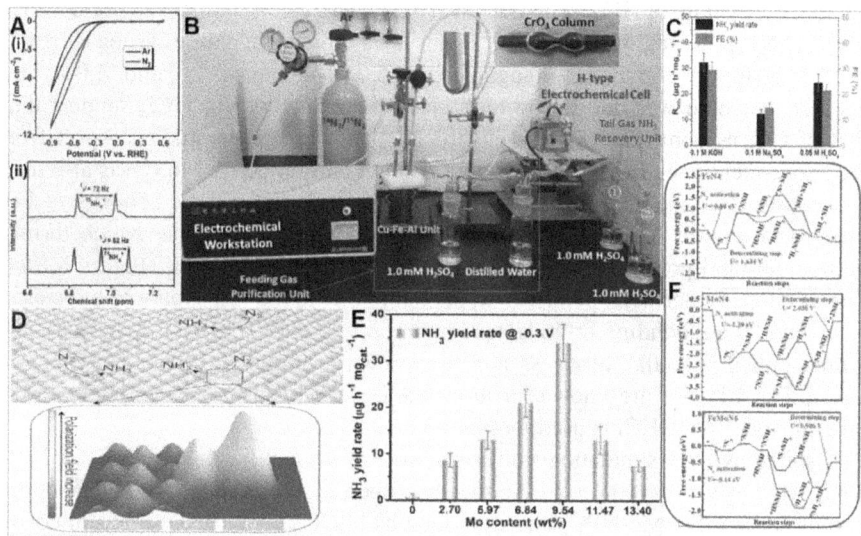

FIGURE 8.10 CV curves of ISAS–Fe/NC in Ar- and N_2-saturated 0.1 M PBS solutions (i), ^{1}H NMR analysis of $^{14}N_2$ and $^{15}N_2$ feeding gas after 2 h of electrolysis at -0.4 V vs. RHE (ii) (A). (Reproduced from Ref.[147]) The experimental setup for feeding gas purification (B). NRR performance of Fe–SAs/LCC in different pH conditions (C). (Reproduced from Ref.[149]) Schematic illustration of NRR performance of protruded Fe SA anchored on MoS_2 (D). (Reproduced from Ref.[125]) The rate of NH_3 formation of SA–Mo/NPC with different Mo loading (E). (Reproduced from Ref.[150]) Free energy diagrams for N_2 reduction on FeN_4, MoN_4, and $FeMoN_6$ (F). (Reproduced from Ref.[153])

is the fingerprint of $^{15}NH_4^+$) confirms that NRR at the catalyst surface is only the source of NH_3 formation. DFT calculations suggested that the electrocatalytic activity of Fe SAC anchored N-doped carbon was due to Fe–N_3 and Fe–N_4 active centers. This assertion was experimentally examined and confirmed using potassium thiocyanate (KSCN). Because SCN$^-$ anions could coordinate to Fe and poison the above-mentioned active sites during NRR. Recently, Zhang et al.[149] illustrated that Fe SAs could anchor onto the oxygen functional groups of nitrogen-free lignocellulose (Fe-SAs/LCC) and could efficiently catalyze NRR. The main advantage of this work was the use of nitrogen-free lignocellulose support which did not interfere with NRR, unlike the N-doped carbon counterpart, which might pose as an interferant. In addition, they employed a fool-proof gas purification system, as given in Figure 8.10C, to eliminate NO_x and NH_3 potential interferants presenting in the N_2 feeding gas. This system consisted of a Cu–Fe–Al catalyst-based NO_x removal unit. If the feeding gas contained any traces of NO converted into water-soluble NO_2, which could be removed by the H_2SO_4 and distilled water adsorbing units before reaching the electrochemical cell. Effect of pH was examined using N_2-saturated 0.05 M H_2SO_4 (pH 1.5), 0.1 M Na_2SO_4 (pH 6.5), and 0.1 M KOH (pH 12.7) at -0.1 V vs. RHE. The highest rate of NH_3 formation and FE were recorded in the alkaline medium, as illustrated in Figure 8.10C. Interestingly, when Fe-SAs/LCC immobilized on carbon cloth (CC), the rate of NH_3 formation and FE were 5.2×10^{-10} mol cm^{-2}s^{-1} and 29.3%, respectively; however, when the same catalyst was coated on glassy carbon electrode (GCE), the former and the latter augmented to 5.0×10^{-9} mol cm^{-2}s^{-1} and 51.0%. This peculiar outcome was due to the high electric conductivity of GCE support than that of CC, as confirmed for the electrochemical impedance studies. Apart from the above-mentioned carbon-based supports, non-carbon-based MoS_2 was also used as a substrate to anchor Fe SAs (Fe–MoS_2).[125] The electrocatalytic behavior of Fe–MoS_2 catalyst was explained based on the interfacial polarization, which formed between N_2 and protrusion-shaped Fe SA immobilized onto MoS_2 (Figure 8.10D). The developed interfacial polarization paves the way to the fast supply of more electrons into N_2 antibonding orbitals, leading to an improved rate of NH_3 formation at a low applied potential. Since K^+ ions are known to promote the most efficient N_2 adsorption, Cl^- ions are known for moderate ion and electron conductivities in the present work, 0.1 M KCl is used for NRR.

Inspired by DFT simulation studies on Mo SAs immobilized on defective BN for NRR, Han et al.[150] synthesized atomically dispersed Mo atoms anchored N-doped porous carbon (SA-Mo/NPC) and examined its NRR activity under ambient conditions. They used NPC as the support instead of BN due to the latter's poor electrical conductivity. The rate of NH_3 formation and FE were increased with increasing applied voltage. Both reached maximum at an applied potential of –0.3 V vs. RHE. Beyond this potential, the increased HER reduction could decrease the FE and NH_3 yield. Potentiostatic experiments for SA-Mo/NPC with different Mo loading were carried out at –0.3 V vs. RHE. As shown in Figure 8.10E, when Mo content is increased from 0 to 13.40 wt%, the rate of NH_3 formation is increased, reaching a maximum at 9.54 wt%, and then decreased. The increase in Mo content paves the way to enhance Mo–N active sites at which N_2 activation occurs. However, when the Mo content is too large, Mo nanoclusters appear, which decreases the Mo–N active sites and thus worsens the rate of NH_3 formation.

The molybdenum carbide (Mo_2C) can give appreciable NRR activity due to its π-back electron donation ability.[151] For example, to exploit Mo_2C intrinsic activity, Mo SAs with Mo_2C nanoparticles were assembled on the N-doped CNTs (MoSAs-Mo_2C/NCNTs) composite catalyst.[152] The rate of NH_3 formation on the composite catalyst was 4.5 and 4 folds higher than those of Mo_2C/NCNTs and Mo SAs/NCNTs individual catalysts. The superior catalytic activity of the composite catalyst was found to be due to a synergistic interaction between Mo SAs and Mo_2C. DFT simulations were carried out to ascertain the individual contribution of Mo_2C and Mo SAs. As per these studies, both Mo_2C and Mo SAs could show selectivity to NRR and HER, respectively. With different selectivity, HER selective Mo SAs could provide a large *H coverage environment around Mo_2C nanoparticles to activate N_2. Thus, by working in tandem, the composite catalyst could improve both selectivity and activity. This work demonstrated that two catalytically different materials could be combined into a single composite that effectively catalyzed complex multistep reactions with ease. Likewise, the heterobimetallic catalytic centers could also show superior NRR.[91] For example, Li et al.[153] synthesized a composite consisting of equal mole ratios of Fe and Mo heterobimetallic SACs on defect-rich N-doped graphene (FeMO@NG). DFT calculations revealed that the N-coordinated FeMo dimer species existed in the form of FeMoN$_6$ due to the lowest formation energy with 0.009 eV, which could serve as an N$_2$ reduction active site. Fe and Mo SAs on defect-rich N-doped graphene were prepared separately (denoted as Fe@NG and Mo@NG), and their NRR activity was examined. The rate of NH$_3$ formation on the composite was more than 14 and 3 folds than Mo and Fe counterparts at an applied potential of −0.4 V vs. RHE. Besides, to gain theoretical insight into the reaction mechanism, they performed DFT calculations on Fe@NG, Mo@NG, and FeMo@NG surfaces based on the model of FeN$_4$, MoN$_4$, and FeMoN$_6$, respectively (Figure 8.10F). Although N$_2$ binding energy on FeMoN$_6$ (−0.14 eV) is lower than FeN$_4$ (−0.84 eV) and MoN$_4$ (−2.39 eV), the former has shown better hydrogenation process. Also, FeMoN$_6$ has the lowest rate-determining step ($U = 0.906$ eV, for NH$_3$ dissociation) than FeN$_4$ ($U = 1.634$ eV, for N$_2$ dissociation) and MoN$_4$ ($U = 2.036$ eV, for NH$_3$ dissociation), suggesting its more feasible N$_2$ activation and the relatively low energy barrier. Liu et al.[154] also prepared Fe and Mo heterobimetallic SACs composite anchored on N-doped carbon (FeMO/NC) using a dissociation-and-carbonization method. The FeMo/NC achieved the maximum FE of 11.8% at −0.25 V and the rate of NH$_3$ formation of 4.3×10^{-10} mol cm^{-2}s^{-1} at −0.3 V vs. RHE in 0.1 M PBS. Because of the abundant FeMo dimer species and a combination of ligand, geometric and synergistic effects, FeMo heterobimetallic catalyst gave a remarkable N$_2$ selectivity and the improved reaction kinetics compared to individual Fe and Mo SACs counterparts. Due to the multifunctional electrocatalytic reduction ability of Co-based catalysts, atomically dispersed Co SAs anchored on N-doped carbon were tested for NRR as well.[155–157] Recently, Liu et al.[155] reported the atomically dispersed Co on N-doped carbon using cobalt-based ZIF (CSA/NPC) for NRR. The CSA/NPC presented the rate of NH$_3$ formation of 1.4×10^{-10} mol cm^{-2}s^{-1} with 10.5% of FE at −0.2 V vs. RHE in 0.05 M Na$_2$SO$_4$. Gao et al.[156] also employed a cobalt-based ZIF-67 to synthesize a Co SAC (Co/NC_500) for NRR. The Co/NC_500 showed the rate of NH$_3$ formation of 8.3×10^{-11} mol cm^{-2}s^{-1} with 10.1% of FE at −0.1 V vs. RHE in 0.1 M KOH. Both studies revealed that the rate of NH$_3$ formation was strongly dependent on the number of Co-N$_4$ active sites present in the catalysts. Zhang et al.[157] adopted a simple carbonization method involving a cobalt acetate, tartaric acid, and polyvinylpyrrolidone mixture for Co SA anchored N-doped carbon (Co-SAs/NC) synthesis. The as-prepared catalyst exhibited a rate of NH$_3$ formation of 1.5×10^{-10} mol cm^{-2}s^{-1} with 7.5% of FE at −0.45 V vs. RHE in 5 mM H$_2$SO$_4$. However, after laser-irritation treatment, the Co-SAs/NC displayed an enhanced rate of NH$_3$ formation (2.7×10^{-10} mol cm^{-2}s^{-1}) and FE (18.8%) at a lower applied potential of −0.25 V. This improved NRR activity was due to Co–N$_x$ sites on Co–SAs/NC after the laser-irritation treatment. The previous Co SACs-based studies stated that Co–N$_4$ sites could act as N$_2$ reduction active centers[155,156]. X-ray absorption spectroscopy and DFT calculations revealed that the Co–N$_3$ coordination sites could serve as the N$_2$ reduction catalytic sites.

Mukherjee et al.[158] showed that atomically dispersed Ni SAs anchored on N-doped carbon (Ni–N$_x$–C) could be used to electrochemically convert N$_2$ to NH$_3$ under ambient conditions using N$_2$ and H$_2$O. Although the rate of NH$_3$ formation was lower in alkaline (0.1 M KOH) medium, a high FE of 21% was achieved at −0.2 V vs. RHE. In the neutral (0.5 M LiClO$_4$) condition, the Ni–N$_x$–C displayed a rate of NH$_3$ formation of 1.8×10^{-9} mol cm^{-2}s^{-1} at −0.8 V vs. RHE. The high activity and selectivity of $Ni-N_x-C$ were due to $Ni-N_3$ active sites. For CO_2 reduction reaction, Cu-based catalysts have been extensively used, but these catalysts have not been exploited in NRR. Zang et al.[159] reported Cu SAs anchored on a porous N-doped carbon network (NC–Cu SA) prepared by a facile surfactant-assisted synthesis approach. In both alkaline and acidic solutions, this NC–Cu SA catalyst was used for NRR. It exhibited FE values of 13.8% and 11.7% and rate of NH_3 formation of 8.7×10^{-10} mol cm^{-2}s^{-1} and 8.0×10^{-10} mol cm^{-2}s^{-1} in 0.1 M NaOH and 0.1 M HCl electrolytes, respectively. These results proved that in a wide range of pH, NRR could be performed by using an NC–Cu SA catalyst. Also, in-depth theoretical calculations in combination with X-ray fine structure analysis showed that the local $Cu-N_2$ coordination active sites were responsible for the elevated NRR activity of the catalyst.

The synthesis of Mn SACs remains a big challenge because Mn atoms are prone to oxidize and aggregate into oxide/carbide species during the thermal treatment at low content.[160] Using the folic acid self-assembly strategy, Wang et al.[160] prepared Mn SAs on ultra-thin N-doped carbon sheets (Mn–N–C SAC). In 0.1 M NaOH electrolyte, this Mn–N–C SAC presented a high FE of 32.02% with a rate of NH_3 of 3.5×10^{-10} mol cm^{-2}s^{-1} at an applied potential of -0.65 V vs. RHE. DFT studies unveiled that Mn atomic sites could significantly promote the adsorption of reacting intermediates and reduce the energy barrier of the first hydrogenation step for efficient NRR.

The rare earth-metal-based catalysts are considered inactive for the room-temperature electrochemical reactions due to their strong reactant adsorption ability. The rich electron orbitals of rare earth elements can form strong coordination bonds with the anions on the supports, make easy to prepare rare earth elements-based SACs. Inspired by this logic, yttrium (Y) and scandium (Sc) rare earth SACs were synthesized on N-doped carbon support (Y_1/NC and Sc_1/NC) and employed as the NRR catalysts under ambient conditions.[161] Unlike TM catalyst's $M-N_4$ structure, Sc and Y atoms with large atomic radium tended to be anchored to the large-sized carbon defects through six coordination bonds of carbon and nitrogen. NRR of these catalysts was explained by the modulation of the local electronic structure of single atoms by N and C coordination. Table 8.1 summarizes non-noble metal-based SACs electrocatalysts employed for NRR.

8.4 SUMMARY, CHALLENGES, AND FUTURE RESEARCH DIRECTIONS

8.4.1 SUMMARY

With the increase of population in the world, the demand for NH_3 production is elevating proportionately. At present, most of the NH_3 production is obtained from the energy-intensive *Haber–Bosch* process. Unfortunately, this process accounts for 1.8% or half a billion metric tons of human-caused global CO_2 emissions each year. The safe and energy-efficient alternative NH_3 production is inevitable. Recently, electrochemical production of NH_3 directly from N_2 and water under ambient conditions using renewable energy has gained significant momentum over the last few years. Especially, the electrocatalytic NRR using SACs has drawn a great deal of attention because of the SACs' maximum atom utilization efficiency, quantum size effect, and low-coordination state. In this book chapter, we have summarized the most important strategies employed for the synthesis of SACs including impregnation and coprecipitation, spatial confinement, coordination sites attachment, defect engineering, ball milling, electrochemical deposition, and atom-trapping strategies. All these different approaches have the following common goals in the synthesis of SACs: (i) uniform atomic dispersion all over the support; (ii) strong interaction between isolated single atoms and atoms of supporting material; (iii) enhanced electron and mass transports as well as better active-site accessibility; and (iv) improved catalytic activity and durability. Apart from these points, the pros and cons of each method are discussed in detail. Recently, with the improvement of various databases and exponential development in supercomputers, DFT computations and high-throughput screening based on algorithms have shown rapid exploration of suitable single-atom electrocatalysts for NRR. As identified, the electrocatalytic activity and stability of SACs strongly depend on interactions between single atoms and support because there is a trade-off between diffusion and aggregation of metal SAs.[86] Nitrogen-doped carbon materials, two-dimensional metal carbides (MXenes), boron nitride (BN), MoS_2, and several other supports have been evaluated theoretically for their suitability to anchor metal-single atoms, also discussed in detail. In the last and most major part, noble metals such as Ru, Pt, Pd, Ru, and Au-based- and non-noble metal-based SACs, which have been employed for NRR to date, are summarized thoroughly. Besides, vital experimental details involved in noble metal-based- and non-noble metal-based SACs for NRR such as type of electrolyte used, NH_3 determination method, rate of NH_3 formation, and FE are also provided in Tables 8.1 and 8.2, respectively.

8.4.2 Challenges

Although significant progress has been achieved in SACs for electrochemical NRR through rational design and facile synthetic strategies, several scientific and technical glitches need to be addressed, as follows:

i. Agglomeration of Metal Single-Atoms: Owing to the large surface free energy, the metal SAs are prone to agglomerate into nanoclusters or nanoparticles during the electrocatalytic process. To solve this issue, most of the SACs synthesis involves less metal loading (≤ 1 wt%). This solution is effective to subdue agglomeration of metal single-atoms but suffers from low active sites. High electrocatalytic NRR activity immensely depends on the number of available active sites. Thus, a suitable strategy to increase single atoms loading on the support is highly desirable. For instance, Han et al.[150] prepared a high-loaded Mo SAs (9.54 wt%) on N-doped highly porous carbon using a high-temperature pyrolysis process. They experimentally proved that by increasing Mo content from 0 to 13.40 wt%, the rate of NH_3 formation was increased to reach a maximum at 9.54 wt%, as given in Figure 8.10E. The higher Mo content could enhance Mo–N active sites, resulting in improved NRR.

ii. Deactivation of SACs: Most of the electrochemical NRR research articles present a limited number of recycling NRR stability of the catalysts. To

TABLE 8.2

Summary of Non-Noble Metal-Based SACs Electrocatalysts and Their Experimental Details for NRR.

SAC Type	Catalyst	Electrolyte	NH$_3$ Determination Method	Rate of NH$_3$ Formation (mol cm^{-2} s^{-1})	FE (%)	References
Non-noble metal-based SACs	Fe$_{SA}$–N–C	0.1 M KOH	Indophenol	1.2×10^{-10}* @ 0 V vs. RHE	56.5	Wang et al.[145]
	Fe$_1$–N–C	0.1 M HCl	Indophenol and ^{1}H NMR	1.5×10^{-11} @ -0.05 V vs. RHE	4.5	Zhang et al.[146]
	ISAS-Fe/NC	0.1 M PBS	Indophenol	1.0×10^{-9}* @ -0.4 V vs. RHE	18.6	Lü et al.[147]
	Fe SAC/N–C	0.1 M KOH	Indophenol	8.6×10^{-10}* @ -0.35 V vs. RHE	39.6	Yang et al.[148]
	Fe–SAs/LCC/CC	0.1 M KOH	Indophenol and ^{1}H NMR	5.2×10^{-10}* @ -0.1 V vs. RHE	29.3	Zhang et al.[149]
	Fe–SAs/LCC/GCE	0.1 M KOH	Indophenol and ^{1}H NMR	5.0×10^{-9}* @ -0.15 V vs. RHE	51.0	Zhang et al.[149]
	Fe-MoS$_2$	0.1 M KCl	Indophenol	1.5×10^{-9} @ -0.1 V vs. RHE	31.6	et al.[125]
	SA–Mo/NPC	0.1 M KOH	Nessler's reagent and chromatography	5.5×10^{-10}* @ -0.3 V vs. RHE	14.6	Han et al.[150]
	MoSAs–Mo$_2$C/NCNTs	5 mM H$_2$SO$_4$ and 0.5 M K$_2$SO$_4$	Indophenol	2.6×10^{-10}* @ -0.25 V vs. RHE	7.1	Ma et al.[152]
	FeMO@NG	0.25 M LiClO$_4$	Indophenol	2.4×10^{-10}* @ -0.4 V vs. RHE	41.7[a]	Li et al.[153]
	FeMO/NC	0.1 M PBS	Indophenol	4.3×10^{-10}* @ -0.3 V vs. RHE	11.8[b]	Liu et al.[154]
	CSA/NPC	0.05 M Na$_2$SO$_4$	Nessler's reagent	1.4×10^{-11} @ -0.2 V vs. RHE	10.5	Liu et al.[155]
	Co/NC_500	0.1 M KOH	Indophenol	8.3×10^{-11}* @ -0.4 V vs. RHE	10.1[c]	Gao et al.[156]
	Co–SAs/NC	5 mM H$_2$SO$_4$	Indophenol	2.7×10^{-10}* @ -0.25 V vs. RHE	18.8	Zhang et al.[157]

(continued)

TABLE 8.2 (Continued)
Summary of Non-Noble Metal-Based SACs Electrocatalysts and Their Experimental Details for NRR.

SAC Type	Catalyst	Electrolyte	NH₃ Determination Method	Rate of NH₃ Formation (mol cm⁻² s⁻¹)	FE (%)	References
	Ni–N$_x$–C	0.5 M LiClO$_4$	Indophenol and Nessler's reagent	1.8×10^{-9} @ -0.8 V vs. RHE	7.6	Mukherjee et al.[158]
	NC–Cu SA	0.1 M KOH	Indophenol	8.7×10^{-10}* @ -0.35 V vs. RHE	13.8	Zang et al.[159]
	Mn–N–C SAC	0.1 M NaOH	Indophenol	3.5×10^{-10}* @ -0.65 V vs. RHE	32.0[d]	Wang et al.[160]
	Y$_1$–NC and Sc$_1$–NC	0.1 M HCl	Indophenol	3.8×10^{-10} @ -0.1 V vs. RHE 3.3×10^{-10} @ -0.1 V vs. RHE	12 11.5	Liu et al.[161]

* Rate of NH₃ formation presented in mol mg$_{cat}^{-1}$ s⁻¹.
[a] FE obtained at an applied potential of -0.2 V vs. RHE.
[b] FE obtained at an applied potential of -0.25 V vs. RHE.
[c] FE obtained at an applied potential of -0.1 V vs. RHE.
[d] FE obtained at an applied potential of -0.45 V vs. RHE.

commercialize NH_3 production, long-term recycling stability of the catalysts is quite essential. Therefore, extensive in-situ/operando studies are required to understand the reason behind the reduction of activity of the catalysts.

iii. Insufficient Understanding of electrocatalytic NRR mechanism on SACs: The comprehensive insight of electrocatalytic NRR mechanism on SACs is an unsolved puzzle due to the complex multistep reaction. Advanced in-situ/ex-situ characterization techniques, such as HAADF-STEM and XAS, need to be employed to Figure 8.8 out the real active sites of the catalyst during NRR. However, these tools have limitations in identifying the active sites during electrocatalysis. This limitation certainly hinders the deep insights into the complicated catalytic mechanism on SACs to a certain extent. Hence, suitable cutting-edge in-situ/operando characterization methods are further needed.

iv. Low Activity and Selectivity: Although SACs exhibit higher FE and rate of NH_3 formation than nanoparticle/nanocluster, it is not sufficient. According to the department of energy of USA, to commercialize the electrocatalyst, it must produce the rate of NH_3 formation of $9.7 \times 10^{-7} mol\ cm^{-2}s^{-1}$ with 90% of FE.[162] Regrettably, current SACs still suffer from low NH_3 yield and FE that are several orders of magnitude lower than the requirements for commercial catalysts.

8.4.3 FUTURE RESEARCH DIRECTIONS

To the best of our knowledge, the following points could be helpful for the development of sustainable and efficient SACs for NRR (Figure 8.11). The first four points are directly related to SACs and the last two for the overall development of electrochemical NH_3 synthesis.

i. Further Fundamental Understanding of NRR Mechanism: NRR on SACs surfaces involves multiple electrons and protons transfer steps. However, it is unclear whether NRR follows a concerted proton-electron transfer pathway or a sequential proton-electron transfer route. Therefore, it is essential to fully decipher the structural performance relationship and NRR catalytic mechanism at the atomic level. it can be done by combining advanced theoretical simulations with in-situ characterization techniques, such as in-situ electron microscopy and in-situ X-ray absorption fine structure.[28] For instance, Wang et al.[145] employed an emerging cutting-edge in-situ XANES operando technique with ultrasensitive to the electronic property of the metal to probe that Fe SA was the actual active phase for the NRR in real-time. Also, they conducted computational studies to evaluate the in-depth analysis of the NRR mechanism. The computed energy profiles revealed that the NRR on the Fe_{SA}–N–C catalyst proceededs through the alternating pathway. Yao et al.[163] adopted surface-enhanced infrared absorption spectroscopy (SEIRAS) to prove that NRR on Au surfaces followed an associated mechanism. Utilizing SEIRAS, they detected the N_2H_y species with bands at 1,453, 1,298, and $1,109\ cm^{-1}$ corresponding to H–N–H

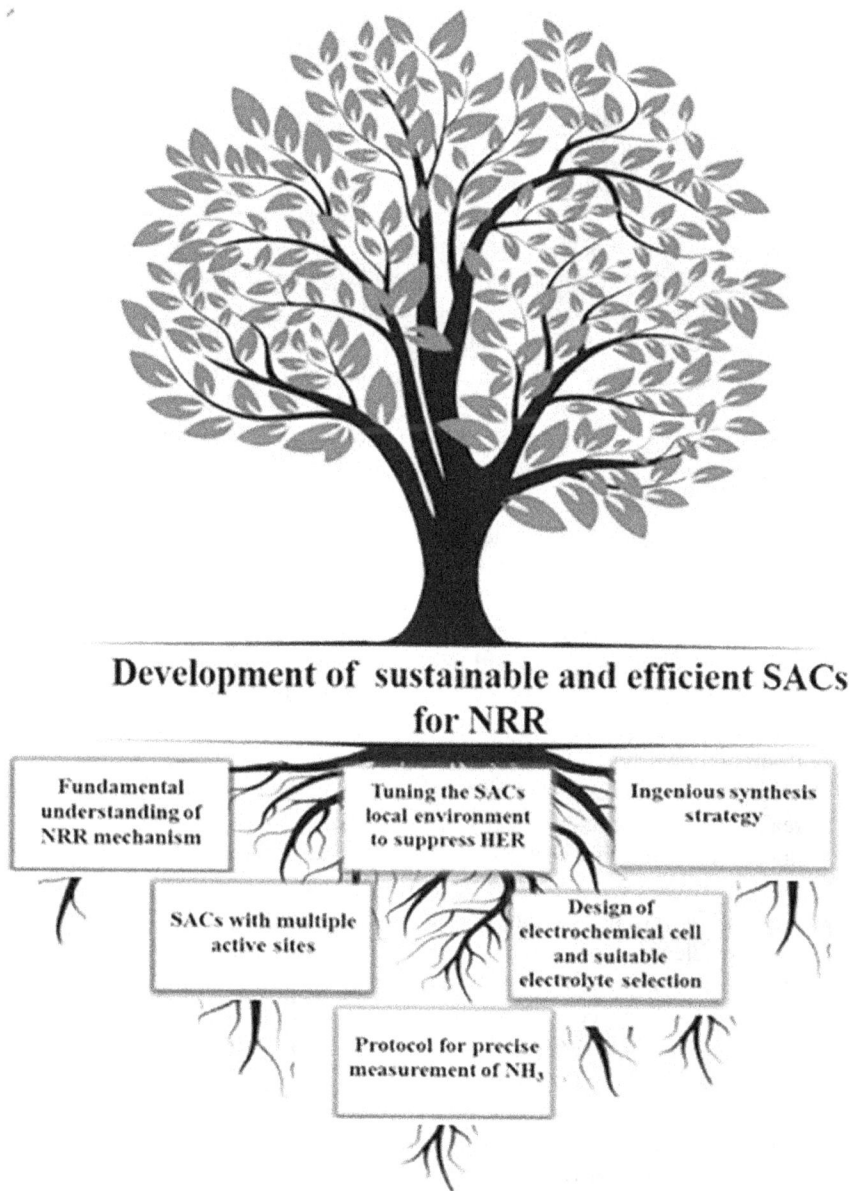

FIGURE 8.11 Schematics for the development of sustainable and efficient SACs for NRR.

bending, NH_2 wagging, and N–N stretching, respectively. Inspired by comprehensive research, more in-situ studies are needed to evaluate the NRR mechanism on different active sites to obtain a holistic perspective.

ii. Tuning the SACs Local Environment to Suppress HER: SACs suffer from undesired but facile kinetics of competing HER that operates under a similar potential window of NRR. Various ingenious strategies have been employed to restrain HER and boost NRR selectivity. For example, Zheng et al.[164] covered hydrophobic polymer with the porous structure on Au catalyst at which N_2 molecules concentrated at the active sites while manipulating the proton activity resulting in improved NH_3 yield and FE. It is better to implement an aerophilic-hydrophilic heterostructure approach for SACs at which diffusion of N_2 and active proton could facilitate near catalyst active sites besides by repelling water molecule HER. Designing SACs which offer a small energy barrier for N_2 molecules and a relatively higher energy barrier for *H adoption is also a popular choice.[145]

iii. Further Designing and Developing Ingenious Synthesis Strategy for SACs: Although the SACs exhibit a high NH_3 yield rate per atomic site, there is still room to improve it by increasing metal loading. Unfortunately, most existing reports suggest that higher metal loading can lead to nanoparticles/ clusters formation because of the aggregation of SAs. On the other hand, several SACs have been proved theoretically as excellent catalysts for NRR. However, only a few SACs have been synthesized due to lack of suitable procedures and differences in the ideal model and actual sample. Therefore, ingenious SACs synthetic procedures need to be developed.

iv. Developing SACs with Multiple Active Sites: As mentioned earlier, NRR involves multiple electrons and protons as well as multiple reaction intermediates. It is difficult for the single-atom center to simultaneously improve the NH_3 yield rate and FE. Thus, the catalyst with multiple active sites is desirable. Few theoretical studies have shown that some SACs can offer more flexible active sites, which can improve the kinetics of the multistep reaction and maximize selectivity.[91,133] A couple of experimental studies have also proved that some SACs exhibit a few fold higher NH_3 yield rate with improved FE compared to single-atom cousins.[153,154]

v. Further Optimizing Design of Electrochemical Cells and Selection of Suitable Electrolytes: Electrochemical cell design plays a significant role in suppressing HER during NRR. For example, by employing either a back-to-back type cell or proton exchange membrane (PEM)-type cell, HER can be minimized to a large extent by controlling the proton supply. Chen et al.[165] showed how back-to-back cell configuration could increase NRR performance by minimizing HER and limiting NH_3 cross-over. Singh et al.[9] suggested that by limiting the proton and electron supply, NRR selectivity could be improved significantly. Proton supply entirely depends on the type of electrolyte used for NRR. Most NRR studies have been carried out in an aqueous solution at which N_2 solubility is limited. The compressive evaluation needs to carry out on non-aqueous solvents ionic liquids as electrolytes since N_2 solubility is high in these solvents.

vi. Optimizing Protocol for Precise Measurement of NH_3. Since the omnipresence nature of NH_3, it can be found almost everywhere, including human breath, latex gloves, Nafion, gas tubing, etc. These all could lead to false-positive NH_3 results if proper care is not taken.[11] Similarly, special attention is required when nitrogen-based materials such as metal nitrides, N-doped carbon, are used as NRR catalysts because their decomposition leads to false-positive results.[166] Therefore, rigorous protocols must be followed for the accurate measurement of NH_3, which is produced only due to NRR.[167]

REFERENCES

1. R. Lan, J. T. S. Irvine and S. Tao, *Sci. Rep.*, 2013, 3, 1–7.
2. C. Zamfirescu and I. Dincer, *Fuel Process. Technol.*, 2009, 90, 729–737.
3. R. Lan, J. T. S. Irvine and S. Tao, *Int. J. Hydrogen Energy*, 2012, 37, 1482–1494.
4. R. Manjunatha, A. Karajić, M. Liu, Z. Zhai, L. Dong, W. Yan, D. P. Wilkinson and J. Zhang, *Electrochem. Energy Rev.*, 2020, 3, 506–540.
5. R. Manjunatha, A. Karajić, V. Goldstein and A. Schechter, *ACS Appl. Mater. Interfaces*, 2019, 11, 7981–7989.
6. N. Cao and G. Zheng, *Nano Res.*, 2018, 11, 2992–3008.
7. J. D. Holladay, J. Hu, D. L. King and Y. Wang, *Catal. Today*, 2009, 139, 244–260.
8. J. H. Montoya, C. Tsai, A. Vojvodic and J. K. Nørskov, *ChemSusChem*, 2015, 8, 2180–2186.
9. A. R. Singh, B. A. Rohr, J. A. Schwalbe, M. Cargnello, K. Chan, T. F. Jaramillo, I. Chorkendorff and J. K. Nørskov, *ACS Catal.*, 2017, 7, 706–709.
10. J. Wang, L. Yu, L. Hu, G. Chen, H. Xin and X. Feng, *Nat. Commun.*, 2018. DOI: 10.1038/s41467-018-04213-9.
11. G.-F. Chen, S. Ren, L. Zhang, H. Cheng, Y. Luo, K. Zhu, L.-X. Ding and H. Wang, *Small Methods*, 2018, 3, 1800337.
12. L. Zhang, G.-F. Chen, L.-X. Ding and H. Wang, *Chem. A Eur. J.*, 2019. DOI:10.1002/chem.201901668.
13. H. Huang, L. Xia, X. Shi, A. M. Asiri and X. Sun, *Chem. Commun.*, 2018, 54, 11427–11430.
14. Z. W. She, J. Kibsgaard, C. F. Dickens, I. Chorkendorff, J. K. Nørskov and T. F. Jaramillo, *Science*, 2017, 355, 1–12.
15. J. Su, R. Ge, Y. Dong, F. Hao and L. Chen, *J. Mater. Chem. A*, 2018, 6, 14025–14042.
16. J. Grunes, J. Zhu and G. A. Somorjai, *Chem. Commun.*, 2003, 3, 2257–2260.
17. H. Duan, D. Wang and Y. Li, *Chem. Soc. Rev.*, 2015, 44, 5778–5792.
18. J. N. Kuhn, W. Huang, C. K. Tsung, Y. Zhang and G. A. Somorjai, *J. Am. Chem. Soc.*, 2008, 130, 14026–14027.
19. S. Cao, F. F. Tao, Y. Tang, Y. Li and J. Yu, *Chem. Soc. Rev.*, 2016, 45, 4747–4765.
20. S. Ji, Y. Chen, X. Wang, Z. Zhang, D. Wang and Y. Li, *Chem. Rev.*, 2020, 120, 11900–11955.
21. Q. Fu, H. Saltsburg and M. Flytzani-Stephanopoulos, *Science*, 2003, 301, 935–938.
22. R. Bashyam and P. Zelenay, *Nature*, 2006, 443, 63–66.
23. B. Qiao, A. Wang, X. Yang, L. F. Allard, Z. Jiang, Y. Cui, J. Liu, J. Li and T. Zhang, *Nat. Chem.*, 2011, 3, 634–641.
24. S. Mitchell and J. Pérez-Ramírez, *Nat. Commun.*, 2020, 11, 10–12.
25. C. C. Hou, H. F. Wang, C. Li, Q. Xu, Q. Xu and Q. Xu, *Energy Environ. Sci.*, 2020, 13, 1658–1693.
26. S. Sun, G. Zhang, N. Gauquelin, N. Chen, J. Zhou, S. Yang, W. Chen, X. Meng, D. Geng, M. N. Banis, R. Li, S. Ye, S. Knights, G. A. Botton, T. K. Sham and X. Sun, *Sci. Rep.*, 2013, 3, 1–9.
27. K. Liu, X. Zhao, G. Ren, T. Yang, Y. Ren, A. F. Lee, Y. Su, X. Pan, J. Zhang, Z. Chen, J. Yang, X. Liu, T. Zhou, W. Xi, J. Luo, C. Zeng, H. Matsumoto, W. Liu, Q. Jiang, K. Wilson, A. Wang, B. Qiao, W. Li and T. Zhang, *Nat. Commun.*, 2020, 11, 1–9.
28. Y. Chen, S. Ji, C. Chen, Q. Peng, D. Wang and Y. Li, *Joule*, 2018, 2, 1242–1264.
29. J. Han, J. Bian and C. Sun, *Research*, 2020, 2020, 1–51.

30. S. Yang, J. Kim, Y. J. Tak, A. Soon and H. Lee, *Angew. Chem. Int. Ed.*, 2016, 55, 2058–2062.
31. H. Shang, W. Chen, Z. Jiang, D. Zhou and J. Zhang, *Chem. Commun.*, 2020, 56, 3127–3130.
32. X. Zhang, Z. Sun, B. Wang, Y. Tang, L. Nguyen, Y. Li and F. F. Tao, *J. Am. Chem. Soc.*, 2018, 140, 954–962.
33. C. H. Choi, M. Kim, H. C. Kwon, S. J. Cho, S. Yun, H. T. Kim, K. J. J. Mayrhofer, H. Kim and M. Choi, *Nat. Commun.*, 2016, 7, 1–9.
34. L. Cao, Q. Luo, J. Chen, L. Wang, Y. Lin, H. Wang, X. Liu, X. Shen, W. Zhang, W. Liu, Z. Qi, Z. Jiang, J. Yang and T. Yao, *Nat. Commun.*, 2019, 10, 1–9.
35. X. Sun, S. R. Dawson, T. E. Parmentier, G. Malta, T. E. Davies, Q. He, L. Lu, D. J. Morgan, N. Carthey, P. Johnston, S. A. Kondrat, S. J. Freakley, C. J. Kiely and G. J. Hutchings, *Nat. Chem.*, 2020, 12, 560–567.
36. J. Li, J. Liu and T. Zhang, *J. Am. Chem. Soc.*, 2013, 135, 15314–15317.
37. S. Zhao, F. Chen, S. Duan, B. Shao, T. Li, H. Tang, Q. Lin, J. Zhang, L. Li, J. Huang, N. Bion, W. Liu, H. Sun, A. Q. Wang, M. Haruta, B. Qiao, J. Li, J. Liu and T. Zhang, *Nat. Commun.*, 2019, 10, 1–9.
38. L. DeRita, S. Dai, K. Lopez-Zepeda, N. Pham, G. W. Graham, X. Pan and P. Christopher, *J. Am. Chem. Soc.*, 2017, 139, 14150–14165.
39. J. Z. Qiu, J. Hu, J. Lan, L. F. Wang, G. Fu, R. Xiao, B. Ge and J. Jiang, *Chem. Mater.*, 2019, 31, 9413–9421.
40. Q. Sun, N. Wang, T. Zhang, R. Bai, A. Mayoral, P. Zhang, Q. Zhang, O. Terasaki and J. Yu, *Angew. Chem. Int. Ed.*, 2019, 58, 18570–18576.
41. T. Zhang, Z. Chen, A. G. Walsh, Y. Li and P. Zhang, *Adv. Mater.*, 2020, 32, 1–29.
42. C. Wang, H. Song, C. Yu, Z. Ullah, Z. Guan, R. Chu, Y. Zhang, L. Zhao, Q. Li and L. Liu, *J. Mater. Chem. A*, 2020, 8, 3421–3430.
43. Y. N. Gong, L. Jiao, Y. Qian, C. Y. Pan, L. Zheng, X. Cai, B. Liu, S. H. Yu and H. L. Jiang, *Angew. Chem. Int. Ed.*, 2020, 59, 2705–2709.
44. X. Han, X. Ling, Y. Wang, T. Ma, C. Zhong, W. Hu and Y. Deng, *Angew. Chem. Int. Ed.*, 2019, 58, 5359–5364.
45. R. Zhao, Z. Liang, S. Gao, C. Yang, B. Zhu, J. Zhao, C. Qu, R. Zou and Q. Xu, *Angew. Chem. Int. Ed.*, 2019, 58, 1975–1979.
46. W. Zhang, P. Jiang, Y. Wang, J. Zhang, Y. Gao and P. Zhang, *RSC Adv.*, 2014, 4, 51544–51547.
47. K. Kamiya, R. Kamai, K. Hashimoto and S. Nakanishi, *Nat. Commun.*, 2014, 5, 1–5.
48. Y. Pan, S. Liu, K. Sun, X. Chen, B. Wang, K. Wu, X. Cao, W. C. Cheong, R. Shen, A. Han, Z. Chen, L. Zheng, J. Luo, Y. Lin, Y. Liu, D. Wang, Q. Peng, Q. Zhang, C. Chen and Y. Li, *Angew. Chem. Int. Ed.*, 2018, 57, 8614–8618.
49. K. Wu, X. Chen, S. Liu, Y. Pan, W. C. Cheong, W. Zhu, X. Cao, R. Shen, W. Chen, J. Luo, W. Yan, L. Zheng, Z. Chen, D. Wang, Q. Peng, C. Chen and Y. Li, *Nano Res.*, 2018, 11, 6260–6269.
50. C. Jin, L. Ma, W. Sun, P. Han, X. Tan, H. Wu, M. Liu, H. Jin, Z. Wu, H. Wei and C. Sun, *Commun. Chem.*, 2019, 2, 1–7.
51. S. Zhou, L. Shang, Y. Zhao, R. Shi, G. I. N. Waterhouse, Y. C. Huang, L. Zheng and T. Zhang, *Adv. Mater.*, 2019, 31, 1–7.
52. M. Tavakkoli, E. Flahaut, P. Peljo, J. Sainio, F. Davodi, E. V. Lobiak, K. Mustonen and E. I. Kauppinen, *ACS Catal.*, 2020, 10, 4647–4658.
53. H. Yin, S. L. Li, L. Y. Gan and P. Wang, *J. Mater. Chem. A*, 2019, 7, 11908–11914.
54. Y. Gao, Z. Cai, X. Wu, Z. Lv, P. Wu and C. Cai, *ACS Catal.*, 2018, 8, 10364–10374.
55. J. Wan, W. Chen, C. Jia, L. Zheng, J. Dong, X. Zheng, Y. Wang, W. Yan, C. Chen, Q. Peng, D. Wang and Y. Li, *Adv. Mater.*, 2018, 30, 1–8.
56. Y. Zhang, L. Guo, L. Tao, Y. Lu and S. Wang, *Small Methods*, 2019, **3**, 1–17.

57. F. Dvořák, M. F. Camellone, A. Tovt, N. D. Tran, F. R. Negreiros, M. Vorokhta, T. Skála, I. Matolínová, J. Mysliveček, V. Matolín and S. Fabris, *Nat. Commun.*, 2019. DOI:10.1038/ncomms10801.

58. J. H. Kwak, J. Hu, D. Mei, C. W. Yi, D. H. Kim, C. H. F. Peden, L. F. Allard and J. Szanyi, *Science,* 2009, 325, 1670–1673.

59. N. Tang, Y. Cong, Q. Shang, C. Wu, G. Xu and X. Wang, *ACS Catal.*, 2017, 7, 5987–5991.

60. K. Jiang and H. Wang, *Chem*, 2018, 4, 194–195.

61. Y. Jia, L. Zhang, A. Du, G. Gao, J. Chen, X. Yan, C. L. Brown and X. Yao, *Adv. Mater.*, 2016, 28, 9532–9538.

62. L. Zhang, Y. Jia, G. Gao, X. Yan, N. Chen, J. Chen, M. T. Soo, B. Wood, D. Yang, A. Du and X. Yao, *Chem*, 2018, 4, 285–297.

63. L. Wu, S. Hu, W. Yu, S. Shen and T. Li, *NPJ Comput. Mater.*, 2020, 6, 1–8.

64. G. Liu, A. W. Robertson, M. M. J. Li, W. C. H. Kuo, M. T. Darby, M. H. Muhieddine, Y. C. Lin, K. Suenaga, M. Stamatakis, J. H. Warner and S. C. E. Tsang, *Nat. Chem.*, 2017, 9, 810–816.

65. R. Q. Zhang, T. H. Lee, B. D. Yu, C. Stampfl and A. Soon, *Phys. Chem. Chem. Phys.*, 2012, 14, 16552–16557.

66. S. L. James, C. J. Adams, C. Bolm, D. Braga, P. Collier, T. Friščic, F. Grepioni, K. D. M. Harris, G. Hyett, W. Jones, A. Krebs, J. Mack, L. Maini, A. G. Orpen, I. P. Parkin, W. C. Shearouse, J. W. Steed and D. C. Waddell, *Chem. Soc. Rev.*, 2012, 41, 413–447.

67. X. He, Y. Deng, Y. Zhang, Q. He, D. Xiao, M. Peng, Y. Zhao, H. Zhang, R. Luo, T. Gan, H. Ji and D. Ma, *Cell Rep. Phys. Sci.*, 2020, 1, 100004.

68. T. Gan, Q. He, H. Zhang, H. Xiao, Y. Liu, Y. Zhang, X. He and H. Ji, *Chem. Eng. J.*, 2020, 389, 124490.

69. D. Deng, X. Chen, L. Yu, X. Wu, Q. Liu, Y. Liu, H. Yang, H. Tian, Y. Hu, P. Du, R. Si, J. Wang, X. Cui, H. Li, J. Xiao, T. Xu, J. Deng, F. Yang, P. N. Duchesne, P. Zhang, J. Zhou, L. Sun, J. Li, X. Pan and X. Bao, *Sci. Adv.*, 2015. DOI:10.1126/sciadv.1500462.

70. X. Cui, J. Xiao, Y. Wu, P. Du, R. Si, H. Yang, H. Tian, J. Li, W. H. Zhang, D. Deng and X. Bao, *Angew. Chem. Int. Ed.*, 2016, 55, 6708–6712.

71. B. Mohanty, B. K. Jena and S. Basu, *ACS Omega*, 2020, 5, 1287–1295.

72. Z. Zhang, C. Feng, C. Liu, M. Zuo, L. Qin, X. Yan, Y. Xing, H. Li, R. Si, S. Zhou and J. Zeng, *Nat. Commun.*, 2020, 11, 1–8.

73. L. Zhang, L. Han, H. Liu, X. Liu and J. Luo, *Angew. Chem. Int. Ed.*, 2017, 56, 13694–13698.

74. Y. Shi, W. M. Huang, J. Li, Y. Zhou, Z. Q. Li, Y. C. Yin and X. H. Xia, *Nat. Commun.*, 2020, 11, 1–9.

75. M. D. Kärkäs, J. A. Porco and C. R. J. Stephenson, *Chem. Rev.*, 2016, 116, 9683–9747.

76. T. Li, J. Liu, Y. Song and F. Wang, *ACS Catal.*, 2018, 8, 8450–8458.

77. P. Liu, Y. Zhao, R. Qin, S. Mo, G. Chen, L. Gu, D. M. Chevrier, P. Zhang, Q. Guo, D. Zang, B. Wu, G. Fu and N. Zheng, *Science*, 2016, 352, 797–801.

78. P. Liu, J. Chen and N. Zheng, *Cuihua Xuebao/Chinese J. Catal.*, 2017, 38, 1574–1580.

79. K. Huang, L. Zhang, T. Xu, H. Wei, R. Zhang, X. Zhang, B. Ge, M. Lei, J. Y. Ma, L. M. Liu and H. Wu, *Nat. Commun.*, 2019. DOI: 10.1038/s41467-019-08484-8.

80. H. Wei, K. Huang, D. Wang, R. Zhang, B. Ge, J. Ma, B. Wen, S. Zhang, Q. Li, M. Lei, C. Zhang, J. Irawan, L. M. Liu and H. Wu, *Nat. Commun.*, 2017, 8, 1–8.

81. H. Wei, H. Wu, K. Huang, B. Ge, J. Ma, J. Lang, D. Zu, M. Lei, Y. Yao, W. Guo and H. Wu, *Chem. Sci.*, 2019, 10, 2830–2836.

82. X. Ge, P. Zhou, Q. Zhang, Z. Xia, S. Chen, P. Gao, Z. Zhang, L. Gu and S. Guo, *Angew. Chem.*, 2020, 132, 238–242.

83. Y. Qu, Z. Li, W. Chen, Y. Lin, T. Yuan, Z. Yang, C. Zhao, J. Wang, C. Zhao, X. Wang, F. Zhou, Z. Zhuang, Y. Wu and Y. Li, *Nat. Catal.*, 2018, 1, 781–786.

84. J. Jones, H. Xiong, A. T. DeLaRiva, E. J. Peterson, H. Pham, S. R. Challa, G. Qi, S. Oh, M. H. Wiebenga, X. I. P. Hernández, Y. Wang and A. K. Datye, *Science*, 2016, 353, 150–154.

85. Y. Qu, L. Wang, Z. Li, P. Li, Q. Zhang, Y. Lin, F. Zhou, H. Wang, Z. Yang, Y. Hu, M. Zhu, X. Zhao, X. Han, C. Wang, Q. Xu, L. Gu, J. Luo, L. Zheng and Y. Wu, *Adv. Mater.*, 2019, 31, 1–7.

86. Y. Zhai, Z. Zhu, C. Zhu, K. Chen, X. Zhang, J. Tang and J. Chen, *Mater. Today*, 2020, 38, 99–113.

87. X. Liu, Y. Jiao, Y. Zheng, M. Jaroniec and S. Z. Qiao, *J. Am. Chem. Soc.*, 2019, 141, 9664–9672.

88. X. Guo, J. Gu, X. Hu, S. Zhang, Z. Chen and S. Huang, *Catal. Today*, 2020, 350, 91–99.

89. C. Ling, X. Bai, Y. Ouyang, A. Du and J. Wang, *J. Phys. Chem. C*, 2018, 122, 16842–16847.

90. M. Riyaz and N. Goel, *ChemPhysChem*, 2019, 20, 1954–1959.

91. T. He, A. R. Puente Santiago and A. Du, *J. Catal.*, 2020, 388, 77–83.

92. T. Deng, C. Cen, H. Shen, S. Wang, J. Guo, S. Cai and M. Deng, *J. Phys. Chem. Lett.*, 2020, 11, 6320–6329.

93. R. Guo, M. Hu, W. Zhang and J. He, *Molecules*, 2019, 24, 1777.

94. M. Qu, G. Qin, J. Fan, A. Du and Q. Sun, *Appl. Surf. Sci.*, 2021, 537, 148012.

95. G. Zheng, L. Li, Z. Tian, X. Zhang and L. Chen, *J. Energy Chem.*, 2021, 54, 612–619.

96. M. Zafari, D. Kumar, M. Umer and K. S. Kim, *J. Mater. Chem. A*, 2020, 8, 5209–5216.

97. D. W. Ma, T. Li, Q. Wang, G. Yang, C. He, B. Ma and Z. Lu, *Carbon N. Y.*, 2015, 95, 756–765.

98. X. Zhai, H. Yan, G. Ge, J. Yang, F. Chen, X. Liu, D. Yang, L. Li and J. Zhang, *Appl. Surf. Sci.*, 2020, 506, 144941.

99. Z. W. Chen, L. X. Chen, M. Jiang, D. Chen, Z. L. Wang, X. Yao, C. V. Singh, Q. Jiang, X. Zhai, H. Yan, G. Ge, J. Yang, F. Chen, X. Liu, D. Yang, L. Li, J. Zhang, T. He, S. K. Matta, A. Du, L. Jasin Arachchige, Y. Xu, Z. Dai, X. Zhang, F. Wang, C. Sun, D. W. Ma, T. Li, Q. Wang, G. Yang, C. He, B. Ma and Z. Lu, *Phys. Chem. Chem. Phys.*, 2020, 21, 1546–1551.

100. L. Jasin Arachchige, Y. Xu, Z. Dai, X. Zhang, F. Wang and C. Sun, *J. Phys. Chem. C*, 2020, 124, 15295–15301.

101. Z. W. Chen, L. X. Chen, M. Jiang, D. Chen, Z. L. Wang, X. Yao, C. V. Singh and Q. Jiang, *J. Mater. Chem. A*, 2020, 8, 15086–15093.

102. Z. Chen, J. Zhao, C. R. Cabrera and Z. Chen, *Small Methods*, 2019, 3, 1–9.

103. C. Ren, Q. Jiang, W. Lin, Y. Zhang, S. Huang, K. Ding and K. Ding, *ACS Appl. Nano Mater.*, 2020, 3, 5149–5159.

104. Y. Cao, Y. Gao, H. Zhou, X. Chen, H. Hu, S. Deng, X. Zhong, G. Zhuang and J. Wang, *Adv. Theory Simulations*, 2018, 1, 1870012.

105. X. Chen, X. Zhao, Z. Kong, W. J. Ong and N. Li, *J. Mater. Chem. A*, 2018, 6, 21941–21948.

106. Y. Cao, S. Deng, Q. Fang, X. Sun, C. X. Zhao, J. Zheng, Y. Gao, H. Zhuo, Y. Li, Z. Yao, Z. Wei, X. Zhong, G. Zhuang and J. Wang, *Nanotechnology*, 2019, 30, 335403.

107. C. C. Leong, Y. Qu, Y. Kawazoe, S. K. Ho and H. Pan, *Catal. Today*, 2021, 370, 2–13.

108. B. Huang, N. Li, W. J. Ong and N. Zhou, *J. Mater. Chem. A*, 2019, 7, 27620–27631.

109. Y. Gao, H. Zhuo, Y. Cao, X. Sun, G. Zhuang, S. Deng, X. Zhong, Z. Wei and J. Wang, *Chinese J. Catal.*, 2019, 40, 152–159.

110. L. Li, X. Wang, H. Guo, G. Yao, H. Yu, Z. Tian, B. Li and L. Chen, *Small Methods*, 2019, 3, 1–7.

111. J. Qu, J. Xiao, H. Chen, X. Liu, T. Wang and Q. Zhang, *Chinese J. Catal.*, 2021, 42, 288–296.

112. Y. Kong, D. Liu, H. Ai, K. H. Lo, S. Wang and H. Pan, *ACS Appl. Nano Mater.*, 2020, 3, 11274–11281.

113. J. Zhao and Z. Chen, *J. Am. Chem. Soc.*, 2017, 139, 12480–12487.
114. Z. Ma, Z. Cui, C. Xiao, W. Dai, Y. Lv, Q. Li and R. Sa, *Nanoscale*, 2020, 12, 1541–1550.
115. X. Mao, S. Zhou, C. Yan, Z. Zhu and A. Du, *Phys. Chem. Chem. Phys.*, 2019, 21, 1110–1116.
116. S. Dai, Q. Ma, M. K. Liu, T. Andersen, Z. Fei, M. D. Goldflam, M. Wagner, K. Watanabe, T. Taniguchi, M. Thiemens, F. Keilmann, G. C. A. M. Janssen, S. E. Zhu, P. Jarillo-Herrero, M. M. Fogler and D. N. Basov, *Nat. Nanotechnol.*, 2015, 10, 682–686.
117. Y. Huang, T. Yang, L. Yang, R. Liu, G. Zhang, J. Jiang, Y. Luo, P. Lian and S. Tang, *J. Mater. Chem. A*, 2019, 7, 15173–15180.
118. L. Chen, Q. Wang, H. Gong and M. Xue, *Appl. Surf. Sci.*, 2021, 546, 149131.
119. C. Fang and W. An, *Nano Res.*, 2021, 12, 1–9.
120. E. G. Da Silveira Firmiano, A. C. Rabelo, C. J. Dalmaschio, A. N. Pinheiro, E. C. Pereira, W. H. Schreiner and E. R. Leite, *Adv. Energy Mater.*, 2014, 4, 1–8.
121. C. Ma, N. Zhai, B. Liu and S. Yan, *Electrochim. Acta*, 2021, 370, 137695.
122. H. Guo, L. Li, X. Wang, G. Yao, H. Yu, Z. Tian, B. Li and L. Chen, *ACS Appl. Mater. Interfaces*, 2019, 11, 36506–36514.
123. T. Yang, T. T. Song, J. Zhou, S. Wang, D. Chi, L. Shen, M. Yang and Y. P. Feng, *Nano Energy*, 2020, 68, 104304.
124. X. Zhai, L. Li, X. Liu, Y. Li, J. Yang, D. Yang, J. Zhang, H. Yan and G. Ge, *Nanoscale*, 2020, 12, 10035–10043.
125. J. Li, S. Chen, F. Quan, G. Zhan, F. Jia, Z. Ai and L. Zhang, *Chem*, 2020, 6, 885–901.
126. H. Zhang, C. Cui and Z. Luo, *J. Phys. Chem. C*, 2020, 124, 6260–6266.
127. W. Y. Li, Y. B. Sun, M. Y. Li, X. Y. Zhang, X. Zhao and J. S. Dang, *Phys. Chem. Chem. Phys.*, 2021, 23, 2469–2474.
128. L. Ge, W. Xu, C. Chen, C. Tang, L. Xu and Z. Chen, *J. Phys. Chem. Lett.*, 2020, 11, 5241–5247.
129. L. Gao, F. Wang, M. A. Yu, F. Wei, J. Qi, S. Lin and D. Xie, *J. Mater. Chem. A*, 2019, 7, 19838–19845.
130. L. Lin, L. Gao, K. Xie, R. Jiang and S. Lin, *Phys. Chem. Chem. Phys.*, 2020, 22, 7234–7240.
131. K. Liu, J. Fu, L. Zhu, X. Zhang, H. Li, H. Liu, J. Hu and M. Liu, *Nanoscale*, 2020, 12, 4903–4908.
132. B. Li, L. Li, B. Li and Q. Guo, *J. Phys. Chem. C*, 2019, 123, 14501–14507.
133. X. Guo, J. Gu, S. Lin, S. Zhang, Z. Chen and S. Huang, *J. Am. Chem. Soc.*, 2020, 142, 5709–5721.
134. S. Wang, F. Ichihara, H. Pang, H. Chen and J. Ye, *Adv. Funct. Mater.*, 2018, 28, 1–26.
135. R. Manjunatha and A. Schechter, *Electrochem. Commun.*, 2018, 90, 96–100.
136. Z. Geng, Y. Liu, X. Kong, P. Li, K. Li, Z. Liu, J. Du, M. Shu, R. Si and J. Zeng, *Adv. Mater.*, 2018, 30, 2–7.
137. B. Yu, H. Li, J. White, S. Donne, J. Yi, S. Xi, Y. Fu, G. Henkelman, H. Yu, Z. Chen and T. Ma, *Adv. Funct. Mater.*, 2020, 30, 1–11.
138. H. Tao, C. Choi, L. X. Ding, Z. Jiang, Z. Han, M. Jia, Q. Fan, Y. Gao, H. Wang, A. W. Robertson, S. Hong, Y. Jung, S. Liu and Z. Sun, *Chem*, 2019, 5, 204–214.
139. L. M. Azofra, N. Li, D. R. Macfarlane and C. Sun, *Energy Environ. Sci.*, 2016, 9, 2545–2549.
140. W. Peng, M. Luo, X. Xu, K. Jiang, M. Peng, D. Chen, T. S. Chan and Y. Tan, *Adv. Energy Mater.*, 2020, 10, 1–9.
141. H. Zou, W. Rong, S. Wei, Y. Ji and L. Duan, *Proc. Natl. Acad. Sci. U. S. A.*, 2020, 117, 29462–29468.
142. Q. Qin, T. Heil, M. Antonietti and M. Oschatz, *Small Methods*, 2018, 2, 1800202.
143. X. Wang, W. Wang, M. Qiao, G. Wu, W. Chen, T. Yuan, Q. Xu, M. Chen, Y. Zhang, X. Wang, J. Wang, J. Ge, X. Hong, Y. Li, Y. Wu and Y. Li, *Sci. Bull.*, 2018, 63, 1246–1253.

144. S. K. Sahoo, J. Heske, M. Antonietti, Q. Qin, M. Oschatz and T. D. Kühne, *ACS Appl. Energy Mater.*, 2020, 3, 10061–10069.
145. M. Wang, S. Liu, T. Qian, J. Liu, J. Zhou, H. Ji, J. Xiong, J. Zhong and C. Yan, *Nat. Commun.*, 2019, 10, 1–8.
146. R. Zhang, L. Jiao, W. Yang, G. Wan and H. L. Jiang, *J. Mater. Chem. A*, 2019, 7, 26371–26377.
147. F. Lü, S. Zhao, R. Guo, J. He, X. Peng, H. Bao, J. Fu, L. Han, G. Qi, J. Luo, X. Tang and X. Liu, *Nano Energy*, 2019, 61, 420–427.
148. H. Yang, Y. Liu, Y. Luo, S. Lu, B, Su and J. Ma, *ACS Sustain. Chem. Eng.*, 2020, 8, 12809–12816.
149. S. Zhang, M. Jin, T. Shi, M. Han, Q. Sun, Y. Lin, Z. Ding, L. R. Zheng, G. Wang, Y. Zhang, H. Zhang and H. Zhao, *Angew. Chem. Int. Ed.*, 2020, 59, 13423–13429.
150. L. Han, X. Liu, J. Chen, R. Lin, H. Liu, L. U. Fang, S. Bak, Z. Liang, S. Zhao, E. Stavitski, J. Luo, R. R. Adzic and H. L. Xin, *Angew. Chem. Int. Ed.*, 2019, 58, 2321–2325.
151. H. Cheng, L. X. Ding, G. F. Chen, L. Zhang, J. Xue and H. Wang, *Adv. Mater.*, 2018, 30, 1–7.
152. Y. Ma, T. Yang, H. Zou, W. Zang, Z. Kou, L. Mao, Y. Feng, L. Shen, S. J. Pennycook, L. Duan, X. Li and J. Wang, *Adv. Mater.*, 2020. DOI:10.1002/adma.202002177.
153. Y. Li, Q. Zhang, C. Li, H. N. Fan, W. Bin Luo, H. K. Liu and S. X. Dou, *J. Mater. Chem. A*, 2019, 7, 22242–22247.
154. W. Liu, L. Han, H. T. Wang, X. Zhao, J. A. Boscoboinik, X. Liu, C. W. Pao, J. Sun, L. Zhuo, J. Luo, J. Ren, W. F. Pong and H. L. Xin, *Nano Energy*, 2020, 77, 105078.
155. Y. Liu, Q. Xu, X. Fan, X. Quan, Y. Su, S. Chen, H. Yu and Z. Cai, *J. Mater. Chem. A*, 2019, 7, 26358–26363.
156. Y. Gao, Z. Han, S. Hong, T. Wu, X. Li, J. Qiu and Z. Sun, *ACS Appl. Energy Mater.*, 2019, 2, 6071–6077.
157. S. Zhang, Q. Jiang, T. Shi, Q. Sun, Y. Ye, Y. Lin, L. R. Zheng, G. Wang, C. Liang, H. Zhang and H. Zhao, *ACS Appl. Energy Mater.*, 2020, 3, 6079–6086.
158. S. Mukherjee, X. Yang, W. Shan, W. Samarakoon, S. Karakalos, D. A. Cullen, K. More, M. Wang, Z. Feng, G. Wang and G. Wu, *Small Methods*, 2020, 4, 1–11.
159. W. Zang, T. Yang, H. Zou, S. Xi, H. Zhang, X. Liu, Z. Kou, Y. Du, Y. P. Feng, L. Shen, L. Duan, J. Wang and S. J. Pennycook, *ACS Catal.*, 2019, 9, 10166–10173.
160. X. Wang, D. Wu, S. Liu, J. Zhang, X. Z. Fu and J. L. Luo, *Nano-Micro Lett.*, 2021, 13, 1–12.
161. J. Liu, X. Kong, L. Zheng, X. Guo, X. Liu and J. Shui, *ACS Nano*, 2020, 14, 1093–1101.
162. I. J. McPherson, T. Sudmeier, J. Fellowes and S. C. E. Tsang, *Dalt. Trans.*, 2019, 48, 1562–1568.
163. Y. Yao, S. Zhu, H. Wang, H. Li and M. Shao, *J. Am. Chem. Soc.*, 2018, 140, 1496–1501.
164. J. Zheng, Y. Lyu, M. Qiao, R. Wang, Y. Zhou, H. Li, C. Chen, Y. Li, H. Zhou, S. P. Jiang and S. Wang, *Chem*, 2019, 5, 617–633.
165. S. Chen, S. Perathoner, C. Ampelli, C. Mebrahtu, D. Su and G. Centi, *ACS Sustain. Chem. Eng.*, 2017, 5, 7393–7400.
166. R. Manjunatha, A. Karajić, H. Teller, K. Nicoara and A. Schechter, *ChemCatChem*, 2020, 12, 438–443.
167. S. Z. Andersen, V. Čolić, S. Yang, J. A. Schwalbe, A. C. Nielander, J. M. McEnaney, K. Enemark-Rasmussen, J. G. Baker, A. R. Singh, B. A. Rohr, M. J. Statt, S. J. Blair, S. Mezzavilla, J. Kibsgaard, P. C. K. Vesborg, M. Cargnello, S. F. Bent, T. F. Jaramillo, I. E. L. Stephens, J. K. Nørskov and I. Chorkendorff, *Nature*, 2019, 570, 504–508.

9 Lithium Batteries Application of Atomically Dispersed Metallic Materials

*Changtai Zhao, Kieran Doyle-Davis,
and Xueliang Sun*
The University of Western Ontario

CONTENTS

9.1 INTRODUCTION

Electrochemical energy storage systems play a crucial role in the use of renewable and environmentally-friendly resources.[1,2] Among them, rechargeable batteries, especially Li batteries, possess high energy density, high voltage, and long cycle life, showing their advantages as power sources for electric vehicles and portable electronic devices. However, the efficiency and kinetics of their electrochemical reactions are relatively low, which depend on the activity of the electrode materials.[3-6] For example, Li–O$_2$ batteries suffer from sluggish redox kinetics and high overpotential due to the insulation and insolubilization of Li$_2$O$_2$, a prominent discharge product, requiring a highly effective electrocatalyst to boost the electrochemical process and further enhance the energy efficiency, rate capability, and cycling stability of Li–O$_2$ batteries.[7] However, most catalysts possess low utilization and low catalytic activity. Besides, these catalysts with a high weight density will undoubtedly decrease the

DOI: 10.1201/9781003153436-9

energy density of batteries when used as a component in batteries.[8] Therefore, it has become more important to design novel catalysts to satisfy the requirements of high catalytic activity, low density, and high utilization.

Single-atom materials with nearly 100% atomic utilization, good interaction with substrates, unsaturated coordination chemistry, and unique electronic structure have been explored in some classical industrial catalytic reactions, such as CO oxidation and selective hydrogenation.[9,10] Recently, the research of single-atom catalysts has been extended to efficiently catalyze electrochemical reactions in fuel cells, solar cells, aqueous secondary batteries, electrolysis of water, CO_2 electrochemical reduction, and N_2 electrochemical reduction.[11–16] Very recently, the applications of single-atom catalysts in Li batteries, including Li-metal batteries, $Li–O_2$ batteries, $Li–CO_2$ batteries, Li–S batteries, and Li–ion batteries, were reported and indicated great potential.[17–20]

In this chapter, we summarize the recent progress and review the working mechanism of single-atom catalysts for Li batteries. First, we discuss the effects of single-atom materials on adjusting the interface interaction and regulating the deposition of Li metal on the Li metal anode. Then, the catalytic functions of single-atom catalysts for regulating the deposition of discharge products (Li_2O_2 and Li_2CO_3) and further promote their decomposition in $Li–O_2$ batteries and $Li–CO_2$ batteries are presented. The application of single-atom catalysts to modify the sulfur cathode and separator to suppress the diffusion of polysulfides and boost the electrochemical conversion reaction are also investigated. The effect of a single-atom catalyst on improving the Li storage ability for Li–ion batteries is also reviewed. Finally, a summary and outlook of single-atom catalyst for Li batteries were present.

9.2 SINGLE ATOMS FOR LI BATTERIES

9.2.1 Single Atoms for Li-Metal Batteries

Li-metal batteries feature high energy density due to the high theoretical capacity (3,860 mAh g^{-1}) and the lowest potential (−3.04 V vs. standard hydrogen electrode) of the Li-metal anode, which are considered ideal candidates for next-generation rechargeable batteries. Unfortunately, Li-metal anodes suffer from severe Li dendrite issues caused by the uneven Li-metal nucleation/deposition during the charging process.[21–23] This will decrease the Coulombic efficiency and cycle lifespan of Li-metal batteries and bring about safety concerns due to the short-circuit risk. Therefore, a safe and high-efficiency Li-metal anode is crucial to the development of Li-metal batteries. Designing advanced battery materials, such as electrolytes,[24,25] host materials,[26] and solid electrolyte interphase (SEI),[27,28] etc., to regulate the nucleation/deposition of Li metal are some effective approaches to overcome these problems. Recently, some works shed light on the controllable Li-metal nucleation by designing the host materials with uniformly distributed metal single atoms.[29] Zhang and colleagues offered fresh insights into guiding Li deposition by introducing single atoms into the Li-metal host to regulate the lithiophilic chemistry.[30] Atomically dispersed CoN$_x$-doped graphene was developed as the host material, in which the coordination between Co atoms and N atoms in the conductive matrix can effectively adjust the local electron structure, thereby enhancing the adsorption of Li ions. This resulted in uniform Li nucleation behavior and further contributed to the smooth Li-metal deposition. These characteristics together led to high Coulombic efficiency of as high as 99.2% at a current density of 2 mA cm^{-2}. In terms of the high nucleation overpotential of Li metal on the surface of lithiophobic matrix, Yan et al. used an N-doped carbon matrix with iron single atoms (Fe SAs–NC) as the uniformly lithiophilic site to guide Li nucleation and deposition and minimize the overpotential (Figure 9.1a).[31] The molecular dynamics simulation quantitatively confirmed the excellent affinity between Li and Fe SAs–NC at the atomic level. This is attributed to the crucial role

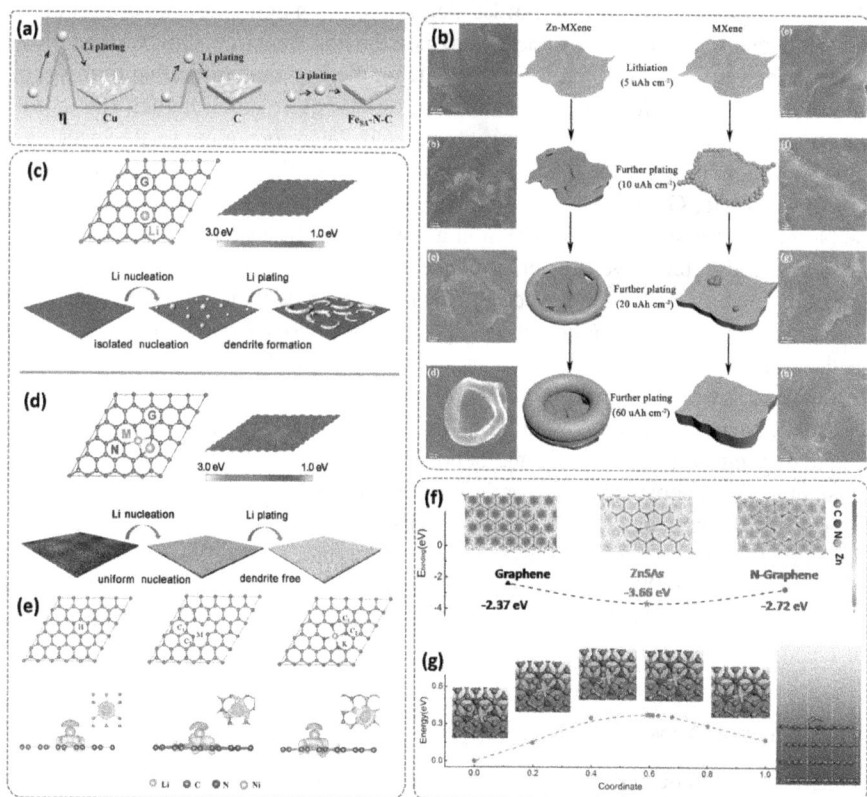

FIGURE 9.1 (a) Schematic diagram of the Li plating process on the substrates of Cu, C@Cu, and Fe SAs–NC@Cu, respectively (η means the nucleation overpotential of Li metal).[31] (b) Nucleation and growth diagram and typical SEM images after Li plating at different current densities on Zn SAs–MXene and MXene.[32] Li adsorption energy distribution and the corresponding schematic diagrams of Li nucleation and deposition process on (c) the surface of pristine graphene and (d) Ni single atoms supported on N-doped graphene. (e) Top view of pristine graphene, N-doped graphene, and Ni single atoms supported on N-doped graphene and side view of electron density differences of pristine graphene, N-doped graphene, and Ni single atoms supported on N-doped graphene with one Li atom adsorbed.[33] (f) Surface binding energy and electron density differences of graphene, Zn single atom, and N-doped graphene. (g) Li migration pathways and barrier energy on the Zn single atom.[34]

of Fe single atoms that can regulate the electron structure around carbon. MXene with high electronic conductivity was also employed as the substrate to immobilize Zn single atoms (Zn SAs-MXene) to achieve the controllable nucleation and growth of Li metal.[32] As shown in Figure 9.1b, Li metal initially tended to uniformly nucleate on the surface of Zn SAs-MXene due to the presence of a lot of Zn single atoms, which could form an alloy phase with Li metal, and then grew vertically on the nucleated sites, resulting in bowl-like Li without Li dendrites. Benefiting from this feature, a low overpotential (11.3 mV), a long cycle life (1,200 h), and a high areal capacity (40 mAh cm^{-2}) were achieved by using this Li-metal host.

Density functional theory calculations are an effective way to explore the effect of single atoms on the regulation of Li-metal deposition and guide the design of the Li-metal host. If the adsorption energy of Li on the host material is higher than the Li cohesive energy, the Li ions prefer to nucleate on the host material rather than aggregate to the Li bulk. However, if the adsorption energy is much higher, the atomic structural stability of host materials is challenged. Therefore, a moderate adsorption energy gradient is better for stable and smooth Li deposition (Figure 9.1c–e).[33] Based on this principle, Gong and colleagues designed metal single atoms supported on the N-doped graphene that could exhibit a low-voltage hysteresis, a high Coulombic efficiency, and a stable cycling performance as the host material for Li-metal anode. Qian and coworkers also employed the first principle calculation to reveal the effect of single atoms on the regulation of Li deposition behaviors (Figure 9.1f and g).[34] The results demonstrated Zn single-atom sites have higher surface energy, better to bind Li ions, and lower Li migration barrier energy, which enables Li to easily migrate toward the surrounding area rather than local aggregation. These advantages contributed to driving the high-dimensional dendrite-free Li deposition.

As mentioned above, the concept of introducing metal single atoms into Li-metal anode shows promising effects, and the results are encouraging. Therefore, this provides a fresh route to develop safe and high-performance Li-metal anode for Li-metal batteries. However, this technique is still in the initial stages and requires some improvements before widespread adoption, namely, a more practical preparation of the electrode surface, as well as a more comprehensive study of the space.

9.2.2 Single Atoms for Li–O$_2$ Batteries

Typical Li–O$_2$ batteries are configured by coupling a Li-metal anode with an O$_2$-breathing cathode via a suitable electrolyte, sitting somewhere between a fuel cell and traditional batteries. The output energy is based on the reversible electrochemical reactions of Li-metal oxidation and oxygen reduction (Li$_2$ + O$_2$ ↔ Li$_2$O$_2$).[35–37] Due to the highest specific capacity (3,860 mAh g^{-1}) and the lowest theoretical potential (−3.04 V) of Li metal and the nature of cathode active materials from surrounding air, thus, they feature the highest theoretical energy density of as high as 3,500 Wh kg^{-1}. This makes Li–O$_2$ batteries attract worldwide research interest and become one of the most promising candidates as the power source for smart electronic devices and electric vehicles.[38,39] However, the study of Li–O$_2$ batteries is early, and the development of Li–O$_2$ batteries is facing several challenges. As previously mentioned, the uncontrollable Li dendrite growth will bring about safety concerns for the practical application of Li–O$_2$ batteries. Similarly, the side reactions of liquid electrolytes at high voltage and the Li-metal corrosion caused by the crossover of air also influence the electrochemical performance of Li–O$_2$ batteries. Developing solid-state Li–O$_2$ batteries by replacing the liquid electrolyte with a solid-state electrolyte is an effective strategy to solve some of these problems.[40] Equally noteworthy, the formed discharge product is insulating and insoluble, which will continuously accumulate in porous air electrodes during the discharging process. The electrochemical performance of the battery will directly depend on the ability of the electrode to decompose the discharge product. In this case, the development of high-efficiency catalysts to accelerate the decomposition of Li$_2$O$_2$ is an effective way to enhance the electrochemical performance of Li–O$_2$ batteries.[41]

So far, various materials have been explored as the electrocatalyst for Li–O$_2$ batteries, such as noble metals,[41,42] transition metal oxides,[43] N (S, B)-doped carbon,[44,45] and soluble catalysts,[46] etc. Recently, metal single atoms were reported as the catalyst for Li–O$_2$ batteries and exhibited unexpected effects and outstanding electrochemical performance due to their unique physical/chemical properties. The high atom utilization, abundant highly active sites, and unique coordination environment of single-atom catalysts are favorable to catalyze the deposition and decomposition of discharge products. In 2018, the concept of the single atom was introduced into the study of Li–O$_2$ batteries, and the effects of single-atom catalyst have been demonstrated.[47] Recently, Xu et al. reported a single-atom cobalt electrocatalyst for Li–O$_2$ batteries, which could provide more active sites and nucleation sites to promote the formation of micrometer-sized flowerlike discharge products (Figure 9.2a).[48] As shown in Figure 9.2b, during charge progress, the binding energy of reduced species on the surface of Co single-atom sites is much lower than those on the surface of Pt/C and N-doped carbon. This is beneficial to spread the reduced species from the cathode surface to the electrolyte and complete the reaction in a one-electron pathway, thus, promoting the decomposition of Li$_2$O$_2$. Benefiting from these features, the Li–O$_2$ battery with single-atom Co catalyst exhibited excellent electrochemical performance, evidenced by the high round-trip efficiency (86.2%) and long cycling stability of more than 250 cycles with a fixed capacity of 1,000 mAh g^{-1} (218 days). Similarly, Yin and workers developed N-rich carbon nanosheets coordinated with Co single atoms by a gas migration-trapping process (Figure 9.2c and d).[49] The Li–O$_2$ battery with this cathode material delivered remarkably decreased overpotential (0.40 V), high discharge specific capacity (20,105 mAh g^{-1}), outstanding rate capability (11,098 mAh g^{-1} at 1 A g^{-1}), and long-term cycle life (260 cycles). Interestingly, the great performance is attributed to the rich Co–N$_4$ moieties as the catalytic sites that can promote to form uniformly distributed nanosized Li$_2$O$_2$ (Figure 9.2e). Further, this accelerated the decomposition kinetics of Li$_2$O$_2$ and decreased side reactions. DFT calculations demonstrated that the formation of nanosized Li$_2$O$_2$ is via a "surface-adsorption pathway". Rich Co–N$_4$ moieties featured strong adsorption ability to LiO$_2$, thus, the immobilized LiO$_2^*$ cannot freely assemble, but homogeneously and intimately deposited around active sites.

Aside from the transition metals,[50] noble metal-based single-atom catalysts were also developed and studied as the component of cathode for Li–O$_2$ batteries. The study of noble metal-based single-atom catalysts may be more significant because

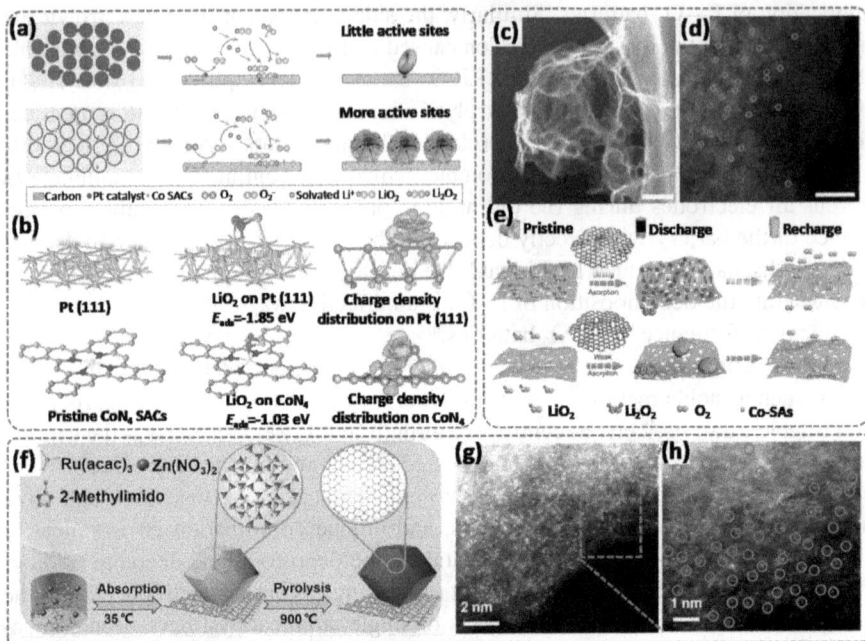

FIGURE 9.2 (a) Schematic diagram of the discharge mechanism with commercial Pt/C and single-atom Co catalyst in Li–O₂ batteries. (b) Optimized structures and the corresponding charge density distribution of the LiO₂ on Pt (111) and CoN₄ site.[48] (c, d) HAADF-STEM images of Co single atoms supported on carbon nanosheet. (e) Schematic diagram of the reaction mechanisms with and without Co single atoms.[49] (f) Schematic illustrations of the preparation process of Ru single atoms on carbon cloth. (g, h) HAADF-STEM images of Ru single atoms supported on carbon cloth.[54]

of their high cost. A hybrid of yolk–shell $Co_3O_4@Co_3O_4$/Ag was synthesized via a facile synchronous reduction strategy by Liu et al.[47] In this work, the introduction of Ag, in terms of surface single atoms and cluster and crystal lattice doping, exhibited the functions of strengthening the interface binding and regulating the electronic structures of Ag and Co species, further providing more active sites. Besides, it showed the important effect on the regulation of discharged product morphology to form the more easily decomposable flowerlike Li_2O_2. As the catalyst for Li–O₂ batteries, it delivered enhanced electrochemical performance such as higher discharge specific capacity (12,000 mAh g⁻¹ at 200 mA g⁻¹), lower overpotential, higher rate capability (4,700 mAh g⁻¹ at 800 mA g⁻¹), and longer cycle life. Ruthenium and ruthenium oxide with various structures have been extensively investigated as catalysts for Li–O₂ batteries and shown enhanced electrochemical performance.[51–53] Xu and coworkers developed a MOF-assisted spatial confinement coupled with ionic substitution strategy to synthesize Ru single atoms, which were used as the catalyst for Li–O₂ batteries and exhibited the function of regulating the size, morphology, and distribution of the discharge product and the enhanced battery performance (Figure 9.2f–h).[54] Besides, Pt single atoms were also prepared and studied for Li–O₂ batteries. Qian and coworkers developed Pt single atoms decorated with holey ultra-thin g-C_3N_4 nanosheets (Pt-CNHS) via a facile liquid-phase reaction process.[55] Benefiting from the porous structure of g-C_3N_4 nanosheets with a large surface area and the single-atom Pt with high dispersibility, stability, and electrocatalytic activity, the Li–O_2 battery with this material showed a high discharge capacity of 17059.5 mAh g^{-1} at the current density of 100 mA g^{-1}, a good rate capability of 5964.7 mAh g^{-1} at 800 mA g^{-1}, and outstanding cycling stability of 100 cycles.

Overall, the study of single-atom catalysts provides a new route to address the issues of low kinetics, high overpotential, and poor cycling stability in Li–O_2 batteries. While the efficacy of single-atom catalysts is impressive, the advantage in reducing the amount of metal catalyst, especially noble metal, is remarkable. However, the electrochemical performance of batteries can be further enhanced by improving the loading of single atoms, optimizing the interaction with substrates, enhancing the catalytic activity via the regulation of electronic structure, and developing new single-atom catalysts.

9.2.3 Single Atoms for Li–CO₂ Batteries

Li–CO_2 batteries can output a high theoretical energy density of 1,876 Wh kg^{-1} which is derived from the reversible redox reaction between Li-metal anode and CO_2 gas to form Li_2CO_3 and carbon ($3CO_2 + 4Li^+ + 4e^- \leftrightarrow 2Li_2CO_3 + C$). Therefore, Li–$CO_2$ batteries can not only achieve energy storage and conversion but also capture and utilize CO_2. It is much important for recycling CO_2 and maintaining carbon neutrality. Besides, Li–CO_2 batteries offer an advantage in providing the power for exploration on the planet Mars, with an atmosphere of 96% CO_2.[56,57] However, similar to Li–O_2 batteries, the sluggish reactions of CO_2 reduction and evolution lower the kinetics of Li–CO_2 batteries, resulting in high overpotential for decomposing the discharge products of Li_2CO_3/C. This will further cause electrolyte decomposition, leading to poor electrochemical performance.[58] Therefore, a high-efficiency catalyst is required to facilitate the reversible formation and decomposition of discharge products during the discharge/charge processes. So far, various kinds of materials have been explored as the catalyst for Li–CO_2 batteries, such as carbon materials,[59–61] noble metals,[62] transition metal oxides,[63] etc.

Very recently, single-atom materials have been proposed as the catalyst to promote the electrochemical reactions of Li–CO_2 batteries. Dai and coworkers reported a microstructure of implanting Fe single atoms into 3D N, S-codoped holey graphene sheets (Fe–SAs/N, S–HG), which can be used as the cathode material for Li–CO_2 batteries.[64] As shown in Figure 9.3a, the 3D porous hierarchical structure with a large surface area, interconnected channels, and sufficient space can not only facilitate electron transfer and mass transportation of CO_2/Li–ion but also allow for high uptake of Li_2CO_3 discharge product. Fe single atoms with the high catalytic activity are in favor of forming the small dense Li_2CO_3 nanoparticles (2–10 nm), which can be readily decomposed on charging. Figure 9.3b and c show the discharge–charge curves and cycling performance of Li–CO_2 batteries with different electrodes. It can be found that the resultant Li–CO_2 batteries delivered a high specific capacity of 23,174 mAh g^{-1} at a current density of 100 mA g^{-1}, high initial Coulombic efficiency of 92.9%, a low overpotential of 1.17 V, and excellent cycling

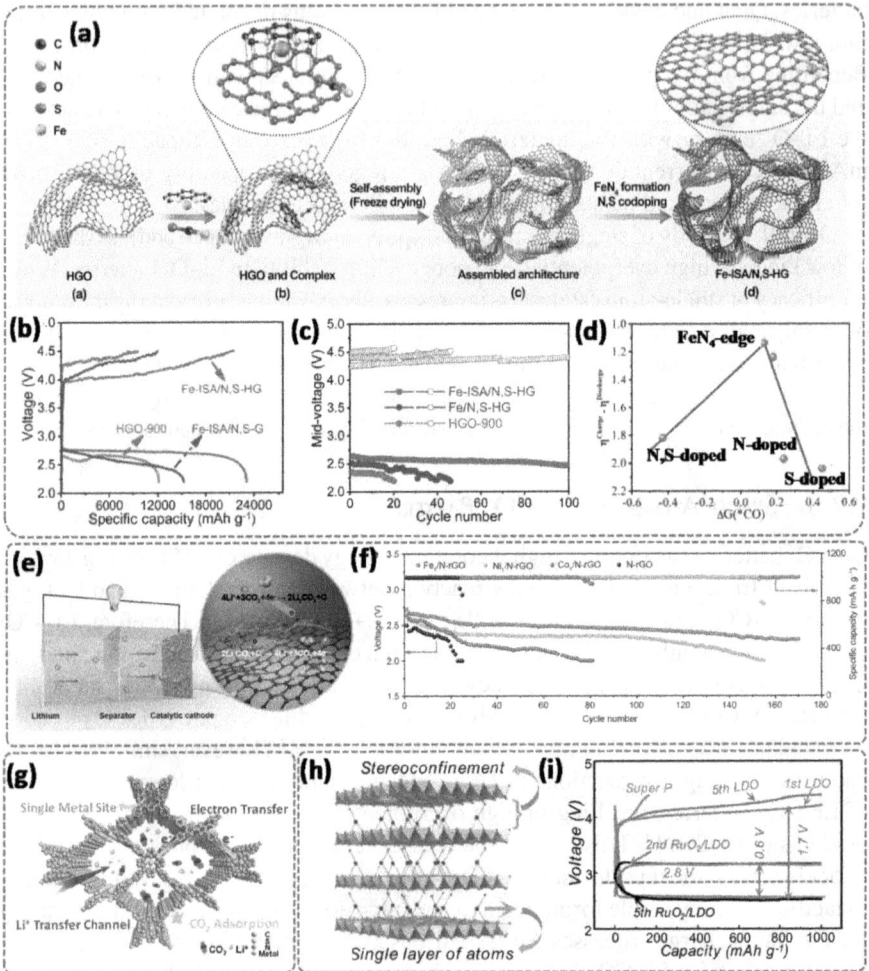

FIGURE 9.3 (a) Schematic illustration of the synthesis process of Fe single atoms supported on 3D N, S-codoped holey graphene sheets. (b) Discharge/charge curves of Li–CO₂ battery with various electrodes at the current density of 100 mA g⁻¹. (c) Cycling performance at the current density of 1 A g⁻¹. (d) The adsorption energy of *CO as a function of the overpotential on the various catalysts.[64] (e) Schematic illustration of Li–CO₂ batteries with single-atom catalysts. (f) Cycling stability of Li–CO₂ batteries with different single-atom catalysts.[65] (g) Schematic diagram of the TTCOF-Mn.[66] (h) Schematic representation for the interlayered stereoconfinement of single-atom Ru. (i) Discharge/charge curves of Li–CO₂ battery with a limited capacity of 1,000 mAh g⁻¹ at a current density of 166 mA g⁻¹.[68]

stability for over 200 cycles with a limited capacity of 1,000 mAh g⁻¹. As shown in Figure 9.3d, DFT calculations indicated Fe–SAs/N, S–HG featured the highest activity to catalyze CO_2 reduction and evolution reactions, which were attributed to the synergistic effects between the metal centers of "Fe–N₄" moieties and the support with charge and spin redistribution resulting from the N, S-codoping. Fe, Co, and Ni single atoms supported on N–rGO (M_1–rGO) were synthesized and compared as the catalyst for Li–CO_2 batteries by evaluating the decomposition ability of Li_2CO_3/C (Figure 9.3e and f).[65] It demonstrated that Fe_1/N–rGO featured the highest catalytic activity and showed the best electrochemical performance of Li–CO_2 batteries. The battery with Fe_1/N–rGO delivered a high discharge capacity of 16,835 mAh g^{-1} and maintained excellent cycling stability for over 170 cycles. A porphyrin-based covalent organic framework with single metal sites (TTCOF-Mn) is a kind of inherent single-atom material, which is used as the catalyst for Li–CO_2 battery by Lan and coworkers.[66] The Li–CO_2 battery with this material exhibited a low potential gap of 1.07 V and long cycling stability for over 180 cycles with a fixed capacity of 1,000 mAh g^{-1}. As shown in Figure 9.3g, the electron-donating properties of TTF and the unique ordered micropores of TTCOF-Mn ensured fast electron transfer, high CO_2 adsorption capacity, and rapid Li–ion transportation. Besides, DFT calculation results demonstrated that the Mn-TAPP site has strong adsorption to CO_2 and can catalyze the CO_2 conversion process in a four-electron reaction pathway. Together, these contributed to the outstanding performance of Li–CO_2 batteries. Apart from single atoms moiety, the interaction between adjacent single atoms also plays an important role in the catalytic action. Qiao and coworkers first investigated the synergy between adjacent single atoms and evaluated the effects for Li–CO_2 batteries.[67] When increasing the Co atom loading to 5.3 wt%, a hybrid of adjacent Co atoms anchored on graphene oxide could be configured, in which the atomic dispersion of Co was maintained and the Co–Co distance was 1.79 Å. This offered sufficient and continuous active sites for CO_2 reduction and evolution reactions. DFT calculation results demonstrated dimer Co/GO featured a unique electronic structure with the synergetic effect of Co–Co and Co–O, which could provide a strong adsorption ability to Li_2CO_3 and were beneficial for the deposition and decomposition of discharge products, thus, leading to higher catalytic activity than that of single-atom Co/GO and Co nanoclusters/GO. As a result, the Li–CO_2 batteries with this catalyst exhibited a high discharge specific capacity of 17,358 mAh g^{-1} and maintained great cycling stability for over 100 cycles.

Precious metals such as Ru and RuO_2 are high-efficient catalysts for Li–CO_2 batteries and have been considerably investigated. Downsizing from nanoparticles to a single atom with low coordination chemistry is a feasible strategy to decrease the cost of catalyst, maximize atomic efficiency, and improve the catalytic activity for Li–CO_2 batteries. Wang et al. proposed an interlayered stereoconfinement approach to dispersing and immobilizing Ru single atoms by intercalating Ru anions into the interlayers of CoTi-layered double oxide (Figure 9.3h).[68] This catalyst not only exhibited high catalytic activity but also overcame the challenging problem of single-atom catalysts that tend to aggregate from metal single atoms to metal clusters due to the high surface energy. As shown in Figure 9.3i, the Li–CO_2 battery with this catalyst exhibited a low charge potential of only 3.2 V at a current density of 166 mA g^{-2} with a limited capacity of 1,000 mAh g^{-1}. Benefiting from the low overpotential that could eliminate side reactions caused by the decomposition of electrolyte at a high voltage, it offered a platform to investigate the reaction mechanism of Li–CO_2 batteries. Based on the differential quantitative mass spectrometry results, the release of CO_2 at about 3.2 V indicated the Li_2CO_3

was decomposed. The ratio of charge-to-mass analysis demonstrated the reaction process of $4Li^+ + 3CO_2 + 4e^- \rightarrow 2Li_2CO_3 + C$ with reversible formation and decomposition of Li_2CO_3.

9.2.4 Single Atoms for Li–S Batteries

Li–S batteries are composed of a Li-metal anode and a sulfur cathode and have attracted considerable attention as a replacement for Li–ion batteries due to their high theoretical capacity density of 2,600 Wh kg^{-1}, high natural abundance, and nontoxicity of sulfur.[39] The discharge reaction is a complicated and multi-step process. First, the solid-state elemental sulfur is reduced to soluble polysulfides (Li_2S_x, $4 \leq x \leq 8$). Then, the liquid polysulfides are converted into solid Li_2S_2 or Li_2S. Finally, the Li_2S_2 is converted to Li_2S via a solid–solid reaction.[69] The working mechanisms of Li–S batteries indicate some necessary considerations. First, one challenging issue is the shuttling effect of highly soluble polysulfides between the Li-metal anode and sulfur cathode, which will worsen the electrode interface and result in the loss of some active material, causing increased internal resistance and rapid capacity fade with cycling. To overcome this issue, confining dissolved polysulfides within the cathode or blocking their diffusion through an interlayer or separator are possible approaches.[70,71] In addition, during the discharge/charge processes, accelerating the transformation of polysulfides with highly efficient electrocatalysts can mitigate their shuttling behavior to some degree.[72,73] Second, the sluggish kinetics of solid-phase reactions caused by the insulativity of Li_2S_2 and Li_2S largely reduces the capacity utilization of the sulfide cathode and rate performance. Besides, at the beginning of the charging process, the activation of Li_2S is difficult, causing a large overpotential.[74,75] Developing a highly efficient electrocatalyst is a promising strategy to enhance redox kinetics by reducing energy barriers.

Single-atom catalysts feature an incomparable atom utilization efficiency of nearly 100%, unsaturated coordination structure, and unique electronic structure, which demonstrates the dual functions of trapping the polysulfides and accelerating the sluggish redox kinetics in Li–S batteries. The strong interaction between single-atom catalysts and their substrate plays an important role in promoting charge transfer. Besides, single-atom catalysts take up less volume and weight in sulfur cathode, being favorable to high energy density.[76] Therefore, single-atom catalysts exhibited tremendous potential for enhancing the performance of Li–S batteries compared to even nanoscale catalysts.[77,78]

9.2.4.1 Single Atoms for the Cathode of Li–S Batteries

The studies of single-atom materials in Li–S batteries consist of the single atom-based sulfur cathode and single atom-based separator.[79] For the sulfur host, Yang et al. first reported the utilization of single-atom catalysts for Li–S batteries in 2018.[80] Iron single atoms supported on porous N-doped carbon were synthesized and further employed as the sulfur host. For Li–S batteries, Fe single atoms featured strong interactions with polysulfides, limiting the diffusion of polysulfides. Besides, Fe single atoms served as the electrocatalyst, promoting the conversion of polysulfides and the nucleation and deposition of Li_2S. Subsequently, it could activate the process of Li_2S oxidation.

Iron single atoms deposited on N, S-codoped porous carbon were also reported as the host material for Li–S batteries.[81] Besides, Fe single atoms were selected as a representative model to carry out a systematic DFT study on the structural stability of FeN_x and the interaction between polysulfides and FeN_x sites.[82] Apart from Fe single atoms, Co single atoms embedded on porous carbon materials were also developed as the catalyst for Li–S batteries.[83–85] Ji et al. reported a hybrid of cobalt single atoms embedded in N-doped graphene, which could be used as the cathode host for Li–S batteries.[86] Experimental and theoretical results demonstrated that Co–N–C coordination centers could play the role of a bifunctional electrocatalyst to decrease the energy barriers of formation and decomposition of Li_2S and facilitate reaction kinetics during discharge and charge processes. As shown in Figure 9.4a and b, the Co–N/G electrode showed the lowest overpotential in the processes of the conversion of soluble Li_2S_4 to solid Li_2S_2 and conversion of solid Li_2S to soluble polysulfides. Besides, the sulfur reduction pathways from S_8 to Li_2S on both N/G and Co–N/G cathodes were

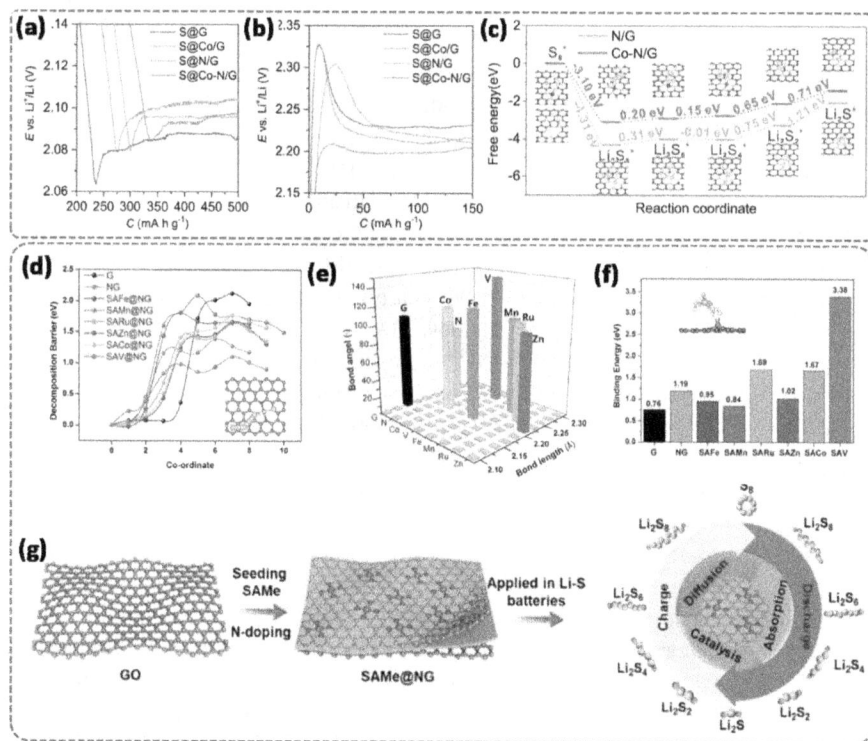

FIGURE 9.4 (a) Discharge and (b) charge profiles of various electrodes showing the overpotentials for conversion reactions. (c) Energy profiles for the reduction of S_8 to Li_2S on N/G and Co-N/G substrates.[86] (d) Decomposition barriers of Li_2S, (e) bond length and bond angle of Li_2S, and (f) binding energy of Li_2S_6 on different substrates. (g) Schematic illustration of the seeding approach of single atoms and the conversion process of Li–S batteries with a single-atom catalyst.[88]

investigated. As shown in Figure 9.4c, the conversion reaction from Li_2S_2 to Li_2S is the rate-determining step with the largest positive Gibbs free energy. The value is 0.71 eV on Co–N/G which is much lower than that on N/G (1.21 eV), indicating that the reduction of sulfur on Co–N/G is thermodynamically more favorable compared with N/G. This hybrid with a high S mass ratio of up to 90 wt% still exhibited a high capacity of 1,210 mAh g^{-1}. With the S loading of 6.0 mg cm^{-2}, it could deliver a high areal capacity of 5.1 mAh cm^{-2} and maintain 100 cycles with a cycle fading rate of just 0.029%. In addition, a single-atom Co catalyst was also used as the model substrate to study how the single-atom catalyst worked in Li–S batteries.[87]

For the research of single-atom catalysts, theoretical modeling is a powerful tool to help search for a stable and highly effective catalyst. Cui et al. screened ten different metal single atoms on N-doped graphene by evaluating the structural stability and the decomposition energy barrier of Li_2S.[88] As shown in Figure 9.4d, the theoretical simulation results indicated vanadium single atoms on N-doped graphene (VSAs@NG) were stable and showed the smallest decomposition energy barrier of 1.10 eV, meaning the best catalytic activity for the decomposition of Li_2S. Figure 9.4e displays that VSAs@NG possessed both the longest bond length (2.28 Å, Li–S) and the largest bond angle (145.83°, Li–S–Li) among all models, meaning the formation of a weaker Li–S bond that is good for the decomposition of Li_2S. Besides, vanadium single atoms also had the highest binding energy to polysulfides, suggesting the best potential on limiting polysulfide dissolution and suppressing their shuttle effect (Figure 9.4f). Under this guidance, vanadium single atoms on N-doped graphene were synthesized by a seeding strategy with controllable loading, adjustable components, and scalable quantities (Figure 9.4g). The batteries with this catalyst delivered great improvement in the capacity, rate capability, and cycling stability, further confirming the merits of this single-atom catalyst in achieving the high utilization of sulfur and rapid redox kinetics. Besides, Mo single atoms were also reported as the catalyst for Li–S batteries. Mo–N_2–C coordination structure in porous N-doped carbon nanosheets was designed by Li and coworkers and used as sulfur host materials.[89] It offered the ability to limit the diffusion of polysulfides and promote the conversion reaction, further leading to high rate capability (a high reversible capacity of 743.9 mAh g^{-1} at the current density of 5 C) and excellent cycling stability (an ultralow capacity fade rate of 0.018% per cycle for 550 cycles).

In addition to carbon materials, many other nanomaterials with unique structural advantages have also been developed as the substrate to support single atoms and configure the sulfur host.[90–92] Graphitic carbon nitride (g-C_3N_4) features ultra-high nitrogen content and low density and shows strong polarity and adsorbability to polysulfide, which is a promising host material for Li–S batteries.[93–95] Chen and coworkers used layer g-C_3N_4 to load Fe single atoms as the sulfur host material for Li–S batteries.[91] The substrate of layered g-C_3N_4 with a high nitrogen content not only provided a lot of coordination sites to realize the high mass loading (8.5%) of Fe single atoms but also suppressed the shuttle effect of polysulfides due to its strong polarity (Figure 9.5a). The implantation of Fe single atoms into the g-C_3N_4 skeleton not only improved the electrical conductivity but also increased the interlayer distance of g-C_3N_4, which facilitated both the electron transfer and ion transportation during discharge/charge processes. The electrochemical and theoretical

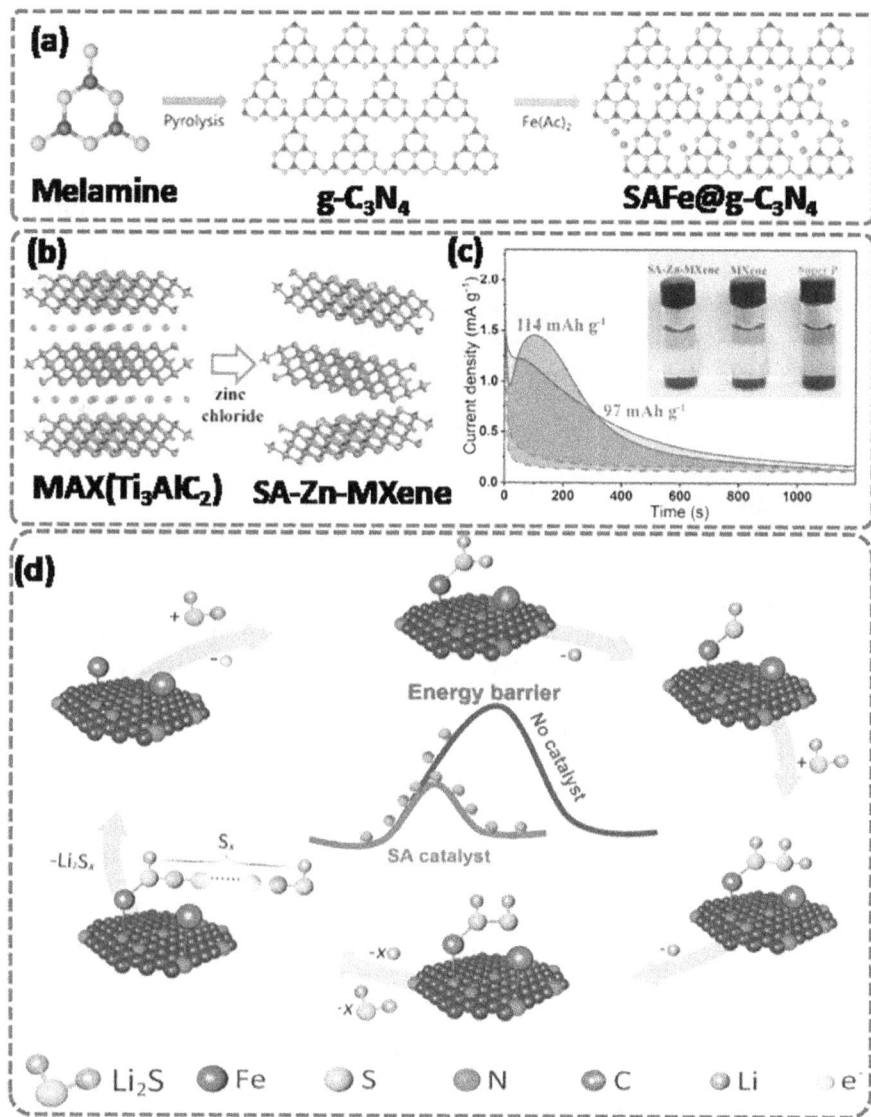

FIGURE 9.5 (a) Schematic illustration for the synthesis process of Fe SAs g-C$_3$N$_4$.[91] (b) Schematic illustration for the synthesis process of Zn-SAs MXene. (c) Potentiostatic nucleation curves on Zn-SAs MXene and MXene at the potential of 2.05 V, the inset is a visual adsorption test of Zn-SAs MXene, MXene, and super P to polysulfides.[90] (d) A proposed mechanism for Fe single atoms catalyzed Li$_2$S delithiation reaction.[99]

results demonstrated Fe single atoms could reduce the energy barriers of electrochemical reactions and further improve the reaction kinetics, thus improving the rate performance of batteries. The achievement of a high sulfur loading cathode with lean electrolyte is crucial for practical applications of Li–S batteries with high energy density. However, these are mainly hindered by the reaction kinetics of polysulfides in a thick electrode with a limited amount of electrolyte.[96] In this case, Fe SAs g-C_3N_4 with high catalytic activity played the role of a booster to facilitate the redox kinetics and improve the sulfur utilization. This idea allowed for a low mass ratio of electrolyte dosage to energy density (5.5 g Ah^{-1}). MXene is one kind of emerging material that features a 2D sheet-shaped structure and high electrical conductivity. Thus, it is considered a promising sulfur host material for Li–S batteries.[97] Yang and coworkers developed a hybrid of zinc single atoms implanted on MXene by etching Al layers of Ti_3AlC_2 MAX in molten zinc chloride (Figure 9.5b).[90] It showed the ability to trap polysulfides and accelerate redox kinetics as the host material for Li–S batteries. As shown in Figure 9.5c, the visualized adsorption experiment and the characterization of UV/vis spectra demonstrated stronger interaction between Zn SAs-MXene and polysulfide compared with the pure MXene and super P. This is in favor of limiting the diffusion of polysulfide and is further good for high reversible capacity and long cycle life. CV test results of the asymmetric cell in 0.2 M Li_2S_6 electrolyte displayed that the cell with Zn SAs-MXene delivered much higher current densities than those of MXene at peak sites, indicating greatly enhanced reaction kinetics. This was also confirmed by the potentiostatic nucleation analysis. As shown in Figure 9.5c, at a voltage of 2.05 V, Zn–SAs MXene exhibited a higher current density of 1.45 mA g^{-1} than that of MXene. DFT calculations were employed to investigate the catalytic activity and chemical affinity to polysulfides with Zn–SAs MXene. The conversion reaction from Li_2S_2 to Li_2S features the highest reaction free-energy, suggesting that this is the rate-determining step and the reason for the sluggish redox kinetics. The reaction free-energy of the rate-determining step on Zn–SAs MXene is 0.71 eV, which is lower than that on MXene (0.92 eV). This demonstrated that the introduction of Zn single atoms efficiently reduced the energy barriers of the reaction of Li_2S_2 reduction and accelerated the reaction kinetics. Similarly, the higher binding energies between polysulfides and Zn SAs Mxene than that with bare MXene further indicated stronger adsorption with polysulfides.

As an alternative to the sulfur cathode, the Li_2S electrode possesses a high specific capacity of 1,166 mAh g^{-1}, which can match with Li-free anodes and minimize volume expansion, leading to a high-energy-density high-safety battery.[98] However, there are also high energy barriers to oxidize the Li_2S and start the battery due to its insulating nature. Catalyzing the solid conversion reaction by single-atom catalysts is an effective strategy to promote the development of the Li_2S electrode. Zhang and coworkers reported an N-doped carbon with Fe single atoms (Fe SAs–NC) was designed and applied as the catalyst to promote the delithiation of relatively-inert Li_2S.[99] As shown in Figure 9.5d, a working mechanism for catalyzing the delithiation reaction of Li_2S on Fe SAs–NC was proposed. The initial coordination between Fe single atom and Li_2S is vital to decrease the reaction energy barrier and speed up the delithiation of Li_2S. The theoretical results demonstrated that the Fe single-atom catalyst not only lowered the activation energy barrier of the first cycle but also enhanced the rate performance and cycling stability of the battery. As a result, the batteries with this hybrid cathode delivered a high reversible capacity of 588 mAh g^{-1} at 12 C and a low capacity decay rate of 0.06% per cycle for 1,000 cycles.

9.2.4.2 Single Atoms for the Separator of Li–S Batteries

The separator is an indispensable component for an electrochemical rechargeable battery which acts as an electronic insulator to separate the anode and cathode. However, channels on the separator for ion transportation cannot limit the diffusion of polysulfides in Li–S batteries. Therefore, apart from the sulfur cathode, the separator can also be carefully designed to block the shuttling effect of polysulfides. It is desired to develop advanced separators modified by functional materials that afford the ability to immobilize the polysulfides and promote their kinetics conversion.[100] Most recently, single-atom materials have been reported as the coating materials for modifying the separator for Li–S batteries.[101,102]

As shown in Figure 9.6a, Fe single-atom catalysts implanted on graphene foam were developed and used as a coating layer to modify the commercial polypropylene separator to catalyze polysulfide conversion and further reduce the overpotential and enhance the cycle life.[103] A freestanding multifunctional interlayer composed of cobalt single atoms-anchored N-doped carbon nanosheets and dual networks of CNTs and cellulose nanofibers was proposed to effectively suppress the shuttle effect of polysulfides and enhance sulfur redox kinetics (Figure 9.6b).[83] As shown in Figure 9.6c, dual networks of CNTs and cellulose nanofibers could not only act as the barriers to block the polysulfide diffusion but also allow the hopping of Li ions, cobalt single atoms on N-doped carbon nanosheets could serve as the electrocatalyst to promote the reversible conversion reactions of polysulfides. Niu et al reported that a hybrid of nickel single atoms supported on N-doped graphene (Ni@NG) was synthesized by a pyrolysis approach with the assistance of sacrificed g-C$_3$N$_4$ templates and used to modify the separator.[104] The ion conductivity test demonstrated the separator coated with Ni@NG layer did not show an obvious influence on the diffusion of Li$^+$. The binding energy of polysulfides on the surface of Ni@NG is larger than that of N-doped graphene, suggesting the ability to immobilize the polysulfides. Additionally, this coating layer is hydrophobic, which plays another important role in restricting the shuttle effect. The decomposition energy barrier of Li$_2$S on Ni@NG is 1.23 eV, which is much lower than the 1.98 eV of N-doped graphene, suggesting that Ni single-atom sites featured higher catalytic activity to break the Li–S bond and achieve rapid reaction kinetics (Figure 9.6d). Besides, Ni single atoms with a Ni–N$_4$ structure show good structural stability even after long-term cycling. Benefiting from the merits of the single-atom coating layer, the batteries delivered an excellent rate capability and stable cycling life. It could still maintain a high specific capacity of 612 mAh g^{-1} even at a current density of 10 C and keep a high specific capacity of 826.2 mAh g^{-1} after 500 cycles at a current density of 1 C (Figure 9.6e). Apart from the coin cell, the performance of the modified separator was further evaluated by assembled pouch cells. The pouch cells with Ni@NG delivered capacities of 1,301 and 580.8 mAh g^{-1} at 0.1 and 3 C, respectively. Recently, Guo and coworkers developed a novel bifunctional separator with a "single atom Co array mimic" on the surface of an ultrathin metal-organic framework nanosheet by the layer-by-layer assembly of bacterial cellulose and ultrathin MOF–Co nanosheets.[105] As shown in Figure 9.6f, it demonstrated simultaneous enhancements on both Li-metal anode and sulfur cathode in Li–S batteries. The periodically arranged Co single atoms coordinated with four oxygen atoms could homogenize the Li–ion flux due to the strong adsorption of Li–ion, leading to stable Li striping/plating

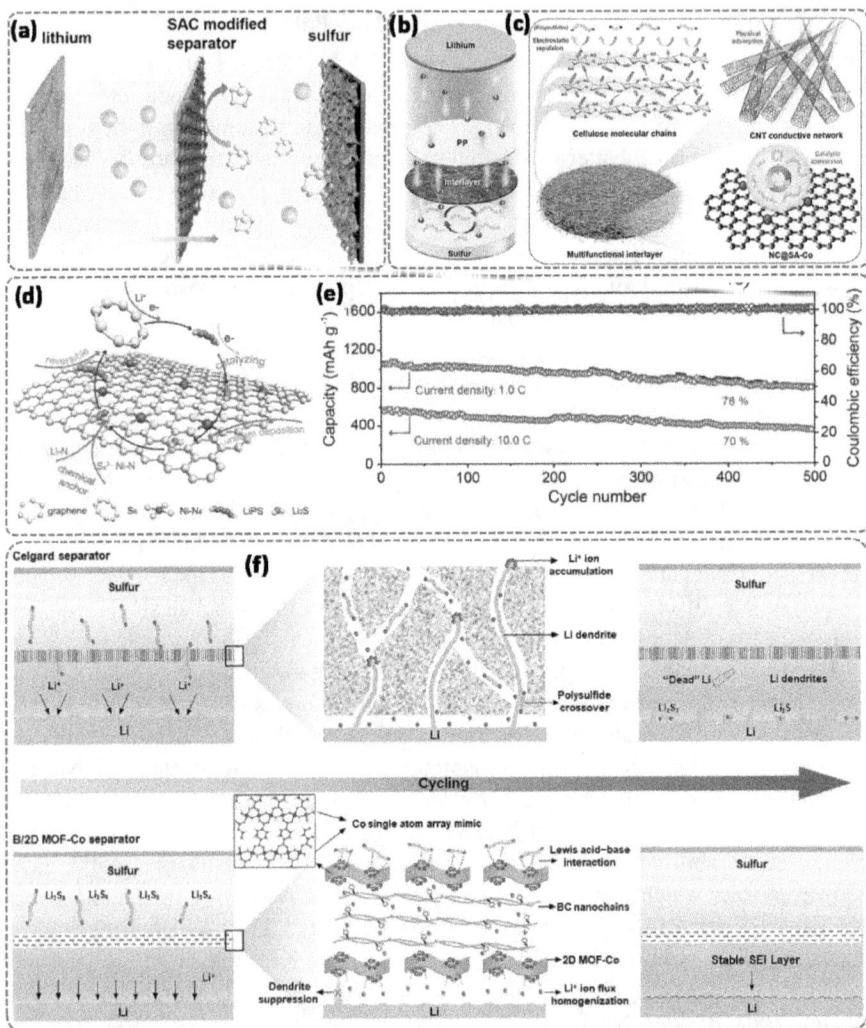

FIGURE 9.6   (a) Schematic illustration of the Li–S battery with the modified separator.[103] (b) Schematic illustration of the Li–S battery with a multifunctional interlayer. (c) Schematic illustration of functions of the multifunctional interlayer.[83] (d) Schematic illustration of the electrochemical process of the polysulfides on the surface of Ni@NG. (e) Cycling performance of the Li–S battery with the Ni@NG modified separator.[104] (f) Schematic illustration of the functions of Celgard separator and B/2D MOF-Co separator in Li–S batteries.[105]

without dendrite-like Li. Besides, the single-atom Co array mimic could effectively immobilize polysulfides due to the Lewis acid-base interaction, suppressing the shuttle effect of polysulfides and improving sulfur utilization. As a result, Li–S batteries with this bifunctional separator exhibited great long-term cycling stability with a low capacity fade rate of 0.07% per cycle for 600 cycles and a high areal capacity of 5.0 mAh cm^{-2} after 200 cycles.

Overall, a series of experiments and theoretical results indicated the introduction of single-atom materials into the sulfur cathode or separators can efficiently suppress the shuttle effect of polysulfides due to the strong adsorbability and catalyze the conversion reaction of polysulfides by reducing the decomposition energy barriers. These are beneficial for the efficient utilization of sulfur species and high rate capability.

9.2.5 Single Atoms for Li–Ion Batteries

Rechargeable Li–ion batteries are one of the most widely used energy storage devices for powering portable electronic devices due to their high energy density, high voltage, and long cycling life. However, the existing Li–ion batteries still need to be enhanced to satisfy the requirements of long-range electric vehicles and high-energy-consumption electronic equipment. The development of anode materials of Li–ion batteries is one of the effective approaches to improving electrochemical performances and meeting these requirements.[106,107] Recently, metal single atoms supported on carbon materials were also reported as the anode materials to enhance the performance of Li–ion batteries. For example, Wang and coworkers developed platinum single atoms supported on biomass-based porous carbon (Pt_1/MC) and used them for Li–ion batteries.[108] The as-synthesized Pt_1/MC featured a high specific surface area and a large pore volume. When used as the anode material for Li–ion batteries, it exhibited a high specific capacity of 846 mAh g^{-1} at 2 A g^{-1} and great cycling stability of remaining 349 mAh g^{-1} at the current density of 5 A g^{-1} after 6,000 cycles. DFT calculation indicated that Pt atoms homogeneously distributed on the porous carbon could serve as Li–ion storage sites. As shown in Figure 9.7a and b, each Pt atom could anchor five Li ions to form Pt_1–Li_5 alloyed on the porous carbon substrate, thus, offering a high specific capacity. Besides, Xu et al. reported a Co single-atom material as the anode for Li–ion batteries. As shown in Figure 9.7c, a 3D porous N-doped carbon with atomically dispersed Co atoms and Co clusters was developed via a template-sacrificed synthesis strategy.[109] The 3D porous structure of this hybrid was favorable for the storage of electrolyte and the quick diffusion of Li ions and provided a large accessible surface area for Li–ion storage. As shown in Figure 9.7d, DFT calculations demonstrated the active sites (Co–N–C) exhibited larger Li adsorption energy than those of N-doped graphene and graphene, indicating stronger Li–ion adsorption ability. Besides, the doping of Co atoms is in favor of electron transfer of the hybrid. The battery with this anode material delivered a high initial specific capacity of 1,587 mAh g^{-1} at a current density of 0.1 C and a reversible capacity of 1,000 mAh g^{-1} after 800 cycles at 5 C. With only a few studies directly on Li–ion batteries, the effect of single atoms is positive, but requires further investigation.

9.3   CONCLUSION AND PERSPECTIVES

In this chapter, we summarized different cases of single atoms supported on different substrates applied in various Li batteries, including Li-metal batteries, Li–O_2 batteries, Li–CO_2 batteries, Li–S batteries, and Li–ion batteries. Single-atom catalysts with unsaturated coordination can adjust the interface interaction, regulate the deposition of discharge products, catalyze the decomposition of discharge products (Li_2O_2, Li_2CO_3, Li_2S), trap polysulfides, decrease the reaction energy barriers, and

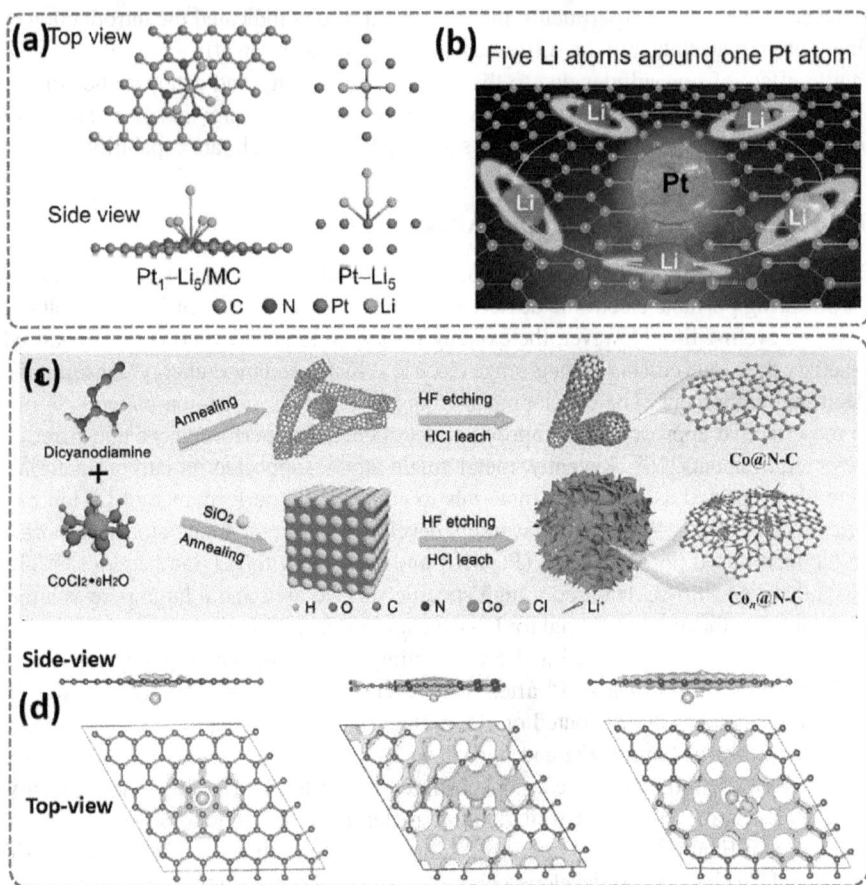

FIGURE 9.7 (a) Optimized geometrical structures of the Pt_1–Li_5 and bulk Pt–Li_5. (b) Schematic illustration of Pt_1–Li_5 structure on the porous carbon support.[108] (c) Schematic illustration for the synthesis process of single-atom Co on 3D porous carbon. (d) The charge density differences of graphene, N-doped graphene, and Co@N-C with one Li adsorbed.[109]

enhance the Li storage ability. Benefiting from these merits, a series of achievements of Li batteries were realized, such as a safe and dendrite-free Li-metal anode, low discharge/charge overpotentials, high specific capacity, high rate capability, and long cycling stability. However, despite these significant enhancements mentioned above, there are still several challenges that should be paid more attention to in the future:

1. Single-atom centers usually present different catalytic activities because of the different metal centers, different substrates, and their complicated coordination environments. Therefore, it is necessary to explore new single-atom catalysts with different metal centers and substrates and regulate their coordination chemistry and interaction with substrates to fit different electrochemical reactions. Besides, dual or multi-atom catalysts possess the potential to

provide continuous catalytic sites and achieve synergistic catalysis to reduce reaction energy barriers and improve catalytic efficiency.

2. The concentration of single-atom sites should be delicately controlled because the metal atoms tend to aggregate under high loading due to the high surface energy. Such aggregation can dramatically decrease the catalytic effect for electrochemical reactions. It is vitally important to develop new synthesis methods to precisely and controllably prepare single-atom catalysts with uniform dispersion and high mass loading. Another thing worth noting is that it is equally important to develop synthesis methods that are suitable for large-scale production. The theoretical calculation is an effective approach to helping screen and forecasts the catalytic activity of single atoms.

3. Aside from the relationship between the chemical/physical structure of the single-atom center and the catalytic activity, it is of great concern to in-depth and comprehensively understand the electrochemical mechanisms of single atoms in Li batteries. It can further guide the design of single-atom catalysts. To gain this information, in-situ and ex-situ chemical character-izations, such as in-situ TEM and in-situ XAS, etc., need to be carried out with the assistance of theoretical simulations.

Overall, the development of novel single-atom catalysts with high mass loading and adjustable coordination chemistry and electronic structure will be promising to fur-ther enhance the electrochemical performance of Li batteries.

REFERENCES

1. S. Chu and A. Majumdar, *Nature*, 2012, **488**, 294.
2. M. S. Dresselhaus and I. L. Thomas, *Nature*, 2001, **414**, 332–337.
3. Y. Liang, C.-Z. Zhao, H. Yuan, Y. Chen, W. Zhang, J.-Q. Huang, D. Yu, Y. Liu, M.-M. Titirici, Y.-L. Chueh, H. Yu and Q. Zhang, *InfoMat*, 2019, **1**, 6–32.
4. X. Zhang, X. Cheng and Q. Zhang, *Journal of Energy Chemistry*, 2016, **25**, 967–984.
5. M. Armand and J. M. Tarascon, *Nature*, 2008, **451**, 652–657.
6. J. M. Tarascon and M. Armand, *Nature*, 2001, **414**, 359.
7. J. Lu, L. Li, J. B. Park, Y. K. Sun, F. Wu and K. Amine, *Chemical Reviews*, 2014, **114**, 5611–5640.
8. N. Feng, P. He and H. Zhou, *Advanced Energy Materials*, 2016, 6, 1502303.
9. B. Qiao, A. Wang, X. Yang, L. F. Allard, Z. Jiang, Y. Cui, J. Liu, J. Li and T. Zhang, *Nature Chemistry*, 2011, **3**, 634–641.
10. P. Liu, Y. Zhao, R. Qin, S. Mo, G. Chen, L. Gu, D. M. Chevrier, P. Zhang, Q. Guo, D. Zang, B. Wu, G. Fu and N. Zheng, *Science*, 2016, **352**, 797–800.
11. C. Zhao, Y. Tang, C. Yu, X. Tan, M. N. Banis, S. Li, G. Wan, H. Huang, L. Zhang, H. Yang, J. Li, X. Sun and J. Qiu, *Nano Today*, 2020, **34**, 100955.
12. A. Wang, J. Li and T. Zhang, *Nature Reviews Chemistry*, 2018, **2**, 65–81.
13. Y. Chen, S. Ji, C. Chen, Q. Peng, D. Wang and Y. Li, *Joule*, 2018, **2**, 1242–1264.
14. H. Fei, J. Dong, M. J. Arellano-Jiménez, G. Ye, N. Dong Kim, E. L. G. Samuel, Z. Peng, Z. Zhu, F. Qin, J. Bao, M. J. Yacaman, P. M. Ajayan, D. Chen and J. M. Tour, *Nature Communications*, 2015, **6**, 8668.
15. Z. Liu, S. Li, J. Yang, X. Tan, C. Yu, C. Zhao, X. Han, H. Huang, G. Wan, Y. Liu, K. Tschulik and J. Qiu, *ACS Nano*, 2020, **14**, 11662–11669.

16. J. Zhang, C. Liu and B. Zhang, *Small Methods*, 2019, **3**, 1800481.
17. Z. Zhang, J. Liu, A. Curcio, Y. Wang, J. Wu, G. Zhou, Z. Tang and F. Ciucci, *Nano Energy*, 2020, **76**, 105085.
18. H. Xia, G. Qu, H. Yin and J. Zhang, *Journal of Materials Chemistry A*, 2020, **8**, 15358–15372.
19. H. Tian, A. Song, H. Tian, J. Liu, G. Shao, H. Liu and G. Wang, *Chemical Science*, 2021, **12**, 7656–7676.
20. C. Lu, R. Fang and X. Chen, *Advanced Materials*, 2020, **32**, e1906548.
21. S. C. Mukul, D. Tikekar, Z. Tu and L. A. Archer, *Nature Energy*, 2016, **1**, 1–7.
22. X.-B. Cheng, R. Zhang, C.-Z. Zhao and Q. Zhang, *Chemical Reviews*, 2017, **117**, 10403–10473.
23. C. Zhao, C. Yu, S. Li, W. Guo, Y. Zhao, Q. Dong, X. Lin, Z. Song, X. Tan, C. Wang, M. Zheng, X. Sun and J. Qiu, *Small*, 2018, **14**, 1803310.
24. F. Ding, W. Xu, G. L. Graff, J. Zhang, M. L. Sushko, X. Chen, Y. Shao, M. H. Engelhard, Z. Nie, J. Xiao, X. Liu, P. V. Sushko, J. Liu and J.-G. Zhang, *Journal of the American Chemical Society*, 2013, **135**, 4450–4456.
25. C. Zhao, J. Liang, Q. Sun, J. Luo, Y. Liu, X. Lin, Y. Zhao, H. Yadegari, M. N. Banis, R. Li, H. Huang, L. Zhang, R. Yang, S. Lu and X. Sun, *Small Methods*, 2019, **3**, 1800437.
26. C. Zhao, Z. Wang, X. Tan, H. Huang, Z. Song, Y. Sun, S. Cui, Q. Wei, W. Guo, R. Li, C. Yu, J. Qiu and X. Sun, *Small Methods*, 2019, **3**, 1800546.
27. K. R. Adair, C. Zhao, M. N. Banis, Y. Zhao, R. Li, M. Cai and X. Sun, *Angewandte Chemie International Edition*, 2019, **58**, 15797–15802.
28. Y. Zhao, M. Amirmaleki, Q. Sun, C. Zhao, A. Codirenzi, L. V. Goncharova, C. Wang, K. Adair, X. Li, X. Yang, F. Zhao, R. Li, T. Filleter, M. Cai and X. Sun, *Matter*, 2019, **1**, 1215–1231.
29. Y. Wang, F. Chu, J. Zeng, Q. Wang, T. Naren, Y. Li, Y. Cheng, Y. Lei and F. Wu, *ACS Nano*, 2021, **15**, 210–239.
30. H. Liu, X. Chen, X. B. Cheng, B. Q. Li, R. Zhang, B. Wang, X. Chen and Q. Zhang, *Small Methods*, 2018, **3**, 1800354.
31. Y. Sun, J. Zhou, H. Ji, J. Liu, T. Qian and C. Yan, *ACS Applied Materials & Interfaces*, 2019, **11**, 32008–32014.
32. J. Gu, Q. Zhu, Y. Shi, H. Chen, D. Zhang, Z. Du and S. Yang, *ACS Nano*, 2020, **14**, 891–898.
33. P. Zhai, T. Wang, W. Yang, S. Cui, P. Zhang, A. Nie, Q. Zhang and Y. Gong, *Advanced Energy Materials*, 2019, **9**, 1804019.
34. K. Xu, M. Zhu, X. Wu, J. Liang, Y. Liu, T. Zhang, Y. Zhu and Y. Qian, *Energy Storage Materials*, 2019, **23**, 587–593.
35. Y. Li and J. Lu, *ACS Energy Letters*, 2017, **2**, 1370–1377.
36. J.-S. Lee, S. Tai Kim, R. Cao, N.-S. Choi, M. Liu, K. T. Lee and J. Cho, *Advanced Energy Materials*, 2011, **1**, 34–50.
37. H.-F. Wang and Q. Xu, *Matter*, 2019, **1**, 565–595.
38. C. Zhao, Q. Sun, J. Luo, J. Liang, Y. Liu, L. Zhang, J. Wang, S. Deng, X. Lin, X. Yang, H. Huang, S. Zhao, L. Zhang, S. Lu and X. Sun, *Chemistry of Materials*, 2020, **32**, 10113–10119.
39. P. G. Bruce, S. A. Freunberger, L. J. Hardwick and J.-M. Tarascon, *Nature Materials*, 2012, **11**, 19–29.
40. C. Zhao, Y. Zhu, Q. Sun, C. Wang, J. Luo, X. Lin, X. Yang, Y. Zhao, R. Li, S. Zhao, H. Huang, L. Zhang, S. Lu, M. Gu and X. Sun, *Angewandte Chemie International Edition*, 2021, **60**, 5821–5826.
41. C. Zhao, C. Yu, M. N. Banis, Q. Sun, M. Zhang, X. Li, Y. Liu, Y. Zhao, H. Huang, S. Li, X. Han, B. Xiao, Z. Song, R. Li, J. Qiu and X. Sun, *Nano Energy*, 2017, **34**, 399–407.

42. C. Zhao, J. Liang, Y. Zhao, J. Luo, Q. Sun, Y. Liu, X. Lin, X. Yang, H. Huang, L. Zhang, S. Zhao, S. Lu and X. Sun, *Journal of Materials Chemistry A*, 2019, **7**, 24947–24952.

43. C. Zhao, C. Yu, J. Yang, S. Liu, M. Zhang and J. Qiu, *Particle & Particle Systems Characterization*, 2016, **33**, 228–234.

44. C. Zhao, C. Yu, S. Liu, J. Yang, X. Fan, H. Huang and J. Qiu, *Advanced Functional Materials*, 2015, **25**, 6913–6920.

45. C. Yu, C. Zhao, S. Liu, X. Fan, J. Yang, M. Zhang and J. Qiu, *Chemical Communications*, 2015, **51**, 13233–13236.

46. H. D. Lim, H. Song, J. Kim, H. Gwon, Y. Bae, K. Y. Park, J. Hong, H. Kim, T. Kim, Y. H. Kim, X. Lepro, R. Ovalle-Robles, R. H. Baughman and K. Kang, *Angewandte Chemie International Edition in English*, 2014, **53**, 3926–3931.

47. R. Gao, Z. Yang, L. Zheng, L. Gu, L. Liu, Y. Lee, Z. Hu and X. Liu, *ACS Catalysis*, 2018, **8**, 1955–1963.

48. L. N. Song, W. Zhang, Y. Wang, X. Ge, L. C. Zou, H. F. Wang, X. X. Wang, Q. C. Liu, F. Li and J. J. Xu, *Nature Communications*, 2020, **11**, 2191.

49. P. Wang, Y. Ren, R. Wang, P. Zhang, M. Ding, C. Li, D. Zhao, Z. Qian, Z. Zhang, L. Zhang and L. Yin, *Nature Communications*, 2020, **11**, 1576.

50. G. Li, C. Dang, C. Hou, F. Dang, Y. Fan and Z. Guo, *Engineered Science*, 2020, **10**, 85–94.

51. D. Wang, X. Mu, P. He and H. Zhou, *Materials Today*, 2019, **26**, 87–99.

52. Z. Chang, J. Xu and X. Zhang, *Advanced Energy Materials*, 2017, **7**, 1700875.

53. L. Ma, T. Yu, E. Tzoganakis, K. Amine, T. Wu, Z. Chen and J. Lu, *Advanced Energy Materials*, 2018, **8**, 1800348.

54. X. Hu, G. Luo, Q. Zhao, D. Wu, T. Yang, J. Wen, R. Wang, C. Xu and N. Hu, *Journal of the American Chemical Society*, 2020, **142**, 16776–16786.

55. W. Zhao, J. Wang, R. Yin, B. Li, X. Huang, L. Zhao and L. Qian, *Journal of Colloid and Interface Science*, 2020, **564**, 28–36.

56. B. Liu, Y. Sun, L. Liu, J. Chen, B. Yang, S. Xu and X. Yan, *Energy & Environmental Science*, 2019, **12**, 887–922.

57. X. Sun, Z. Hou, P. He and H. Zhou, *Energy & Fuels*, 2021, **35**, 9165–9186.

58. A. Hu, C. Shu, C. Xu, R. Liang, J. Li, R. Zheng, M. Li and J. Long, *Journal of Materials Chemistry A*, 2019, **7**, 21605–21633.

59. Y. Xiao, F. Du, C. Hu, Y. Ding, Z. L. Wang, A. Roy and L. Dai, *ACS Energy Letters*, 2020, **5**, 916–921.

60. X. Li, J. Zhou, J. Zhang, M. Li, X. Bi, T. Liu, T. He, J. Cheng, F. Zhang, Y. Li, X. Mu, J. Lu and B. Wang, *Advanced Materials*, 2019, **31**, e1903852.

61. Z. Zhang, Q. Zhang, Y. Chen, J. Bao, X. Zhou, Z. Xie, J. Wei and Z. Zhou, *Angewandte Chemie International Edition in English*, 2015, **54**, 6550–6553.

62. S. Yang, Y. Qiao, P. He, Y. Liu, Z. Cheng, J.-J. Zhu and H. Zhou, *Energy & Environmental Science*, 2017, **10**, 972–978.

63. S. Li, Y. Liu, J. Zhou, S. Hong, Y. Dong, J. Wang, X. Gao, P. Qi, Y. Han and B. Wang, *Energy & Environmental Science*, 2019, **12**, 1046–1054.

64. C. Hu, L. Gong, Y. Xiao, Y. Yuan, N. M. Bedford, Z. Xia, L. Ma, T. Wu, Y. Lin, J. W. Connell, R. Shahbazian-Yassar, J. Lu, K. Amine and L. Dai, *Advanced Materials*, 2020, **32**, e1907436.

65. L. Zhou, H. Wang, K. Zhang, Y. Qi, C. Shen, T. Jin and K. Xie, *Science China Materials*, 2021, **64**, 2139–2147.

66. Y. Zhang, R. L. Zhong, M. Lu, J. H. Wang, C. Jiang, G. K. Gao, L. Z. Dong, Y. Chen, S. L. Li and Y. Q. Lan, *ACS Central Science*, 2021, **7**, 175–182.

67. B. W. Zhang, Y. Jiao, D. L. Chao, C. Ye, Y. X. Wang, K. Davey, H. K. Liu, S. X. Dou and S. Z. Qiao, *Advanced Functional Materials*, 2019, **29**, 1904206.

68. S.-M. Xu, Z.-C. Ren, X. Liu, X. Liang, K.-X. Wang and J.-S. Chen, *Energy Storage Materials*, 2018, **15**, 291–298.
69. R. Fang, S. Zhao, Z. Sun, D.-W. Wang, H.-M. Cheng and F. Li, *Advanced Materials*, 2017, **29**, 1606823.
70. L. Fan, M. Li, X. Li, W. Xiao, Z. Chen and J. Lu, *Joule*, 2019, **3**, 361–386.
71. W. Ren, W. Ma, S. Zhang and B. Tang, *Energy Storage Materials*, 2019, **23**, 707–732.
72. D. Liu, C. Zhang, G. Zhou, W. Lv, G. Ling, L. Zhi and Q.-H. Yang, *Advanced Science*, 2018, **5**, 1700270.
73. J. He and A. Manthiram, *Energy Storage Materials*, 2019, **20**, 55–70.
74. W.-G. Lim, S. Kim, C. Jo and J. Lee, *Angewandte Chemie International Edition*, 2019, **58**, 18746–18757.
75. Y. Song, W. Cai, L. Kong, J. Cai, Q. Zhang and J. Sun, *Advanced Energy Materials*, 2020, **10**, 1901075.
76. L. Fang, Z. Feng, L. Cheng, R. E. Winans and T. Li, *Small Methods*, 2020, **4**, 2000315.
77. Z. Shi, L. Wang, H. Xu, J. Wei, H. Yue, H. Dong, Y. Yin and S. Yang, *Chemical Communications*, 2019, **55**, 12056–12059.
78. J. Wang, B. Ding, X. Lu, H. Nara, Y. Sugahara and Y. Yamauchi, *Advanced Materials Interfaces*, 2021, **8**, 2002159.
79. F. Wang, J. Li, J. Zhao, Y. Yang, C. Su, Y. L. Zhong, Q.-H. Yang and J. Lu, *ACS Materials Letters*, 2020, **2**, 1450–1463.
80. Z. Liu, L. Zhou, Q. Ge, R. Chen, M. Ni, W. Utetiwabo, X. Zhang and W. Yang, *ACS Applied Materials and Interfaces*, 2018, **10**, 19311–19317.
81. H. Zhao, B. Tian, C. Su and Y. Li, *ACS Applied Materials and Interfaces*, 2021, **13**, 7171–7177.
82. Q.-W. Zeng, R.-M. Hu, Z.-B. Chen and J.-X. Shang, *Materials Research Express*, 2019, **6**, 095620.
83. Y. Li, P. Zhou, H. Li, T. Gao, L. Zhou, Y. Zhang, N. Xiao, Z. Xia, L. Wang, Q. Zhang, L. Gu and S. Guo, *Small Methods*, 2020, **4**, 1900701.
84. C. Wang, H. Song, C. Yu, Z. Ullah, Z. Guan, R. Chu, Y. Zhang, L. Zhao, Q. Li and L. Liu, *Journal of Materials Chemistry A*, 2020, **8**, 3421–3430.
85. J. Xie, B. Q. Li, H. J. Peng, Y. W. Song, M. Zhao, X. Chen, Q. Zhang and J. Q. Huang, *Advanced Materials*, 2019, **31**, 1903813.
86. Z. Du, X. Chen, W. Hu, C. Chuang, S. Xie, A. Hu, W. Yan, X. Kong, X. Wu, H. Ji and L. J. Wan, *Journal of the American Chemical Society*, 2019, **141**, 3977–3985.
87. B. Q. Li, L. Kong, C. X. Zhao, Q. Jin, X. Chen, H. J. Peng, J. L. Qin, J. X. Chen, H. Yuan, Q. Zhang and J. Q. Huang, *InfoMat*, 2019, **1**, 533–541.
88. G. Zhou, S. Zhao, T. Wang, S. Z. Yang, B. Johannessen, H. Chen, C. Liu, Y. Ye, Y. Wu, Y. Peng, C. Liu, S. P. Jiang, Q. Zhang and Y. Cui, *Nano Letters*, 2020, **20**, 1252–1261.
89. F. Ma, Y. Wan, X. Wang, X. Wang, J. Liang, Z. Miao, T. Wang, C. Ma, G. Lu, J. Han, Y. Huang and Q. Li, *ACS Nano*, 2020, **14**, 10115–10126.
90. D. Zhang, S. Wang, R. Hu, J. Gu, Y. Cui, B. Li, W. Chen, C. Liu, J. Shang and S. Yang, *Advanced Functional Materials*, 2020, **30**, 2002471.
91. C. Lu, Y. Chen, Y. Yang and X. Chen, *Nano Letters*, 2020, **20**, 5522–5530.
92. J. Liang, Z.-H. Sun, F. Li and H.-M. Cheng, *Energy Storage Materials*, 2016, **2**, 76–106.
93. Z. Jia, H. Zhang, H. Yu, Y. Chen, J. Yan, X. Li and H. Zhang, *Journal of Energy Chemistry*, 2020, **43**, 71–77.
94. J. Zhang, J.-Y. Li, W.-P. Wang, X.-H. Zhang, X.-H. Tan, W.-G. Chu and Y.-G. Guo, *Advanced Energy Materials*, 2018, **8**, 1702839.
95. J. Wu, J. Chen, Y. Huang, K. Feng, J. Deng, W. Huang, Y. Wu, J. Zhong and Y. Li, *Science Bulletin*, 2019, **64**, 1875–1880.
96. H.-J. Peng, J.-Q. Huang, X.-B. Cheng and Q. Zhang, *Advanced Energy Materials*, 2017, **7**, 1700260.

97. C. Zhang, L. Cui, S. Abdolhosseinzadeh and J. Heier, *InfoMat*, 2020, **2**, 613–638.
98. J. Wang, L. Jia, S. Duan, H. Liu, Q. Xiao, T. Li, H. Fan, K. Feng, J. Yang, Q. Wang, M. Liu, J. Zhong, W. Duan, H. Lin and Y. Zhang, *Energy Storage Materials*, 2020, **28**, 375–382.
99. J. Wang, L. Jia, J. Zhong, Q. Xiao, C. Wang, K. Zang, H. Liu, H. Zheng, J. Luo, J. Yang, H. Fan, W. Duan, Y. Wu, H. Lin and Y. Zhang, *Energy Storage Materials*, 2019, **18**, 246–252.
100. Y. C. Jeong, J. H. Kim, S. Nam, C. R. Park and S. J. Yang, *Advanced Functional Materials*, 2018, **28**, 1707411.
101. R. Xiao, K. Chen, X. Zhang, Z. Yang, G. Hu, Z. Sun, H.-M. Cheng and F. Li, *Journal of Energy Chemistry*, 2021, **54**, 452–466.
102. Z. Zhuang, Q. Kang, D. Wang and Y. Li, *Nano Research*, 2020, **13**, 1856–1866.
103. K. Zhang, Z. Chen, R. Ning, S. Xi, W. Tang, Y. Du, C. Liu, Z. Ren, X. Chi, M. Bai, C. Shen, X. Li, X. Wang, X. Zhao, K. Leng, S. J. Pennycook, H. Li, H. Xu, K. P. Loh and K. Xie, *ACS Applied Materials and Interfaces*, 2019, **11**, 25147–25154.
104. L. Zhang, D. Liu, Z. Muhammad, F. Wan, W. Xie, Y. Wang, L. Song, Z. Niu and J. Chen, *Advanced Materials*, 2019, **31**, e1903955.
105. Y. Li, S. Lin, D. Wang, T. Gao, J. Song, P. Zhou, Z. Xu, Z. Yang, N. Xiao and S. Guo, *Advanced Materials*, 2020, **32**, e1906722.
106. J. Lu, Z. Chen, F. Pan, Y. Cui and K. Amine, *Electrochemical Energy Reviews*, 2018, **1**, 35–53.
107. V. Etacheri, R. Marom, R. Elazari, G. Salitra and D. Aurbach, *Energy & Environmental Science*, 2011, **4**, 3243–3262.
108. T. Li, D. Yu, J. Liu and F. Wang, *ACS Applied Materials & Interfaces*, 2019, **11**, 37559–37566.
109. S. Liu, Z. Lin, F. Xiao, J. Zhang, D. Wang, X. Chen, Y. Zhao and J. Xu, *Chemical Engineering Journal*, 2020, **389**, 124377.

10 Stabilization of Atomically Dispersed Metallic Catalysts for Electrochemical Energy Applications

Junjie Li, Xiaozhang Yao, Yi Guan, Kieran Doyle-Davis, and Xueliang Sun
University of Western Ontario

CONTENTS

10.1 INTRODUCTION

With the development of human society, the consumption of fossil fuels is rapidly increasing. However, these fossil fuels, such as coal, natural gas, and oil, are non-renewable energy sources that could cause future energy shortages and further damage our environment. Toxic gases, such as SO_x and NO_x, are other harmful side products of the use of fossil fuels, which will further affect our health. Besides that, the generated CO_2 will further lead to global warming. The effects are far-reaching. In this case, pursuing renewable energy sources, such as hydrogen, is important to

DOI: 10.1201/9781003153436-10

replace traditional fossil fuels. Therefore, finding a suitable energy conversion technique to cleanly and efficiently convert these sources to energy is essential. Most conversion devices operate through electrochemical reactions, for example, the oxygen reduction reaction (ORR) at the cathode used in proton-exchange membrane fuel cells (PEMFCs), oxygen evolution reaction (OER) in rechargeable metal-air batteries, hydrogen evolution reaction (HER), formic acid oxidation reaction (FAOR), and methanol oxidation reaction (MOR).[1–5] Among the components in these devices, the electrocatalyst is one of the most important parts, as the rate and efficiency at which these reactions occur are of critical importance. Pt-based noble metal electrocatalysts are widely used because of their superior catalytic performance. However, for traditional electrocatalysts, the metal particles used have a wide size distribution. Because the electrochemical reactions are more related to the surface of electrocatalysts, only the surface metal atoms participate in the reaction; thus, most of the metal atoms in the bulk phase are wasted. If many atoms in the bulk of the metal particle are not contributing to the reaction, they could be seen as a tremendous wasted expense, particularly in the case of Pt. Therefore, an electrocatalyst with high atom utilization efficiency is in demand.

In recent years, single-atom catalysts (SAC) have attracted considerable interest due to their remarkable catalytic performance, which seems to be the best scenario to solve the issue mentioned above.[6–9] When the size of metal species decreases from nanoparticle (NP) to single atom, the atom utilization efficiency reaches the maximum as all atoms act as the active center with high dispersion and uniformity. More importantly, as the atoms are atomically dispersed, the single atoms always display a highly unsaturated coordination environment. The unique local structure and electronic properties lead to their outstanding catalytic activity and selectivity in some reactions, which showed much improved catalytic performance over NPs-based catalysts. However, with a decrease in size from NPs to a single atom, it simultaneously increases the surface free energy that improves the mobility of supported single-atom thus leading to the formation of clusters/NPs through aggregation.[10,11] Therefore, increasing the stability of SACs by some efficient strategies is important. In this chapter, we first briefly describe the challenges and opportunities for SACs and then highlight the recent progress to achieve stable SACs. Further, we will summarize the recent work regarding the electrochemical energy applications of SACs. Finally, the summary and perspective for single-atom catalysis materials are discussed.

10.2 SACs: OPPORTUNITIES AND CHALLENGES

SACs, with singly dispersed metal atoms, have shown their promising applications in various chemical reactions.[7,9,12] The most significant difference between SACs and traditional catalysts is the single-atom site where the catalytic active center is a single atom rather than the two-dimensional metal clusters or the three-dimensional metal NPs. This unique single-atom-site not only increases the atom utilization efficiency to almost 100% but also exhibits much improved catalytic performance compared to traditional catalysts. For example, Yan and coworkers successfully developed the Pd_1/graphene SAC which exhibited a high selectivity to the 1-butadiene in the selective hydrogenation of 1,3-butadiene, compared with the Pd NPs catalysts. They attributed

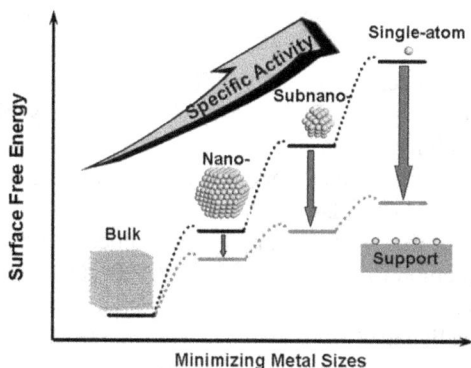

FIGURE 10.1 Schematic illustration of the changes of surface free energy and specific activity per metal atom with metal particle size and the support effects on stabilizing single atoms. (Reproduced with permission from Reference,[12] Copyright 2013, American Chemical Society.)

such a high selectivity to the mono-π-adsorption of 1,3-butadiene molecular on the Pd single atoms, which differs from the di-π-adsorption on the Pd NPs.[13]

Although SACs show outstanding catalytic performance in catalysis, the stability of SACs is still a big challenge for their practical application. When the size of the metal species is down to a single atom, the surface free energy dramatically increases (Figure 10.1), thus leading to the much improved mobility of single atoms on the support.[12] As a result, the single atoms easily aggregate and form clusters/NPs. This phenomenon could be even worse under chemical reactions, as the reactants could be adsorbed by the single atoms, which could weaken the interactions between single atoms and the support, leading to the migration and detachment of the catalyst. This process not only decreases the number of active sites but also leads to the deactivation of SACs.[11,14] Therefore, increasing the stability of SACs is one of the most emergent requirements for single-atom catalysis.[7–9,12] However, finding a suitable and effective method to stabilize single atoms is not an easy task, especially for those SACs with high metal loadings and applied in harsh reaction conditions. As such, several synthesis strategies have been developed to improve the stability of SACs.

10.3 SYNTHETIC STRATEGY FOR THE STABILIZATION OF SACs

10.3.1 SPATIAL CONFINEMENT

The high surface free energy induced high mobility of single atoms is one of the main reasons behind the low stability of SACs. Based on this, researchers developed several strategies that can specifically confine the single atoms in special structures. This spatial confinement method provides limited space that single atoms can move around, thus strongly decreasing the possibility of single atoms aggregating. Moreover, it can also promote the catalytic performance of SACs through the confinement effect.

Using atomic layer deposition (ALD) to deposit nano oxide cages around metal species is found to be a promising strategy to improve the stability of catalysts. For instance, Cheng et al. successfully used the ZrO_2 nanocages to stabilize the Pt NPs using ALD. The achieved ZrO_2–Pt catalysts show extremely high stability in the ORR and showed much less decreased catalytic activity than the pristine Pt sample after stability testing.[15] In their later work, Song and coworkers further demonstrated that the TaO_x nano cages can also promote the stability of Pt catalysts.[15] Besides the NPs-based catalyst, this strategy can also be applied to SACs. Liu and coworkers used a grafting method to chemisorb the (^{Ph}PCP)Pt–OH precursor on the Al_2O_3 support and then used ALD to deposit the metal oxide, including TiO_2, ZnO, and Al_2O_3. Because of the ligand effect, the oxide deposited by ALD can only be located around the Pt species, thus resulting in an oxide nanocage structure. After repeating several cycles of oxide coatings, differently sized oxide nanocages would be formed, which can capture the Pt single atoms (Figure 10.2a). Later on, the ligands of the Pt precursor were removed after thermal treatment. With the coatings of Al_2O_3, TiO_2, and ZnO, the sizes of Pt species are 1.7 ± 1.4, 0.6 ± 0.3, and 1.4 ± 1.1 nm, respectively (Figure 10.2b). The samples without the protection of oxide nanocage were found to have a much larger Pt particle size after thermal treatment. In the CO diffuse reflectance Infrared Fourier transform spectroscopy (DRIFTS) measurements, they found that the TiO_2 and ZnO coated samples only displayed a CO linear adsorption peak, indicating the formation of Pt single atoms is dominant. However, in addition to the CO liner adsorption peak on Pt single atoms, another shoulder peak at a lower frequency region can also be observed on the bare and Al_2O_3 coated samples, demonstrating the existence of Pt NPs. The X-ray photoelectron spectroscopy (XPS) results further confirm that the TiO_2 coated sample has much more Pt single atoms than the Al_2O_3 coated one which shows the metallic state Pt^0. Combined with the high-angle annular dark-field scanning transmission electron microscopy (HAADF-STEM), DRIFTS, and XPS results, they demonstrated that the oxide nanocage using TiO_2 and ZnO can help to stabilize Pt single atoms.[16]

In addition to the nano cage stabilization strategy, some materials naturally have a special structure such as a cavity that can help to stabilize single atoms. Carbon nitride (C_3N_4), a two-dimension material that has N-coordinating "sixfold cavities" is found to be acted as suitable support to stabilize single atoms. Vilé and coworkers are the first to successfully use mesoporous polymeric graphitic carbon nitride (mpg-C_3N_4) to support Pd single atoms. In their work, the isolated Pd species were strongly anchored into the host cavity from mpg-C_3N_4, and the loading of Pd is 0.5 wt%. The singly dispersed Pd atoms were confirmed by HAADF-STEM and extended X-ray absorption fine structure (EXAFS) results. Density functional theory (DFT) calculations were used to understand the nature of Pd single atoms into mpg-C_3N_4, which indicated that the trapped Pd single atoms in the sixfold cavities have strong interactions with the nitrogen atoms and result in a large amount of homogeneously distributed anchoring centers. Hence, the stability of Pd single atoms could be strongly improved. In the hydrogenation of alkynes and nitroarenes, the Pd SAC exhibited much improved catalytic activity compared to the NPs-based catalysts and maintained high selectivity and stability. The theoretical calculation results demonstrated that the high activity and stability are due to the fact that the Pd

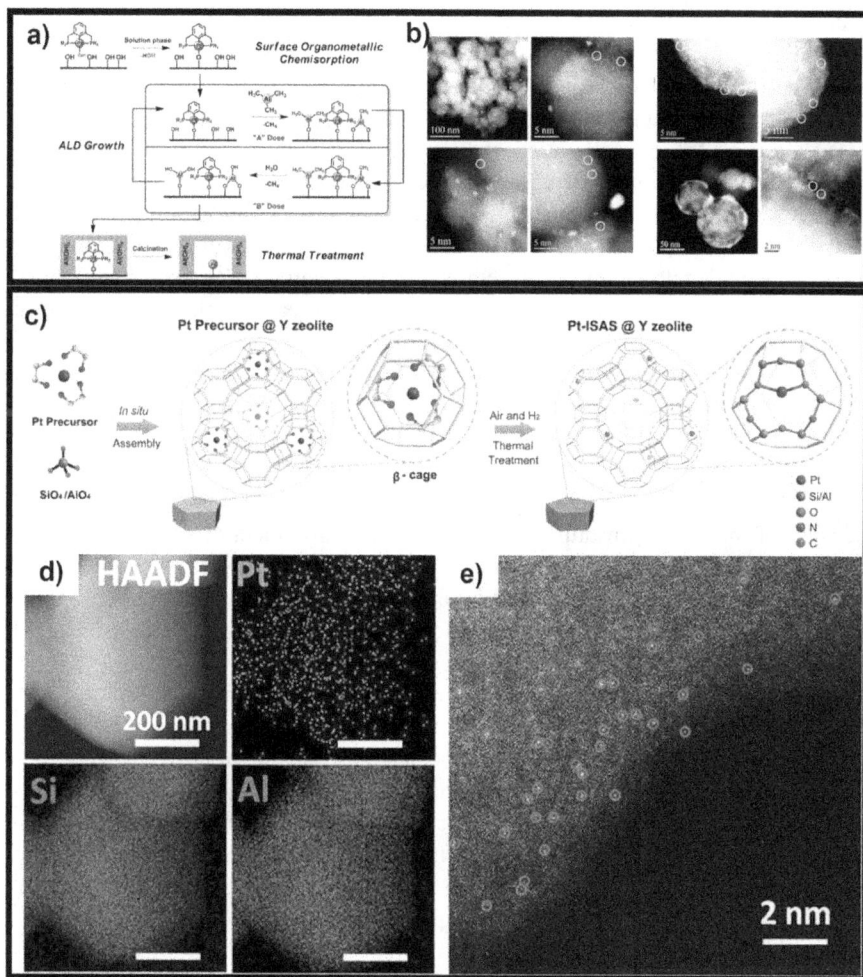

FIGURE 10.2 (a) Overall schematic for proposed single Pt atom synthetic methodology combining surface organometallic chemisorption and ALD approaches. (b) HAADF-STEM images of (PhPCP)Pt–Al$_2$O$_3$–400 cal, 20Al–(PhPCP)Pt–Al$_2$O$_3$–400 cal, 40Ti–(PhPCP)Pt–Al$_2$O$_3$–400 cal, and 20Zn–(PhPCP)Pt–Al$_2$O$_3$–400 cal. (Reproduced with permission from Reference,[16] Copyright 2017, American Chemical Society.) (c) Schematic illustration of the in-situ separation and confinement of a platinum precursor in a β-Cage. (d) Elements mapping, and (e) AC-HAADF-STEM image of Pt-ISAS@NaY. (Reproduced with permission from Reference,[23] Copyright 2019, American Chemical Society.)

single atoms facilitate hydrogen activation and alkyne adsorption. More importantly, the special cavities of the support make the active sites homogeneously distributed, which prohibit the adsorption of poisons such as CO. In the stability test, the Pd SAC exhibited no decrease in both the activity and selectivity, indicating the high stability.[17] With this concept, the C$_3$N$_4$ was reported to be a suitable support to stabilize various types of single atoms, such as Pt and Co.[18–20] Besides the C$_3$N$_4$, some other materials such as zeolite, which has a high surface area and porous structure, also can stabilize single atoms.[21-23] For instance, Liu et al. developed a general method to synthesize single atoms in Y zeolite, including Pt, Pd, Ru, Rh, Co, Ni, and Cu single atoms by in situ separating and confining a metal-ethanediamine complex into β-cages during the crystallization and thermal treatment (Figure 10.2c–e).[23] Also using zeolite as the support, Kistler et al. used the zeolite KLTL and $[Pt(NH_4)](NO_3)_2$ as the reactant, the supported $[Pt(NH_4)]^{2+}$ complex which were oxidized under high temperature. From the HAADF STEM results, they found the $[Pt(NH_4)]^{2+}$ complexes predominantly located in the D sites with the largest pores (66%), and the proportion decreased to 56% after oxidation. In the CO oxidation reaction, the achieved Pt SAC exhibits comparable catalytic performance among the reported catalysts. This work demonstrated the location of the Pt single atoms and indicated using zeolite as the support could be an efficient way to achieve stable single-site catalysts.[21]

10.3.2 NEIGHBORING SITE MODIFICATIONS

Because of the highly unsaturated coordination environment of single atoms, their chemical properties are highly dependent on the neighboring sites. Therefore, modifying the neighboring sites of SACs is important to tailor the electronic properties and local coordination environment, which may increase both the catalytic activity and stability of SACs.

Carbon-based materials are widely used in electrochemistry due to their low price and high conductivity.[24] However, for the conventional carbon-based catalysts, the interactions between metal species and carbon support are weak, thus leading to the aggregation and detachment of metal NPs. Carbon corrosion also occurs under high overpotentials. These factors result in the low stability of the catalyst, especially for SACs. Doping other elements into the carbon materials is one of the most promising ways to both increase the strength of carbon materials and generate stable anchor sites for single atoms. Nitrogen-doped carbon materials are found to be suitable for supporting single atoms. The strong interactions between N and metal single atoms could strongly enhance the stability. Cheng et al. successfully synthesized Pt single atoms on N-doped graphene using the ALD method and firstly applied it in the HER (Figure 10.3a–c). The achieved Pt SAC exhibited more than 37 times higher catalytic activity than the Pt/C and high stability without showing any obvious activity decrease after 1,000 cycles stability test. The theoretical calculations results show that there is a 0.25 e⁻ charge transfer from Pt to the neighboring N atom, indicating the strong interactions between Pt single atoms the N-doped graphene support. The analysis of X-ray absorption spectroscopy (XAS) further demonstrated that the partially unoccupied density of states of Pt $5d$ orbitals is the reason for its high catalytic performance.[25] Similar to the N dopant, S-doped materials can also help to stabilize single atoms. For instance, Choi and coworkers prepared sulfur-doped zeolite-templated carbon materials which have 17 wt% S. Later on, they found that these high S content materials can achieve a high loading (5 wt%) of Pt SAC. Based on the EXAFS analysis results, the Pt single atom is coordinated with four S atoms (Figure 10.3d and e). In the ORR, the Pt SAC showed high selectivity to the H_2O_2.[26]

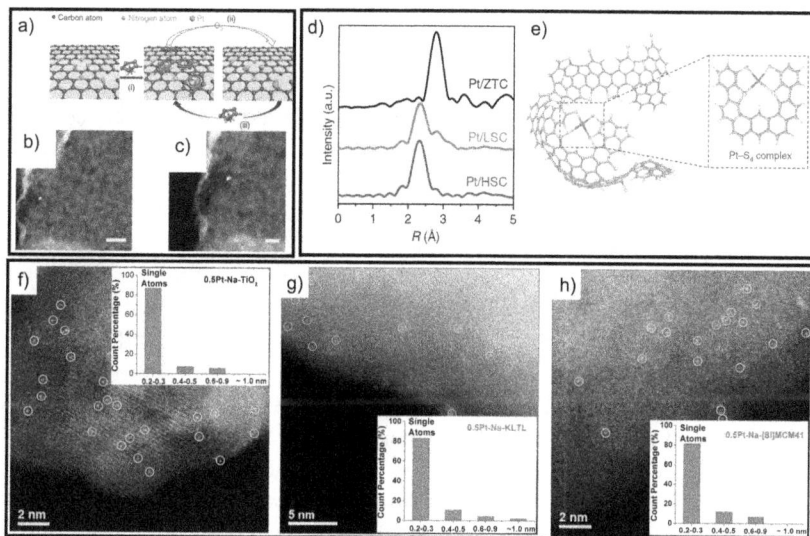

FIGURE 10.3 (a) Schematic illustration of the Pt ALD mechanism on NGNs. (b and c) ADF-STEM images of 50ALDPt/NGNs sample. Scale bars, 10 nm (b); 5 nm (c). (Reproduced with permission from Reference,[25] Copyright 2016, Nature Publishing Group.) (d) Fourier transforms of k^3-weighted Pt L_{III}-edge EXAFS. (e) Proposed atomistic structure of the Pt/HSC. (Reproduced with permission from Reference,[26] Copyright 2016, Nature Publishing Group.) (f–h) AC-HAADF-STEM images of the Na-containing Pt catalysts. (Reproduced with permission from Reference,[28] Copyright 2015, American Chemical Society.)

Besides the N- and S-doping method, the alkali ions are also confirmed to have the ability to achieve SACs.[27,28] Zhai and coworkers demonstrated that the alkali ions can strongly suppress the formation of Pt NPs. In the Na-modified silica support, the size of Pt NPs is much smaller than the one without Na. The theoretical calculation results show that the partially oxidized Pt-alkali-$O_x(OH)_y$ is the active site in the water-gas shift (WGS) reaction.[27] In their following work, they extend this synthesis strategy to other materials to achieve stable SACs.[28,29] In the sample preparation, they made the atomic ratio of Na/Pt as 10:1 from the $Pt(NH_3)_4(NO_3)_2$ and NaOH precursors using a solid-state impregnation method. They have applied this method to various supports, including the high-surface-area anatase, the microporous K-type L-zeolite, and mesoporous silica MCM-41. After the synthesis process, all of the three samples show the dominantly existed Pt single atoms (Figure 10.3f–h). The Pt single atoms are displayed as Pt (II), and they believe the Pt–$O(OH)_x$-species are the active sites that catalyze the WGS reaction. This work provides a general method to prepare stable SACs and the understanding of the active center for SACs.[28]

10.3.3 STRONG METAL-SUPPORT INTERACTIONS

Because of the high surface free energy of SACs, the single atoms tend to aggregate into clusters/NPs through Ostwald ripening, especially under harsh reaction conditions such as high temperature or in acidic/basic media. However, some reported work shows that some special supports have very strong interactions with the single atoms, which can reduce the degree of metal migration and further trap the single atoms. The metal NPs may evaporate and redisperse to single atoms under high-temperature conditions.

Recently, several works show that the SACs can be achieved using bulk metal or NPs. Jones and coworkers successfully achieved the Pt SACs through an atomic migration trapping method. In their work, they found that CeO_2 materials with different shapes show distinct abilities to trap the Pt single atoms. When the Pt NPs on the La-Aluminum sample were mixed with the ceria cubes and further treated under high temperature (800°C) in air, larger Pt NPs could be found, which is similar to the result when directly treating the Pt NPs. However, when the Pt NPs sample was mixed with ceria rods or polyhedral ceria, the Pt single atoms will be obtained.[30] Lang et al. successfully achieved a thermally stable Pt SAC through a strong covalent metal-support interaction. They first supported colloidal Pt NPs on Fe_2O_3 support and then calcinated in the air at 500°C to remove the stabilizer. HAADF-STEM result indicated that the size of Pt NPs is around 2–3 nm. However, when the Pt–NPs sample was treated at a high temperature (800°C) in the air, the Pt NPs disappeared but formed a very high density of Pt single atoms (Figure 10.4a). The singly dispersed Pt_1 atoms were confirmed by HAADF-STEM and EXAFS characterizations. However, if the Pt–NPs/Fe_2O_3 was calcinated in Ar, the size of the Pt NPs is even larger, indicating the thermal treatment conditions are critically important. They further performed in-situ HAADF-STEM to confirm their assumption and found that the total number of Pt NPs decreased from 300 to 200 after heating to 800°C under O_2 for 20 min. With the higher resolution, time-resolved imaging, they found the small Pt clusters disintegrated in <35 s (Figure 10.4b–d). This method provides a new way to synthesize Pt SACs by simply introducing the iron oxide.[31] This high

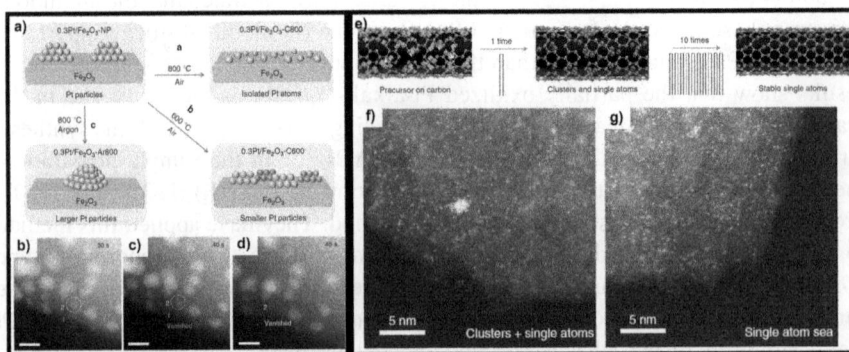

FIGURE 10.4 (a) Illustration of thermally induced Pt NPs restructuring. (b–d) Sequential HAADF-STEM images from the same area showing the dissociation of small particles during in-situ calcination: 5 nm scale bars. (Reproduced with permission from Reference,[31] Copyright 2019, Nature Publishing Group.) (e) The schematic diagram shows the synthesis and dispersion process. Typical HAADF images of Pt HT-SAs after one (f) and ten (g) cycles of the thermal shock (0.01 μmol cm⁻²). (Reproduced with permission from Reference,[32] Copyright 2019, Nature Publishing Group.)

temperature-induced formation of single atoms can also be found elsewhere using other methods. For example, Yao and coworkers used a high-temperature shockwave to achieve SACs (Figure 10.4e–g). More importantly, this method can be general and achieve various types of SACs such as Pt, Ru, and Co.[32]

Similar to the thermal treatment method, the pyrolysis process using metal-organic frameworks (MOFs) is widely used to obtain SACs, including the direct pyrolysis from MOFs with precursors, pyrolysis of MOFs with metal precursors, and followed acid washing to remove the metal NPs.[33] Among the reported literature, most of the SACs achieved by MOFs display in a metal-N_4 structure. Strong interactions between single atoms and the four N atoms enhance the stability of the catalysts. Very recently, Wei and coworkers demonstrated that noble metal NPs can be transformed into single atoms above 900°C in an inert atmosphere. More importantly, they directly observed this transformation. In the HAADF-STEM results, they found at the beginning of heating, the Pd NPs grew larger with an inhomogeneous size distribution. However, after 1.5 h, they found the total number of Pd NPs decreased, and the crystalline Pd NPs became an amorphous state. When the pyrolysis time went to 3 h, the Pd NPs vanished and formed the Pd single atoms (Figure 10.5a and b). They further carried out theoretical calculations to understand the mechanism of these NPs to single atoms transformation. DFTs results showed that this NPs-to-single atom conversion was driven by more thermodynamically stable Pd–N_4 sites and the Pd atoms were trapped during the thermal treatment (Figure 10.5c). In the semi-hydrogenation of acetylene, the achieved Pd SAC exhibited much improved catalytic activity and selectivity than the Pd NPs.[34]

10.4 SACs FOR EFFICIENT ELECTROCATALYSIS

Because of the unique properties of SACs, they showed outstanding catalytic performance in electrochemical reactions.[7,9,35] In the following section, we highlight some examples that demonstrated the promising applications of SACs as electrocatalysts.

10.4.1 OXYGEN REDUCTION REACTION

Global warming and other environmental problems strongly urge our society to use clean energy to substitute for fossil fuels. Fuel cells have high efficiency, are environmental-friendly, and have high power density, which is considered a next-generation energy conversion system. PEMFCs are promising energy conversion devices in applying practical electric vehicles due to their high power density, simple structure, and low operating temperature.[36] Currently, the commercialization of PEMFCs still has several key problems. In addition to problems of production, transportation, and storage of hydrogen, the high cost of PEMFCs is the main obstacle to its practical application. Noble metal catalysts such as platinum in cathode would account for 41% in whole fuel cell stack since the slow low reaction rate of the ORR in cathode needs a large amount of Pt catalysts. Thus, improving atom utilization efficiency to the single-atom level would be significant for reducing the whole cost and propelling the development of PEMFC. Currently, tremendous efforts had been devoted to prepare single-atom electrocatalysts for ORR.

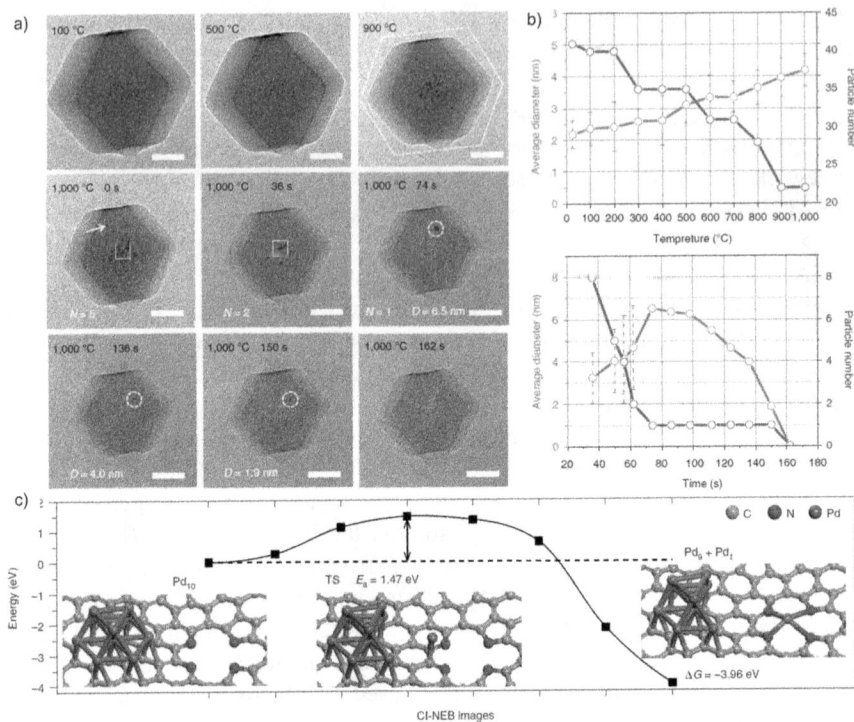

FIGURE 10.5 (a) Frames of Pd-NPs@ZIF-8 pyrolyzed in-situ TEM under an Ar atmosphere, scale bars are 50 nm. (b) Average diameter and number of particles as a function of heating temperatures and average diameter and number of particles versus pyrolyzing time at 1,000°C. (c) Calculated energies along the stretching pathway of the Pd atom from the Pd_{10} cluster to Pd–N_4 defect. (Reproduced with permission from Reference,[34] Copyright 2018, Nature Publishing Group.)

Before introducing the recent progress of single-atom electrocatalysts for ORR, understanding the ORR mechanism is critical to rationally design high-performance single-atom electrocatalysts. The first step is the adsorption of oxygen molecular by the active site, as shown in Figure 10.6a.[37,38] Then based on the absorbed oxygen molecular, the mechanism divides into the associated mechanism and disassociated mechanism. In the associated mechanism, the absorbed oxygen molecular would combine with an electron and a proton to form the *OOH intermediate. After that, *OOH will combine another proton to form H_2O_2 or producing a water molecular and the *O intermediate, then the *O would be further oxidized into water molecular. In the disassociated mechanism, the O–O bond would be broken and the *O bonds with adjacent active sites. In summary, the final product of ORR is water or hydrogen peroxide, and there are two pathways: two-electron pathway or four-electron pathway.

Furthermore, the mechanism on Pt nanocatalysts is shown in Figure 10.6b. Since Pt nanocatalysts exhibit high performance for ORR both in the acidic and alkaline environment and very low selectivity toward hydrogen peroxide, so the mechanism

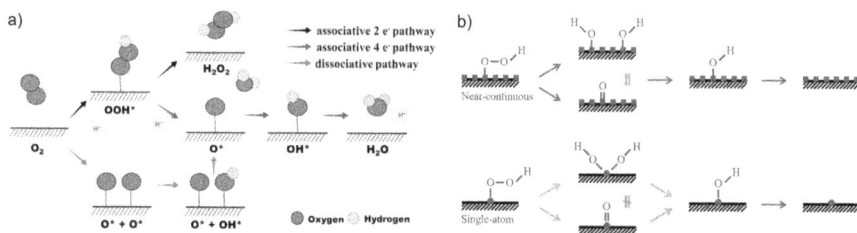

FIGURE 10.6 (a) The mechanism of the ORR process. (Reproduced with permission from Reference,[38] Copyright 2020, American Chemical Society.) (b) The ORR mechanism on near continuous sites and single-atom catalysts. (Reproduced with permission from Reference,[37] Copyright 2019, American Chemical Society.)

on Pt nanocatalysts is preferred to be the four-electron pathway. However, the associated mechanism or disassociated mechanism is still elusive. Recently, Li and coworkers found that the absorbed intermediate of different facets of Pt nanocatalysts is various based on surface-enhanced Raman spectroscopy, so this might indicate that mechanism of ORR would change on different facets.[39] Notably, the ORR pathway of Pt single atom is still controversial. For instance, Liu et al. synthesized Pt single atom supported on N-doped carbon by the impregnation method, and these as-prepared electrocatalysts exhibited excellent performance and durability. The Tafel plot and hydrogen peroxide experiment indicated that the four-electron pathway is the main mechanism for this type of catalyst.[40] However, other Pt single-atom electrocatalysts are believed to follow a two-electron pathway.[26,41]

For non-noble metal single-atom electrocatalysts, Fe SACs are the most attractive electrocatalysts since they show the highest activity toward ORR among non-noble metal catalysts, so it's critical to identify the structure of its active sites, which further guide the preparation of high-performance Fe SACs.[42–45] Fe SACs as the representative one, most studies show the mechanism of Fe SACs is through the four-electron pathway due to the low yield of hydrogen peroxide. The associated mechanism is generally considered an ideal process. Figure 10.7 shows the ORR mechanism on FeN_x ($x=4$, 5) by Mossbauer spectroscopy.[46] It indicates that both FeN_4 or FeN_5 follow the four-electron pathway and associated mechanism. However, some researchers suggest that Fe SACs would absorb two *OH as intermediate rather than a water molecular and intermediate *O.[37] The more detailed process still needs to be further explored by advanced characterization techniques and computational modeling. Notably, during the preparation of Fe SACs, it is unavoidable to introduce N-doped carbon materials, which show high selectivity for hydrogen peroxide.[47]

First, we discuss the structure of Fe, one type is crystal-phase iron NP, and another is atomically isolated iron atoms, which are usually located on graphene or micropores as Fe–N_x structure. Notably, the amorphous structure of atomically isolated iron cannot be characterized by common diffraction or microscopy techniques, which need a more advanced characterization method. These two types of structures co-existed in Fe–NC catalysts and can be tolerated for high temperatures. However, during the synthesis procedure, these two types of structures are competitive and closely related to the iron local environment. Commonly, the excessive iron

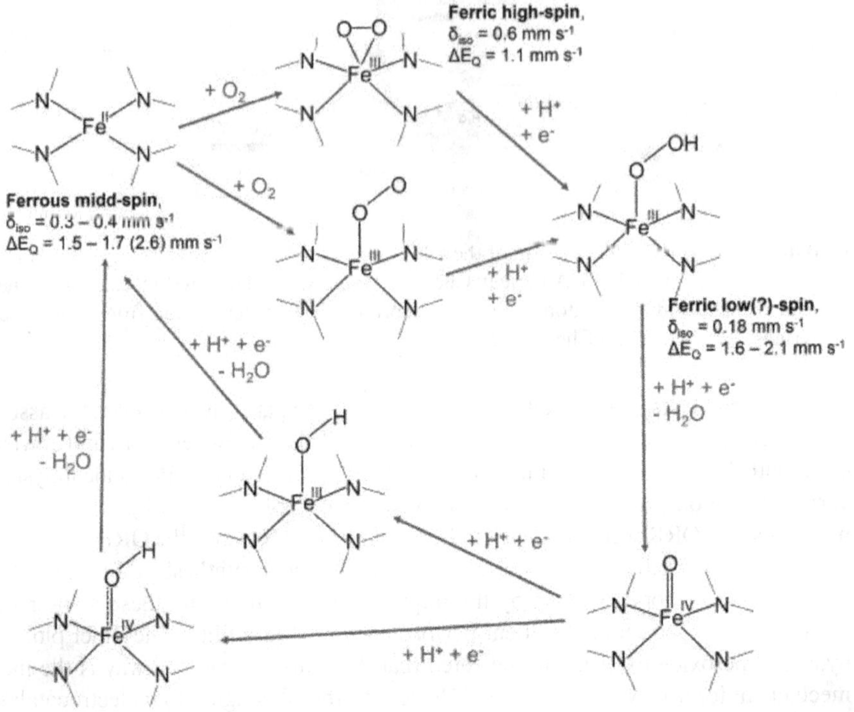

FIGURE 10.7 The ORR catalytic process of FeN_4 or FeN_5. (Reproduced with permission from Reference,[46] Copyright 2019, Wiley-VCH.)

precursor would lead to the increment of the crystal iron phase and reduce the density of a single iron structure.[48] This indicates that the atomically isolated iron needs a suitable environment to stabilize otherwise agglomerate into Fe nanoparticles.

Currently, the atomically isolated Fe single atoms rather than crystal iron NPs are considered active sites for ORR but still lack a comprehensive understanding. The iron nitrogen-coordinated structure could divide into three types: low spin state, medium spin state, and high spin state.[49] The spin state of Fe is determined by its coordination environment. Only low spin state and high spin state of Fe single atom show catalytic activity toward ORR in the acidic environment and high spin state Fe single atom show higher activity. This might be from the $3d_z^2$ orbit of medium spin state iron single atom is fully occupied and thus cannot interact with oxygen molecular while the orbit of the high spin state is half occupied and the orbit of the low spin state is unoccupied, so $3d_z^2$ orbit of high spin state and low spin state Fe singe atom catalysts could absorb oxygen molecular and catalyze ORR.

To achieve high performance, the most effective way is to increase the intrinsic activity of Fe SACs by tuning the coordination environment. Li and coworkers synthesized S and P co-doped Fe SAC by mixing S, P-contained precursors with metal precursors.[35] In addition, the doped sulfur and the evaporated zinc atoms would produce the Kirkendall effect, which facilitates the hollow structure of the dodecahedron shape. Then, based on DFT calculation, they found the doped S and P could exert electron effects on Fe active sites that with doped S and P, the Fe SAC shows a lower reaction free energy barrier during the catalytic process, thus improving its activity toward ORR. In the practical PEMFC test, FeNC with S- and P-doped SACs show high performance that its peak power density reached $0.3\,mW\,cm^{-2}$, which is closed to the commercial Pt/C. Furthermore, doping other metal atoms such as Co, Mn, and others also enhances its catalytic activity toward ORR. For example, Fe, Co dual metal active sites SACs were prepared by MOF-assisted method, the results indicated that Co-doped to Fe SACs would lead to synergistic effects and reduce the free energy to weakening the O=O bond. The ORR test showed these dual active sites Fe, Co catalysts exhibited high performance which is like commercial Pt/C.[50]

Additionally, optimizing support, especially the pore structure is another way to improve its performance. Pore structure could divide into three types: micropores (< 2 nm), mesopores (2–50 nm), and macropores (>50 nm).[51] The effects of three types of pore structure on FeNC catalytic performance had been investigated. Active sites of FeNC are mainly located on micropores, so the increasing density of active sites needs to increase the micropore on the carbon support. The mesopores structure would propel the transfer reactants to active sites, which fewer mesopores would lead to most active sites located in micropores can't be accessed. Macropores determine the rate of the oxygen and reactants that emerged in the overall catalysts layer. If lacking enough macropores, during PEMFC operation, the active sites in the deep catalysts layer cannot have enough oxygen to react. For example, Shui and coworkers used the SiO_2 coating method to synthesize concave-shaped FeNC single atoms to increase mesopores and surface area, the procedure is shown in Figure 10.8a.[52] Figure 10.8a

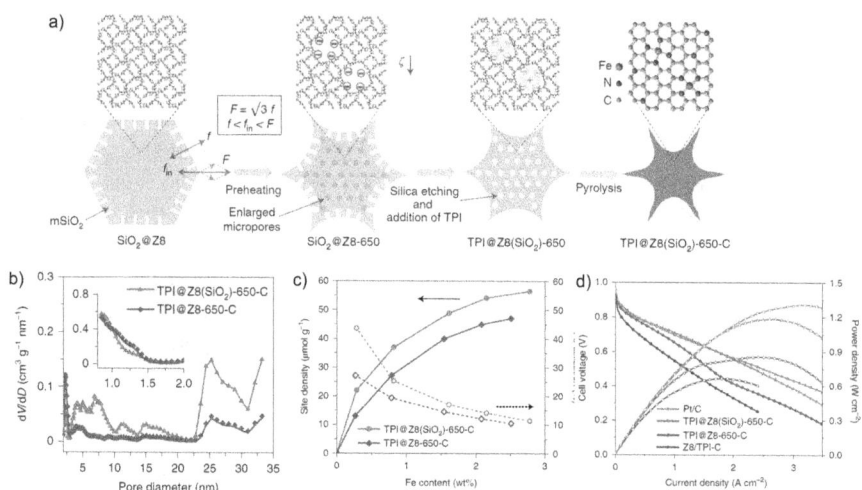

FIGURE 10.8 (a) The illustration of synthesis procedure. (b) The comparison of differential pore volume distribution of two catalysts. (c) The comparation of active sites density. (d) The polarization curve and power density of FeNC single-atom catalysts. (Reproduced with permission from Reference,[52] Copyright 2019, Nature Publishing Group.)

indicated that SiO_2 coating would be etched to destroy the smooth dodecahedron surface and formed a concave shape, which increases the surface area and meso-pores. Deferential pore volume results showed that after SiO_2 coating, the mesopores of FeNC SACs were enhanced from 5 to 10 nm and 22 to 30 nm (Figure 10.8b). In addition, the comparison of the density of active sites between SiO_2 coating FeNC catalysts and uncoating FeNC catalysts corroborated that the density of active sites of SiO_2 coating FeNC catalysts increased by nearly 40% (Figure 10.8c). Finally, they tested the ORR performance of these catalysts as the cathode in PEMFC. Its peak power density reached 1.18 W cm^{-2} at 0.47 V, which far surpassed the department of energy target and is one of the highest peak power densities in the current literature (Figure 10.8d).

Other researchers also optimize pore structure to improve active sites utilization and mass transport to gain high-performance FeNC SACs in PEMFC. Xu et al. used the templated-assisted method to synthesize overhang-eave shape catalysts and it shows high performance in PEMFC, which resulted from more three-phase react places on its surface.[53] Another highly ordered porous carbon support structure is prepared in two steps: the first is solvent-induced nucleation and the second is poly-styrene sphere template-assisted.[54] The highly ordered macropores and mesopores improve mass transport and active sites utilization at the same time, thus it shows high performance in PEMFC compared with random carbon-supported FeNC cata-lysts. Many ways have been developed for rational design pore structure of FeNC catalysts to gain high performance in PEMFC. However, the optimum distribution of the three-type pore structure still needs to be further explored.

However, currently, the most challenging task for non-noble metal single-atom electrocatalysts is stability. Currently, four mechanisms have been proposed for the degradation mechanism of FeNC SACs. Because the Fe single atoms serve as active sites for ORR, the dissolution of Fe atoms in fuel cell operation would reduce the density of active sites and thus decrease catalytic activity. In addition, the Fe ion in the electrolyte would block the proton-exchange membrane and accelerate the decomposition of Nafion, which further lowers fuel cell performance. Dodelet and coworkers found that the initial current density loss during the stability test is similar to the decrease of FeN_x active sites density.[55] They pointed out that though FeN_x is stable in the acidic environment, there is flowable water containing proton and oxy-gen in the practical fuel cell which destroys the balance of Fe/Fe^{2+} and leads to the dissolution of Fe atoms. Other researchers conceived that FeN_x is stable in an acidic environment and it is Fe ions dissolved from Fe NPs that react with high energy radi-cal to attack the Fe atoms of active sites.[56] The actual working potential of PEMFC is around 0.6 V, which far surpasses the oxidation potential of carbon support in an acidic environment, thus carbon support is inevitable to be oxidized during opera-tion, which would be detrimental to fuel cell performance. The oxidation equation of carbon support is as follows:

$$C + 2H_2O \rightarrow CO_2 + 4H^+ + 4e^- \quad E = 0.207\,V$$

Frederic explored the stability of ZIF-based FeNC SACs in the fuel cell by in-situ dif-ferential electrochemical mass spectroscopy (DEMS) with other in-situ techniques.

Results indicated that under 0.7 V, only Fe ions in electrolyte were detected when the potential is higher than 0.9 V. Carbon support would be oxidized and dissolved in electrolyte.[57] The stable operation window is during 0.7–0.9 V. Notably, when the Fe dissolves in electrolyte, the performance is still retained, but when carbon support started to be oxidized, the catalytic performance quickly decreased. Furthermore, the oxidation rate of carbon support would be accelerated at higher temperatures and potentials. Thus, it's important to improve carbon support stability. Furthermore, in addition to electrochemical oxidation, the chemical oxidation induced by high active radicals produced by the Fenton reaction also needs to be considered. Among the two mechanisms mentioned above, one factor is strong oxidized free radicals produced by the Fenton reaction. In general, during the practical application, the by-product hydrogen peroxide would react with Fe ion and produced strong oxidized free radicals, such as $^*OH$, and the equation is as follows:

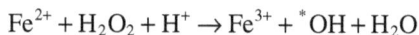

$$Fe^{2+} + H_2O_2 + H^+ \rightarrow Fe^{3+} + {}^*OH + H_2O$$

The free radicals would be detrimental to the Fe atom of active sites and carbon support, which decreases catalytic performance. For example, after H_2O_2 treatment, researchers found the performance suffered a severe loss after the stability test.[58] They conceived that the hydrogen peroxide would change the nitrogen coordination rather than the dissolution of the Fe atom. However, other researchers believed that adding hydrogen peroxide would directly dissolve the Fe atoms from FeN sites and thus causing performance loss.[59] These disagreements might result from different synthesis methods and different FeN_x configurations. In addition, the free radicals $^*OH$ could be further electro-oxidized with carbon support and form $^*OOH$ or $^*COOH$. These functional groups further destroy the active sites' structure and carbon support, which causes performance loss. Overall, hydrogen peroxide in fuel cells would exert its oxidation and electro-oxidation effects on FeNC catalysts, which severely affects its stability. Other two mechanisms are: micropores flooding and active sites protonation. The micropores flooding points out that actives sites mainly locate in micropores, and during operation, the hydrophilic of micropores would increase, and then micropores would be submerged by water, thus active sites in micropores would not access reactants and cause a dramatical performance loss.[60] The active sites protonation mechanism elucidate that the doped N such as pyridine N would be protonation in the acidic environment and further combine with anion in the electrolyte, thus leading to performance loss.[61] These two mechanisms are still in debate because they contradict other experiment results.[62]

10.4.2 Oxygen Evolution Reaction

Increasing energy consumption and severe environmental pollution motivated new and sustainable energy to substitute for fossil fuels. Hydrogen energy is considered the ultimate solution for human society because it is environmentally friendly and renewable. Currently, steam reforming of gas, such as methane, is the main pathway to produce hydrogen gas, but its high cost and relatively low purity restrict its large-scale production. Electrochemical water splitting is an efficient and clean technology to large-scale yield high purity hydrogen gas, with a theoretical efficiency of 100%. Water electrolysis is involved with two half-reactions: HER and OER.[63–66] HER is a two-electron transfer pathway and relatively fast reaction rate kinetically. However, the OER is a four-electron transfer pathway and requires higher energetic activation. Thus, the main obstacle of water splitting occurs at the anode, with a high overpotential of OER.

The mechanism of OER proposed by Norskov from computational modeling is generally accepted.[67,68] In acidic conditions, the reaction pathway is below:

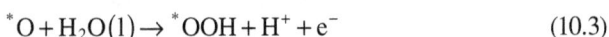

$$^* + H_2O(l) \rightarrow {}^*OH + H^+ + e^- \tag{10.1}$$

$$^*OH \rightarrow {}^*O + {}^*H + e^- \tag{10.2}$$

$$^*O + H_2O(l) \rightarrow {}^*OOH + H^+ + e^- \tag{10.3}$$

$$^*OOH \rightarrow O_2(g) + H^+ + e^- + {}^* \tag{10.4}$$

In alkaline conditions, there are four steps:

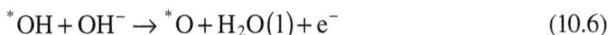

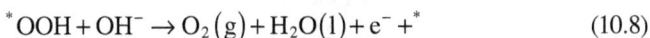

$$^* + OH^- \rightarrow {}^*OH + e^- \tag{10.5}$$

$$^*OH + OH^- \rightarrow {}^*O + H_2O(l) + e^- \tag{10.6}$$

$$^*O + OH^- \rightarrow {}^*OOH + e^- \tag{10.7}$$

$$^*OOH + OH^- \rightarrow O_2(g) + H_2O(l) + e^- + {}^* \tag{10.8}$$

The * represents the active site of catalysts.

Since the OER is a multiple-step reaction, the intermediates, such as $^*O$, $^*OH$, and $^*OOH$, directly affect the reaction rate. Among these steps, the slowest step ultimately determines the whole reaction velocity, named the rate-determining step. From Figure 10.9a, the whole reaction Gibbs energy of ideal catalysts is 1.23 V, and under this circumstance, the overpotential is zero.[68] However, the relationship of the absorption energy of three intermediates is linearly correlated, and some researchers suggested that the discrepancy of $^*OOH$ and $^*OH$ absorption energy stabilized at 3.2 V, as shown in Figure 10.9b.[68] Thus, we can calculate one intermediate absorption energy rather than calculating every single one. Additionally, the reaction rate could be simplified to two intermediates: $^*O$ and $^*OH$ absorption energy. Figure 10.9c indicates the volcano shape of metal oxide catalysts for OERs. We found that IrO_2 and RuO_2 exhibit the best performance for OER due to their medium bonding energy.

To rationally design electrocatalysts for OER, understanding the catalytic activity descriptor is critically important. Currently, there are several activity descriptors proposed by different research groups such as E_g descriptor,[69,70] M–O bonding theory,[71] d band center theory,[72] and so on. This chapter mainly introduces E_g descriptor and M–O bonding theory. E_g descriptor for OER is proposed by the Shao-horn Yang group, and they investigated the relationship between the E_g occupancy of the surface

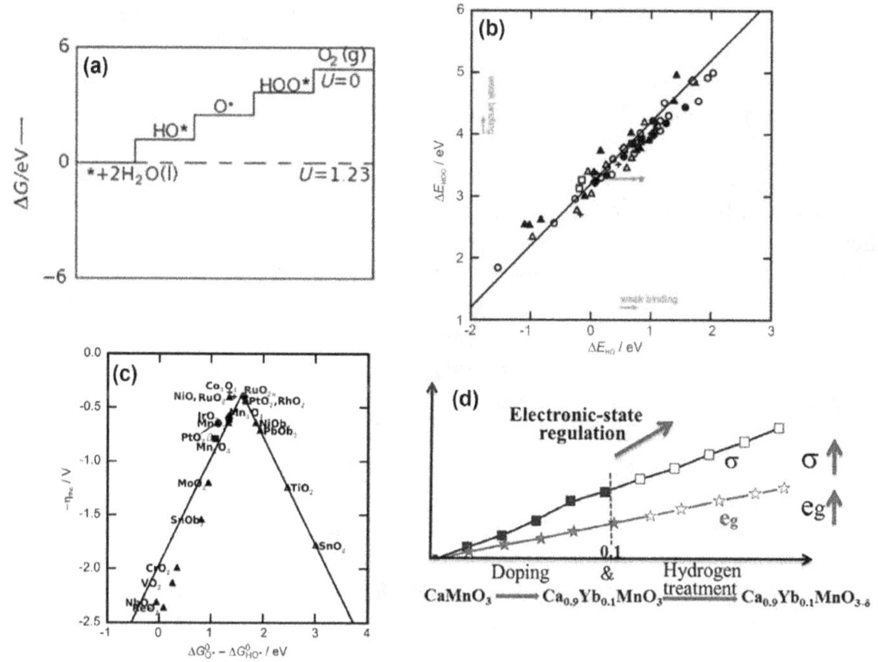

FIGURE 10.9 (a) The Gibbs energy of ideal catalysts in each reaction coordinate. (b) The relationship between reaction intermediate *OOH and *OH. (c) The volcano shape plot-based electrocatalytic activity and Gibbs energy difference. (Reproduced with permission from Reference,[68] Copyright 2019, Wiley-VCH.) (d) Electronic state regulation by the combination of lightly doping and hydrogen treatment for ABO_3 perovskite. σ, conductivity, e_g, number of electrons at Mn e_g orbit. (Reproduced with permission from Reference,[74] Copyright 2015, Wiley-VCH.)

metal atom and OER activity from tens of metal oxide electrocatalysts.[73] The results indicated that the OER activity of these electrocatalysts exhibited a volcano shape plot depending on the occupancy of 3d electron with an E_g symmetry of surface transition metal atom because the electron in E_g orbits would interact with oxygen species during the OER process and thus affected electron transfer and absorption/desorption intermediate. For instance, Xie et al. applied element doping and hydrogen treatment into perovskite calcium–manganese oxide (Figure 10.9d) and modulated the E_g filling state to 0.81. The prepared electrocatalyst activity was 100 fold higher than the pristine one at an overpotential of 500 mV and excellent stability.[74]

Another universal performance descriptor is the M–O bonding theory, proposed by the Markovic group.[75-77] During OER, the metal atom is generally considered as an active site, and reaction intermediate is absorbed on the metal atom surface. The coordinated oxygen atom plays an important role in the OER process for strong metal-oxygen p-d interaction electrocatalysts as well. Increasing metal-oxygen hybridization/bonding would enhance the performance since high metal-oxygen binding boosted the electron transfer. Thus, transition metal oxide, such as Ni, Co, and Fe, exhibited superior performance for OER due to their strong metal-oxygen bonding. For instance, Xu et al. introduced the third element, such as Fe into $ZnCoO_4$, to produce more Zn vacancies to improve the Co–O bonding because the formation energy of Fe is lower than Zn. The enhanced Co–O bonding facilitated its OER performance by accelerating the electron transfer process between metal and oxygen species.[78]

Currently, IrO_2 and RuO_2 are considered as benchmark catalysts for oxygen evolution reaction, but their low abundance and high cost restrict their practical applications. Developing low-cost, high-efficient electrocatalysts for OER can pave the way to the commercialization of water splitting. Single-atom electrocatalysts attracted great attention for oxygen evolution reactions due to their superior electrocatalytic performance, ultrahigh atom utilization, and low cost. In this chapter, we introduce the recent progress of single-atom electrocatalysts in oxygen evolution reactions from two aspects: noble metal single atom and non-noble metal single-atom electrocatalysts.

Before presenting recent progress, elucidating some measurement criteria for OER performance is helpful to evaluate different electrocatalysts. Four factors including overpotential, Tafel plot, mass activity/specific activity, and stability are usually employed to assess the performance of the catalyst. Overpotential is to determine electrocatalysts' performance for OER. Overpotential under $10 \, mA \, cm^{-2}$ is commonly used to assess electrocatalysts' performance and catalysts with overpotentials <300 mV are believed to be excellent electrocatalysts for OER. A Tafel plot could be used to identify the rate-determining step under the whole reaction. Mass activity/specific activity is another useful parameter for comparing electrocatalytic performance per unit weight or area. Stability is a significantly important factor for electrocatalysts since it represents the whole working period. It could be measured using chronopotentiometry at a constant potential or current density.

Since IrO_2 and RuO_2 are the benchmark electrocatalysts for OER, there are many studies to prepare noble metal single-atom electrocatalysts such as Ir, Ru, Pt for OER.[79–82] First, we introduce carbon materials supported noble metal single-atom electrocatalysts because of low-cost and high electron conductivity of the carbon-based substrate. For instance, Ru single atom embedded into nitrogen-doped carbon support (Ru–N–C) are synthesized by a defect-driven impregnation method. Ru coordinated with four neighboring pyridine N structures is precisely identified by XAS, XPS, and STEM. This special Ru–N–C electrocatalyst exhibited excellent performance toward OER with only 267 mV overpotential under $10 \, mA \, cm^{-2}$ and 5% degradation after 30 h operation, shown in Figure 10.10a. Specifically, the mass activity of Ru–N–C displays $3,571 \, A \, g^{-1}$ at 267 mV overpotential, which is 322 times higher than those of RuO_2/C (Figure 10.10b). Interestingly, the operando XAS study and FTIR revealed that during the OER process, intermediate O species would absorb on Ru–N_4 moiety to form an O–Ru–N_4 structure. Furthermore, X-ray absorption near edge structure (XANES) results indicated that the electronic structure of Ru would be distorted by intermediate O species through electron transfer and the valence state of Ru slightly increased, indicated by Figure 10.10c, thus favoring bonding with reaction intermediates such as *O, *OH.[83]

In addition to carbon-based support, transition metal oxides also could be used as the substrate, as strong metal–oxide interaction would stabilize noble single atom. Plenty of transition metal oxide materials are promising electrocatalysts for OER due

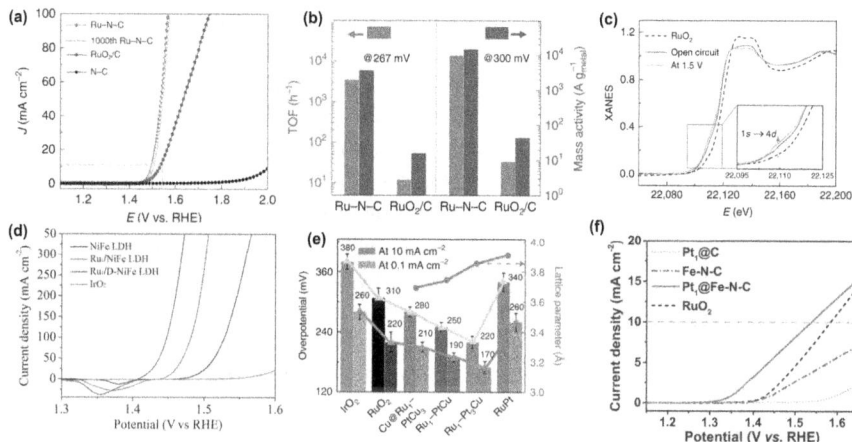

FIGURE 10.10 (a) The CV curves of four types of electrocatalysts. (b) The comparison of mass activity and TOF. (c) The operando XAS results in different potential. (Reproduced with permission from Reference,[83] Copyright 2019, Nature Publishing Group.) (d) The CV curves of NiFe-LDH, Ru/NiFe LDH, Ru/D-NiFe LDH and IrO₂. (Reproduced with permission from Reference,[87] Copyright 2021, Nature Publishing Group.) (e) The relationship between lattice amount and overpotential. (Reproduced with permission from Reference,[92] Copyright 2019, Nature Publishing Group.) (f) The OER performance of prepared electrocatalysts. (Reproduced with permission from Reference,[93] Copyright 2017, Wiley-VCH.)

to their special structure and many active sites.[19,84–86] Thus, anchoring a single atom on a transition metal oxide support is an ideal way to optimize its performance. Zhai et al. electrodeposited Ru single atom NiFeAl layered double hydroxides (LDH) supported by a 3-D Ni foam and subsequently etching Al to form defected NiFe LDH. STEM images and XAFS results revealed the existence of Ru single atom and its corresponding electronic structure. Figure 10.10d shows the as-prepared catalysts exhibited an outstanding performance for HER and OER that only 189 mV overpotential at 10 mA cm⁻². Then, the author tested this electrocatalyst in a practical water electrolyzer for overall water splitting. The Ru/D–NiFe LDH catalysts only require 1.44 V to reach 10 mA cm⁻² and 1.72 V for a practical operation current density of 500 mA cm⁻². DFT calculation proposed that the synergistic effects between Ru single atom and defected NiFe LDH boost its electrocatalytic performance for OER. More specifically, the formation of Ru–O active sites from the interaction of Ru single atom and LDH support showed an ultralow overpotential for OER.[87] Other researchers are devoted to preparing super high metal loading Ir single atom on Ni oxide matrix for OER. It exhibited an overpotential of 215 mV at 10 mA cm⁻² from increased active sites and cooperative interaction between a single atom and NiO matrix. DFT calculation and XAS results demonstrated the valence state evolution of the Ir single atom and its surrounding Ni atoms, accounting for its excellent performance.[88] However, some other noble metal elements, such as Au, are also used to dope in NiFe LDH for OER, but single atoms are not served as active sites. The incorporation of Au single atom into NiFe LDH induces electron transfer from Au single atom to surround transition metal atom and stabilize surface LDH structure under operation, thus modulating the absorption energy of oxygen intermediate to enhance OER performance, verified by DFT calculation and experiment results.[89]

The third type of support for noble metal single-atom electrocatalysts is a metal alloy. Since stability is an important factor to assess OER electrocatalysts, but for Ru-based electrocatalysts, frequent change of redox state under operation is the main reason for its degradation.[90,91] Thus, tuning the electronic structure of Ru-based electrocatalysts could improve its stability and simultaneously enhance activity. Li's group synthesized atomically dispersed Ru atom on a PtCu/Pt core–shell structure alloy support, precisely modulating the electronic structure and redox state of Ru atom by compressive strain. Through a series of leaching processes, the surface composition of PtCu alloy is modulated to achieve different lattice parameters, thus leading to different compressive strains for deposited Ru atoms. Then, as depicted in Figure 10.10e, a volcano-shaped relationship between electrocatalytic performance and lattice parameter is found, with the OER activity located on the optimum value being only 220 mV overpotential under 10 mA cm^{-2} when a surface composition is Pt$_3$Cu.[92] Notably, PtCu alloy exhibited poor OER performance, so Ru single atoms play a critical role in OER activity. In-situ XAS studies indicated that the electronic structure of the Ru atom is constant during OER measurement. DFT calculations revealed that compressive strain for Ru atoms modulates its electronic structure to optimize its OER activity, more importantly, suppressing the electron transfer between Ru atoms and oxygen species during operation to improve its stability. Moreover, to further enhance electrocatalytic performance, researchers design a single atom to single atom moiety to boost its OER activity. A Pt–O$_2$–Fe–N–C special structure is prepared by pyrolysis method and subsequent impregnation method, corroborated by STEM and XAS characterization. Figure 10.10f describes that its onset potential only 1.33 V and overpotential 310 mV under 10 mA cm^{-2}, surpass the commercial RuO$_2$ catalyst under alkaline conditions. Calculation and experimental results indicated that the synergistic effects between Pt atom and Fe atoms enhance its OER performance. Additionally, the low-coordinated Pt atoms would facilitate the dissociation of the water molecule to promote the OER performance.[93]

These findings provide valuable insight into the rational design of noble metal single atom for OER, but some underlying mechanisms are still elusive such as the role of synergistic effects in OER performance. Therefore, exploring the relationship between structure or coordination configuration and catalytic activity is significant for noble metal single-atom electrocatalysts. Meanwhile, substantial efforts are still required to devote to the preparation of high performance in activity and stability single-atom electrocatalysts, especially in practical water electrolyzes.

Non-noble metal single-atom electrocatalysts are widely investigated by many computational models that suggested transition metal single-atom electrocatalysts would exhibit excellent performance for OER in alkaline conditions.[94–97] For instance, Ni single-atom electrocatalysts had been extensively explored since Ni contained oxide and derivatives showed great activity toward OER (Figure 10.11a).[94] A 3d transition metal anchored into a nitrogen-doped carbon framework is firstly prepared since the M–N–C structure displayed excellent electrocatalytic activity for the ORR.[98–100] Through XAS analysis and STEM characterization (Figure 10.11b), the

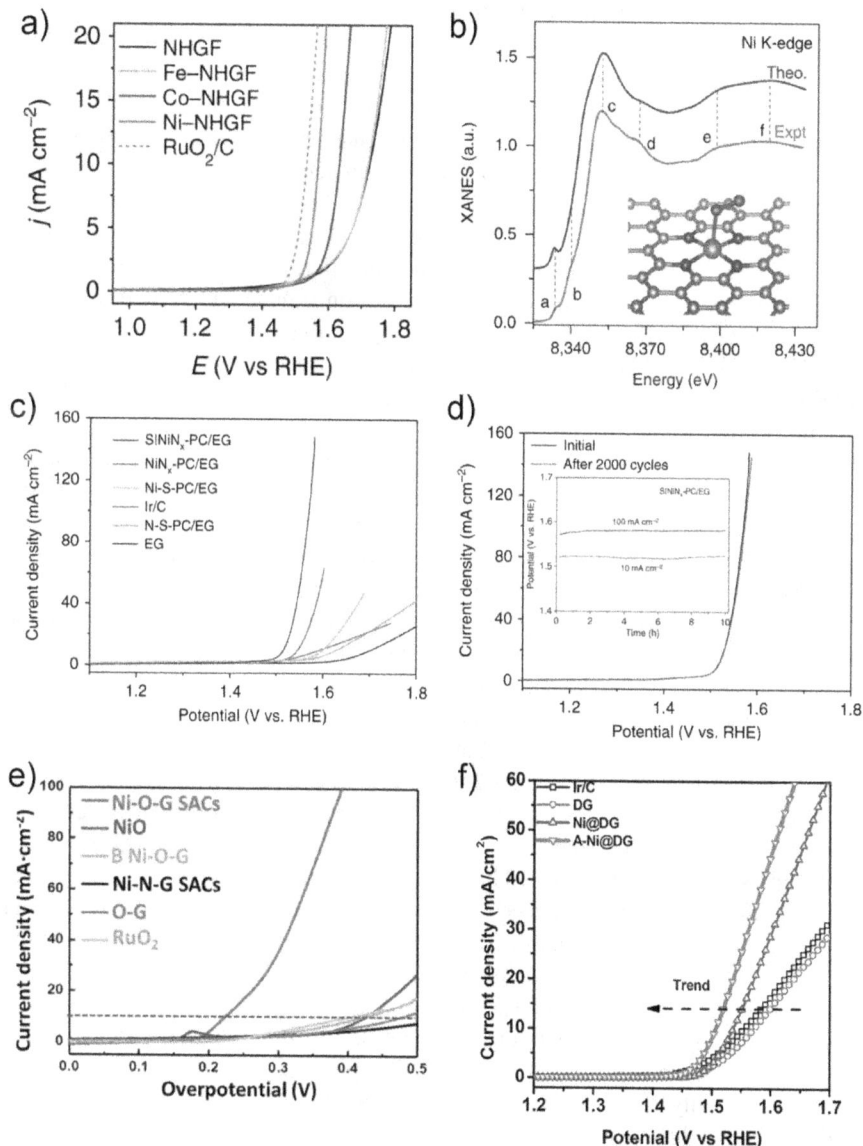

FIGURE 10.11 (a) The OER curves of prepared electrocatalysts and commercial RuO_2. (b) Ni K-edge XANES spectra. (Reproduced with permission from Reference,[94] Copyright 2018, Nature Publishing Group.) (c) LSV of S/NiN$_x$-PCEG, NiN$_x$-PCEG, Ni-S-PECG, Ir/C, N-S-PECG and CG. (d) The OER stability test of S/NiN$_x$-PECG electrocatalysts. (Reproduced with permission from Reference,[101] Copyright 2019, Nature Publishing Group.) (e) The OER performance of different Ni single-atom electrocatalysts and benchmark electrocatalyst. (Reproduced with permission from Reference,[102] Copyright 2020, Wiley-VCH.) (f) LSV curves of A-Ni@DG, Ni@DG, DG and Ir/C. (Reproduced with permission from Reference,[103] Copyright 2018, Elsevier.)

center transition metal atom coordinated with four nitrogen atoms and one axial oxygen atom with fours carbon atoms on the second shell is identified. Before testing these three types of catalysts' performance, DFT calculation had been used to predict its activity based on electronic structure and coordination configuration. For Fe and Co elements, the OER process prefers to occur on center metal atoms to favor the single active site mechanism since the M–O bonding is much stronger than C–O bonding. However, *O and *OH would absorb on coordinated carbon atoms, and *OOH would absorb on Ni atoms to carry on dual active sites mechanism and thus, facilitating the OER process. Practical electrochemical evaluation for these samples indicated that Ni supported on nitrogen-doped hollow carbon framework (NHCF) displayed 331 mV overpotential at 10 mA cm^{-2}, much higher than Co–NHCF (402 mV) and Fe–NHCF (488 mV), shown in Figure 10.11a. More importantly, turnover frequency (TOF) of these active sites results showed the TOF of Ni single atom is two orders of magnitude higher than Ni-based oxide.[94] Additionally, to further tune the electronic structure of Ni single atom to improve OER kinetics, other researchers developed nitrogen sulfur co-doped carbon support for Ni single-atom electrocatalysts. By the pyrolysis of ternary dicyandiamide–thiophene–nickel precursor, Figure 10.11c indicates that the as-prepared electrocatalysts exhibited 1.51 V overpotential at 10 mA cm^{-2}. More importantly, after 2,000 cycles, NiN$_x$/S also exhibited excellent stability, with nearly no loss (Figure 10.11d). From experimental results and DFT calculation, Ni single atom coordinated with three N atoms and one sulfur atom embedded into carbon matrix had been revealed. Furthermore, the incorporation of a sulfur atom changes the local electron distribution and reduces the energy barrier of elementary reaction, thus enhancing OER performance.[101]

In addition to nitrogen-coordinated structure, other coordinated elements, such as oxygen and carbon, are also used to stabilize single atoms to modulate their electronic structure. Li et al. synthesized Ni single atom bonded with oxygen atoms on carbon matrix by NaCl template method. The detailed structure and electronic environment are elucidated by various advanced characterization techniques such as STEM and XAS. The experimental results verified the Ni single atom and demonstrated that the valence state of Ni single atom is higher than Ni(II). The simulation results showed that a Ni single atom coordinated with four oxygen atoms and two OH groups and this saturated coordination configuration decreases the energy barrier of elementary reaction and thus, enhances its OER activity. This prepared Ni single-atom electrocatalyst achieved only 214 mV overpotential at 10 mA cm^{-2}, as shown in Figure 10.11e.[102] Additionally, designing Ni coordinated with carbon atoms is also a promising pathway to accelerate electron transfer during OER though the Ni–C bonding is much weaker than Ni–N bonding. Researchers reported a defect-assisted impregnation method to prepare Ni single atom embedded in defected graphene. HAADF-STEM images and XANES confirmed the coordinated configuration of Ni–C with different types. Precise modeling and analysis indicated that there are three types of carbon coordination: divacancy, D5775, and the perfect hexagons. Then, the electrochemical performance is evaluated (Figure 10.11f), showing the overpotential of Ni on defects carbon is only 270 mV at 10 mA cm^{-2} and demonstrates excellent stability as well. Systematic DFT calculation indicated that the interaction between absorbed species and Ni single atom anchored in divacancy is favorable for the OER process.[103]

Moreover, other transition metal single-atom electrocatalysts, such as cobalt and iron, also exhibited promising performance for OER.2,[104–109] Qiao's group prepared Co single atom on the g-C_3N_4 substrate, and modeling results showed that the most stable location for Co single atom is void space. Advanced characterization from STEM and XAS results verified the Co single atom structure and indicated that one Co atom coordinated with two N atoms to form a Co–N_2 moiety. Notably, the valence state of Co atoms is closed to Co(II). The electrocatalytic activity evaluation of Co–N–C showed a 1.61 V overpotential under 10 mA cm^{-2}.[110] As the electrocatalytic performance of single-atom electrocatalysts is needed to further enhance, dual atom electrocatalysts for OER are also explored. Through the high-temperature polymerization method, researchers prepared Ni–Fe dual metal embedded in a g-C_3N_4 substrate. The STEM and XAS results corroborated the Ni–Fe dual metal active sites and revealed more oxidation states of Ni atoms, which improve its OER kinetics. The as-prepared Ni–Fe dual metal electrocatalysts exhibited outstanding OER activity at only 326 mV overpotential under 10 mA cm^{-2}. The synergistic effects from Ni and Fe facilitate OER kinetics from exposure to more active sites, reasonable pore structure, and charge transfer between Ni and Fe.[111]

Overall, tremendous efforts have been devoted to develop single-atom electrocatalysts for OER, but there are still some challenges. The first is the overpotential of almost all electrocatalysts is still relatively high, especially under high current density to meet practical operation. In addition, underlying mechanisms are still elusive, especially how to depict the correlation between a single-atom active site and its corresponding performance. The last one is the stability of a single atom electrocatalyst. Since severe operation environments for OER, like high potential and acid or alkaline environment, the stability of single-atom electrocatalysts remains a huge challenge.

10.4.3 HYDROGEN EVOLUTION REACTION

As another half-reaction of water splitting systems, the HER is also very attractive. Besides, hydrogen gas (H_2) as a high value and no emission fuel to be used in the fuel cell, is considered as the most promising candidate to replace fossil fuel.[112] Normally, HER is a multistep electrochemical process, which may follow two different pathways depending on the electrolyte pH.[113] The first step is electrochemical hydrogen adsorption (Volmer reaction):

$$\text{In acid solution}: \text{H} + {}^{*} + e^{-} \rightarrow \text{H}^{*}$$

$$\text{In alkaline solution}: \text{H}_2\text{O} + {}^{*} + e^{-} \rightarrow \text{H}^{*} + \text{OH}^{-}$$

The following step has two different pathways to generate the final product of H_2. When the H* coverage is low, the adsorbed hydrogen atom prefers to couple with a new electron and another proton in the electrolyte to evolve H_2. This is the so-called Heyrovsky reaction,

$$\text{In acid solution}: \text{H} + e^{-} + \text{H}^{*} \rightarrow \text{H}_2 + {}^{*}$$

$$\text{In alkaline solution}: \quad H_2O + e^- + H^* \rightarrow^* + OH + H_2$$

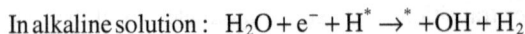

Nowadays, Pt and its derivatives still are the best choice for HER and its commercial counterpart Pt/C show a low overpotential in high current density. However, the scarcity and high cost are still the biggest obstacles to the commercialization of precious metal catalysts. Herein, to reduce the cost and maximize the activity, a huge effort has been made to develop SACs. For example, after functionalizing the graphene via π–π interactions, Ye et al.[114] successfully dropped single Pt atoms on graphene (denoted as Pt SASs/AG). After being functionalized by aniline molecules via π–π interactions, the graphene became hydrophilic and positively charged, which successfully make $[PtCl_6]^{2-}$ ions anchored on AG. Finally, the anchored $[PtCl_6]^{2-}$ ions were reduced to single Pt atomic sites by using microwave irradiation (Figure 10.12a). Pt SASs/AG presents excellent HER activity with $\eta = 12\,mV$ at $10\,mA\,cm^{-2}$ and a mass current density of 22,400 $AgPt^{-1}$ at $\eta = 50\,mV$, which is 46 times higher than that of commercial 20 wt% Pt/C (Figure 10.12b). DFT calculations show that the coordination of atomically separated Pt and aniline nitrogen optimizes the electronic structure and hydrogen adsorption energy of Pt, and ultimately promotes HER activity (Figure 10.12c).

In another study, Kuang et al.[115] focused on the electronic metal-support interaction (EMSI) in catalysis to modulate the electronic state of the supported metal, and optimize the reduction of intermediate species. They reported the synthesis of N-doped mesoporous hollow carbon spheres loaded with Pt single atoms (Pt_1/NMHCS) with a strong EMSI effect by carbonization of polydopamine (PDA) through a SiO_2-templated strategy. SEM and TEM results demonstrate that NMHCS not only can effectively immobilize metal atoms but also provides an efficient three-phase region for electron transfer, ion transport, and gaseous product diffusion. The XAF and theoretical simulations reveal that the strong EMSI effect in a unique N_1–Pt_1–C_2 coordination structure significantly tailors the electronic structure of Pt 5d states in Pt_1/NMHCS, facilitated H–H coupling, and improve HER activity.

Ru, as another noble metal, its derivatives also show excellent performance for HER and are often compared with Pt.[116,117] On the basis that the original price of Ru is only 5% of the price of Pt, Ru SAC can further reduce the cost while increasing the catalytic activity.[118] Chen et al.[119] prepared a nanocomposite based on ruthenium and nitrogen co-doped carbons by coating melamine-formaldehyde (MF) polymer on the tellurium nanowires (Te NWs) to adsorb an appropriate amount of ruthenium (III) chloride, and then pyrolyze to synthesize Ru, N-co-doped carbon nanowires. It must be pointed out that both ruthenium nanoparticles and ruthenium single atoms were embedded within the carbon matrix to form Ru–N/Ru–C moieties, while Ru nanoparticles made only minor contributions. In addition, consistent results were obtained through first-principles calculations, in which it was found that the RuC_xN_y part showed much lower hydrogen binding energy than ruthenium nanoparticles and a lower water dissociation kinetic barrier than platinum. Among them, RuC_2N_2 stands out as the most active catalytic center, in which ruthenium and adjacent carbon atoms are possible active sites.

As an emerging 2D transition metal carbide/nitride, MXenes have high conductivity, many exposed active sites on the base surface, and hydrophilic surface functional

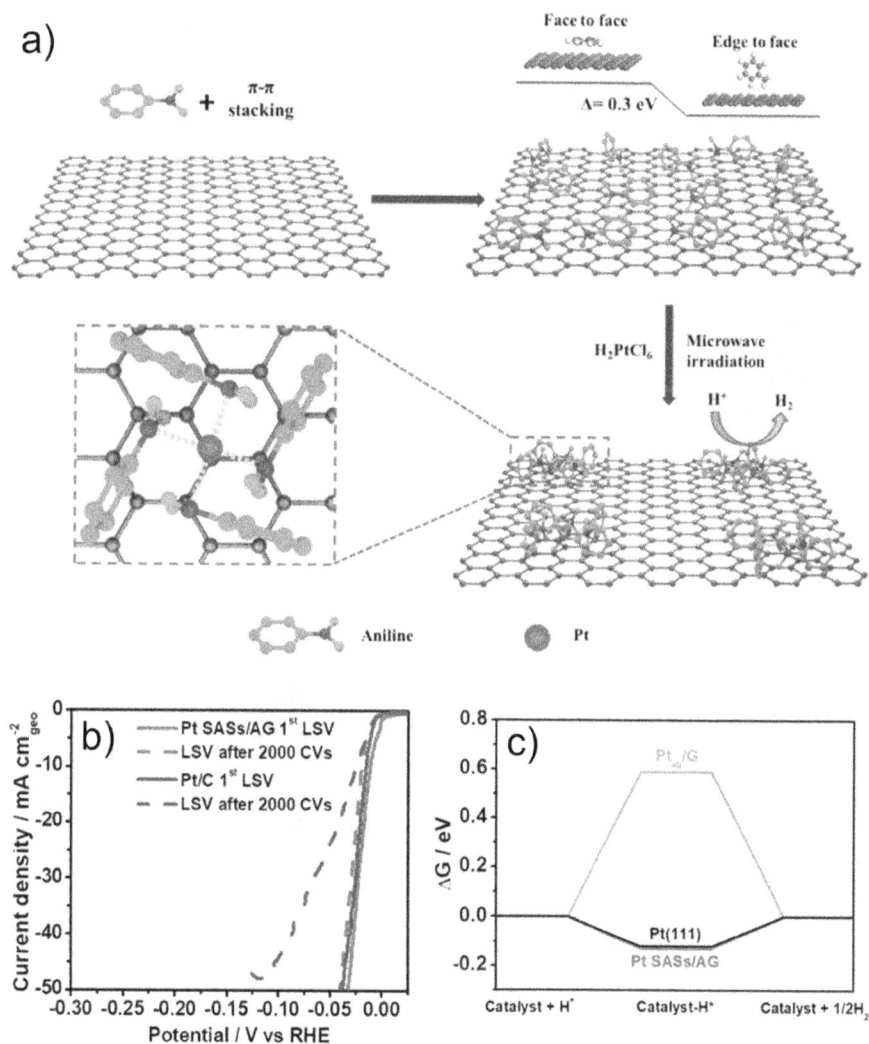

FIGURE 10.12 (a) Synthetic scheme of Pt SASs/AG. (b) LSV curves of Pt SASs/AG before and after 2000 catalytic cycles. (c) Calculated free energy diagrams of HER for Pt(111), Pt_{ab}/G and Pt SASs/AG. (Reproduced with permission from Reference,[114] Copyright 2019, Royal Society of Chemistry.)

groups (such as −O, −OH, and −F groups), are seen as promising candidates for various electrochemical applications. He et al.[120] reported the coordination interaction between isolated ruthenium single atoms (Ru_{sa}) and a 2D titanium carbide ($Ti_3C_2T_x$) MXene support through the nitrogen (N) and sulfur (S) heteroatom dopants. Using the XAFS and AC-STEM, they found that the atomic dispersion of Ru on the $Ti_3C_2T_x$ MXene support and the successful coordination of Ru_{SA} with the N and S species on the $Ti_3C_2T_x$ MXene. DFT calculations suggest that RuSA coordinated with N and S sites on the $Ti_3C_2T_x$ MXene support is the origin of this enhanced activity, which also alters the electronic structure of RuSA with optimal ΔG_{H*} to effectively facilitate the HER process.

Even the precious metal-based SACs could be successfully prepared, their high cost is still a great challenge. Herein, great effort must be made to design non-precious metal. the ion-base SACs hold great potential to replace the noble metal. Recently, Xue et al.[121] have used the graphdiyne as support to anchor Fe SACs through a two-step strategy, including first using hexaethynylbenzene (HEB) as a precursor, preparing three-dimensional (3D) GD foam (GDF) on the surface of carbon cloth (CC) through an acetylenic cross-coupling reaction, and then anchoring Ni/Fe atom by a facile electrochemical reduction method (Figure 10.13a). STEM and XAF tests identify the very narrow size distributions of both nickel (1.23 Å) and iron (1.02 Å), typical sizes of single-atom nickel and iron. DFT simulation also reveals that the strong interaction between the active site and substrate not only enhances the charge transfer but also modulates the electronic environment of the catalyst, generating superior ORR activity (Figure 10.13).

Co, as another transition metal, also demonstrates outperform activity in various electrochemical reactions as SAC. Co–SAC also shows a more unique HER performance than Fe–SAC. For, example, Chen et al.[122] reported a cheap, concise, and scalable method to disperse metallic cobalt onto nitrogen-doped graphene (denoted as Co–NG) by simply heat-treating graphene oxide (GO) and a small amount of cobalt salt in gaseous NH_3. HAADF-STEM imaging in an aberration-corrected STEM consistency with EXAFS analysis data shows that in Co–NG, Co atoms are dispersed in the nitrogen-doped graphene matrix and are in an ionic state with the nitrogen atoms in the first coordination sphere of cobalt. Furthermore, the results show that the HER activity does not increase linearly with the Co content, but there is a saturation point of the Co content, beyond which the HER activity begins to decrease. This trend may be due to excessive Co content; additional Co atoms will not be incorporated into the C–N lattice of graphene. On the contrary, excessive Co will form Co-containing particles or clusters, such as cobalt oxide, which can be evidenced by the higher oxygen content with the highest Co content in the Co–NG sample.

Carbon is not the only choice for high-performance electrocatalysts, Co SAC based on transition metal sulfides can also exhibit excellent HER performance. Cui et al.[123] reported the electrochemical cyclic voltammetry (CV) leaching of Co nanodisks (NDs) through Co–S bonds through Co–S bonds to covalently bond monoatomic cobalt (Co) arrays to twisted 1T MoS_2 nanosheets (representing it as the process of SA Co–D 1T MoS_2)–MoS_2 nanosheet hybrid. The EXAFS and XANES have been used to verify the coordination environment of atomic Co dispersed on SA Co–D 1T MoS_2 that a single Co atom is coordinated to three adjacent sulfur atoms and located at the site directly above the Mo atom. During the assembly process, the strain caused by lattice mismatch and the formation of covalent Co–S bonds between Co ND and 2H MoS_2 nanosheets is the main reason for the 2H to D 1T phase transition. The two-dimensional morphology of Co NDs is the key to the preparation of SA Co–D 1T MoS_2 because the large-area contact between Co and MoS_2 basal surface is essential to generate sufficient compressive strain for phase transition. DFT calculations confirm that the activity of HER for the catalysts is mainly due to the overall effect of the

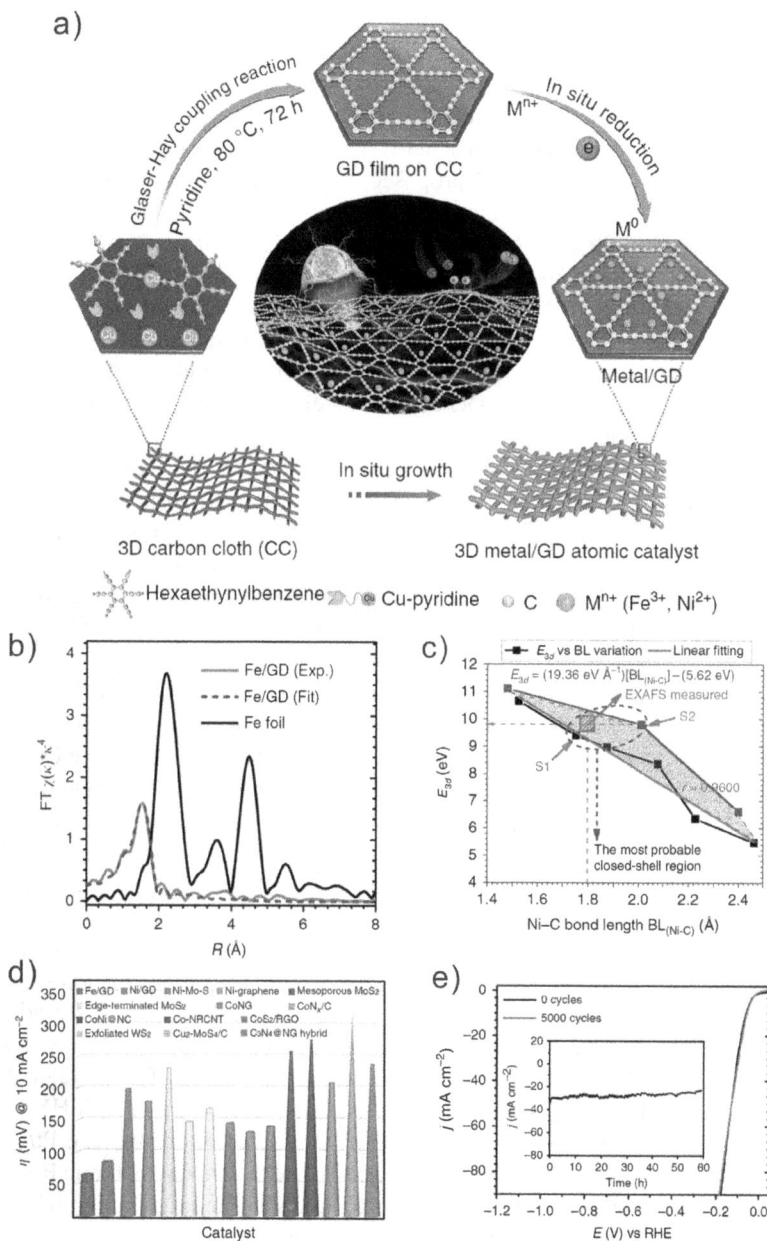

FIGURE 10.13 (a) Protocols for the synthesis of Ni/GD and Fe/GD. (b) Ex-situ EXAFS spectra of Fe/GD and Fe foil at the Fe K-edge. (c) The variation of orbital energy variation with related to the newly formed Ni–C. (d) Overpotentials at $10\,mA\,cm^{-2}$ of Ni/GD and Fe/GD along with other non-precious single-atom HER catalysts and several bulk catalysts. The overpotentials of catalysts from left to right on the horizontal axis is followed as the order of legends accordingly. (e) Fe/GD (insets: respective time-dependent current density curves). (Reproduced with permission from Reference,[121] Copyright 2018, Nature Publishing Group.)

synergistic effect of Co adsorption atoms and MoS_2 (111) carrier S through adjusting the hydrogen bonding mode at the interface.

As the most abundant metal on the fourth earth (after Fe, Ti, and Zr), nickel has also been extensively studied for the preparation of Ni-based SAC. Due to its excellent HER activity, Ni SAC is particularly promising as a substitute for Pt-based electrocatalysts.[124] Zhang et al.[125] proposed a simple and inexpensive strategy for the production of highly stable atom-dispersed Ni catalysts on DG (A–Ni@DG) by incipient wetness impregnation and subsequent acid leaching, with Ni loading up to 1.24 wt%. DFT calculations clarify by constructing three possible Ni–C configuration models that different Ni–C coordination conditions show different preferences for HER and OER. This is a kind of comparison with traditional metal-NC catalysis. It is a completely different catalytic mechanism. The results indicate that aNi@defect is the active site of HER. A–Ni@DG exhibits an excellent turnover frequency (TOF) calculated as $5.7\,s^{-1}$ at 100 mV.

Both 4d and 5d non-precious metal SACs have a certain HER potential. Sun et al.[123] prepared a metal cobalt sulfide (Co_9S_8) precursor by using a simple solvothermal method. In this process, Co_9S_8 may first be oxidized to CoSOH and then to CoSO. At this time, oxygen species derived from surface oxidation are used as anchor sites to efficiently capture single-atom metal sites. Effect derives from the O species to anchor single-site Mo, and the strong interaction between Mo SAs species and Co_9S_8-support surface/interface make a great contribution to the superior electro performance. $Mo–Co_9S_8@C$ exhibits unique catalytic activity, with a low onset potential of about 41 mV and an overpotential of 98 mV at a current density of $10\,mA\,cm^{-2}$, indicating that its catalytic activity is better than that of bare Co_9S_8 (η is 310 mV), $Co_9S_8@C$ (η is 230 mV) and $Mo–Co_9S_8$ (η is 140 mV) at a current density of $10\,mA\,cm^{-2}$. At a higher current density ($180\,mA\,cm^{-2}$), $Mo–Co_9S_8@C$ still provides better performance, despite showing a greater overpotential (140 mV) than Pt/C.[126]

In addition, to further improve the performance of the catalyst, dual-site SACs have also attracted the attention of researchers. Zhang et al.[127] successfully prepared high-quality one-to-one A-B bimetallic dimer structures (PtRu dimers) through the ALD process (Figure 10.14a). Both XAS spectra and first-principles calculations indicate that the Pt–Ru dimer complex contains a Pt–Ru bond. In addition, first-principles calculations show that Pt–Ru dimer can be easily changed from metal to semiconductor by adsorption, leaving unoccupied orbitals. Pt can regulate the inertia between H and Ru through a synergistic effect, resulting in high HER activity (more than 50 times) and excellent stability compared to commercial Pt/C catalysts. Besides, Lei et al.[128] fabricated a bimetallic SAC where monoatomic Ru and Ni co-modify MoS_2 using a mild method (Ru/Ni–MoS_2). The introduction of the heteroatom Ni with strong electronegativity anchors the single-atom Ru and enhances the stability of the catalyst. The experimental results confirm that a single Ru atom is mainly distributed on the Ni atom at the top of the Ru/Ni–MoS_2 site. Thanks to this unique structure, the synthesized catalyst exhibits a very low overpotential of 32 mV, the corresponding Tafel curve is 41 mV·dec^{-1}, and has excellent long-term stability within 20 h, far exceeding Ni–MoS_2 and Ru–MoS_2. DFT calculations show that the prepared Ru/Ni–MoS_2 synergistically promotes the chemical adsorption of H adsorbed on S atoms bound to Ni and OH bound to Ru atoms, thereby effectively

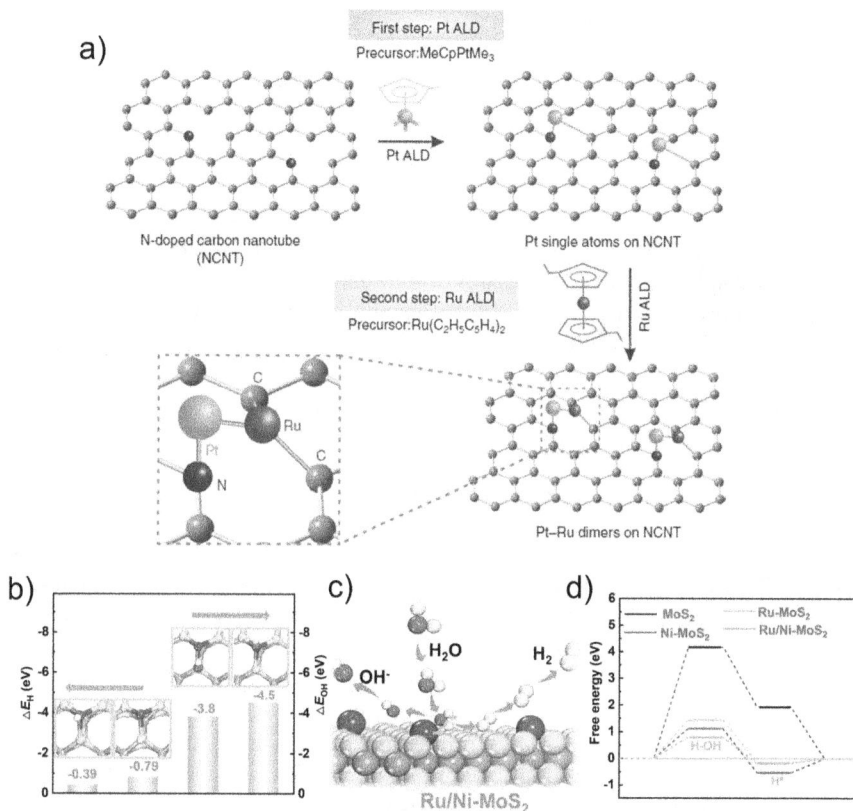

FIGURE 10.14 (a) Schematic illustration of ALD synthesis of Pt–Ru dimers on nitrogen-doped carbon nanotubes (NCNTs). (Reproduced with permission from Reference,[127] Copyright 2019, Nature Publishing Group.) (b) DFT-calculated adsorption energies of H and OH at different positions on the surfaces of Ru/Ni–MoS$_2$, respectively. (c) The illustration of mechanism for the electrocatalytic HER under alkaline conditions. (d) Free energy diagrams on the surface of these catalysts in alkaline solution. (Reproduced with permission from Reference,[128] Copyright 2019, Elsevier.)

accelerating the dissociation of water and HER process (Figure 10.14b–d). Besides, Cheng[126] developed a general and simple room-temperature impregnation strategy to construct Ru–C$_5$ single atoms and Ru oxide nanoclusters ($\approx$1.5 nm) Ru atomic dispersion catalyst (Ru ADC), which can also be extended to the Preparation of Ir, Rh, Pt, Au, and Mo atom dispersion catalyst (ADC). It is found that the obtained Ru ADC greatly promotes the precipitation of basic hydrogen through the synergistic catalysis between single atoms and sub-nano clusters, and only requires an overpotential of 18 mV at 10 mA cm^{-2}. Further mechanism studies have shown that in a catalyst, Ru–C$_5$ single atom and Ru oxide nanoclusters with Ru–O$_4$ configuration can synergistically promote water molecule capture, hydrolysis, and hydrogen release.

In summary, SACs represent a new class of low-cost, high-efficient HER electrocatalysts. The dispersion of metal atoms in these SAC materials is usually achieved by doping and coordinating the metal atoms with heteroatoms on the substrate or anchoring the defect sites. SAEC based on precious metals and non-precious metals can show significant HER activity in a wide pH range. The electronic structure of the metal center usually depends on the type of metal element, the coordination environment, and the interaction between the substrate to determine the bonding strength of *H, which determines the catalytic performance of HER.

It can be seen from the above discussion that it is relatively easy to adjust the metal-substrate coordination to achieve the desired adsorption characteristics (ie minimize $|\Delta G_{H*}|$) to reduce the proton desorption/adsorption energy. However, to promote the dissociation of water and the recombination of hydrogen, that is, water reduction and HER in an alkaline environment, a synergy between the two active sites is usually required. As a result, more and more researchers focus on how to prepare SAC with multi-metal active sites, which is also a major trend in the future.

10.5   SUMMARY AND PERSPECTIVES

In this chapter, we have summarized the synthesis strategies that can be applied to stabilize the SACs, including spatial confinement strategy, neighboring sites modifications, and strong metal-support interactions. We also highlighted the electrochemical energy applications for noble-metal and non-noble metal SACs in the ORR, OER, and HER. With the maximum atom utilization efficiency and unique coordination environment, SACs show superior catalytic performance in the catalysis fields.

SACs are promising catalysts for future applications, but some challenges still limited their practical applications. The first one is the low loadings of SACs. Due to the high surface free energy, the loadings of SACs are normally controlled at a low level (<1 wt%). Although these SACs with low metal loadings may have high mass activity in some reactions, they may require higher working conditions to catalyze such as higher temperature and higher pressure than the traditional catalysts. Therefore, developing some general and efficient synthesis methods to achieve high-loading SACs is highly demanded. Another issue for SACs is the unclear nature of active sites under reactions. SACs can indeed be well-characterized by several techniques such as HAADF-STEM, DRIFTS, XAS and et al. Most of the characterizations are conducted under ex-situ conditions, thus leading to the unambiguous conclusion about the nature of single atoms. Thus, *in-situ/operando* techniques would be very helpful for understanding the mechanism and more importantly the behaviors of SACs under realistic reaction conditions.

ACKNOWLEDGMENTS

This research was supported by the Natural Sciences and Engineering Research Council of Canada (NSERC), Canada Research Chair (CRC) Program, Canada Foundation for Innovation (CFI), Ontario Research Fund (ORF), Automotive Partnership of Canada, and the University of Western Ontario. J. Li and Y. Guan were supported by the Chinese Scholarship Council.

REFERENCES

1. Y. Sun, J. Wang, Q. Liu, M. Xia, Y. Tang, F. Gao, Y. Hou, J. Tse and Y. Zhao, *J. Mater. Chem. A*, 2019, **7**, 27175–27185.
2. C. Du, Y. Gao, J. Wang and W. Chen, *J. Mater. Chem. A*, 2020, **8**, 9981–9990.
3. S. Mukherjee, X. Yang, W. Shan, W. Samarakoon, S. Karakalos, D. A. Cullen, K. More, M. Wang, Z. Feng, G. Wang and G. Wu, *Small Methods*, 2020, **4**, 1900821.
4. Z. G. Geng, Y. Liu, X. D. Kong, P. Li, K. Li, Z. Y. Liu, J. J. Du, M. Shu, R. Si and J. Zeng, *Adv. Mater.*, 2018, **30**, 1803498.
5. X. Zou and Y. Zhang, *Chem. Soc. Rev.*, 2015, **44**, 5148–5180.
6. D. Liu, B. Wang, H. Li, S. Huang, M. Liu, J. Wang, Q. Wang, J. Zhang and Y. Zhao, *Nano Energy*, 2019, **58**, 277–283.
7. D. Zhao, Z. Zhuang, X. Cao, C. Zhang, Q. Peng, C. Chen and Y. Li, *Chem. Soc. Rev.*, 2020, **49**, 2215–2264.
8. L. Zhang, K. Doyle-Davis and X. Sun, *Energy Environ. Sci.*, 2019, **12**, 492–517.
9. A. Han, B. Wang, A. Kumar, Y. Qin, J. Jin, X. Wang, C. Yang, B. Dong, Y. Jia, J. Liu and X. Sun, *Small Methods*, 2019, **3**, 1800471.
10. X. Shi, Y. Lin, L. Huang, Z. Sun, Y. Yang, X. Zhou, E. Vovk, X. Liu, X. Huang and M. Sun, *ACS Catal.*, 2020, **10**, 3495–3504.
11. H. Xiong, S. Lin, J. Goetze, P. Pletcher, H. Guo, L. Kovarik, K. Artyushkova, B. M. Weckhuysen and A. K. Datye, *Angew. Chem. Int. Ed.*, 2017, **56**, 8986–8991.
12. X. F. Yang, A. Wang, B. Qiao, J. Li, J. Liu and T. Zhang, *Acc. Chem. Res.*, 2013, **46**, 1740–1748.
13. H. Yan, H. Cheng, H. Yi, Y. Lin, T. Yao, C. Wang, J. Li, S. Wei and J. Lu, *J. Am. Chem. Soc.*, 2015, **137**, 10484–10487.
14. E. Bayram, J. Lu, C. Aydin, N. D. Browning, S. Özkar, E. Finney, B. C. Gates and R. G. Finke, *ACS Catal.*, 2015, **5**, 3514–3527.
15. N. Cheng, M. N. Banis, J. Liu, A. Riese, X. Li, R. Li, S. Ye, S. Knights and X. Sun, *Adv. Mater.*, 2015, **27**, 277–281.
16. S. Liu, J. M. Tan, A. Gulec, L. A. Crosby, T. L. Drake, N. M. Schweitzer, M. Delferro, L. D. Marks, T. J. Marks and P. C. Stair, *Organometallics*, 2017, **36**, 818–828.
17. G. Vilé, D. Albani, M. Nachtegaal, Z. Chen, D. Dontsova, M. Antonietti, N. López and J. Pérez-Ramírez, *Angew. Chem. Int. Ed.*, 2015, **54**, 11265–11269.
18. Y. Cao, S. Chen, Q. Luo, H. Yan, Y. Lin, W. Liu, L. Cao, J. Lu, J. Yang and T. Yao, *Angew. Chem. Int. Ed.*, 2017, **56**, 12191–12196.
19. X. Huang, Y. Xia, Y. Cao, X. Zheng, H. Pan, J. Zhu, C. Ma, H. Wang, J. Li, R. You, S. Wei, W. Huang and J. Lu, *Nano Res.*, 2017, **10**, 1302–1312.
20. P. Zhou, F. Lv, N. Li, Y. Zhang, Z. Mu, Y. Tang, J. Lai, Y. Chao, M. Luo and F. Lin, *Nano Energy*, 2019, **56**, 127–137.
21. J. D. Kistler, N. Chotigkrai, P. Xu, B. Enderle, P. Prasertdham, C. Y. Chen, N. D. Browning and B. C. Gates, *Angew. Chem. Int. Ed.*, 2014, **53**, 8904–8907.
22. Y. Chen, S. Ji, W. Sun, W. Chen, J. Dong, J. Wen, J. Zhang, Z. Li, L. Zheng and C. Chen, *J. Am. Chem. Soc.*, 2018, **140**, 7407–7410.
23. Y. Liu, Z. Li, Q. Yu, Y. Chen, Z. Chai, G. Zhao, S. Liu, W.-C. Cheong, Y. Pan and Q. Zhang, *J. Am. Chem. Soc.*, 2019, **141**, 9305–9311.
24. Z. Lu, J. Wang, S. Huang, Y. Hou, Y. Li, Y. Zhao, S. Mu, J. Zhang and Y. Zhao, *Nano Energy*, 2017, **42**, 334–340.
25. N. Cheng, S. Stambula, D. Wang, M. N. Banis, J. Liu, A. Riese, B. Xiao, R. Li, T. K. Sham, L. M. Liu, G. A. Botton and X. Sun, *Nat. Commun.*, 2016, **7**, 13638.
26. C. H. Choi, M. Kim, H. C. Kwon, S. J. Cho, S. Yun, H. T. Kim, K. J. J. Mayrhofer, H. Kim and M. Choi, *Nat. Commun.*, 2016, **7**, 10922.

27. Y. Zhai, D. Pierre, R. Si, W. Deng, P. Ferrin, A. U. Nilekar, G. Peng, J. A. Herron, D. C. Bell, H. Saltsburg, M. Mavrikakis and M. Flytzani-Stephanopoulos, *Science*, 2010, **329**, 1633–1636.

28. M. Yang, J. Liu, S. Lee, B. Zugic, J. Huang, L. F. Allard and M. Flytzani-Stephanopoulos, *J. Am. Chem. Soc.*, 2015, **137**, 3470–3473.

29. M. Yang, L. F. Allard and M. Flytzani-Stephanopoulos, *J. Am. Chem. Soc.*, 2013, **135**, 3768–3771.

30. J. Jones, H. Xiong, A. T. DeLaRiva, E. J. Peterson, H. Pham, S. R. Challa, G. Qi, S. Oh, M. H. Wiebenga, X. I. P. Hernández, Y. Wang and A. K. Datye, *Science*, 2016, **353**, 150–154.

31. R. Lang, W. Xi, J.-C. Liu, Y.-T. Cui, T. Li, A. F. Lee, F. Chen, Y. Chen, L. Li, L. Li, J. Lin, S. Miao, X. Liu, A.-Q. Wang, X. Wang, J. Luo, B. Qiao, J. Li and T. Zhang, *Nat. Commun.*, 2019, **10**, 234.

32. Y. Yao, Z. Huang, P. Xie, L. Wu, L. Ma, T. Li, Z. Pang, M. Jiao, Z. Liang, J. Gao, Y. He, D. J. Kline, M. R. Zachariah, C. Wang, J. Lu, T. Wu, T. Li, C. Wang, R. Shahbazian-Yassar and L. Hu, *Nat. Nanotechnol.*, 2019, **14**, 851–857.

33. Z. Song, L. Zhang, K. Doyle-Davis, X. Fu, J.-L. Luo and X. Sun, *Adv. Energy Mater.*, 2020, **10**, 2001561.

34. S. Wei, A. Li, J.-C. Liu, Z. Li, W. Chen, Y. Gong, Q. Zhang, W.-C. Cheong, Y. Wang, L. Zheng, H. Xiao, C. Chen, D. Wang, Q. Peng, L. Gu, X. Han, J. Li and Y. Li, *Nat. Nanotechnol.*, 2018, **13**, 856–861.

35. Y. Chen, S. Ji, C. Chen, Q. Peng, D. Wang and Y. Li, *Joule*, 2018, **2**, 1242–1264.

36. M. K. Debe, *Nature*, 2012, **486**, 43–51.

37. L. Zhong and S. Li, *ACS Catal.*, 2020, **10**, 4313–4318.

38. X. Guo, S. Lin, J. Gu, S. Zhang, Z. Chen and S. Huang, *ACS Catal.*, 2019, **9**, 11042–11054.

39. Y. H. Wang, J. B. Le, W. Q. Li, J. Wei, P. M. Radjenovic, H. Zhang, X. S. Zhou, J. Cheng, Z. Q. Tian and J. F. Li, *Angew. Chem. Int. Ed.*, 2019, **58**, 16062–16066.

40. J. Liu, M. Jiao, L. Lu, H. M. Barkholtz, Y. Li, Y. Wang, L. Jiang, Z. Wu, D.-J. Liu and L. Zhuang, *Nat. Commun.*, 2017, **8**, 15938.

41. S. Yang, J. Kim, Y. J. Tak, A. Soon and H. Lee, *Angew. Chem. Int. Ed.*, 2016, **55**, 2058–2062.

42. D. Xia, X. Yang, L. Xie, Y. Wei, W. Jiang, M. Dou, X. Li, J. Li, L. Gan and F. Kang, *Adv. Funct. Mater.*, 2019, **29**, 1906174.

43. S. G. Peera, J. Balamurugan, N. H. Kim and J. H. Lee, *Small*, 2018, **14**, 1800441.

44. J. Liu, X. Wan, S. Liu, X. Liu, L. Zheng, R. Yu and J. Shui, *Adv. Mater.*, 2021, **33**, 2103600.

45. J. Qin, Z. Liu, D. Wu and J. Yang, *Appl. Catal. B Environ.*, 2020, **278**, 119300.

46. U. I. Kramm, L. Ni and S. Wagner, *Adv. Mater.*, 2019, **31**, e1805623.

47. R. A. Sidik, A. B. Anderson, N. P. Subramanian, S. P. Kumaraguru and B. N. Popov, *J. Phys. Chem. B*, 2006, **110**, 1787–1793.

48. H. Zhang, H. T. Chung, D. A. Cullen, S. Wagner, U. I. Kramm, K. L. More, P. Zelenay and G. Wu, *Energy Environ. Sci.*, 2019, **12**, 2548–2558.

49. U. I. Kramm, J. Herranz, N. Larouche, T. M. Arruda, M. Lefevre, F. Jaouen, P. Bogdanoff, S. Fiechter, I. Abs-Wurmbach, S. Mukerjee and J. P. Dodelet, *Phys. Chem. Chem. Phys.*, 2012, **14**, 11673–11688.

50. J. Wang, W. Liu, G. Luo, Z. Li, C. Zhao, H. Zhang, M. Zhu, Q. Xu, X. Wang, C. Zhao, Y. Qu, Z. Yang, T. Yao, Y. Li, Y. Lin, Y. Wu and Y. Li, *Energy Environ. Sci.*, 2018, **11**, 3375–3379.

51. S. H. Lee, J. Kim, D. Y. Chung, J. M. Yoo, H. S. Lee, M. J. Kim, B. S. Mun, S. G. Kwon, Y. E. Sung and T. Hyeon, *J. Am. Chem. Soc.*, 2019, **141**, 2035–2045.

52. X. Wan, X. Liu, Y. Li, R. Yu, L. Zheng, W. Yan, H. Wang, M. Xu and J. Shui, *Nat. Catal.*, 2019, **2**, 259–268.

53. C. C. Hou, L. Zou, L. Sun, K. Zhang, Z. Liu, Y. Li, C. Li, R. Zou, J. Yu and Q. Xu, *Angew. Chem. Int. Ed.*, 2020, **59**, 7384–7389.
54. M. Qiao, Y. Wang, Q. Wang, G. Hu, X. Mamat, S. Zhang and S. Wang, *Angew. Chem. Int. Ed.*, 2020, **59**, 2688–2694.
55. R. Chenitz, U. I. Kramm, M. Lefèvre, V. Glibin, G. Zhang, S. Sun and J.-P. Dodelet, *Energy Environ. Sci.*, 2018, **11**, 365–382.
56. C. H. Choi, C. Baldizzone, G. Polymeros, E. Pizzutilo, O. Kasian, A. K. Schuppert, N. Ranjbar Sahraie, M.-T. Sougrati, K. J. J. Mayrhofer and F. Jaouen, *ACS Catal.*, 2016, **6**, 3136–3146.
57. C. H. Choi, C. Baldizzone, J. P. Grote, A. K. Schuppert, F. Jaouen and K. J. Mayrhofer, *Angew. Chem. Int. Ed.*, 2015, **54**, 12753–12757.
58. H. Schulenburg, S. Stankov, V. Schünemann, J. Radnik, I. Dorbandt, S. Fiechter, P. Bogdanoff and H. Tributsch, *J. Phys. Chem. B*, 2003, **107**, 9034–9041.
59. M. Lefèvre and J.-P. Dodelet, *Electrochim. Acta*, 2003, **48**, 2749–2760.
60. G. Zhang, R. Chenitz, M. Lefèvre, S. Sun and J.-P. Dodelet, *Nano Energy*, 2016, **29**, 111–125.
61. G. Liu, X. Li and B. Popov, *ECS Trans.*, 2009, **25**, 1251.
62. J.-Y. Choi, L. Yang, T. Kishimoto, X. Fu, S. Ye, Z. Chen and D. Banham, *Energy Environ. Sci.*, 2017, **10**, 296–305.
63. X. Li, X. Hao, A. Abudula and G. Guan, *J. Mater. Chem. A*, 2016, **4**, 11973–12000.
64. I. Roger, M. A. Shipman and M. D. Symes, *Nat. Rev. Chem.*, 2017, **1**, 1–13.
65. B. Zhang, Y. Zheng, T. Ma, C. Yang, Y. Peng, Z. Zhou, M. Zhou, S. Li, Y. Wang and C. Cheng, *Adv. Mater.*, 2021, **33**, 2006042.
66. C. Zhu, Q. Shi, S. Feng, D. Du and Y. Lin, *ACS Energy. Lett.*, 2018, **3**, 1713–1721.
67. J. Rossmeisl, Z.-W. Qu, H. Zhu, G.-J. Kroes and J. K. Nørskov, *J. Electroanal. Chem.*, 2007, **607**, 83–89.
68. I. C. Man, H. Y. Su, F. Calle-Vallejo, H. A. Hansen, J. I. Martínez, N. G. Inoglu, J. Kitchin, T. F. Jaramillo, J. K. Nørskov and J. Rossmeisl, *ChemCatChem*, 2011, **3**, 1159–1165.
69. A. Vojvodic and J. K. Nørskov, *Science*, 2011, **334**, 1355–1356.
70. C. Wei, Z. Feng, G. G. Scherer, J. Barber, Y. Shao-Horn and Z. J. Xu, *Adv. Mater.*, 2017, **29**, 1606800.
71. H. Li, S. Sun, S. Xi, Y. Chen, T. Wang, Y. Du, M. Sherburne, J. W. Ager, A. C. Fisher and Z. J. Xu, *Chem. Mater.*, 2018, **30**, 6839–6848.
72. B. Hammer and J. K. Nørskov, *Adv. Catal.*, 2000, **45**, 71–129.
73. J. Suntivich, K. J. May, H. A. Gasteiger, J. B. Goodenough and Y. Shao-Horn, *Science*, 2011, **334**, 1383–1385.
74. Y. Guo, Y. Tong, P. Chen, K. Xu, J. Zhao, Y. Lin, W. Chu, Z. Peng, C. Wu and Y. Xie, *Adv. Mater.*, 2015, **27**, 5989–5994.
75. R. Subbaraman, D. Tripkovic, K.-C. Chang, D. Strmcnik, A. P. Paulikas, P. Hirunsit, M. Chan, J. Greeley, V. Stamenkovic and N. M. Markovic, *Nat. Mater.*, 2012, **11**, 550–557.
76. D. A. Kuznetsov, B. Han, Y. Yu, R. R. Rao, J. Hwang, Y. Román-Leshkov and Y. Shao-Horn, *Joule*, 2018, **2**, 225–244.
77. A. Grimaud, W. Hong, Y. Shao-Horn and J. Tarascon, *Nat. Mater.*, 2016, **15**, 121–126.
78. Y. Zhou, S. Sun, J. Song, S. Xi, B. Chen, Y. Du, A. C. Fisher, F. Cheng, X. Wang and H. Zhang, *Adv. Mater.*, 2018, **30**, 1802912.
79. X. Luo, X. Wei, H. Zhong, H. Wang, Y. Wu, Q. Wang, W. Gu, M. Gu, S. P. Beckman and C. Zhu, *ACS Appl. Mater. Interfaces*, 2019, **12**, 3539–3546.
80. L. Zhuang, Y. Jia, H. Liu, X. Wang, R. K. Hocking, H. Liu, J. Chen, L. Ge, L. Zhang and M. Li, *Adv. Mater.*, 2019, **31**, 1805581.
81. P. Li, M. Wang, X. Duan, L. Zheng, X. Cheng, Y. Zhang, Y. Kuang, Y. Li, Q. Ma and Z. Feng, *Nat. Commun.*, 2019, **10**, 1711.

82. D. D. Babu, Y. Huang, G. Anandhababu, X. Wang, R. Si, M. Wu, Q. Li, Y. Wang and J. Yao, *J. Mater. Chem. A*, 2019, **7**, 8376–8383.
83. L. Cao, Q. Luo, J. Chen, L. Wang, Y. Lin, H. Wang, X. Liu, X. Shen, W. Zhang and W. Liu, *Nat. Commun.*, 2019, **10**, 4849.
84. F. Tang, T. Liu, W. Jiang and L. Gan, *J. Electroanal. Chem.*, 2020, **871**, 114282.
85. Y. Wang, Y. Zhang, Z. Liu, C. Xie, S. Feng, D. Liu, M. Shao and S. Wang, *Angew. Chem. Int. Ed.*, 2017, **56**, 5867–5871.
86. K. L. Nardi, N. Yang, C. F. Dickens, A. L. Strickler and S. F. Bent, *Adv. Energy Mater.*, 2015, **5**, 1500412.
87. P. Zhai, M. Xia, Y. Wu, G. Zhang, J. Gao, B. Zhang, S. Cao, Y. Zhang, Z. Li and Z. Fan, *Nat. Commun.*, 2021, **12**, 4587.
88. Q. Wang, X. Huang, Z. L. Zhao, M. Wang, B. Xiang, J. Li, Z. Feng, H. Xu and M. Gu, *J. Am. Chem. Soc.*, 2020, **142**, 7425–7433.
89. J. Zhang, J. Liu, L. Xi, Y. Yu, N. Chen, S. Sun, W. Wang, K. M. Lange and B. Zhang, *J. Am. Chem. Soc.*, 2018, **140**, 3876–3879.
90. N. Hodnik, P. Jovanovič, A. Pavlišič, B. Jozinović, M. Zorko, M. Bele, V. S. Šelih, M. Šala, S. Hočevar and M. Gaberšček, *J. Phys. Chem. C*, 2015, **119**, 10140–10147.
91. C. Iwakura, K. Hirao and H. Tamura, *Electrochim. Acta*, 1977, **22**, 329–334.
92. Y. Yao, S. Hu, W. Chen, Z.-Q. Huang, W. Wei, T. Yao, R. Liu, K. Zang, X. Wang, G. Wu, W. Yuan, T. Yuan, B. Zhu, W. Liu, Z. Li, D. He, Z. Xue, Y. Wang, X. Zheng, J. Dong, C.-R. Chang, Y. Chen, X. Hong, J. Luo, S. Wei, W.-X. Li, P. Strasser, Y. Wu and Y. Li, *Nat. Catal.*, 2019, **2**, 304–313.
93. X. Zeng, J. Shui, X. Liu, Q. Liu, Y. Li, J. Shang, L. Zheng and R. Yu, *Adv. Energy Mater.*, 2018, **8**, 1701345.
94. H. Fei, J. Dong, Y. Feng, C. S. Allen, C. Wan, B. Volosskiy, M. Li, Z. Zhao, Y. Wang and H. Sun, *Nat. Catal.*, 2018, **1**, 63–72.
95. D. M. Morales, M. A. Kazakova, S. Dieckhöfer, A. G. Selyutin, G. V. Golubtsov, W. Schuhmann and J. Masa, *Adv. Funct. Mater.*, 2020, **30**, 1905992.
96. J. W. D. Ng, M. García-Melchor, M. Bajdich, P. Chakthranont, C. Kirk, A. Vojvodic and T. F. Jaramillo, *Nat. Energy*, 2016, **1**, 1–8.
97. C. A. Kent, J. J. Concepcion, C. J. Dares, D. A. Torelli, A. J. Rieth, A. S. Miller, P. G. Hoertz and T. J. Meyer, *J. Am. Chem. Soc.*, 2013, **135**, 8432–8435.
98. J. Li, M. Chen, D. A. Cullen, S. Hwang, M. Wang, B. Li, K. Liu, S. Karakalos, M. Lucero and H. Zhang, *Nat. Catal.*, 2018, **1**, 935–945.
99. S. Liu, M. Wang, X. Yang, Q. Shi, Z. Qiao, M. Lucero, Q. Ma, K. L. More, D. A. Cullen and Z. Feng, *Angew. Chem. Int. Ed.*, 2020, **59**, 21698–21705.
100. X. X. Wang, V. Prabhakaran, Y. He, Y. Shao and G. Wu, *Adv. Mater.*, 2019, **31**, 1805126.
101. Y. Hou, M. Qiu, M. G. Kim, P. Liu, G. Nam, T. Zhang, X. Zhuang, B. Yang, J. Cho and M. Chen, *Nat. Commun.*, 2019, **10**, 1392.
102. Y. Li, Z. S. Wu, P. Lu, X. Wang, W. Liu, Z. Liu, J. Ma, W. Ren, Z. Jiang and X. Bao, *Adv. Sci.*, 2020, **7**, 1903089.
103. L. Zhang, Y. Jia, G. Gao, X. Yan, N. Chen, J. Chen, M. T. Soo, B. Wood, D. Yang and A. Du, *Chem*, 2018, **4**, 285–297.
104. X. Sun, S. Sun, S. Gu, Z. Liang, J. Zhang, Y. Yang, Z. Deng, P. Wei, J. Peng and Y. Xu, *Nano Energy*, 2019, **61**, 245–250.
105. X. Lv, W. Wei, H. Wang, B. Huang and Y. Dai, *Appl. Catal. B Environ.*, 2020, **264**, 118521.
106. W. Xie, Y. Song, S. Li, J. Li, Y. Yang, W. Liu, M. Shao and M. Wei, *Adv. Funct. Mater.*, 2019, **29**, 1906477.
107. W. H. Lai, L. F. Zhang, W. B. Hua, S. Indris, Z. C. Yan, Z. Hu, B. Zhang, Y. Liu, L. Wang and M. Liu, *Angew. Chem. Int. Ed.*, 2019, **58**, 11868–11873.

108. D. Lyu, Y. B. Mollamahale, S. Huang, P. Zhu, X. Zhang, Y. Du, S. Wang, M. Qing, Z. Q. Tian and P. K. Shen, *J. Catal.*, 2018, **368**, 279–290.

109. Z. Zhang, C. Feng, X. Li, C. Liu, D. Wang, R. Si, J. Yang, S. Zhou and J. Zeng, *Nano Lett.*, 2021.

110. Y. Zheng, Y. Jiao, Y. Zhu, Q. Cai, A. Vasileff, L. H. Li, Y. Han, Y. Chen and S.-Z. Qiao, *J. Am. Chem. Soc.*, 2017, **139**, 3336–3339.

111. D. Liu, S. Ding, C. Wu, W. Gan, C. Wang, D. Cao, Z. ur Rehman, Y. Sang, S. Chen and X. Zheng, *J. Mater. Chem. A*, 2018, **6**, 6840–6846.

112. J. Luo, J.-H. Im, M. T. Mayer, M. Schreier, M. K. Nazeeruddin, N.-G. Park, S. D. Tilley, H. J. Fan and M. Grätzel, *Science*, 2014, **345**, 1593–1596.

113. X. Tian, P. Zhao and W. Sheng, *Adv. Mater.*, 2019, **31**, 1808066.

114. S. Ye, F. Luo, Q. Zhang, P. Zhang, T. Xu, Q. Wang, D. He, L. Guo, Y. Zhang, C. He, X. Ouyang, M. Gu, J. Liu and X. Sun, *Energy Environ. Sci.*, 2019, **12**, 1000–1007.

115. P. Kuang, Y. Wang, B. Zhu, F. Xia, C.-W. Tung, J. Wu, H. M. Chen and J. Yu, *Adv. Mater.*, 2021, **33**, 2008599.

116. M. Gobbi, S. Bonacchi, J. X. Lian, Y. Liu, X.-Y. Wang, M.-A. Stoeckel, M. A. Squillaci, G. D'Avino, A. Narita, K. Müllen, X. Feng, Y. Olivier, D. Beljonne, P. Samorì and E. Orgiu, *Nat. Commun.*, 2017, **8**, 14767.

117. Q. Qin, H. Jang, L. Chen, G. Nam, X. Liu and J. Cho, *Adv. Energy Mater.*, 2018, **8**, 1801478.

118. S. Ye, F. Luo, T. Xu, P. Zhang, H. Shi, S. Qin, J. Wu, C. He, X. Ouyang, Q. Zhang, J. Liu and X. Sun, *Nano Energy*, 2020, **68**, 104301.

119. B. Lu, L. Guo, F. Wu, Y. Peng, J. E. Lu, T. J. Smart, N. Wang, Y. Z. Finfrock, D. Morris, P. Zhang, N. Li, P. Gao, Y. Ping and S. Chen, *Nat. Commun.*, 2019, **10**, 631.

120. V. Ramalingam, P. Varadhan, H.-C. Fu, H. Kim, D. Zhang, S. Chen, L. Song, D. Ma, Y. Wang, H. N. Alshareef and J.-H. He, *Adv. Mater.*, 2019, **31**, 1903841.

121. Y. Xue, B. Huang, Y. Yi, Y. Guo, Z. Zuo, Y. Li, Z. Jia, H. Liu and Y. Li, *Nat. Commun.*, 2018, **9**, 1460.

122. H. Fei, J. Dong, M. J. Arellano-Jiménez, G. Ye, N. Dong Kim, E. L. G. Samuel, Z. Peng, Z. Zhu, F. Qin, J. Bao, M. J. Yacaman, P. M. Ajayan, D. Chen and J. M. Tour, *Nat. Commun.*, 2015, **6**, 8668.

123. L. Wang, X. Duan, X. Liu, J. Gu, R. Si, Y. Qiu, Y. Qiu, D. Shi, F. Chen, X. Sun, J. Lin and J. Sun, *Adv. Energy Mater.*, 2020, **10**, 1903137.

124. S. Sultan, J. N. Tiwari, A. N. Singh, S. Zhumagali, M. Ha, C. W. Myung, P. Thangavel and K. S. Kim, *Adv. Energy Mater.*, 2019, **9**, 1900624.

125. L. Zhang, Y. Jia, G. Gao, X. Yan, N. Chen, J. Chen, M. T. Soo, B. Wood, D. Yang, A. Du and X. Yao, *Chem*, 2018, **4**, 285–297.

126. D. Cao, J. Wang, H. Xu and D. Cheng, *Small*, 2021, **17**, 2101163.

127. L. Zhang, R. Si, H. Liu, N. Chen, Q. Wang, K. Adair, Z. Wang, J. Chen, Z. Song, J. Li, M. N. Banis, R. Li, T. K. Sham, M. Gu, L. M. Liu, G. A. Botton and X. Sun, *Nat. Commun.*, 2019, **10**, 4936.

128. J. Ge, D. Zhang, Y. Qin, T. Dou, M. Jiang, F. Zhang and X. Lei, *Appl. Catal. B Environ.*, 2021, 298, 120557.

Index

Note: **Bold** page numbers refer to tables and *italic* page numbers refer to figures.

For Product Safety Concerns and Information please contact our EU
representative GPSR@taylorandfrancis.com
Taylor & Francis Verlag GmbH, Kaufingerstraße 24, 80331 München, Germany

* 9 7 8 0 3 6 7 7 2 1 0 0 8 *